Hig ... ineering

The I

This book is dedicated to my wife,
Ann
for her help and encouragement in the writing of this book

High Frequency and Microwave Engineering

E. da Silva
The Open University

OXFORD AUCKLAND BOSTON JOHANNESBURG MELBOURNE NEW DELHI

Butterworth–Heinemann
Linacre House, Jordan Hill, Oxford OX2 8DP
225 Wildwood Avenue, Woburn, MA 01801-2041
A division of Reed Educational and Professional Publishing Ltd

A member of the Reed Elsevier plc group

First published 2001

British Library Cataloguing in Publication Data
A catalogue record for this book is available from the British Library

ISBN 0 7506 5646 X

Typeset in 10/12 pt Times by Cambrian Typesetters, Frimley, Surrey
Printed and bound by MPG Books Ltd, Bodmin, Cornwall

Contents

Preface

This book was started while the author was Professor and Head of Department at Etisalat College which was set up with the technical expertise of the University of Bradford, England. It was continued when the author returned to the Open University, England. Many thanks are due to my colleagues Dr David Crecraft and Dr Mike Meade of the Open University, Dr L. Auchterlonie of Newcastle University and Dr N. McEwan and Dr D. Dernikas of Bradford University. I would also like to thank my students for their many helpful comments.

High Frequency and Microwave Engineering has been written with a view to ease of understanding and to provide knowledge for any engineer who is interested in high frequency and microwave engineering. The book has been set at the third level standard of an electrical engineering degree but it is eminently suitable for self-study. The book comprises standard text which is emphasised with over 325 illustrations. A further 120 examples are given to emphasise clarity in understanding and application of important topics.

A software computer-aided-design package, PUFF 2.1 produced by California Institute of Technology (CalTech) U.S.A., is supplied free with the book. PUFF can be used to provide scaled layouts and artwork for designs. PUFF can also be used to calculate the scattering parameters of circuits. Up to four scattering parameters can be plotted simultaneously and automatically on a Smith chart as well as in graphical form. In addition to the PUFF software I have also included 42 software application examples. These examples have been chosen to calculate and verify some of the examples given in the text, but many are proven designs suitable for use in practical circuits. The confirmation of manual design and CAD design is highly gratifying to the reader and it helps to promote greater confidence in the use of other types of software. An article 'Practical Circuit Design' explaining how PUFF can be used for producing layout and artwork for circuits is explained in detail. There is also a detailed microwave amplifier design which uses PUFF to verify circuit calculations, match line impedances, and produce the artwork for amplifier fabrication. There is also a copy of CalTech's manual on disk. This will prove useful for more advanced work.

The book commences with an explanation of the many terms used in radio, wireless, high frequency and microwave engineering. These are explained in Chapter 1. Chapter 2 provides a gentle introduction to the subject of transmission lines. It starts with a gradual introduction of transmission lines by using an everyday example. Diagrams have been used to illustrate some of the characteristics of transmission lines. Mathematics has been kept to a minimum. The chapter ends with some applications of transmission lines especially in their use as inductors, capacitors, transformers and couplers.

Chapter 3 provides an introduction to Smith charts and scattering parameters. Smith charts are essential in understanding and reading manufacturers' data because they also provide a 'picture' of circuit behaviour. Use of the Smith chart is encouraged and many examples are provided for the evaluation and manipulation of reflection coefficients, impedance, admittance and matching circuits. For those who want it, Smith chart theory is presented, but it is stressed that knowledge of the theory is not essential to its use.

The installation of PUFF software is introduced in Chapter 4. The chapter goes on to deal with the printing and fabrication of artwork and the use and modification of templates. Particular attention is paid to circuit configurations including couplers, transformers and matching of circuits. Scattering parameters are re-introduced and used for solving scattering problems. Many of the examples in this chapter are used to confirm the results of the examples given in Chapters 2 and 3.

Amplifier circuitry components are dealt with in Chapter 5. Particular attention is paid to the design of Butterworth and Tchebyscheff filters and their uses as low pass, bandpass, high pass and bandstop filters. Impedance matching is discussed in detail and many methods of matching are shown in examples.

Chapter 6 deals with the design of amplifiers including transistor biasing which is vitally important for it ensures the constancy of transistor parameters with temperature. Examples are given of amplifier circuits using unconditionally stable transistors and conditionally stable amplifiers. The use of the indefinite matrix in transistor configurations is shown by examples.

The design of microwave amplifiers is shown in Chapter 7. Design examples include conjugately matched amplifiers, constant gain amplifiers, low noise amplifiers, broadband amplifiers, feedback amplifiers and r.f. power amplifiers.

Oscillators and frequency synthesizers are discussed in Chapter 8. Conditions for oscillation are discussed and the Barkhausen criteria for oscillation is detailed in the early part of the chapter. Oscillator designs include the Wien bridge, phase shift, Hartley, Colpitts, Clapp, crystal and the phase lock loop system. Frequency synthesizers are discussed with reference to direct and indirect methods of frequency synthesis.

Chapter 9 is a discussion of topics which will prove useful in future studies. These include signal flow diagrams and the use of software particularly the quasi-free types. Comments are made regarding the usefulness of Hewlett Packard's AppCAD and Motorola's impedance matching program, MIMP.

Finally, I wish you well in your progress towards the fascinating subject of high frequency and microwave engineering.

Ed da Silva

1

Basic features of radio communication systems

1.1 Introduction

This chapter describes communication systems which use radio waves and signals. Radio signals are useful for two main reasons. They provide a relatively cheap way of communicating over vast distances and they are extremely useful for mobile communications where the use of cables is impractical.

Radio signals are generally considered to be electromagnetic signals which are broadcast or radiated through space. They vary in frequency from several kilohertz[1] to well over 100 GHz (10^{11} Hz). They include some well known public broadcasting bands: long-wave (155–280 kHz), medium-wave (522–1622 kHz), short-wave (3–30 MHz), very high frequency FM band (88–108 MHz), ultra high frequency television band (470–890 MHz) and the satellite television band (11.6 to 12.4 GHz). The frequencies[2] quoted above are approximate figures and are only provided to give an indication of some of the frequency bands used in Radio and TV broadcasting.

1.1.1 Aims

The aims of this chapter are to introduce you to some basic radio communications principles and methods. These include **modulation** (impressing signal information on to radio carrier waves), **propagation** (transmission of radio carrier waves) and **demodulation** (detection of radio carrier waves) to recover the original signal information.

The method we use here is to start with an overview of a communication system. The system is then divided to show its sub-systems and the sub-systems are then expanded to show individual circuits and items.

1.1.2 Objectives

The general objectives of this chapter are:

- to help you understand why certain methods and techniques are used for radio frequency and high frequency communication circuits;

[1] One hertz (Hz) means 1 cyclic vibration per second: 1 kHz = 1000 cyclic vibrations per second, 1 MHz = 1 000 000 cyclic vibrations per second, and 1 GHz = 1 000 000 000 cyclic vibrations per second. The word Hertz is named after Heinrich Hertz, one of the early pioneers of physics.

[2] The frequencies quoted are for Europe. Other countries do not necessarily follow the exact same frequencies but they do have similar frequency bands.

- to appreciate the need for modulation;
- to understand the basic principles of modulation and demodulation;
- to understand the basic principles of signal propagation using antennas;
- to introduce radio receivers;
- to introduce you to the requirements of selectivity and bandwidth in radio communication circuits.

1.2 Radio communication systems

1.2.1 Stages in communication

Let's commence with a simple communications example and analyse the important stages necessary for communication. This is shown diagramatically in Figure 1.1. We start by writing a *letter-message*, putting it in an *envelope*, and sending it through a *post-carrier* (postal carrier system) to our destination. At the other end, our recipient *receives* the letter from the post office, *opens* the envelope and *reads* our message. Why do we carry out these actions?

We write a letter because it contains the information we want to send to our recipient. In radio communications, we do the same thing; we use a message signal, which is an electrical signal derived from analogue sound or digitally encoded sound and/or video/data signals, as the information we want to convey. The process of putting this information into an 'envelope' for transmission through the carrier is called modulation and circuits designed for this purpose are known as modulation circuits or **modulators**.

We use the post office as the carrier for our letters because the post office has the ability to transmit messages over long distances. In radio communications, we use a **radio frequency carrier** because a radio carrier has the ability to carry messages over long distances. A radio frequency carrier with an enveloped message impressed on it is often called an **enveloped carrier wave** or a **modulated carrier wave**.

When the post office delivers a letter to a destination, the envelope must be opened to enable the message to be read. In radio communications when the enveloped carrier wave

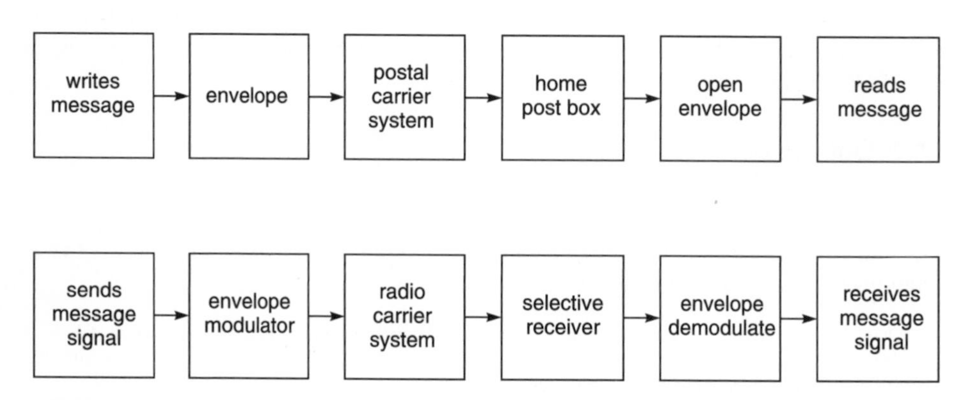

Fig. 1.1 Analogy between the postal system and a radio system

arrives at its destination, the enveloped carrier must be 'opened' or **demodulated** to recover the original message from the carrier. Circuits which perform this function are known as demodulation circuits or **demodulators**.

The post office uses a system of postal codes and addresses to ensure that a letter is selected and delivered to the correct address. In radio communications, **selective** or **tuned circuits** are used to select the correct messages for a particular receiver. Amplifiers are also used to ensure that the signals sent and received have sufficient amplitudes to operate the message reading devices such as a loudspeaker and/or a video screen.

In addition to the main functions mentioned above, we need a post box to send our letter. The electrical equivalent of this is the **transmitting antenna**. We require a letter box at home to receive letters. The electrical equivalent of this is the **receiving antenna**.

1.2.2 Summary of radio communications systems

A pictorial summary of the above actions is shown in Figure 1.1. There are three main functions in a radio communications system. These are: modulation, transmission and demodulation. There are also supplementary functions in a radio communications system. These include transmitting antennas,[3] receiving antennas, selective circuits, and amplifiers. We will now describe these methods in the same order but with more detail.

1.3 Modulation and demodulation

Before discussing modulation and demodulation, it is necessary to clarify two points: the modulation information and the modulation method.

In the case of a letter in the postal system, we are free to write our messages (modulation information) in any language, such as English, German, French, pictures, data, etc. However, our recipient must be able to read the language we use. For example it is useless to write our message in Japanese if our recipient can only read German. Hence the modulation information system we use at the transmitter must be **compatible** with the demodulation information system at the receiver.

Secondly, the method of putting information (modulation method) on the letter is important. For example, we can type, use a pencil, ultra violet ink, etc. However, the reader must be able to decipher (demodulate) the information provided. For example, if we use ultra violet ink, the reader must also use ultra violet light to decipher (demodulate) the message. Hence the modulation and demodulation methods must also be compatible.

In the discussions that follow we are only discussing modulation and demodulation methods; not the modulation information. We also tend to use sinusoidal waves for our explanation. This is because a great mathematician, Joseph Fourier,[4] has shown that periodic waveforms of any shape consist of one or more d.c. levels, sine waves and cosine waves. This is similar to the case in the English language, where we have thousands of words but, when analysed, all come from the 26 letters of the alphabet. Hence, the sinusoidal wave is a useful tool for understanding modulation methods.

[3] Antennas are also known as aerials.
[4] Fourier analysis will be explained fully in a later section.

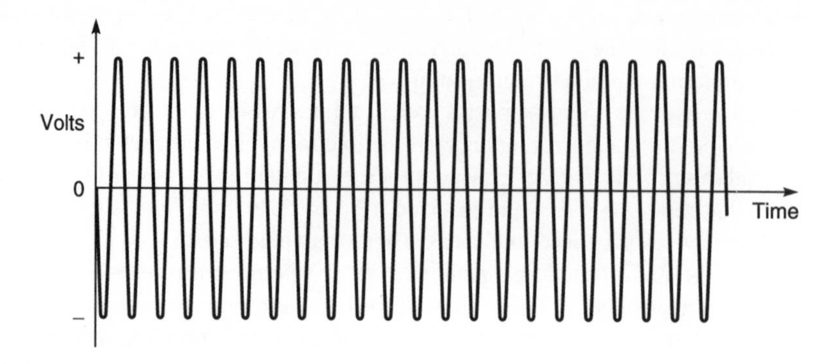

Fig. 1.2 A sinusoidal radio carrier wave

We now return to our simple radio carrier wave which is the sinusoidal wave[5] shown in Figure 1.2.

A sinusoidal wave can be described by the expression

$$v_c = V_c \cos(\omega_c t + \phi_c) \tag{1.1}$$

where

v_c = instantaneous carrier amplitude (volts)
V_c = carrier amplitude (peak volts)
ω_c = angular frequency in radians and $\omega_c = 2\pi f_c$ where
f_c = carrier frequency (hertz)
ϕ_c = carrier phase delay (radians)

If you look at Figure 1.2, you can see that a sinusoidal wave on its own provides little information other than its presence or its absence. So we must find some method of modulating our information on to the radio carrier wave. We can change:

- its *amplitude* (V_c) according to our information – this is called **amplitude modulation** and will be described in Section 1.3.1;
- its *frequency* (ω_c) according to our information – this is called **frequency modulation** and will be described in Section 1.3.2;
- its *phase* (ϕ_c) according to our information – this is known as **phase modulation** and will be described in Section 1.3.3;
- or we can use a combination of one or more of the methods described above – this method is favoured by **digital modulation**.

1.3.1 Amplitude modulation (AM)

This is the method used in medium-wave and short-wave radio broadcasting. Figure 1.3 shows what happens when we apply amplitude modulation to a sinusoidal carrier wave.

[5] A sinusoidal wave is a generic name for a sine or cosine wave. In many cases, cosine waves are used because of ease in mathematical manipulation.

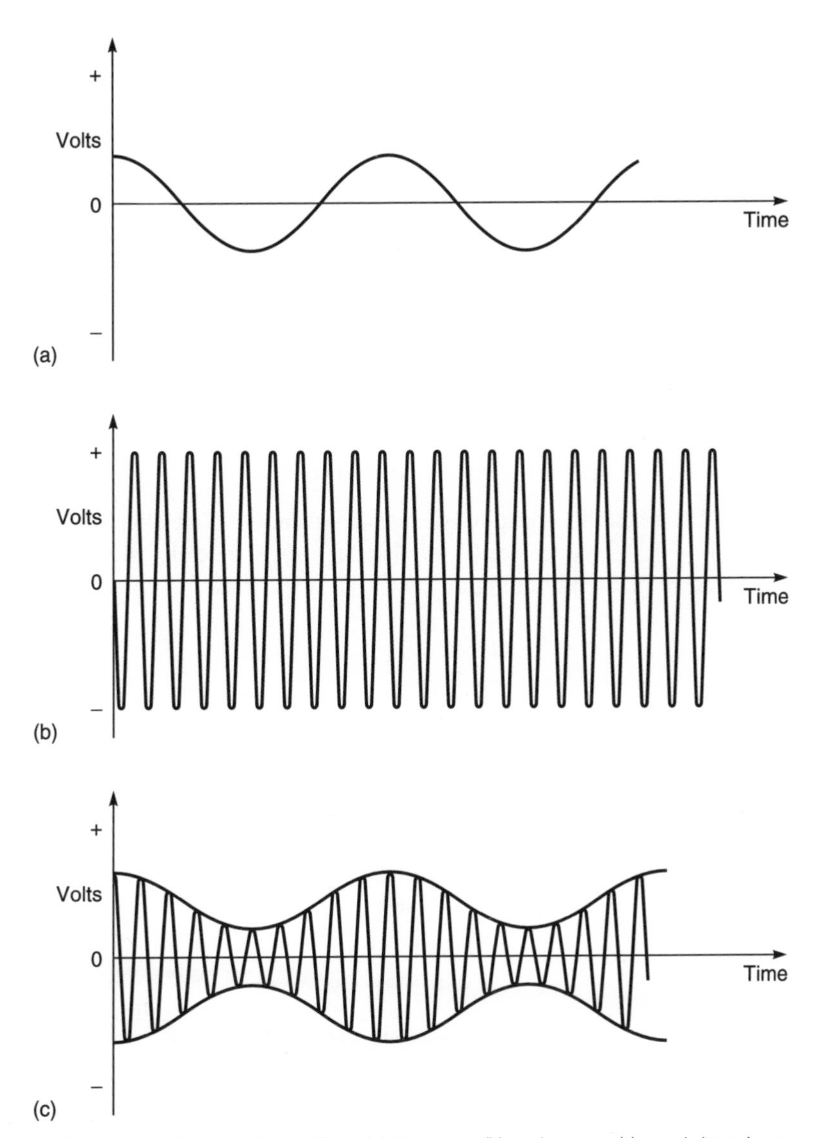

Fig. 1.3 Amplitude modulation waveforms: (a) modulating wave; (b) carrier wave; (c) modulated wave

Figure 1.3(a) shows the modulating wave on its own.[6] Figure 1.3(b) shows the carrier wave on its own. Figure 1.3(c) shows the resultant wave. The resultant wave shape is due to the fact that at times the modulating wave and the carrier wave are adding (in phase) and at other times, the two waves are opposing each other (out of phase).

Amplitude modulation can also be easily analysed mathematically. Let the sinusoidal modulating wave be described as

[6] I have used a cosine wave here because you will see later when we use Fourier analysis that waveforms, no matter how complicated, can be resolved into a series of d.c., sine and cosine terms and their harmonics.

$$v_m = V_m \cos(\omega_m t) \tag{1.2}$$

where

v_m = instantaneous modulating amplitude (volts)
V_m = modulating amplitude (peak volts)
ω_m = angular frequency in radians and $\omega_m = 2\pi f_m$ where
f_m = modulating frequency (hertz)

When the amplitude of the carrier is made to vary about V_c by the message signal v_m, the modulated signal amplitude becomes

$$[V_c + V_m \cos(\omega_m t)] \tag{1.3}$$

The resulting envelope AM signal is then described by substituting Equation 1.3 into Equation 1.1 which yields

$$[V_c + V_m \cos(\omega_m t)] \cos(\omega_c t + \phi_c) \tag{1.4}$$

It can be shown that when this equation is expanded, there are three frequencies, namely $(f_c - f_m)$, f_c and $(f_c + f_m)$. Frequencies $(f_c - f_m)$ and $(f_c + f_m)$ are called sideband frequencies. These are shown pictorially in Figure 1.4.

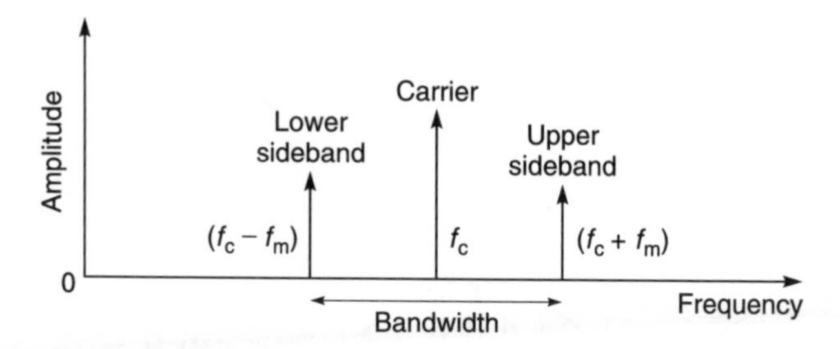

Fig. 1.4 Frequency spectrum of an AM wave

The modulating information is contained in one of the sideband frequencies which must be present to extract the original message. The bandwidth (bw) is defined as the highest frequency minus the lowest frequency. In this case, it is $(f_c + f_m) - (f_c - f_m) = 2f_m$ where f_m is the highest modulation frequency. Hence, a radio receiver must be able to accommodate the bandwidth of a signal.[7]

1.3.2 Frequency modulation (FM)

Frequency modulation is the modulation method used in VHF radio broadcasting. Figure 1.5 shows what happens when we apply frequency modulation to a sinusoidal carrier wave. Figure 1.5(a) shows the modulating wave on its own. Figure 1.5(b) shows the carrier wave on its own. Figure 1.5(c) shows the resultant wave. The resultant wave shape is due to the

[7] This is not unusual because speech or music also have low notes and high notes and to hear them our own ears (receivers) must be able to accommodate their bandwidth. Older people tend to lose this bandwidth and often are unable to hear the high notes.

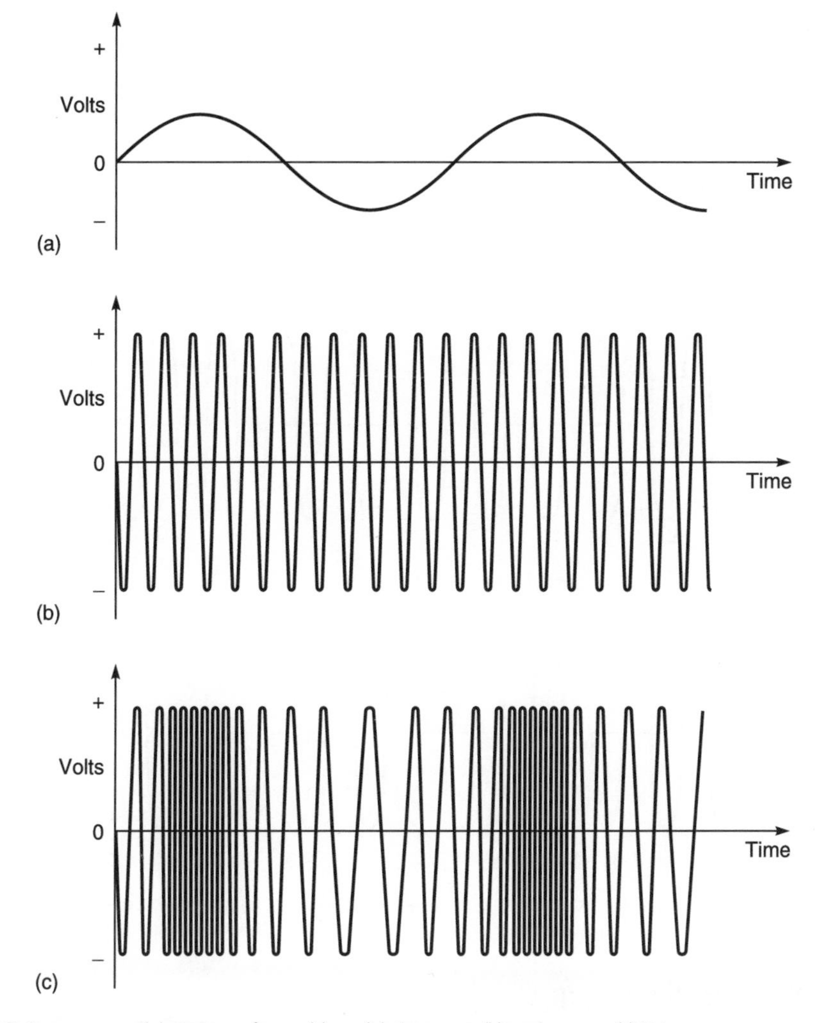

Fig. 1.5 Frequency modulation waveforms: (a) modulating wave; (b) carrier wave; (c) FM wave

fact that the carrier wave frequency increases when the modulating signal is positive and decreases when the modulating signal is negative. Note that in pure FM, the amplitude of the carrier wave is not altered.

The frequency deviation (Δf_c) of the carrier is defined as $[f_{c\ (max)} - f_{c\ (min)}]$ or

$$\Delta f_c = f_{c\ (max)} - f_{c\ (min)} \tag{1.5}$$

According to Carson's rule, the frequency bandwidth required for wideband FM is approximately 2 × (maximum frequency deviation + highest frequency present in the message signal) or

$$\text{bw} = 2\ [\Delta f_c + f_{m\ (max)}] \tag{1.6}$$

In FM radio broadcasting, the allocated channel bandwidth is about 200 kHz.

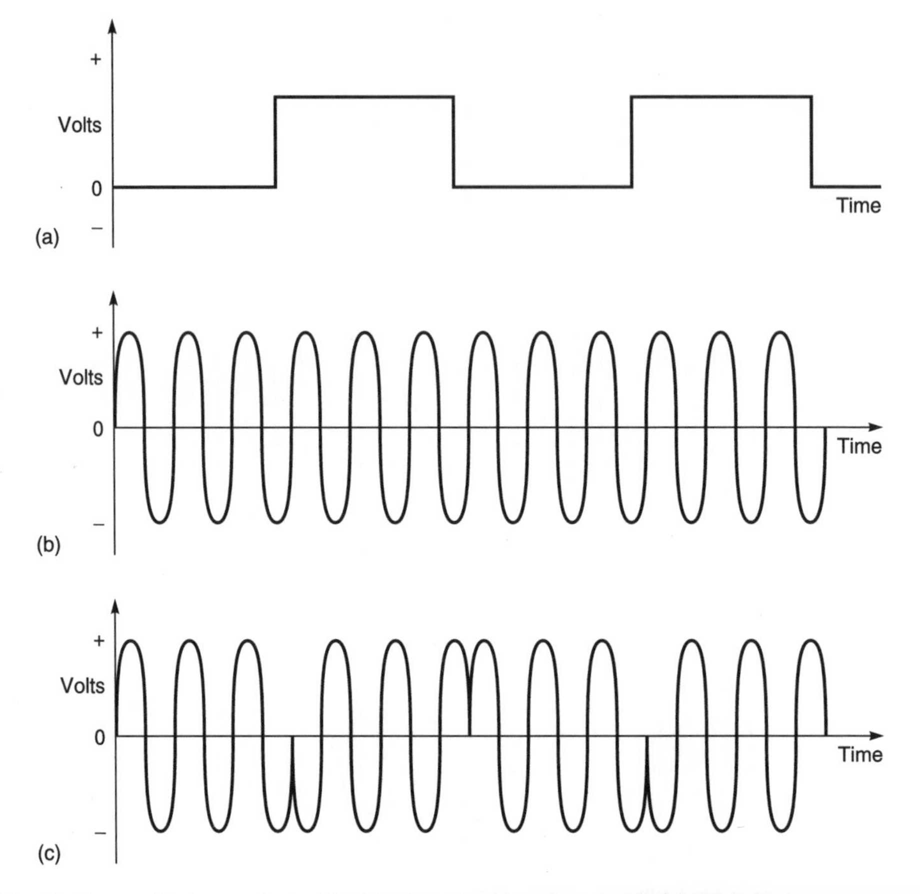

Fig. 1.6 Phase modulation waveforms: (a) modulating wave; (b) carrier wave; (c) modulated wave

1.3.3 Phase modulation (PM)

Phase modulation is particularly useful for digital waveforms. Figure 1.6 shows what happens when we apply phase modulation to a sinusoidal carrier wave. Figure 1.6(a) shows a digital modulating wave on its own. We have used a pulse waveform as opposed to a sine wave in this instance because it demonstrates phase modulation more clearly. Figure 1.6(b) shows the carrier wave on its own. Figure 1.6(c) shows the resultant wave. Note particularly how the phase of the carrier waveform changes when a positive modulating voltage is applied. In this particular case, we have shown you a phase change of 180°, but smaller phase changes are also possible.

Phase modulation is popularly used for digital signals. Phase modulation is synonymous with frequency modulation in many ways because an instantaneous change in phase[8] is also an instantaneous change in frequency and vice-versa. Hence, much of what is said about FM also applies to PM.

[8] Phase (ϕ) = angular velocity (ω) multiplied by time (t). Hence $\phi = \omega t$. Note this equation is similar to that of distance = velocity × time. This is because ϕ = amount of angle travelled = velocity (ω) × time (t).

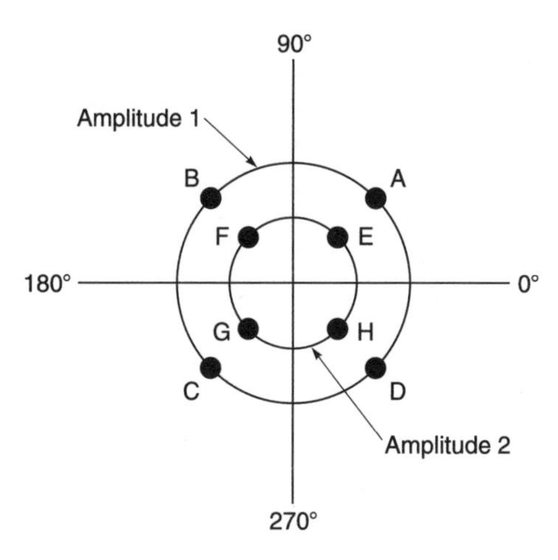

Fig. 1.7 An eight level coded signal modulated on to a radio carrier

1.3.4 Combined modulation methods

Digital signals are often modulated on to a radio carrier using both phase and amplitude modulation. For example, an eight level coded digital signal can be modulated on to a carrier by using distinct 90° phase changes and two amplitude levels. This is shown diagrammatically in Figure 1.7 where eight different signals, points A to H, are encoded on to a radio carrier. This method is also known as quadrature amplitude modulation (QAM).

1.3.5 Summary of modulation systems

In this section, we have shown you four methods by which information signals can be modulated on to a radio carrier.

1.4 Radio wave propagation techniques

1.4.1 Properties of electromagnetic waves

In Figure 1.8 we show the case of a radio generator feeding energy into a load via a two wire transmission line. The radio generator causes voltage and current waves to flow towards the load. A voltage wave produces a voltage or electric field. A current wave produces a cuurent or magnetic field. Taken together these two fields produce an **electro-magnetic field** which at any instant varies in intensity along the length of the line.

The electromagnetic field pattern is, however, far from stationary. Like the voltage on the line, it propagates from end to end with finite velocity which – for an air spaced line – is close to the velocity of light in free space.[9] The flow of power from source to

[9] Strictly speaking 'free space' is a vacuum. However, the velocity of propagation of electro-magnetic waves in the atmosphere is practically the same as that in a vacuum and is approximately 3×10^8 metres per second. Wavelength (λ) is defined as the ratio, velocity/frequency.

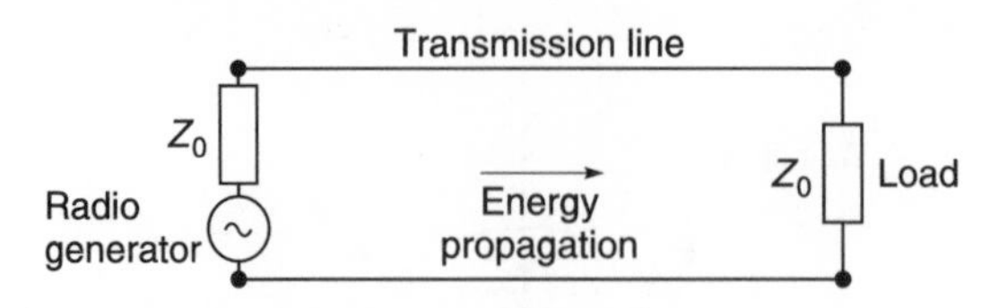

Fig. 1.8 Energy propagation in a transmission line

load is then regarded as that of an **electromagnetic wave** propagating between the conductors.

The equivalence between the circuit and field descriptions of waves on transmission lines is demonstrated by the fact that at any point in the electromagnetic field the instantaneous values of the electric field (E) (volts/metre) and the magnetic field (H) (amperes/metre) are related by

$$\frac{E(\text{V/m})}{H(\text{A/m})} = Z_0 \text{ (ohms)} \tag{1.7}$$

where Z_0 is the characteristic impedance of the transmission line.[10] It can also be shown that both approaches give identical results for the power flow along a matched line.

In the two wire transmission line shown in Figure 1.8, the parallel conductors produce electromagnetic fields which overlap and cancel in the space beyond the conductors. The radio frequency energy is thus confined and guided by the conductors from the source to its destination. If, however, the conductor spacing is increased so that it becomes comparable with the wavelength of operation the line will begin to radiate r.f. energy to its surroundings. The energy is lost in the form of **free-space electromagnetic waves** which radiate away from the line with the velocity of light.

The 19th century mathematician James Clerk Maxwell was the first to recognise that electromagnetic waves can exist and transport energy quite independently of any system of conductors. We know now that radio waves, heat waves, visible light, X-rays are all electromagnetic waves differing only in frequency. Figure 1.9 shows the range of frequencies and the regions occupied by the different types of radiation. This is known as the **electromagnetic spectrum**.

Hz	Power and audio signals	Radio signals	Microwaves	Infra red and visible	UV X-rays X-rays->
10^0	10^1 10^4	10^5 10^8	10^9 10^{11}	10^{12} ~ 10^{15}	10^{15} ~ 10^{17}

Fig. 1.9 The electromagnetic frequency spectrum

[10] Transmission lines have impedances because they are constructed from physical components which have resistance, self inductance, conductance and capacitance.

1.4.2 Free-space radiation

Introduction

At operational frequencies, where the operational wavelengths are comparable in size to circuit components,[11] any circuit consisting of components connected by conductors will tend to act as an imperfect transmission line. As a result, there will always be some loss of r.f. energy by way of radiation. In other words, the circuit will tend to behave like a crude radio transmitter antenna.

It follows that for minimal radiation, components should be small with respect to their operational wavelengths. Conversely, if radiation is desired, then the physical components should be large, approximately 1/4 wavelength for optimum radiation. This is why antennas are physically large in comparison with their operational wavelength.

Energy radiates from an r.f. source or transmitter in all directions. If you imagine a spherical surface surrounding the transmitter, then the interior of the surface would be 'illuminated' with radiated energy, just like the inside of a globular lamp-shade. The illumination is not necessarily uniform, however, since all transmitters are, to some extent, directional.

If the r.f. source is sinusoidal, then the electric and magnetic fields will also be varying sinusoidally at any point in the radiation field. Now it is difficult to depict a propagating electromagnetic field but some of its important properties can be identified. To do this we consider propagation in a particular direction on a straight line connecting a transmitter to a distant receiver as shown in Figure 1.10. You will see that this line coincides with the z-direction in Figure 1.10. Measurements at the radio receiver would then indicate that the oscillating electric field is acting all in one direction, the x-direction in Figure 1.10. The magnetic field is in-phase with the electric field but acts at right-angles to the electric field, in the y-direction. The two fields are thus at right-angles to each other and to the direction of propagation. An electromagnetic wave with these characteristics is known as a **plane wave**.

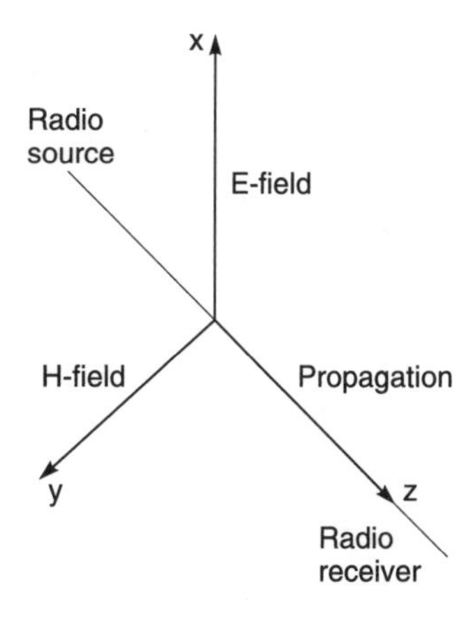

Fig. 1.10 Electric and magnetic field directions for an electromagnetic wave propagating in the z-direction

[11] Generally taken to be the case when the operational wavelength is about 1/20 of the physical size of components.

Polarisation

Provided there is no disturbance in the propagation path, the electric and magnetic field orientations with respect to the earth's surface will remain unchanged. By convention, the orientation of the electric field with respect to the earth's surface is called the **polarisation** of the electromagnetic wave. If the electric field is vertical, the wave is said to be **vertically polarised**; if horizontal, the wave is **horizontally polarised**. A wave is **circularly polarised** if its electric field rotates as the wave travels. Circular polarisation can be either *clockwise* or *anti-clockwise*.

Polarisation is important because antennas must be mounted in the correct plane for optimum signal reception.[12] Terrestrial broadcasting stations tend to use either vertical or horizontal polarisation. Satellite broadcasting stations use circular polarisation. The polarisation of a wave is sometimes 'twisted' as it propagates through space. This twisting is caused by interfering electric or magnetic fields. It is particularly noticeable near steel-structured buildings where aerials are mounted at odd angles to the vertical and horizontal planes to compensate for these effects.

Field strength

The strength of a radio wave can be expressed in terms of the strength of its electric field or by the strength of its magnetic field. You should recall that these are measured in units of volts per metre and amperes per metre respectively. For a sinusoidally varying field it is customary to quote r.m.s. values E_{rms} and H_{rms}. What is the physical significance of E_{rms}? This is numerically equal to the r.m.s. voltage induced in a conductor of length 1 m when a perpendicular electromagnetic wave sweeps over the conductor with the velocity of light.

As stated earlier, the electric and magnetic fields in a plane wave are everywhere in phase. The ratio of the field strengths is always the same and is given by

$$\frac{\text{electric field strength } E_{rms}(\text{V/m})}{\text{magnetic field strength } H_{rms}(\text{A/m})} = 377\,\Omega \tag{1.8}$$

This ratio is called the free-space **wave impedance**. It is analogous to the characteristic impedance of a transmission line.

Example 1.1

The electric field strength at a receiving station is measured and found to have an r.m.s value of 10 microvolts/m. Calculate (a) the magnetic field strength; (b) the amount of power incident on a receiving aerial with an effective area of 5 m^2.

Given: Electric field strength = 10 microvolts/m.
Required: (a) Magnetic field strength, (b) incident power on a receiving aerial with effective area of 5 m^2.

[12] You can see this effect by looking at TV aerials mounted on houses. In some districts, you will see aerials mounted horizontally whilst in other areas you will find aerials mounted vertically. As a general rule, TV broadcasting authorities favour horizontal polarisation for main stations and vertical polarisation for sub or relay stations.

Solution. Using equation 1.8

(a) H_{rms} = 10 μV/m/377 Ω = 2.65×10^{-8} A/m
(b) Power density is given by

$E_{rms} \times H_{rms} = 10 \times 10^{-6} \times 2.65 \times 10^{-8}$ W/m^2 = 2.65×10^{-13} W/m^2

This is the amount of power incident on a surface of area 1 m^2. For an aerial with area 5 m^2, the total incident power will be

$P = 2.65 \times 10^{-13}$ W/m^2 $\times$ 5 m^2 = 1.33 pW

Power density

The product $E_{rms} \times H_{rms}$ has the dimensions of 'volts per metre' times 'amps per metre', giving watts per square metre. This is equivalent to the amount of r.f. power flowing through one square metre of area perpendicular to the direction of propagation and is known as the **power density** of the wave. The power density measures the intensity of the 'illumination' falling on a receiving aerial.

A plane wave expands outwards as it travels through space from a point source. As a result, the power density falls off with increasing distance from the source. If you have studied any optics then you will be familiar with the idea that the power density falls off as the *square of the distance* from the source, i.e.

$$\frac{P_{D2}}{P_{D1}} = \left[\frac{D_1}{D_2}\right]^2 \qquad (1.9)$$

where P_{D1}, P_{D2} = power densities at distances D_1 and D_2 respectively.

Example 1.2

If the data in Example 1.1 applies to a receiver located 10 km from the transmitter, what will be the values of E_{rms} and H_{rms} at a distance of 100 km?

Given: Data of Example 1.1 applied to a receiver at 10 km from transmitter.
Required: (a) E_{rms} at 100 km, (b) H_{rms} at 100 km.

Solution. Using Equation 1.9 at a distance of 100 km, the power density will be reduced by a factor $(10/100)^2 = 0.01$, so power density = 2.65×10^{-15} W/m^2. Now, power density = $E_{rms} \times H_{rms}$ and since $H_{rms} = E_{rms}/377$ (Equation 1.7)

$$\frac{E_{rms}^2}{377} = 2.65 \times 10^{-15} \text{ W/m}^2$$

Hence

$$E_{rms} = \sqrt{2.65 \times 10^{-15} \times 377} = 1\,\mu\text{V}/m$$

and

$$H_{rms} = 1\ \mu\text{V/m}/377\ \Omega = 2.65 \times 10^{-9}\ \text{A/m}$$

Summary of propagation principles

Several important points have been established in Section 1.4.

- R.F. energy is radiated by way of travelling electric and magnetic fields which together constitute an electromagnetic wave propagating in free space with the velocity of light.
- In a plane wave, the electric and magnetic fields vary in phase and act at right-angles to each other. Both fields are at right-angles to the direction of propagation.
- The direction of the electric field determines the polarisation of a plane wave.
- At any point, the ratio of the electric and magnetic fields is the same and equal to the wave impedance. This impedance is 377 Ω approximately.
- The product $E_{rms} \times H_{rms}$ gives the power density of the wave.
- The power density falls off as the square of the distance from the r.f. source.
- To obtain optimum signal reception from free space a receiving aerial should be set for the correct polarisation and be suitably located with regard to height and direction.

1.5 Antennas and aerials

1.5.1 Introduction

An **antenna** or **aerial** is a structure, usually made from good conducting material, that has been designed to have a shape and size so that it will provide an efficient means of transmitting or receiving electromagnetic signals through free space. Many of the principles used in the construction of antennas can be easily understood by analogy to the headlamp of your car (see Figure 1.11).

An **isotropic** light source is a light source which radiates light equally in all directions. The radiation pattern from an isotropic light source can be altered by placing a reflecting mirror on one side of the light source. This is carried out in car headlamps where a quasi-parabolic reflecting mirror (reflector) is placed behind a bulb to increase the light intensity of the lamp in the forward direction. The reflector has therefore produced a change in the **directivity** of the light source. The increase or 'gain' of light intensity in the forward direction has been gained at the expense of losing light at the back of the lamp. This gain is not a 'true gain' because total light energy from the lamp has not been increased; light energy has only been *re-directed* to produce an intensity gain in the forward direction.

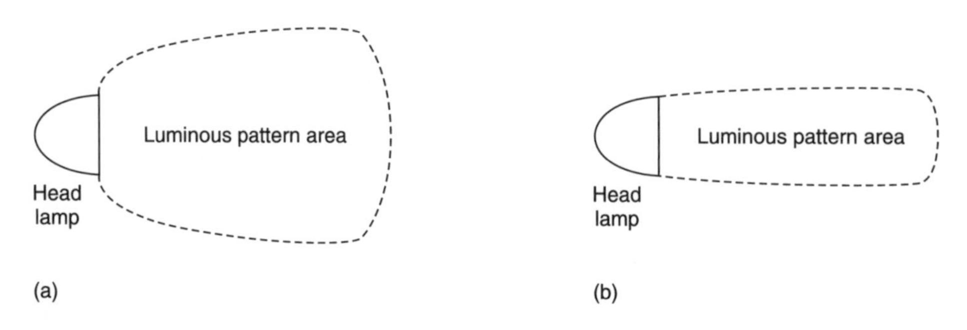

Fig. 1.11 Radiation patterns from a car headlamp: (a) top view; (b) side view

The forward light intensity of a car lamp can be further improved by using one or more lenses to concentrate its forward light into a main beam or **main lobe**. Again, this 'gain' in light intensity has been achieved by confining the available light into a narrower beam of illumination; there has been no overall gain in light output from the bulb.

There are also optimum sizes and distances for the placement of reflectors and lenses. These are dictated by the physical size of the bulb, the desired gain intensity of the main beam or main lobe, the required width of the main beam and the requirement to suppress minor or spurious light lobes which consume energy and cause unnecessary glare to on-coming motorists.

A car headlamp (Figure 1.11) has two main light-emitting patterns; a **horizontal** pattern and a **vertical** pattern. The horizontal pattern (Figure 1.11(a)) is a bird's eye view of the illumination pattern. A plot of the horizontal pattern is called a **polar diagram**. The vertical or **azimuth** pattern (Figure 1.11(b)) is the pattern seen by an observer standing to one side of the lamp. The vertical pattern is sometimes called the **end-fire** pattern. Both light patterns must be considered because modern headlamp reflectors tend to be elliptical and affect emitted light in the horizontal and vertical planes differently.

In the above description, light has been assumed to travel from bulb to free space but the effect is equally true for light travelling in the opposite direction, i.e. the system is *bi-directional.* It can be used either for *transmitting* light from the bulb or for *receiving* external light at the point source usually occupied by the bulb filament. This can be easily verified by shining an external light source through the lens and the reflector in the opposite direction from which light had emerged, and seeing it converge on the bulb source.[13]

Many of the principles introduced above apply to antennas as well. Because of its bi-directional properties, a radio antenna can be used for transmitting or receiving signals.

1.5.2 Radiating resistance

The relationship, power (watts) = (volts^2/ohms), is used for calculating power loss in a circuit. It is not always possible to apply this law directly to a radiating circuit because a physical resistor does not always exist. Yet we cannot deny that there is a radiated power loss when a voltage is applied across a radiating circuit. To overcome this problem, engineers postulate an 'equivalent' resistor to represent a physical resistor which would absorb the same radiated power loss. This equivalent resistor is called the **radiating resistance** of the circuit.

The radiating resistance of an antenna should not be confused with its input impedance. The input impedance is the value used when considering the connection of an antenna to a transmission line with a specified characteristic impedance. Antennas are bi-directional and it is not uncommon to use the same antenna for transmitting and receiving signals.

Example 1.3

A transmitter with an output resistance of 72 Ω and an r.m.s. output of 100 V is connected via a matched line to an antenna whose input resistance is 72 Ω. Its radiation resistance is also 72 Ω. Assuming that the antenna is 100% efficient at the operating frequency, how much power will be transmitted into free space?

[13] If you have any doubts about the system being bi-directional, you should visit a lighthouse which uses a similar reflector and lens system. Curtains must be drawn around the system during daylight hours because sunlight acting on the system has been known to produce such high light and heat intensities that insulation melt-down and fires have been caused.

Given: Transmitter output = 100 V, transmitter output impedance = 72 Ω, antenna input impedance = 72 Ω, radiation resistance = 72 Ω, antenna efficiency = 100%.
Required: Power radiated into free space.

Solution. The antenna has an input impedance Z_{in} = 72 Ω and provides a matched termination to the 72 Ω line. The r.f. generator then 'sees' an impedance of 72 Ω, so the r.m.s. voltage applied to the line will be 100/2 = 50 V. The amount of power radiated is calculated using

$$\text{radiated power} = \frac{50^2}{R}$$

where R = 72 Ω is the radiation resistance. The radiated power is therefore 34.7 W. Notice that, because in this case $R = Z_{in}$, maximum power is radiated into free space.

1.5.3 The half-wave dipole antenna

Most antennas can be analysed by considering them to be transmission lines whose configurations and physical dimensions have been altered to present easy energy transfer from transmission line to free space. In order to do this effectively, most antennas have physical sizes comparable to their operational wavelengths.

Figure 1.12(a) shows a two wire transmission line, open-circuited at one end and driven by a sinusoidal r.f. generator. Electromagnetic waves will propagate along the line until it reaches the open-circuit end of the line. At the open-circuit end of the line, the wave will be reflected and travel back towards the sending end. The forward wave and the reflected wave then combine to form a voltage standing wave pattern on the line. The voltage is a maximum at the open end. At a distance of one quarter wavelength from

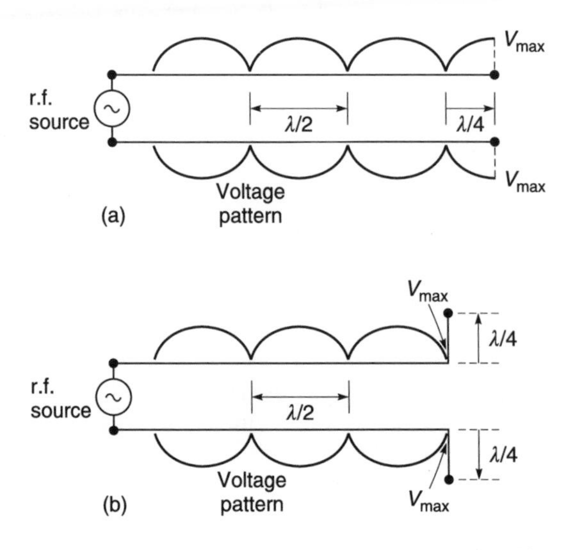

Fig. 1.12 (a) Voltage standing-wave pattern on an open-circuited transmission line; (b) open-circuited line forming a dipole

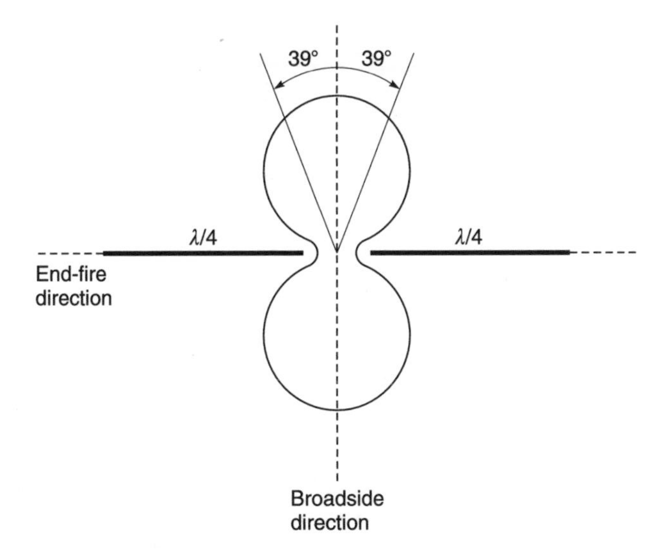

Fig. 1.13 Polar pattern of a half-wave dipole

the end, the voltage standing wave is at a minimum because the sending wave and the reflected wave oppose each other. Suppose now that the wires are folded out from the $\lambda/4$ points, as in Figure 1.12(b). The resulting arrangement is called a **half-wave dipole antenna**.

Earlier we said that the electromagnetic fields around the parallel conductors overlap and cancel outside the line. However, the electromagnetic fields along the two ($\lambda/4$) arms of the dipole are now no longer parallel. Hence there is no cancellation of the fields. In fact, the two arms of the dipole now act in series and are additive. They therefore reinforce each other. Near to the dipole the distribution of fields is complicated but at a distance of more than a few wavelengths electric and magnetic fields emerge in phase and at right-angles to each other which propagate as an electromagnetic wave.

Besides being an effective radiator, the dipole antenna is widely used as a VHF and TV receiving antenna. It has a polar diagram which resembles a figure of eight (see Figure 1.13). Maximum sensitivity occurs for a signal arriving broadside on to the antenna. In this direction the 'gain' of a dipole is 1.5 times that of an isotropic antenna. An isotropic antenna is a theoretical antenna that radiates or receives signals uniformly in all directions.

The gain is a minimum for signals arriving in the 'end-fire' direction. Gain decreases by 3 dB from its maximum value when the received signal is ±39° off the broadside direction. The maximum gain is therefore 1.5 and the half-power beam-width is 78°. The input impedance of a half-wave dipole antenna is about 72 Ω. It turns out that the input impedance and the radiation resistance of a dipole antenna are about the same.

1.5.4 Folded dipole antenna

The folded dipole (Figure 1.14) is a modified form of the dipole antenna. The antenna is often used for VHF FM receivers. The impedance of a folded $\lambda/2$ dipole is approximately 292 Ω. This higher input impedance is advantageous for two main reasons:

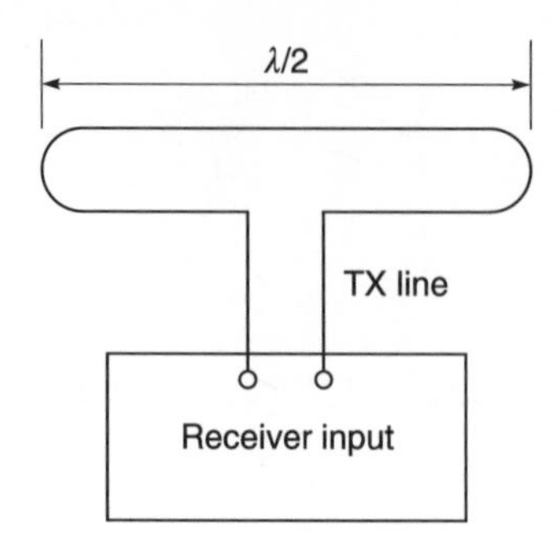

Fig. 1.14 Folded dipole antenna

- it allows easy connection to 300 Ω balanced lines.
- its higher impedance makes it more compatible for use in directive aerials (particularly Yagi arrays) which will be described in Section 1.6.

1.5.5 The monopole or vertical rod antenna

The monopole or vertical rod antenna (Figure 1.15) is basically a coaxial cable[14] whose outer conductor has been removed and connected to earth. It is usually about $\lambda/4$ long except in cases where space restrictions or other electrical factors restrict its length. At high frequencies, the required $\lambda/4$ length is short and the antenna can be made self-supporting by the use of hollow metal tubing. At low frequencies where a greater length is required, the antenna is often supported by poles.

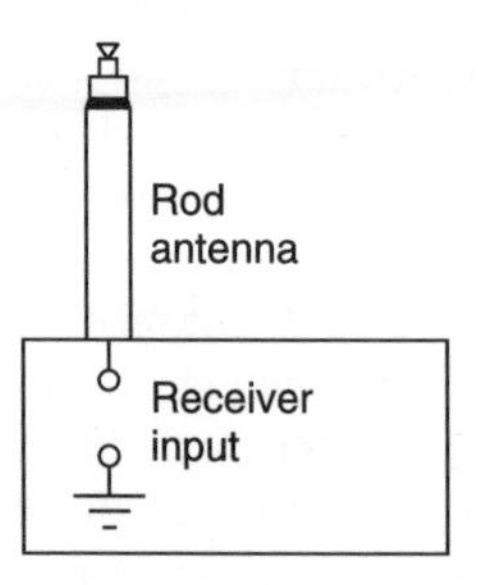

Fig. 1.15 Rod or monopole antenna

This antenna is favoured for use in low frequency transmitting stations, in portable radio receivers, in mobile radio-telephones, and for use on motor vehicles because it has a circular polar receiving pattern, i.e. it transmits and receives signals equally well in all directions around its circumference. This is particularly important in mobile radio-phones and in motor vehicles because a motor vehicle may be moving in any direction with respect to a transmitting station. To minimise interference from the engine of the vehicle and for

[14] A typical example of a coaxial cable is the TV lead which connects your television set to the antenna.

maximum receiving height, rod aerials are frequently mounted on the roofs of vehicles. These aerials are also often mounted at an angle of about 45° to the horizon to enable them to be receptive to both horizontal and vertical polarisation transmissions.

1.5.6 Single loop antennas

Another type of antenna which is frequently used for TV reception is the single loop antenna shown in Figure 1.16. This loop antenna usually has an electrical length equal to approximately $\lambda/2$ at its operating frequency. It is popular with TV manufacturers because it is comparatively cheap and easy to manufacture. The antenna's input impedance is approximately 292 Ω and it is easily coupled to 300 Ω balanced transmission lines. The antenna is directive and has to be positioned for maximum signal pick-up.

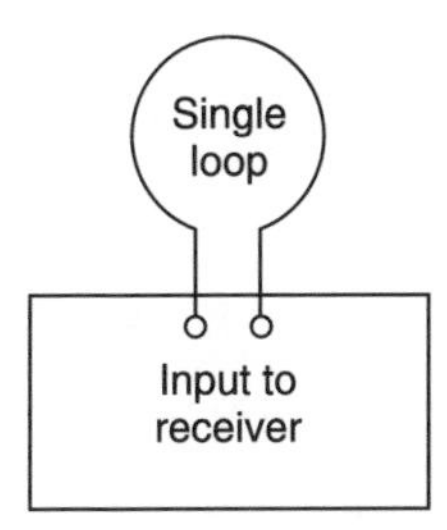

Fig. 1.16 Single loop antenna

1.5.7 Multi-loop antennas

At low frequencies, particularly at frequencies in the medium wave band where wavelengths are long, single loop $\lambda/2$ length antennas are not practical; multi-loop antennas (Figure 1.17) are used instead. The multi-loop antenna can be reduced even further in size if a ferrite rod is inserted within the loop.

The open-circuit voltage induced in multiple loop antennas can be calculated by making use of Faraday's Law of Electromagnetic Induction which states that the voltage induced in a coil of n turns is proportional to the rate of change of magnetic flux linkage. For simplicity in derivation, it will be assumed that the incident radiation is propagating along the axis of the coil (see Figure 1.18).

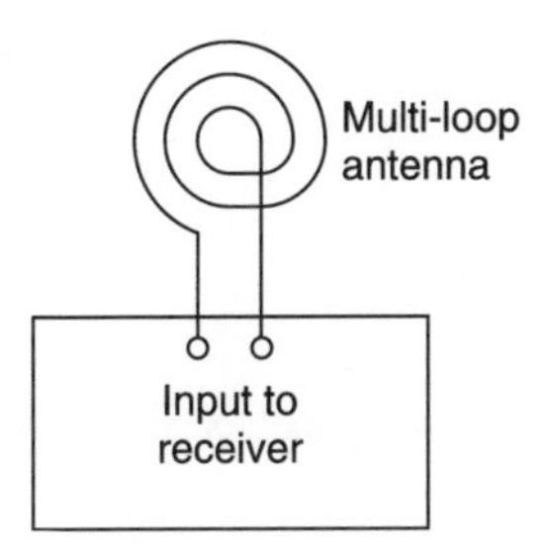

Fig. 1.17 Multi-loop antenna

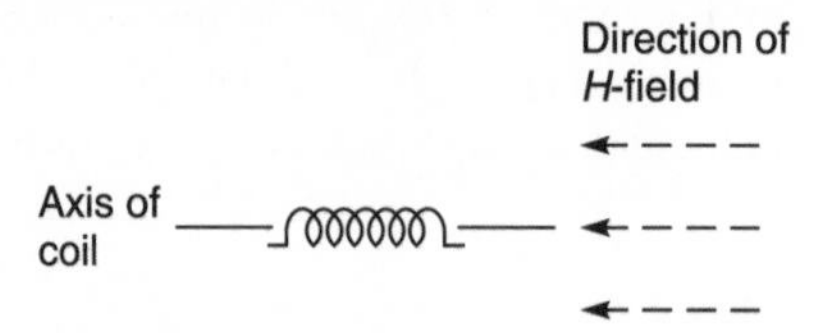

Fig. 1.18 Multi-looped antenna aligned for maximum flux linkage

Expressing Faraday's Law mathematically,

$$e = n\frac{d\phi}{dt} \tag{1.10}$$

where

e = open-circuit voltage in volts
n = number of turns on coil
$d\phi/dt$ = rate of change of magnetic flux linkage (ϕ = webers and t = seconds)

Some fundamental magnetic relations are also required. These include:

total flux ϕ = flux density (B) per unit area × area (A)

or

$$\phi_{\text{webers}} = B_{\text{tesla}} \times A_{\text{square metres}} \tag{1.11}$$

By definition flux density in air cored coil (B_{tesla}) is given by

free-space permeability (μ_0) × magnetic field strength (H)

$$B_{(\text{tesla})} = \mu_{0\,(\text{henry/metre})} \times H_{(\text{ampere/metre})} \tag{1.12}$$

or
Suppose that the incident wave has a magnetic field strength

$$H = H_{\max} \sin \omega t \tag{1.13}$$

where ω is the angular frequency of the r.f. signal. Then substituting Equations 1.12 and 1.13 in Equation 1.11 yields

$$\phi = BA = \mu_0 H_{\max} \sin \omega t \times A \tag{1.14}$$

Taking the rate-of-change[15] in Equation 1.14, then the induced voltage is

$$\begin{aligned} e &= n\frac{d\phi}{dt} \\ &= n\omega\mu_0 A H_{\max} \cos \omega t \end{aligned} \tag{1.15}$$

[15] If you do not know how to differentiate to get the rate of change of a value, then please refer to a maths book.

For a coil with a ferrite core, the flux density is increased by the relative effective permeability (μ_r), giving

$$e = n\omega\mu_0\mu_r AH_{max} \cos \omega t \tag{1.16}$$

You will see that the ferrite core has increased the effective area of the coil by a factor μ_r. Ferrite cores with effective relative permeabilities of 100–300 are readily available but even with these values, the effective area of the aerial is relatively small when compared with a $\lambda/2$ aerial length. The ferrite rod aerial is therefore very inefficient when compared to an outdoor aerial but it is popular because of its convenient size and portability. At medium wave frequencies, the inherent poor signal pick-up is acceptable because broadcast stations radiate large signals.

In the foregoing derivation, it has been assumed that the magnetic field has been cutting the coil along its axis. Occasions arise when the incident magnetic field arrives at an angle α with respect to the axis of the coil. This is shown in Figure 1.19. In this case the effective core area is reduced by cos α, and the induced voltage becomes

$$e = n\omega\mu_0 AH_{max} \cos \omega t \cos \alpha \tag{1.17}$$

This expression shows that the induced open-circuit voltage, e, is dependent on the axial direction of the aerial coil with respect to the direction of the propagation. It is maximum when $\cos \alpha = 1$, i.e. $\alpha = 0°$, and minimum when $\cos \alpha = 0$, i.e. $\alpha = 90°$. This explains why it is necessary to position a loop aerial to receive maximum signal from a particular broadcasting station and this is done in a portable radio receiver by orienting its direction.

The above reasons apply equally well to ferrite rod aerials and for these cases we have an induced voltage

$$e = n\omega\mu_r\mu_0 AH_{max} \cos \omega t \cos \alpha \tag{1.18}$$

If the magnetic field strength is given as an r.m.s. value (H_{rms}), then the r.m.s. value of the induced voltage is

$$e_{rms} = n\omega\mu_r\mu_0 AH_{rms} \cos \alpha \tag{1.19}$$

Finally, ferrite aerials are seldom used at the higher frequencies because ferrite can be extremely lossy above 10 MHz.

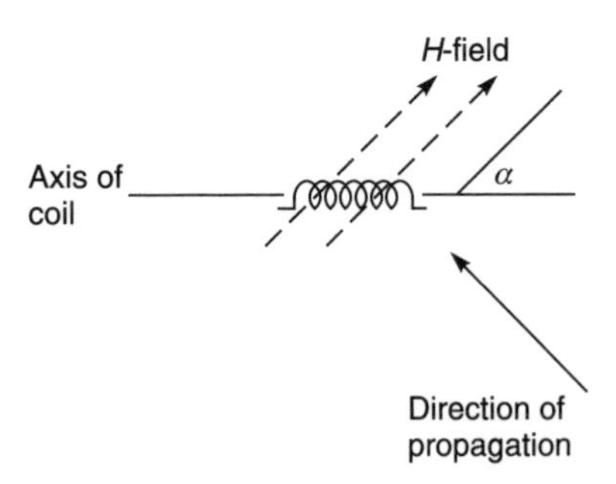

Fig. 1.19 *H* field arriving at an angle α

Example 1.4

A coil of 105 turns is wound on a ferrite rod with an effective cross-sectional area of 8×10^{-5} m^2. The relative permeability of the ferrite is 230 and the permeability of air is $4\pi \times 10^{-7}$ henry/m. The r.m.s. field strength is 10 μA/m. If the magnetic field is incident along the axis of the coil and the frequency of operation is 1 MHz, what is the r.m.s. open-circuit voltage induced in the coil?

Given: No. of coil turns = 105, effective cross-sectional area of ferrite rod = 8×10^{-5} m^2, relative permeability (μ_r) = 230, permeability of air (μ_0) = $4\pi \times 10^{-7}$ Henry/metre, r.m.s. field strength = 10 μA/m, frequency = 1 MHz.
Required: r.m.s. open-circuit voltage induced in coil.

Solution. Using Equation 1.19

$$\begin{aligned} e_{\text{rms}} &= n\omega\mu_r\mu_0 AH_{\text{rms}} \cos\alpha \\ &= 105 \times 2\pi \times 1 \times 10^6 \times 230 \times 4\pi \times 10^{-7} \times 10 \times 10^{-6} \times 8 \times 10^{-5} \times \cos 0° \\ &= 152.5\ \mu\text{V} \end{aligned}$$

Broadcasting authorities tend to quote electric field strengths rather than magnetic field strengths for their radiated signals. This creates no problems because the two are related by the wave impedance formula given earlier as Equation 1.8. This is repeated below:

$$\frac{\text{electric field strength } (E)}{\text{magnetic field strength } (H)} = 377\,\Omega$$

Example 1.5

A coil of 100 turns is wound on a ferrite rod with an effective cross-sectional area of 8×10^{-5} m^2 . The relative permeability of the ferrite is 200 and the permeability of air is $4\pi \times 10^{-7}$ henry/m. The magnetic field is incident at an angle of 60° to the axis of the coil and the frequency of operation is 1 MHz. If the electric field strength is 100 μV/m, what is the r.m.s. open-circuit voltage induced in the coil?

Given: No. of coil turns = 105, effective cross-sectional area of ferrite rod = 8×10^{-5} m^2, relative permeability (μ_r) = 200, permeability, of air (μ_0) = $4\pi \times 10^{-7}$ henry/metre, incidence of magnetic field = 60°, frequency = 1 MHz, electric field strength = 100 μV/m.
Required: Open-circuit voltage (e_{rms}).

Solution. Substituting Equation 1.8 in Equation 1.19 yields

$$\begin{aligned} e_{\text{rms}} &= n\omega\mu_r\mu_0 A \frac{E_{\text{rms}}}{377} \cos\alpha \\ &= 100 \times 2\pi \times 1 \times 10^6 \times 200 \times 4\pi \times 10^{-7} \times 8 \times 10^{-5} \times \frac{100 \times 10^{-6}}{377} \times \cos 60° \\ &= 1.68\ \mu\text{V} \end{aligned}$$

1.6 Antenna arrays

1.6.1 Introduction

Antenna arrays are used to shape and concentrate energy in required patterns. One of the more common domestic arrays is the **Yagi-Uda array** used for the reception of television signals.

1.6.2 Yagi-Uda array

The Yagi-Uda aerial array shown in Figure 1.20 is one of the most commonly used antenna arrays. It is used extensively for the reception of TV signals and can be seen on the roofs of most houses. The Yagi array is an antenna system designed with very similar principles to the car headlamp system described in Section 1.5.1. Its main elements are a folded dipole, a reflector, and directivity elements which serve as 'electrical lenses' to concentrate the signal into a more clearly defined beam. The number of directors per array varies according to the gain required from the aerial. The length of directors and the spacing between them are also dependent on the number of elements used in the array. In general, gain increases with the number of directors, but greater gain needs more careful alignment with the transmitting station and requires that the antenna be more sturdily mounted otherwise its pointing direction will waver in high winds which can cause fluctuations in the received signal strength.

The Yagi array is usually designed to be connected to a 75 Ω transmission line.[16] Yagi

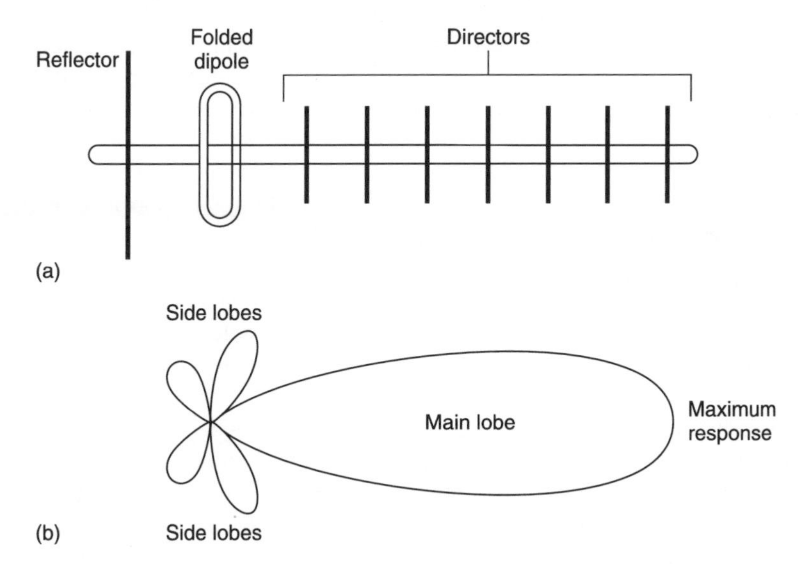

Fig. 1.20 Yagi-Uda array: (a) physical arrangement;(b) radiation pattern

[16] Earlier on, we said that the impedance of a folded dipole aerial was 292 Ω, yet now we say that this antenna is designed to operate with a 75 Ω system. This apparent discrepancy arises because the use of reflector and directors loads the folded dipole and causes its impedance to fall. Judicious director spacing is then used to set the array to the required impedance.

arrays suitable for operation over the entire TV band can be obtained commercially, but these broadband arrays are usually designed to 'trade off' bandwidth against aerial gain. Broadband Yagi arrays are extremely useful for mobile reception where minimum space and convenience are of importance. (You often see them on top of mobile caravans.) Domestic Yagi arrays are usually designed to provide greater gain but with a more restricted operational frequency band. The latter is not a disadvantage because TV stations operating from a common transmitting site confine their broadcasts to well defined frequency bands. The common practice for domestic Yagi arrays[17] is to use three or more designs (scaled in size) to provide reception for the complete TV band.

Typical values for Yagi arrays operating in the TV band are shown in Table 1.1. These figures have been taken from a well known catalogue but some of the terms need explanation.

Table 1.1 Typical values for Yagi arrays operating in the TV band

No. of elements	Forward gain (±0.5 dB)	Front/back ratio (± 2 dB)	Acceptance angle (±3°)
10	12	27	21
13	13	28	19
14	17	30	14
21	19	31	12

- 'Number of elements' means the total number of directors, folded dipoles and reflectors used in the array. For example, if the number of elements in an array is 10, the array includes eight directors, one folded dipole and one reflector.
- 'Forward gain' is the maximum 'gain' which the antenna can provide with respect to an isotropic aerial. A maximum aerial gain of 10 dB means that the antenna will provide 10 times the 'gain' you would get from an isotropic aerial when the array is pointed in its maximum gain direction.
- 'Front to back ratio' is the difference in gain between the direction of maximum antenna gain and the minimum direction of gain which is usually in the opposite direction. This ratio is important because it provides a measure of how the array behaves towards interfering signals arriving from different directions. It is particularly useful in confined areas such as cities where interfering signals 'bounce' off high buildings and interfere with a strong desired signal. In such cases, it is often better to select an antenna with a large front to back ratio to provide rejection to the interfering signal than trying to get maximum antenna gain.
- 'Acceptance angle' is the beamwidth angle in degrees where antenna gain remains within 3 dB of its stated maximum gain. An acceptance angle of 20° and a maximum array gain of 10 dB means that for any signal arriving within ±10° of the maximum gain direction the antenna will provide at least (10 – 3) dB, i.e. 7 dB of gain. However, you should be aware that the acceptance angle itself is not accurate and that it can vary by ±3° as well.

[17] There is a class of Yagi arrays known as Log Periodic Yagis. These have greater bandwidths because the directors are spaced differently. They do cover the entire TV bands but their gain is a compromise between frequency bandwidth and gain.

The values given in the table are representative of the middle range of commercially available Yagi arrays. The figures quoted above have been measured by manufacturers under ideal laboratory conditions and proper installation is essential if the specification is to be achieved in practice.

1.7 Antenna distribution systems

Occasions often arise where it is desired to have one antenna supply signal to several television and radio receivers. A typical example is that of an apartment block, where a single aerial on the roof supplies signals to all the apartments. Another possible use for such a system is in your own home where you would like to distribute signals to all rooms from a single external aerial. In such cases, and for maximum efficiency, an aerial distribution system is used. There are many ways of designing such a system but before discussing them, it is best to understand some of the terms used.

1.7.1 Balanced and unbalanced systems

Examples of balanced and unbalanced aerials and distribution lines are shown in Figures 1.21 and 1.22. You should refer to these figures while you are reading the descriptions given below.

A **balanced antenna** (Figure 1.21(a) and (b)) is an aerial which has neither conductor connected directly to earth; it is balanced because the impedance between earth and each conductor is the same. A folded dipole is a typical example of a balanced antenna because the impedance from each end of the antenna to earth is equal and balanced. An **unbalanced**

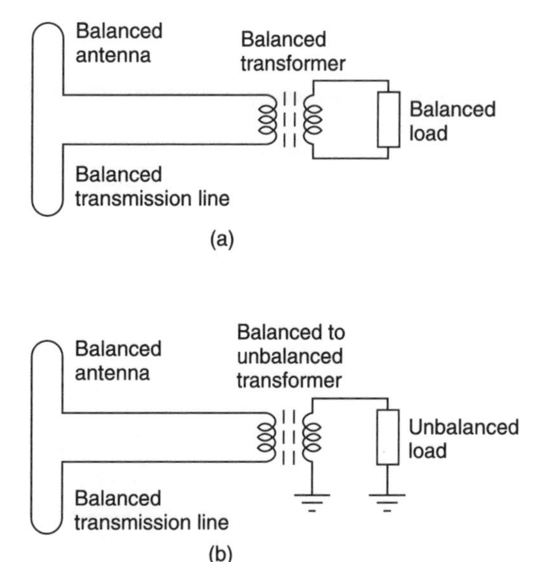

Fig. 1.21 Balanced antenna system and (a) balanced distribution system; (b) unbalanced distribution system

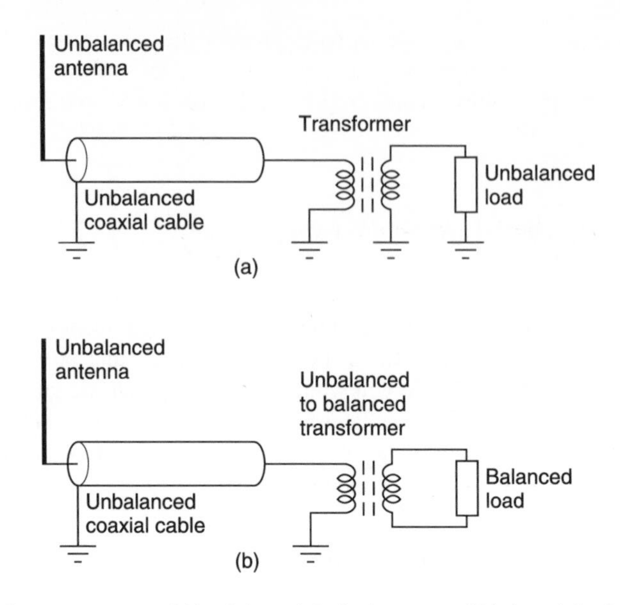

Fig. 1.22 Unbalanced antenna system and (a) unbalanced distribution system; (b) balanced distribution system

antenna (Figure 1.22(a) and (b)) is an aerial which has one of its conductors connected directly to earth. The impedance between earth and each conductor is not the same. A monopole aerial is a typical example of an unbalanced aerial because its other end (see Figure 1.15) is connected to earth.

A **balanced line** (Figure 1.21(a) and (b)) is a transmission line where the impedance between earth and each conductor is identical. A twin pair cable is an example of a balanced line because the impedance between earth and each conductor is the same. An **unbalanced line** (Figures 1.22(a), 1.22(b) is a transmission line where the impedance between earth and each conductor is not equal. A coaxial cable is an example of an unbalanced line because the impedance between earth and the outer shield is different to the impedance between earth and the inner conductor.

The key to the connections in Figures 1.21 and 1.22 is the **balanced/unbalanced transformer**. These transformers are carefully wound to produce maximum energy transfer by magnetic coupling. Coil windings are designed to have minimum self-capacitance, minimum inter-winding capacitance and minimum capacity coupling between each winding and earth. No direct connection is used between input and output circuits. The above conditions are necessary, otherwise balanced circuits will become unbalanced when parts of the circuit are connected together. The balanced/unbalanced transformer is bi-directional; it can be used to pass energy in either direction.

As the operational frequencies become higher and higher (above 2 GHz), it becomes increasingly difficult to make such a good transformer and a transformer is simply not used and antennas and transmission lines are connected directly. In such cases, the systems resolve to either an unbalanced antenna and distribution system or a balanced antenna and distribution system. The unbalanced system is almost always used because of convenience and costs.

1.7.2 Multi-point antenna distribution systems

In the design of antenna distribution systems, transmission lines connecting signal distribution points must function efficiently; they must carry signal with minimum loss, minimum interference and minimum reflections. Minimum loss cables are made by using good conductivity materials such as copper conductors and low loss insulation materials. Minimum interference is obtained by using coaxial cables whose outer conductor shields out interference signals. Reflections in the system are minimised by proper termination of the cables. For proper termination and no reflections in the system, two conditions must be fulfilled:

- the antenna and cable must be terminated in its characteristic impedance Z_0;
- the source impedance (Z_s) feeding each receiver must be matched to the input impedance of the receiver (Z_{in}), i.e. $Z_s = Z_{in}$, otherwise there will be signal reflections and minimum cable transmission loss will not be obtained.

In Figure 1.23, an aerial of characteristic impedance (Z_0) is used to feed a transmission (TX) line with a characteristic impedance Z_0. The output of the line is fed to a number (n) of receivers, each of which is assumed to have an input impedance (Z_{in}) equal to Z_0. Resistors R represent the matching network resistors which must be evaluated to ensure properly terminated conditions.

For the system to be properly terminated, it is essential that the aerial and cable system be terminated with Z_0, i.e. the impedance to the right of the plane 'AE' must present an impedance Z_0 to the antenna and cable system. It is also essential that each receiver be energised from a source impedance (Z_s) matched to its own input impedance (Z_{in}), i.e. $Z_s = Z_{in}$. For ease of analysis we will assume the practical case, $Z_s = Z_{in} = Z_0$.

Now for the transmission line in Figure 1.23 to be properly terminated:

$$R + [R + Z_0]/n = Z_0$$

Multiplying both sides by n:

$$nZ_0 = nR + R + Z_0$$

Collecting and transposing terms gives:

$$R = \frac{(n-1)}{(n+1)} Z_0 \tag{1.20}$$

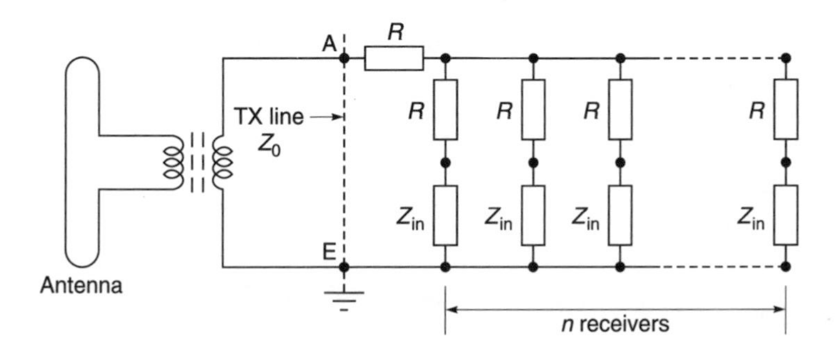

Fig. 1.23 Aerial distribution system for n receivers, each with an input impedance of Z_{in}

This equation is all we need to calculate the value of the matching resistors in Figure 1.23.

Example 1.6

A 75 Ω aerial system is used to supply signals to two receivers. Each receiver has an input impedance of 75 Ω. What is the required value of the matching resistor?

Given: 75 Ω aerial system, input impedance of each receiver = 75 Ω, no. of receivers = 2.
Required: Value of matching resistor.

Solution. Using Equation 1.20 with $n = 2$, we obtain

$$R = \frac{(n-1)}{(n+1)} Z_0 = \frac{(2-1)}{(2+1)} 75 = 25\ \Omega$$

Example 1.7

A 50 Ω aerial receiving system is to be used under matched conditions to supply signal to four receivers, each of input impedance 50 Ω. If the configuration shown in Figure 1.23 is used, calculate the value of the resistor, R, which must be used to provide matching conditions.

Given: 50 Ω aerial system, input impedance of each receiver = 50 Ω, no. of receivers = 4.
Required: Value of matching resistor.

Solution. From Equation 1.20

$$R = \frac{(n-1)}{(n+1)} Z_0 = \frac{(4-1)}{(4+1)} 50 = 30\ \Omega$$

From the answers above, it would appear that an aerial system can be matched to any number of receivers. This is true only within limits because the signal level supplied to individual receivers decreases with the number of distribution points. With large numbers of receivers, network losses become prohibitive.

Transmission losses associated with the matching network of Figure 1.23 can be calculated by reference to Figure 1.24. The network has been re-drawn for easier derivation of circuit losses but Z_0, R and n still retain their original definitions.

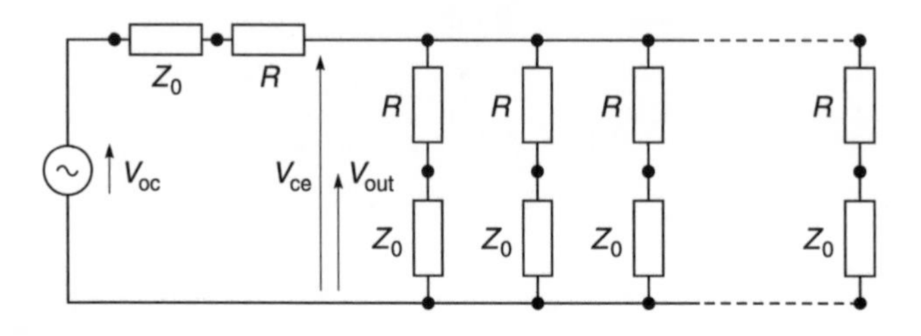

Fig. 1.24 Calculating the signal loss in an antenna distribution system

In Figure 1.24

V_{oc} = open-circuit source voltage from the aerial
V_{ce} = terminated voltage at an intermediate point in the network
V_{out} = terminated voltage at the input to a receiver

By inspection

$$V_{out} = \frac{Z_0}{R + Z_0} V_{ce} \quad \text{and} \quad V_{ce} = \frac{\left\{\dfrac{R + Z_0}{n}\right\}}{\left\{\dfrac{R + Z_0}{n}\right\} + R + Z_0} V_{oc}$$

Therefore

$$V_{out} = \frac{Z_0}{R + Z_0} \times \frac{R + Z_0}{(n+1)(R + Z_0)} V_{oc} = \frac{Z_0}{(n+1)(R + Z_0)} V_{oc}$$

Using Equation 1.20 and substituting $R = [(n - 1)/(n + 1)]Z_0$ in the above equation

$$V_{out} = \frac{Z_0}{(n+1)\left[\dfrac{(n-1)}{(n+1)} Z_0 + Z_0\right]} V_{oc} = \frac{V_{oc}}{2n}$$

Transposing, we find that

$$\text{voltage transmission loss} = \frac{V_{out}}{V_{oc}} = \frac{1}{2n} \qquad (1.21)$$

or

$$\text{voltage transmission loss} = 20 \log\left[\frac{1}{2n}\right] \text{dB}$$ [18]

Example 1.8

A broadcast signal induces an open-circuit voltage of 100 μV into a rod aerial. The aerial system has a characteristic impedance of 50 Ω and it is used to supply signal to three identical receivers each of which has an input impedance of 50 Ω. If the matching network type shown in Figure 1.23 is used, calculate (a) the value of the resistance (R) required for the matching network and (b) the terminated voltage appearing across the input terminals of the receiver.

[18] dB is short for decibel. The Bel is a unit named after Graham Bell, the inventor of the telephone. 1 Bel = $\log_{10}$ [power 1(P_1)/power 2(P_2)]. In practice the unit Bel is inconveniently large and another unit called the decibel is used. This unit is 1/10 of a Bel. Hence 1 Bel = 10 dB or dB = 10 $\log_{10} [P_1/P_2]$ = 10 $\log_{10} [(V_1^2/R)/(V_2^2/R)]$ = 20 $\log_{10} [V_1/V_2]$

Given: 50 Ω aerial system, input impedance of each receiver = 50 Ω, no. of receivers = 3, open-circuit voltage in aerial = 100 μV.
Required: (a) Value of matching resistor, (b) terminated voltage at receiver input terminal.

Solution

(a) For the matching network of Figure 1.23

$$R = \frac{(n-1)}{(n+1)} Z_0 = \frac{(3-1)}{(3+1)} \times 50 = 25\Omega$$

(b) Using Equation 1.20

$$V_{\text{receiver}} = \frac{1}{2n} V_{\text{antenna}} = \frac{1}{6} \times 100\ \mu\text{V} = 16.67\ \mu\text{V}$$

1.7.3 Other aerial distribution systems

The matching network shown in Figure 1.23 is only one type of matching network. Figure 1.25 shows a commercially available matching network for two outlets. This network is sometimes called a **two way splitter** because it splits the signal from a single input port into two output ports. The circuit has been designed for low insertion loss and it does this by trading off proper matching against insertion loss.

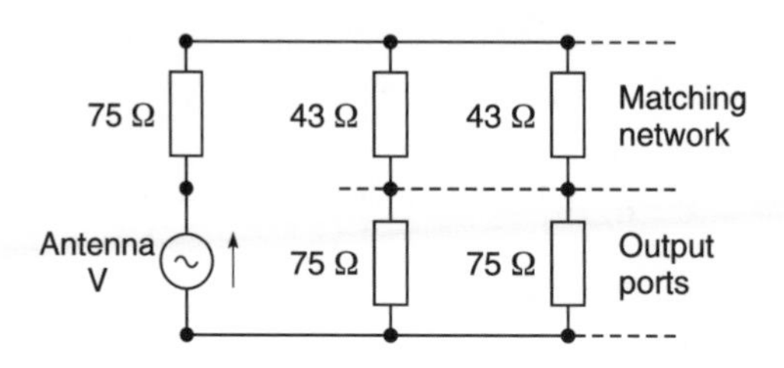

Fig. 1.25 Two way splitter

Example 1.9

Figure 1.25 shows a commercially available 75 Ω matching network. Calculate: (a) the ratio V_{out}/V_{oc} when all ports are each terminated with 75 Ω, (b) the input impedance to the matching network when the output ports are each terminated with 75 Ω and (c) the source impedance to either receiver when the remaining ports are each terminated in 75 Ω.

Given: 75 Ω network splitter of Figure 1.25 with 75 Ω terminations.
Required: (a) Ratio V_{out}/V_{oc}, (b) input impedance of matching network, (c) source impedance to either receiver.

Solution. By inspection:

$$\text{(a)}\quad V_{out} = \frac{75}{43 + 75} \times \frac{(43 + 75)/2}{(43 + 75)/2 + 75} \times V_{oc} = 0.28$$

(b) input impedance to the network = (43 + 75)/2 = 59 Ω

(c) receiver source impedance $= 43 + \dfrac{(43 + 75)(75)}{(43 + 75) + (75)} = 89\ \Omega$

From the answers to Example 1.9, it can be seen that the insertion loss is slightly reduced but this has been carried out at the expense of system match. The manufacturer is fully aware of this but relies on the fact that the reflected signal will be weak and that it will not seriously affect signal quality. The manufacturer also hopes that the cable system will be correctly matched by the antenna and that any reflections set up at the receiver end will be absorbed by the hopefully matched antenna termination to the cable system. This design is popular because the installation cost of an additional resistor is saved by the manufacturer.

1.7.4 Amplified antenna distribution systems

Amplified aerial distribution systems are aerial distribution systems which incorporate amplifiers to compensate for signal transmission, distribution and matching losses. Two systems will be discussed here. The first concerns a relatively simple distribution system where indoor amplifiers are used. The second system deals with a more elaborate system using amplifiers mounted on the aerial (masthead amplifiers) to compensate for distribution and matching losses.

1.7.5 Amplified aerial distribution systems using amplifiers

The block diagram of an amplified aerial distribution system using an amplifier is shown in Figure 1.26. This system is often used in domestic environments. Outdoor aerials provide the incoming signals, UHF for TV and/or VHF for FM-radio to the input of an amplifier. The gain of this amplifier is nominally greater than 10 dB but this varies according to the particular amplifier used. The amplifier is usually placed on the antenna masthead or near the aerial down-lead cables and a power supply point in the attic.

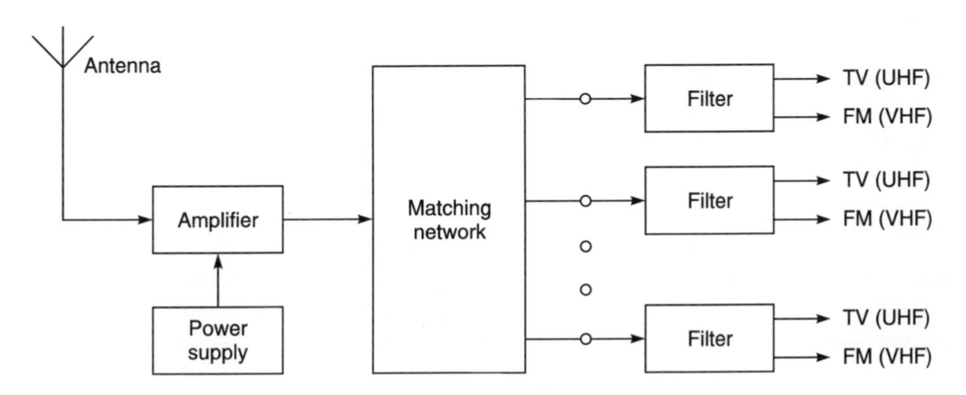

Fig. 1.26 Antenna amplifier distribution network

Output signals from the amplifier are fed into matching networks for distribution to individual terminals. To save cabling costs, both UHF and VHF signals are often carried on the same cables. Filters (a high pass filter for UHF and a low pass filter for VHF) are installed at individual terminals to feed the signals to their designated terminals.

The main advantage of such a system is that it is relatively easy to install especially if wiring for the distribution already exists. It also compensates for signal loss in the distribution network. The amplifier casing is relatively cheap when the amplifier is used indoors as it does not have to be protected from extreme weather conditions. The main disadvantage of indoor mounting is that signals are attenuated by the aerial down-lead cables before amplification. This signal loss decreases the available signal before amplification and therefore a poorer signal-to-noise ratio is available to the distribution points than if the amplifier was to be mounted on the masthead.

1.8 Radio receivers

1.8.1 Aims

The aims of this section are to introduce you to:

- the tuned radio frequency receiver
- the superhet receiver
- the double superhet receiver
- selectivity requirements in receivers
- sensitivity requirements in receivers
- concepts of signal-to-noise and sinad ratios
- noise figures of receivers

1.8.2 Objectives

After reading this section you should be able to understand:

- the basic principles of tuned radio frequency receivers
- the basic principles of superhet receivers
- the basic principles of satellite receivers
- the concepts of selectivity
- the concepts of sensitivity
- the concepts of signal-to-noise and sinad ratios
- the concepts of noise figures

1.8.3 Introduction

Radio receivers are important because they provide a valuable link in communications and entertainment. Early receivers were insensitive, inefficient, cumbersome, and required large power supplies. Modern designs using as little as one integrated circuit have overcome most of these disadvantages and relatively inexpensive receivers are readily available.

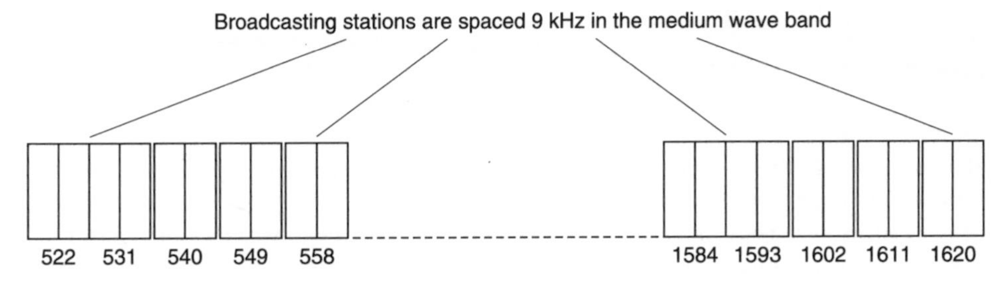

Fig. 1.27 Spacing of broadcast stations in the medium wave band

1.8.4 Fundamental radio receiver requirements

In the AM medium wave band, broadcasting stations transmit their signals centred on assigned carrier frequencies. These carrier frequencies are spaced 9 kHz apart from each other as in Figure 1.27 and range from 522 kHz to 1620 kHz. The information bandwidth allocated for each AM transmission is 9 kHz. This means that modulation frequencies greater than 4.5 kHz are not normally used. To receive information from a broadcast signal, an AM broadcast receiver must be tuned to the correct carrier frequency, have a bandwidth that will pass the required modulated signal, and be capable of extracting information from the required radio signal to operate desired output devices such as loudspeakers and earphones.

Discussions that follow pertain mainly to receivers operating in this band. This is not a limitation because many of the principles involved apply equally well to other frequency bands. When the need arises, specific principles applying to a particular frequency band will be mentioned but these occasions will be clearly indicated.

1.9 Radio receiver properties

A radio receiver has three main sections (see Figure 1.28).

- A **radio frequency section** to *select* and if necessary to *amplify* a desired radio frequency signal to an output level sufficient to operate a demodulator.
- A **demodulator section** to *demodulate* the required radio signal and *extract* its modulated information.
- A **post-demodulation section** to amplify demodulated signals to the required level to *operate* output devices such as loudspeakers, earphones and/or TV screens.

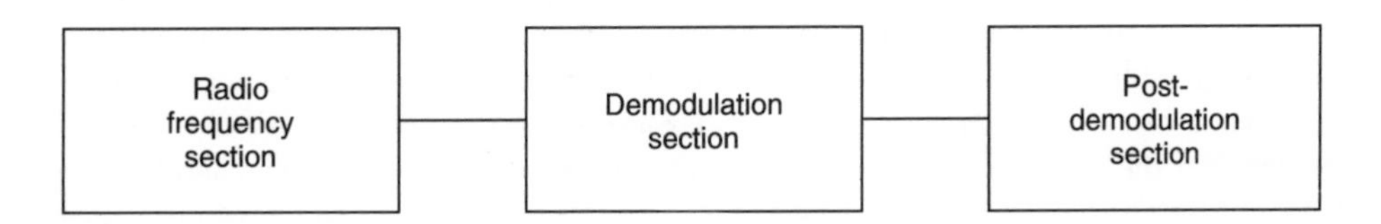

Fig. 1.28 Three main sections of a radio receiver

1.9.1 Radio frequency section

A radio frequency section is designed to have the following properties.

Selectivity

Receiver selectivity is a measure of the ability of a radio receiver to select the desired transmitted signal from other broadcast signals. An ideal selectivity response curve for an AM broadcast receiver centred on a desired carrier frequency (f_0) is shown in Figure 1.29. Two main points should be noted about the ideal selectivity response curve. First, it should have a wide enough passband (9 kHz approx.) to pass the entire frequency spectrum of the desired broadcast signal. Second, the passband should present equal transmission characteristics to all frequencies of the desired broadcast signal. In addition, the bandwidth should be no wider than that required for the desired signal because any additional bandwidth will allow extraneous signals and noise from adjacent channels to impinge on the receiver. Notice that the skirts of the ideal selectivity curve are vertical, so that the attenuation of any signal outside the passband is infinitely high.

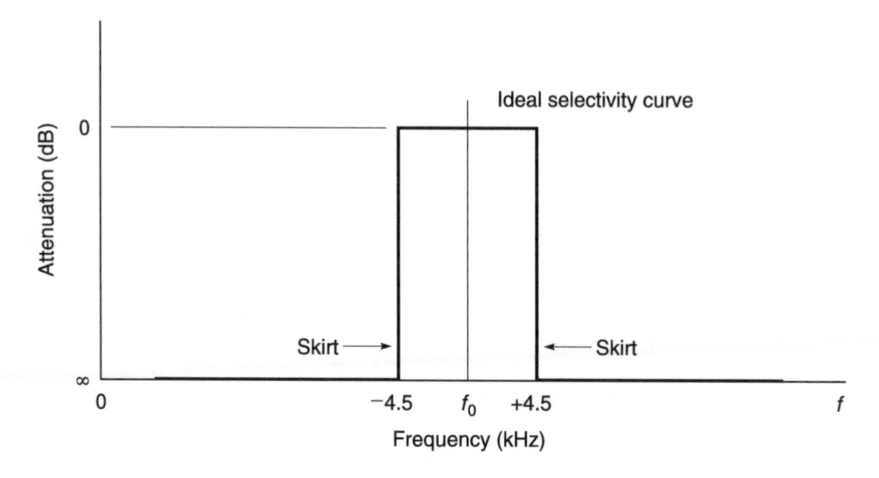

Fig. 1.29 Ideal selectivity curve for an AM medium wave broadcast receiver

In practice, costs and stability constraints prevent the ideal selectivity response curve from ever being attained and it is more rewarding to examine what is achieved by commercial receivers.

An overall receiver selectivity response curve for a typical domestic transistor receiver for the reception of AM broadcast signals is shown in Figure 1.30. In the table supplied with Figure 1.30, you should note that the selectivity curve is not symmetrical about its centre frequency. This is true of most tuned circuits because the effective working quality factor (Q_w) of components, particularly inductors, varies with frequency. Note also that the 3 dB bandwidth is only 3.28 kHz and that the 6 dB bandwidth points are approximately 4.82 kHz apart. The 60 dB points are 63.1 kHz apart.

Consider the case of a carrier signal (f_0) modulated with two inner sideband frequencies f_{1L}, f_{1U} (±1.64 kHz) and two outer sideband frequencies f_{2L}, f_{2U} (±2.4 kHz) away from the carrier. The frequency spectrum of this signal is shown in Figure 1.31(a). When

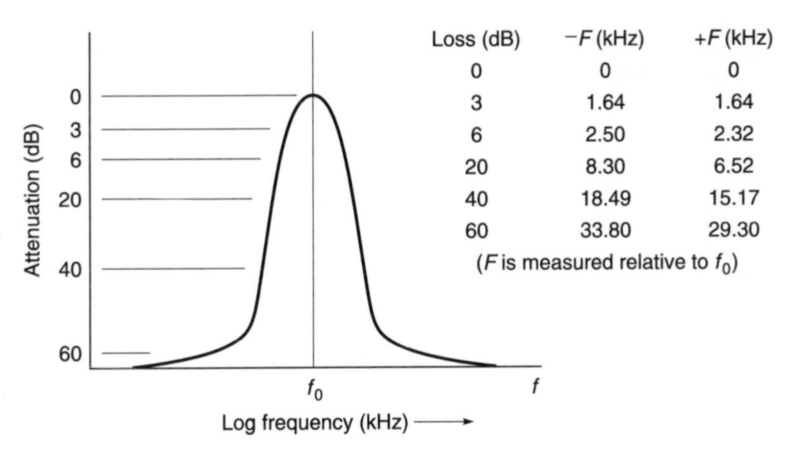

Loss (dB)	−F (kHz)	+F (kHz)
0	0	0
3	1.64	1.64
6	2.50	2.32
20	8.30	6.52
40	18.49	15.17
60	33.80	29.30

(F is measured relative to f_0)

Fig. 1.30 Typical selectivity curve of a commercial AM six transistor receiver

this frequency spectrum is passed through a receiver with the selectivity response shown in Figure 1.30, the inner sidebands f_{1L}, f_{1U} (±1.64 kHz) and the outer sidebands f_{2L}, f_{2U} (±2.4 kHz) will suffer attenuations of 3 dB and 6 dB approximately with respect to the carrier. (See the table in Figure 1.30.) The new spectrum of the signal is shown in Figure 1.31(b).

Comparison of Figures 1.31(a) and (b) shows clearly that amplitude distortion of the sidebands has occurred but what does this mean in practice? If the transmitted signal had been music, there would have been an amplitude reduction in high notes. If the transmitted signal had been speech, the speaker's voice would sound less natural.

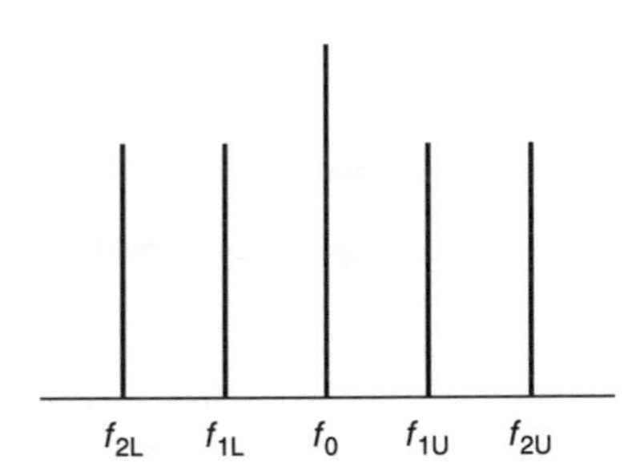

Fig. 1.31(a) Transmitted spectrum

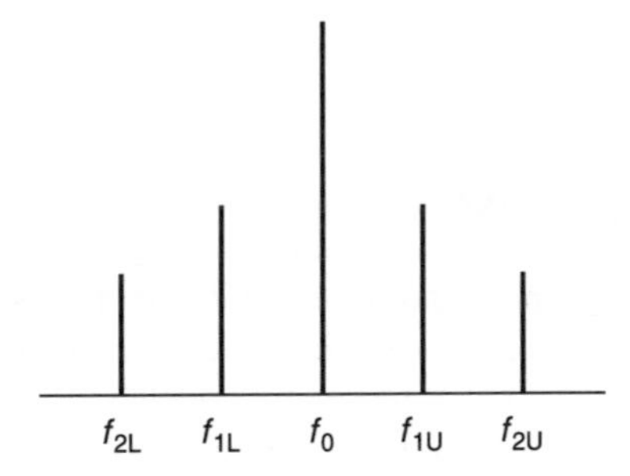

Fig. 1.31(b) Distorted spectrum

From the above discussion, it should be noted that for good quality reproduction the selectivity curve of a receiver should be wide enough to pass all modulation frequencies without discriminatory frequency attenuation.

Adjacent channel selectivity

A graphical comparison of the selectivity curves of Figures 1.29 and 1.30 is shown in Figure 1.32. From these curves, it can be seen that the practical selectivity curve does not provide complete rejection of signals to stations broadcasting on either side of the desired

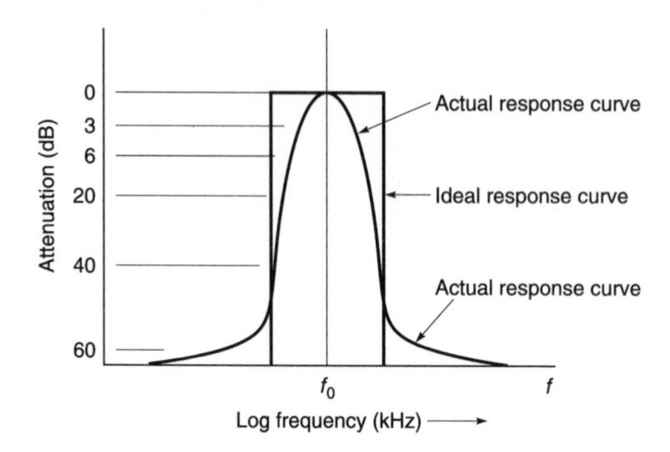

Fig. 1.32 A comparison between the ideal and practical selectivity curve

response channel. The breakthrough of signal from adjacent channels into the desired channel is known as **adjacent channel interference**.

Adjacent channel interference causes signals from adjacent channels to be heard in the desired channel. It is particularly bad when strong adjacent channel signals are present relative to the desired station. What does this mean in practice? It means that you will obtain interference from the unwanted station. In a broadcast receiver, you often hear signals from both channels simultaneously.

Broadcasting authorities minimise adjacent channel interference by forbidding other transmitters situated near the desired station to broadcast on an adjacent channel. Stations geographically distant from the desired station are allowed to operate on adjacent channels because it is likely that their signals will have suffered considerable transmission loss by the time they impinge on the desired channel.

Sensitivity

The sensitivity of a radio receiver is a measure of the modulated signal input level which is required to produce a given output level. A receiver with good sensitivity requires a smaller input signal than a receiver with poor sensitivity to produce a given output level. The sensitivity of a small portable receiver (audio output rated at 250 mW) may be quoted as 200 μV/m. What this means is that a modulated AM carrier (modulated with a 400 Hz tone and with an AM modulation depth of 30%) will produce an audio output of 50 mW under its maximum gain conditions when the input signal is 200 μV/m.

1.9.2 Signal-to-noise ratios

Any signal transmitted through a communications system suffers attenuation in the passive (non-amplifying) parts of the system. This is particularly true for radio signals propagating between transmitting and receiving aerials. Attenuation is compensated for by subsequent amplification, but amplifiers add their own inherent internally generated random noise to the signal. Noise levels must always be less than the required signal, otherwise the required signal will be lost in noise. Some means must be provided to specify the level of the signal above noise. This means is called the **Signal-to-*N*oise ratio**. It is defined as:

$$\text{signal-to-noise ratio} = \frac{S}{N} = \frac{\text{signal power}}{\text{noise power}} \qquad (1.23)$$

Notice that *S*/*N* is specified as a ratio of power levels.

An alternative way of specifying signal-to-noise ratios is to quote the ratio in decibels. This is defined by Equation 1.24:

$$\text{signal-to-noise ratio} = \frac{S}{N}(\text{dB}) = 10\log_{10}\left[\frac{\text{signal power}}{\text{noise power}}\right] \text{dB} \qquad (1.24)$$

A strong signal relative to the noise at the receiver input is essential for good reception. In practice, we require an *S*/*N* of 10–20 dB to distinguish speech, an *S*/*N* of 30 dB to hear speech clearly, and an *S*/*N* of 40 dB or better for good television pictures.

Noise figure

Certain amplifiers have more inherent electrical noise than others. Manufacturers usually produce a batch of transistors, then classify and name the transistors according to their inherent electrical noise levels. The inherent noise produced by a transistor is dependent on its general operating conditions, particularly frequency, temperature, voltage and operating current, and these conditions must be specified when its noise level is measured. Engineers use the ratio term **noise figure** to specify noise levels in transistors.

Noise figure is defined as

$$\text{noise figure (N.F.)} = \left[\frac{(S/N)_{\text{in}}}{(S/N)_{\text{out}}}\right] \text{at 290 K} \qquad (1.25)$$

If a transistor introduces no noise, then its *S*/*N* at both the input and output is the same, therefore from Equation 1.25, N.F. = 1 or in dB N.F. = 10 log 1 = 0 dB. Hence a 'perfect' or 'noiseless' amplifier has a noise figure of 0 dB. An imperfect amplifier has a noise figure greater than 0 dB. For example an amplifier with a noise figure of 3 dB (= 2 ratio) means that it is twice as bad as a perfect amplifier.

1.10 Types of receivers

There are many types of radio receivers. These include:

- tuned radio frequency receivers (TRF)
- superheterodyne receivers (superhets)
- double superheterodyne receivers (double superhets)

1.10.1 Tuned radio frequency receiver

A tuned radio frequency receiver (Figure 1.33) has three main sections, a radio frequency amplifier section, a detector section, and an audio amplifier section.

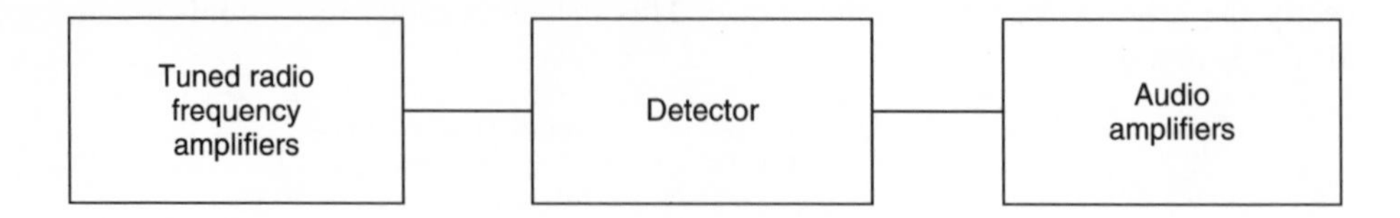

Fig. 1.33 Main sections of a tuned radio frequency receiver

The radio frequency section consists of one or more r.f. amplifiers connected in cascade.[19] For efficient operation, all tuned circuit amplifiers must be tuned to exactly the same broadcast frequency and to ensure that this is the case, all tuning adjusters are fixed on to a common tuning shaft. Tuning capacitors which are connected in this manner are said to be 'ganged' and two and three stage ganged tuning capacitors are common.

The detector is usually a conventional AM diode type detector. This type of detector is usually a diode which detects the positive peaks of the modulated carrier and filters the r.f. out, so that the remaining signal is the inital low frequency modulation frequency.

The audio section uses audio amplifiers which serve to amplify the signals to operate a loudspeaker. This section is similar to the amplifier in your home which is used for playing compact disks (CD) and cassettes.

Advantages

The main advantages of TRF receivers are that they are relatively simple, easy to construct, and require a minimum of components. A complete TRF receiver can be constructed using a single integrated circuit such as a ZN414 type chip.

Disadvantages

TRF receivers suffer from two main disadvantages, gain/bandwidth variations and poor selectivity. The inevitable change of gain and bandwidth as the receiver is tuned through its frequency range is due to changes in the selectivity circuits.

Circuit instability can be a problem because it is relatively easy for any stray or leaked signal to be picked up by one of the many r.f. amplifiers in the receiver. R.F. signal can also be easily coupled from one r.f. stage to another through the common power supply. To minimise these risks, r.f. amplifiers are usually shielded and de-coupled from the common power supply.

1.10.2 Superheterodyne receiver

Block diagram

A block diagram of a superheterodyne (commonly called superhet) receiver is shown in Figure 1.34.

This receiver features an r.f. section which selects the desired signal frequency (f_{rf}). This signal is then mixed with a local carrier at frequency (f_o) in a frequency changer to produce an intermediate frequency (f_{if}) which retains the modulated information initially

[19] Here, cascade is meant to imply one amplifier following another amplifer and so on.

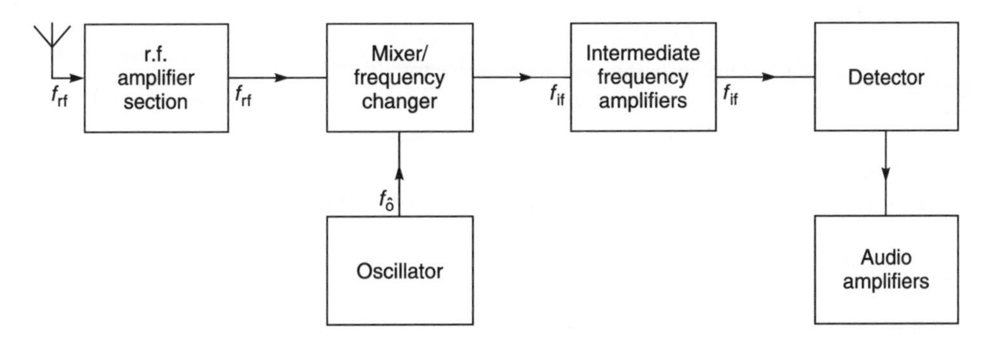

Fig. 1.34 Block diagram of a superhet radio receiver

carried by f_{rf}. The intermediate frequency (f_{if}) then undergoes intensive amplification (60–80 dB) in the intermediate frequency amplifiers to bring the signal up to a suitable level for detection and subsequent application to the post-detection (audio) amplifiers.

Radio frequency amplifiers are sometimes included in the r.f. section in order to make the noise figure of the receiver as small as possible. Frequency changers have comparatively larger noise figures (6–12 dB) than r.f. amplifiers.

The frequency of the local oscillator (f_o) is always set so that its frequency differs from the desired frequency (f_{rf}) by an amount equal to the intermediate frequency (f_{if}), i.e.

$$f_o - f_{rf} = f_{if} \tag{1.26}$$

or

$$f_{rf} - f_o = f_{if} \tag{1.27}$$

Equation 1.26 is more usual for medium-wave receivers. Typical tuning ranges for a medium-wave receiver with f_{if} = 465 kHz are 522–1620 kHz for f_{rf} and 987–2085 kHz for f_o.

Advantages

The main advantages of the superhet receiver are as follows.

- Better selectivity because fixed bandpass filters with well defined cut-off frequency points can be used in the i.f. stages of a superhet. Filters and tuned circuits are also less complex because they need only operate at one frequency, namely the intermediate frequency.
- In a superhet, tuning is relatively simple. A two ganged capacitor can be used to tune the r.f. and oscillator sections simultaneously to produce the intermediate frequency for the i.f. amplifiers.
- R.F. circuit bandwidths are not critical because receiver selectivity is mainly determined by the i.f. amplifiers.

Disadvantages

The main disadvantages of superhets are as follows.

- **Image channel interference** is caused by the local oscillator (f_o) combining with an undesired frequency (f_{im}) which is separated from the desired frequency (f_{rf}) by twice the i.f. frequency (f_{if}). Expressed mathematically

$$f_{im} = f_{rf} \pm 2f_{if} \qquad (1.28)$$

The term **2nd channel interference** is another name for image channel interference.

Image channel interference is more easily understood by substituting some arbitrary values into Equations 1.26 and 1.27. For example, assume that the local oscillator of a superhet is set to 996 kHz and that its intermediate frequency amplifiers operate at 465 kHz. Then, either of two input frequencies, 996 – 465 = 531 kHz (Equation 1.26) or 996 + 465 = 1461 kHz (Equation 1.27), will mix with the local oscillator to produce a signal in the i.f. amplifiers. If the desired frequency is 531 kHz, then the undesired frequency of 1461 kHz is $2(f_{if})$ or 930 kHz away, i.e. it forms an image on the other side of the oscillator frequency. This condition is shown graphically in Figure 1.35.

- There is the possibility that any strong signal or sub-harmonics of 465 kHz (f_{if}) might impinge directly on the i.f. amplifiers and cause interference.
- Any harmonic of the oscillator (f_o) could mix with an unwanted signal to produce unwanted responses. For example

$$2 \times 996(f_o)\text{ kHz} - 1527\text{ kHz} = 465\text{ kHz}$$

The spurious responses stated above are minimised in superhets by using tuned circuits in the r.f. section of the receiver to select the desired signal and to reject the undesired ones. The local oscillator is also designed to be 'harmonic free'.

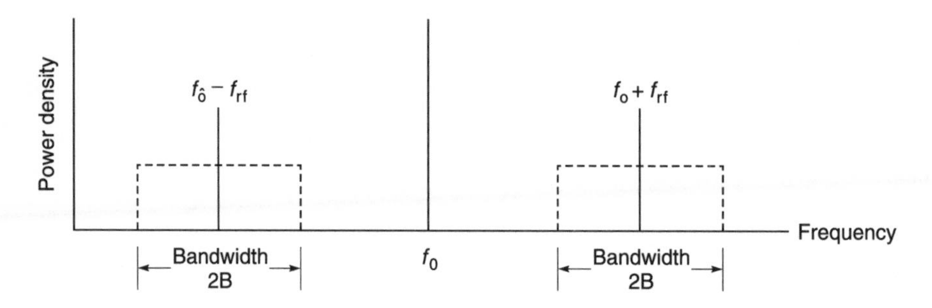

Fig. 1.35 Image response in superhet receivers

1.10.3 Double superheterodyne receivers

A block diagram of a double conversion superhet used for receiving direct broadcast signals (DBS) from satellites is shown in Figure 1.36. Direct broadcasting satellites for the United Kingdom region transmit in the 11.6–12.4 GHz band. Each TV channel uses a 26 MHz bandwidth.

The double superhet is basically a superhet receiver with two i.f. sections. The first i.f. section operates at a much higher i.f. frequency than the second i.f. section. This choice is deliberate because a higher 1st i.f. frequency gives better image channel rejection. You have already seen this in the calculations of p. 40. The 2nd i.f. section is made to operate at a lower frequency because it gives better adjacent channel selectivity.

In this receiver, the input signal, f_1, is selected, mixed with a local oscillator carrier, f_x, and frequency translated to form the first i.f. frequency (f_{if1}). This signal is applied to the

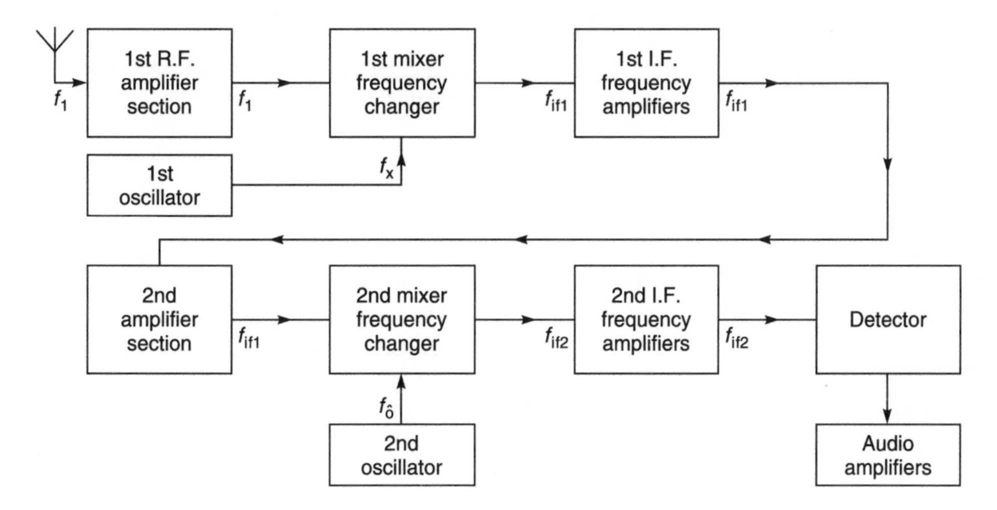

Fig. 1.36 Block diagram of a double conversion superhet receiver

1st i.f. amplifier section, then mixed with f_o to produce a second intermediate frequency (f_{if2}) and amplified prior to detection and low frequency amplification.

In a typical direct broadcast satellite receiver, the first r.f. amplifier section operates in the band 11.6–12.4 GHz. The first local oscillator (f_x) is operated at a fixed frequency of 10.650 GHz. The resultant first i.f. frequency bandwidth range is 950 to 1750 MHz and is really the r.f. band translated to a lower frequency band. This i.f. is then amplified by the first set of 1st i.f. amplifiers. All the foregoing action takes place in a masthead unit which is mounted directly on the antenna. The total gain including r.f. amplification, frequency conversion and i.f. amplification is about 55 dB. This high order of gain is necessary to compensate for the losses which occur in the down-lead coaxial cable to the satellite receiver which is situated within the domestic environment. The satellite receiver treats the 1st i.f. frequency band (950–1750 MHz) as a tuning band and f_o is varied to select the required TV channel which is amplified by the 2nd i.f. section before signal processing.

1.11 Summary

The main purpose of Chapter 1 has been to introduce you to the radio environment in your home. The knowledge you have gained will assist you in understanding basic radio propagation and reception principles. It will also help you to remedy some of the simpler radio and TV problems which you are likely to encounter in your home.

In Sections 1.1–1.3, we started with the necessity for modulation and demodulation and you were introduced to the basic principles of modulation, demodulation and radio propagation. You should now understand the meaning of terms such as amplitude modulation (AM), frequency modulation (FM), phase modulation (PM) and digital modulation.

In Section 1.4, you were introduced to radio propagation, wave polarisation, field strength and power density of radio waves.

In Sections 1.5 and 1.6, you learned about the properties of several antennas. These included $\lambda/2$ dipole, folded dipole, monopole, loop antennas and the Yagi-Uda array.

Section 1.7 dealt with various antenna distributions and matching systems.

In Section 1.8, you encountered some basic concepts concerning the reception of radio signals. You should now be able to carry out simple calculations with regard to selectivity, adjacent channel selectivity, sensitivity, S/N ratio and noise figure ratio as applied to radio receivers.

Sections 1.9 and 1.10 described the main functions required in a radio receiver, and also the main advantages and disadvantages of three basic radio receiver types, namely the TRF, superhet and double superhet receivers. The first type is used in very simple receivers, the second type is used extensively in domestic receivers and the last type is used for direct broadcast reception from satellites.

You have now been provided with an overview of a basic radio communication system. Having established this overview, we will now be in a position to deal with individual sub-systems and circuits in the next chapters.

2

Transmission lines

2.1 Introduction

At this stage, I would like to prepare you for the use of the software program called PUFF which accompanies this book. PUFF (Version 2.1) is very useful for matching circuits, and the design of couplers, filters, line transformers, amplifiers and oscillators. Figure 2.1 shows what you see when you first open the PUFF program. Figure 2.2 shows you how the program can be used in the design of a filter. In Figure 2.1 you can see for yourself that

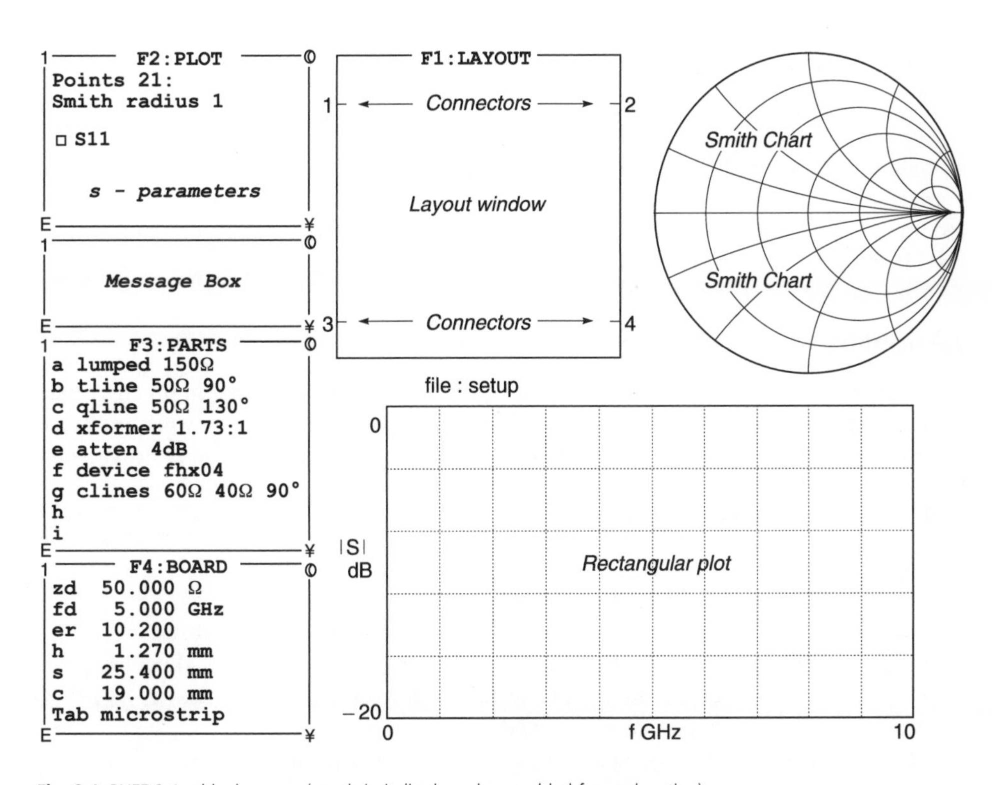

Fig. 2.1 PUFF 2.1 – blank screen (words in italics have been added for explanation)

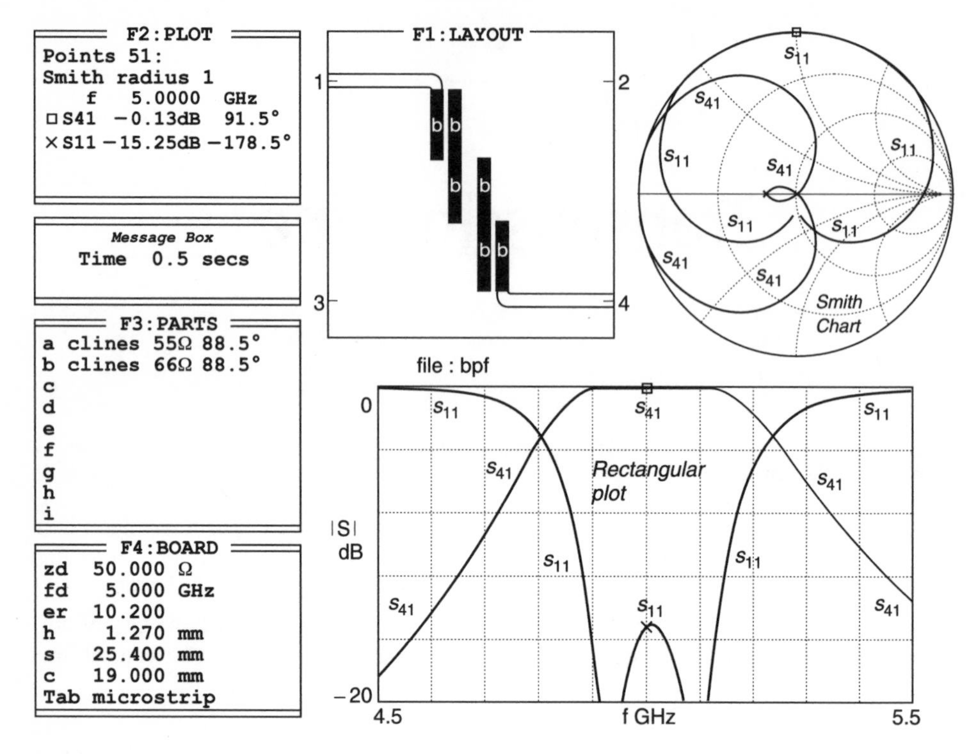

Fig. 2.2 Bandpass filter design using PUFF

to understand and use the program, you must be familiar with Smith charts (top right hand corner) and scattering or '*s*-parameters' (top left hand corner), transmission lines and the methods of entering data (F3 box) into the program. Within limits, the layout window (F1 box) helps to layout your circuit for etching.

2.1.1 Aims

We shall cover the basic principles of transmission lines in this part, and Smith charts and *s*-parameters in Part 3. We will then be in a position to save ourselves much work and avoid most of the tedious mathematical calculations involved with radio and microwave engineering.

The main aims of this chapter are:

- to introduce you to various types of transmission lines;
- to explain their characteristic impedances from physical parameters;
- to provide and also to derive expressions for their characteristic impedances;
- to explain their effects on signal transmission from physical and electrical parameters;
- to explain and derive expressions for reflection coefficients;
- to explain and derive expressions for standing wave ratios;
- to explain and derive the propagation characteristics of transmission lines;
- to provide an understanding of signal distortion, phase velocity and group delay;

- to show how transmission lines can be used as inductors;
- to show how transmission lines can be used as capacitors;
- to show how transmission lines can be used as transformers.

2.1.2 Objectives

This part is mainly devoted to transmission lines. Knowledge of transmission lines is necessary in order to understand how high frequency engineering signals can be efficiently moved from one location to another. For example, the antenna for your domestic TV receiver is usually mounted on the roof and it is therefore necessary to find some means of efficiently transferring the received signals into your house. In the commercial world, it is not unusual for a radio transmitter to be situated several hundred metres from a mast-mounted transmitting antenna. Here again, we must ensure that minimal loss occurs when the signal is transferred to the antenna for propagation.

2.2 Transmission line basics

2.2.1 Introduction to transmission lines

In this discussion we shall start off using some basic terms which are easily understood with sound waves. We will then use these terms to show that these properties are also applicable to electrical transmission systems. Much of the explanation given in these sections will be based on examples using sinusoids because they are easier to understand. But this information applies equally well to digital waveforms because digital signals are composed of sinusoid components combined in a precise amplitude and phase manner. Therefore, it is vitally important that you do not form the mistaken idea that transmission line theory only applies to analogue waveforms.

2.2.2 General properties of transmission systems

Transmission systems are used to transfer energy from one point to another. The energy transferred may be sound power or electrical power, or digital/ analogue/optical signals or any combination[1] of the above.

One easy way of refreshing your memory about signal transmission is to imagine that you are looking into a deep long straight tunnel with walls on either side of you. When you speak, you **propagate** sound energy along a **transmission path** down the length of the tunnel. Your voice is restricted to propagation along the length of the tunnel because walls on either side act as **waveguides**.

Waves emerging directly from the sender are known as **incident waves**. As your vocal cords try to propagate incident waves along the tunnel, they encounter an opposition or **impedance** caused by the air mass in the tunnel. The impedance is determined by the physical characteristics of the tunnel such as its width and height and the manner in which

[1] For example, the coaxial cable connecting the domestic satellite receiver to the low noise amplifier on the satellite dish often carries d.c. power up to the low noise amplifier, radio frequency signals down to the receiver, and in some cases even digital control signals for positioning the aerial.

it impedes air mass movement within the tunnel. This impedance is therefore called the **characteristic impedance** (Z_0) of the tunnel.

Bends or rock protrusions along the tunnel walls cause a change in the effective dimensions of the tunnel. These **discontinuities** in effective dimensions can cause minor **reflections** in the signal propagation path. They also affect the characteristic impedance of the transmission channel.

You should note that the walls of the tunnel do not take part in the main propagation of sound waves. However, they do absorb some energy and therefore weaken or **attenuate** the propagated sound energy. Amplitude attenuation per unit length is usually represented by the symbol α.

Moss, lichen and shrubs growing on the walls will tend to absorb high frequency sound better than low frequency sound, therefore your voice will also suffer **frequency attenuation**. Frequency attenuation is known as **dispersion**.

There is also a speed or **propagation velocity** with which your voice will travel down the tunnel. This velocity is dependent on the material (air mixture of gases), its density and temperature within the tunnel. With sound waves, this velocity is about 331 metres per second.

If the tunnel is infinitely long, your voice will propagate along the tunnel until it is totally attenuated **or absorbed**. If the tunnel is not infinitely long, your voice will be **reflected** when it reaches the end wall of the tunnel and it will return to you as an **echo** or **reflected wave**. The ratio reflected wave/incident wave is called the **reflection coefficient**. You can prevent this reflection if it were physically possible to put some good sound absorption material at the end of the tunnel which absorbs all the incident sound. In other words, you would be creating a **matching termination** or **matched load** impedance (Z_L) which **matches** the **propagation characteristics** of an infinitely long tunnel in a tunnel of finite length. The ratio of the **received sound** relative to the **incident sound** is known as the **transmission coefficient.**

A signal travelling from a point A to another point B takes time to reach point B. This **time delay** is known as **propagation time delay** for the signal to travel from point A to point B. In fact, any signal travelling over any distance undergoes a propagation time delay.

Time propagation delay can be specified in three main ways: (i) **seconds**, (ii) **periodic time** (T) and (iii) **phase delay**. The first way is obvious, one merely has to note the time in seconds which it has taken for a signal to travel a given distance. Periodic time (T) is an interval of time; it is equal to [1/(frequency in Hz)] seconds. For example, if a 1000 Hz sinusoid requires four periodic times ($4T$) to travel a certain distance, then the time delay is $4 \times (1/1000)$ seconds or four milliseconds. Phase delay can be used to measure time because there are 2π radians in a period time (T). For the example of a 1000 Hz signal, a phase delay of $(4 \times 2\pi)$ radians is equivalent to four periodic times (T) or four milliseconds. Phase delay per unit length is usually represented by the symbol (β). It is measured in radians per metre.

Hence if we were to sum up propagation properties, there would be at least three properties which are obvious:

- **attenuation** of the signal as it travels along the line;
- the time or **phase delay** as the signal travels along the line;
- **dispersion** which is the different attenuation experienced by different frequencies as it travels along a line.

Finally, if you walked along a tunnel which produces echoes, while a friend whistled at a constant amplitude and pitch, you would notice the reflected sound **interfering** with the incident sound. In some places, the whistle will sound louder (addition of incident and reflected signal); in other places the whistle will sound weaker (subtraction of incident and reflected signal). Provided your friend maintains the whistle at a constant amplitude and pitch, you will find that louder and weaker sounds always occur at the same locations in the tunnel. In other words, the pattern of louder and weaker sounds remains stationary and appears to be standing still. This change in sound intensity levels therefore produces a **standing wave pattern** along the length of the tunnel. The ratio of the maximum to minimum sound is known as the **standing wave ratio** (SWR). It will be shown later that the measurement of standing wave patterns is a very useful technique for describing the properties of transmission line systems.

In the above discussions, you have used knowledge gained from the university of life to understand the definitions of many transmission line terms. These definitions are not trivial because you will soon see that many of the above principles and terms also relate to electrical transmission lines. In fact, if you can spare the time, re-read the above paragraphs again just to ensure that you are fully cognisant of the terms shown in bold print.

2.3 Types of electrical transmission lines

Many of the terms introduced in the last section also apply to electrical transmission lines. However, you should be aware of the great difference in the velocity of sound waves (331 ms^{-1}) and the velocity of electrical waves (3×10^8 ms^{-1} in air). There is also a great difference in frequency because audible sound waves are usually less than 20 kHz whereas radio frequencies are often in tens of GHz. For example, satellite broadcasting uses frequencies of about 10–12 GHz. Since wavelength = velocity/frequency, it follows that there will be a difference in wavelength and that in turn will affect the physical size of transmission lines. For example, the dimensions of a typical waveguide (Figure 2.3(a)) for use at frequencies between 10 GHz and 15 GHz are A = 19 mm, B = 9.5 mm, C = 21.6 mm, D = 12.1 mm.

There are many types of transmission lines.[2] These range from the two wire lines which you find in your home for table lamps, and three wire lines used for your electric kettle. Although these cables work efficiently at power frequencies (50~60 Hz), they become very inefficient at high frequencies because their inherent construction blocks high frequency signals and encourages radiation of energy.

2.3.1 Waveguides and coplanar waveguides

Other methods must be used and one method that comes readily to mind is the tunnel or waveguide described in Section 2.2.2. This waveguide is shown in Figure 2.3(a). It works efficiently as a high frequency transmission line because of its low attenuation and radiation losses but it is expensive because of its metallic construction (usually copper). It is also relatively heavy and lacks flexibility in use because special arrangements must be

[2] The term 'transmission line' is often abbreviated to 'tx lines'.

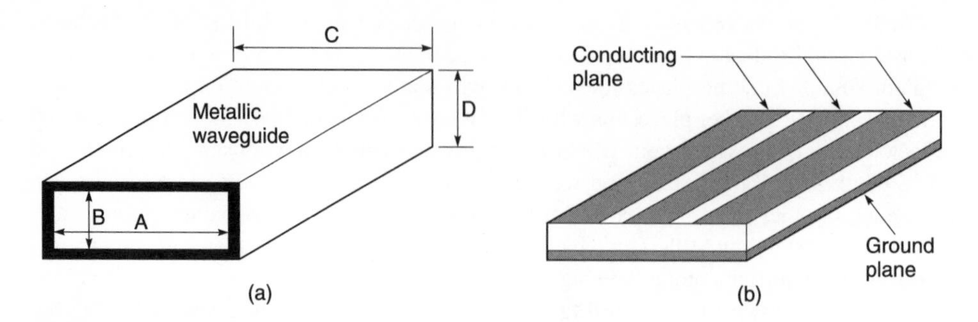

Fig. 2.3 (a) Metallic waveguide; (b) coplanar waveguide

used to bend a transmission path. One variant of the waveguide is known as the coplanar waveguide (Figure 2.3(b)).

2.3.2 Coaxial and strip lines

Another way of carrying high frequency signals is to use a coaxial transmission line similar to the one that connects your TV set to its antenna. The coaxial line is shown in Figure 2.4(a). This is merely a two wire line but the outer conductor forms a circular shield around the inner conductor to prevent radiation.

One variation of the coaxial line appears as the strip line (Figure 2.4(b)). The strip line is similar to a 'flattened' coaxial line. It has the advantage that it can be easily constructed with integrated circuits.

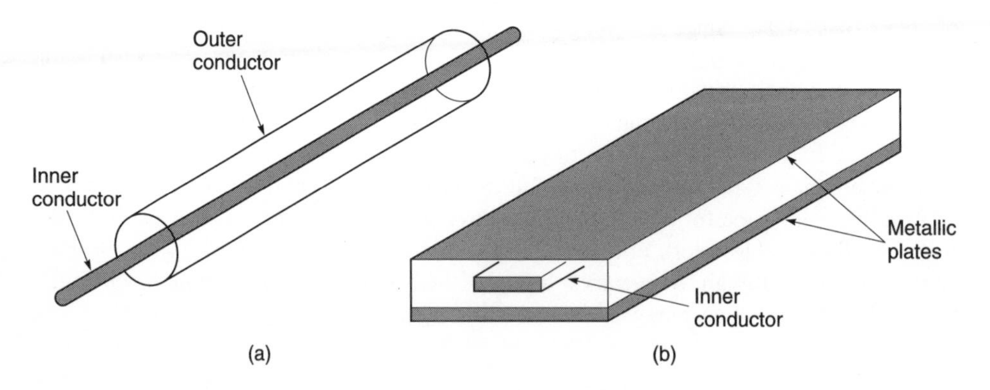

Fig. 2.4 (a) Coaxial cable; (b) strip line

2.3.3 Microstrip and slot lines

The microstrip line (Figure 2.5(a)) is a variant of the stripline with part of the 'shield' removed. The slot line (Figure 2.5(b)) is also a useful line for h.f. transmission.

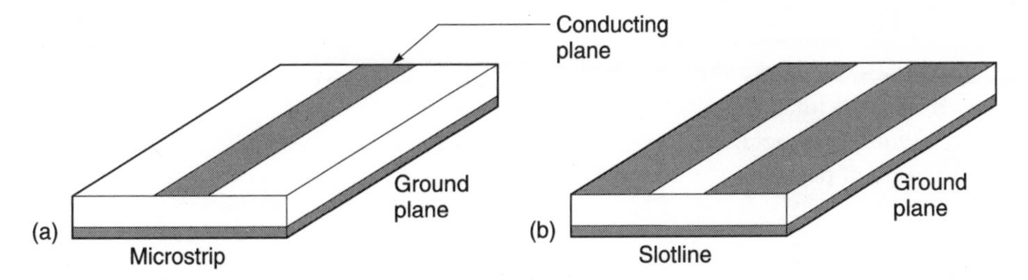

Fig. 2.5 (a) Microstrip line; (b) slot line

2.3.4 Twin lines

In Figure 2.6, we show a sketch of a twin line carefully spaced by a polyethylene dielectric. This is used at relatively low frequencies. This twin cable is designed to have a characteristic impedance (Z_0) of approximately 300 Ω and it is frequently used as a VHF cable or as a dipole antenna for FM radio receivers in the FM band. The parallel wire line arrangement of Figure 2.6 without a dielectric support can also be seen mounted on poles as overhead telephone lines, overhead power lines, and sometimes as lines connecting high power, low and medium frequency radio transmitters to their antennas.

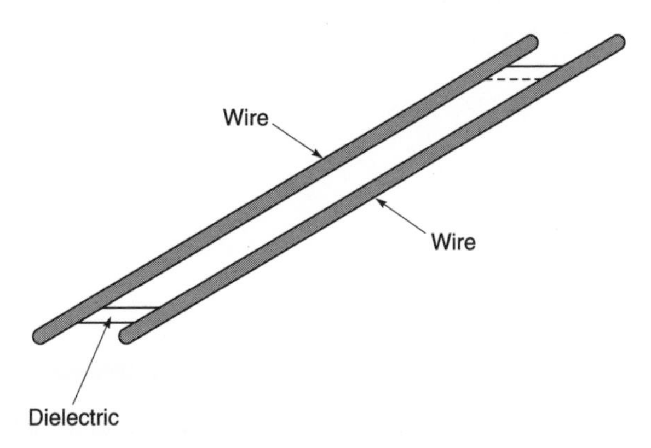

Fig. 2.6 Twin parallel wire VHF cable

All seven transmission lines shown in Figures 2.3–2.6 have advantages and disadvantages. For minimum loss, you would use the waveguide, the coaxial line and the strip line in integrated circuits. However, the latter two lines present difficulties in connecting external components to the inner conductor. The coplanar waveguide is better in this respect and finds favour in monolithic microwave integrated circuits (MMIC) because it allows easy series and parallel connections to external electrical components. The microstrip line is also useful for making series connections but not parallel connections because the only way through to the ground plane is either through or around the edge of the substrate. This is particularly true when a short circuit is required between the upper conductor and the ground plane; holes have to be drilled through the substrate. Microstrip also suffers from radiation losses. Nevertheless, microstrip can be made easily and conveniently and it is therefore used extensively.

2.3.5 Coupled lines

Coupled lines are lines which are laid alongside each other in order to permit coupling between the two lines. One example of microstrip coupled lines is shown in the F1 layout box of Figure 2.2 where three sets of coupled lines are used to couple energy from input port 1 to output port 4.

2.4 Line characteristic impedances and physical parameters

The characteristic impedance of transmission lines is calculated in two main ways:

- from physical parameters and configuration;
- from distributed electrical parameters of the line.

Some relevant expressions for calculating the impedance of these lines from physical parameters are given in the following sections.

2.4.1 Coaxial line characteristic impedance (Z_0)

The expression for calculating the characteristic impedance of the coaxial transmission line shown in Figure 2.4(a) is:

$$Z_0 = \frac{138}{\sqrt{\varepsilon}} \log_{10} \frac{D}{d} \tag{2.1}$$

where

d = outer diameter of the inner conductor
D = inner diameter of the outer conductor
ε = dielectric constant of the space between inner and outer conductor ($\varepsilon = 1$ for air)

Example 2.1

You will often find two types of flexible coaxial cables: one with a characteristic impedance Z_0 of 50 Ω which is used mainly for r.f. instrumentation and the other has a characteristic impedance of 75 Ω used mainly for antennas. The inner diameter of the outer conductor is the same in both cables. How would you distinguish the impedance of the two cables using only your eye?

Solution. In general, to save money, both cables are normally made with the same outer diameter. This is even more evident when the cables are terminated in a type of r.f. connector known as BNC.[3] Since these connectors have the same outer diameter, by using Equation 2.1 you can deduce that for $Z_0 = 75$ Ω, the inner conductor will be smaller than that of the 50 Ω cable. In practice, you will be able to recognise this distinction quite easily.

[3] BNC is an abbreviation for 'baby N connector'. It is derived from an earlier, larger threaded connector, the Type N connector, named after Paul Neill, a Bell Laboratories engineer. BNC uses a bayonet type fixing. There is also a BNC type connector which uses a thread type fixing; it is called a TNC type connector.

2.4.2 Twin parallel wire characteristic impedance (Z_0)

The expression for calculating the characteristic impedance of the type of parallel transmission line shown in Figure 2.6 is:

$$Z_0 \approx \frac{276}{\sqrt{\varepsilon}} \log 10 \frac{2D}{d} \tag{2.2}$$

where

d = outer diameter of one of the identical conductors
D = distance between the centres of the two conductors
ε = relative dielectric constant = 1 for air

Example 2.2

The twin parallel transmission line shown in Figure 2.6 is separated by a distance (D) of 300 mm between the centre lines of the conductors. The diameter (d) of the identical conductors is 4 mm. What is the characteristic impedance (Z_0) of the line? Assume that the transmission line is suspended in free space, i.e. $\varepsilon = 1$.

Given: $D = 300$ mm, $d = 4$ mm, $\varepsilon = 1$.
Required: Z_0.

Solution. Using Equation 2.2

$$Z_0 \approx \frac{276}{\sqrt{1}} \log_{10} \frac{2D}{d} = \frac{276}{\sqrt{1}} \log_{10} \frac{2 \times 300}{4} \approx 600\ \Omega$$

2.4.3 Microstrip line characteristic impedance (Z_0)

Before we start, it is best to identify some properties that are used in the calculations on microstrip. These are shown in Figure 2.7 where w = width of the microstrip, h = thickness of the substrate, t = thickness of the metallisation normally assumed to approach zero in these calculations, ε_r = dielectric constant of the substrate. Note that there are two dielectric constants involved in the calculations, the relative bulk dielectric constant ε_r and the effective dielectric constant ε_e. The effective dielectric is inevitable because some of the electric field passes directly from the bottom of the strip width to the ground plane whereas some of the electric field travels via air and the substrate to the ground plate.

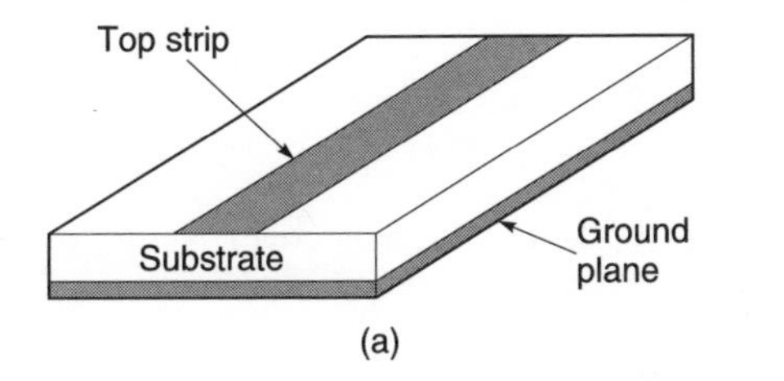

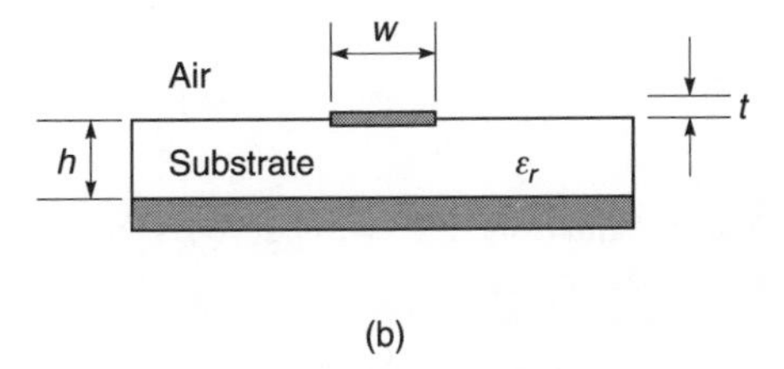

Fig. 2.7 (a) Microstrip line; (b) end view of microstrip line

There are many expressions for calculating microstrip properties[4] but we will use two main methods. These are:

- an *analysis* method when we know the width/height (w/h) ratio and the bulk dielectric constant (ε_r) and want to find Z_0;
- a *synthesis* method when we know the characteristic impedance Z_0 and the bulk dielectric constant (ε_r) and want to find the w/h ratio and the effective dielectric constant (ε_e).

Analysis formulae

In the analysis case we know w/h and ε_r and want to find Z_0. The expressions which follow are mainly due to H. Wheeler's work.[5]

For *narrow* strips, i.e. $w/h < 3.3$

$$Z_0 = \frac{119.9}{\sqrt{2(\varepsilon_r + 1)}} \left\{ \ln\left[\frac{4h}{w} + \sqrt{16\left(\frac{h}{w}\right)^2 + 2} \right] - \frac{1}{2}\left(\frac{\varepsilon_r - 1}{\varepsilon_r + 1}\right)\left(\ln\frac{\pi}{2} + \frac{1}{\varepsilon_r}\ln\frac{4}{\pi} \right) \right\} \tag{2.3}$$

For *wide* strips, i.e. $w/h > 3.3$

$$Z_0 = \frac{119.9}{2\sqrt{\varepsilon_r}} \left\{ \frac{w}{2h} + \frac{\ln 4}{\pi} + \frac{\ln(e\pi^2/16)}{2\pi}\left(\frac{\varepsilon_r - 1}{\varepsilon_r^2}\right) + \frac{\varepsilon_r + 1}{2\pi\varepsilon_r}\left[\ln\frac{\pi\varepsilon_r}{2} + \ln\left(\frac{w}{2h} + 0.94\right) \right] \right\}^{-1} \tag{2.4}$$

Synthesis formulae

In the synthesis case we know Z_0 and ε_r and want to find w/h and ε_e.

For *narrow* strips, i.e. $Z_0 > (44 - 2\varepsilon_r)\ \Omega$

$$\frac{w}{h} = \left(\frac{\exp^H}{8} - \frac{1}{4\exp^H} \right)^{-1} \tag{2.5}$$

where

$$H = \frac{Z_0\sqrt{2(\varepsilon_r + 1)}}{119.9} + \frac{1}{2}\frac{\varepsilon_r - 1}{\varepsilon_r + 1}\left(\ln\frac{\pi}{2} + \frac{1}{\varepsilon_r}\ln\frac{4}{\pi} \right) \tag{2.6}$$

and

$$\varepsilon_e = \frac{\varepsilon_r + 1}{2} + \left[1 - \frac{1}{2H}\frac{\varepsilon_r - 1}{\varepsilon_r + 1}\left(\ln\frac{\pi}{2} + \frac{1}{\varepsilon_r}\ln\frac{4}{\pi} \right) \right]^{-2} \tag{2.7}$$

Note: Equation 2.7 was derived under a slightly different changeover value of $Z_0 > (63 - \varepsilon_r)\ \Omega$.

[4] In practice these are almost always calculated using CAD/CAE programmes.

[5] Wheeler, H.A. Transmission lines properties of parallel wide strips separated by a dielectric sheet, *IEEE Trans*, MTT-13 No. 3, 1965.

For *wide* strips, i.e. $Z_0 < (44 - 2\varepsilon_r)\Omega$

$$\frac{w}{h} = \frac{2}{\pi}\left[(d_e - 1) - \ln(2d_e - 1)\right] + \frac{\varepsilon_r - 1}{\pi\varepsilon_r}\left[\ln(d_e - 1) + 0.293 - \frac{0.517}{\varepsilon_r}\right] \quad (2.8)$$

where

$$d_e = \frac{59.95\pi^2}{Z_0\sqrt{\varepsilon_r}} \quad (2.9)$$

and under a slightly different value of $Z_0 > (63 - 2\varepsilon_r)\Omega$

$$\varepsilon_e = \frac{\varepsilon_r + 1}{2} + \frac{\varepsilon_r - 1}{2}\left(1 + 10\frac{h}{w}\right)^{-0.555} \quad (2.10)$$

Equations 2.3 to 2.10 are accurate up to about 2 GHz. For higher frequencies, the effect of frequency dependence of ε_e has to be taken into account. An expression often used to evaluate $\varepsilon_e(f)$ as frequency (f) varies is

$$\varepsilon_e(f) = \varepsilon_r - \frac{\varepsilon_r - \varepsilon_e}{1 + (h/Z_0)^{1.33}(0.43f^2 - 0.009f^3)} \quad (2.11)$$

where h is in millimetres, f is in gigahertz, and ε_e is the value calculated by either Equation 2.7 or 2.10.

Example 2.3

Two microstrip lines are printed on the same dielectric substrate. One line has a wider centre strip than the other. Which line has the lower characteristic impedance? Assume that there is no coupling between the two lines.

Solution. If you refer to Equation 2.3 and examine the *h*/*w* ratio, you will see that Z_0 varies as a function of *h*/*w*. Therefore, the line with the lower characteristic impedance will have a wider centre conductor.

As you can see for yourself, Equations 2.3 to 2.11 are rather complicated and should be avoided when possible. To avoid these types of calculations, we have included with this book a computer software program called PUFF. With this program, it is only necessary to decide on the characteristic impedance of the microstrip or stripline which we require and PUFF will do the rest. We will return to PUFF when we have explained the basic terms for using it.

Expressions also exist for calculating the characteristic impedance of other lines such as the strip line, coplanar waveguide, slot line, etc. These are equally complicated but details of how to calculate them have been compiled by Gupta, Garg and Chadha.[6] There is also a software program called AppCAD[7] which calculates these impedances.

[6] Gupta, K.C., Garg, R. and Chadha, R., *Computer-Aided Design of Microwave Circuits*, Artech House Inc, Norwood MA 02062 USA, ISBN: 0–89006–105–X.

[7] AppCAD is a proprietary software program from the Hewlett Packard Co, Page Mill Road, Palo Alto CA, USA.

Note: In the previous sections, I have produced equations which are peculiar to types of different transmission lines. From now on, and unless stated otherwise, all the equations in the sections that follow apply to all types of transmission lines.

2.5 Characteristic impedance (Z_0) from primary electrical parameters

A typical twin conductor type transmission line is shown in Figure 2.8. Each wire conductor has resistance and inductance associated with it. The resistance is associated with the material of the metal conductors, effective conductor cross-sectional area and length. The inductance is mainly dependent on length and type of material. In addition to these, there is capacitance between the two conductors. The capacitance is mainly dependent on the dielectric type, its effective permittivity, the effective cross-sectional area between conductors, the distance between the conductors and the length of the transmission line. When a voltage is applied, there is also a leakage current between the two conductors caused by the non-infinite resistance of the insulation between the two conductors. This non-infinite resistance is usually expressed in terms of a shunt resistance or parallel conductance.

Therefore, transmission lines possess inherent resistance, inductance, capacitance and conductance. It is very important to realise that these properties are distributed along the length of the line and that they are not physically lumped together. The lumped approach is only applicable when extremely short lengths of line are considered and as a practical line is made up of many short lengths of these lines, the lumped circuit equivalent of a transmission line would look more like that shown in Figure 2.8. This is an approximation but nevertheless it is an extremely useful one because it allows engineers to construct and simulate the properties of transmission lines.

2.5.1 Representation of primary line constants

In Figure 2.8, let:

R represent the resistance per metre (ohms/metre)
L represent the inductance per metre (henry/metre)
G represent the conductance per metre (siemen/metre)
C represent the capacitance per metre (farad/metre)

It follows that for a short length δl, we would obtain $R\delta l$, $L\delta l$, $G\delta l$, and $C\delta l$ respectively. Hence

$$\frac{Z_1}{4}\delta l = \left(\frac{R}{4} + \frac{\mathrm{j}\omega L}{4}\right)\delta l \tag{2.12}$$

$$Z_1\delta l = (R + \mathrm{j}\omega L)\delta l \tag{2.13}$$

$$Y\delta l = (G + \mathrm{j}\omega C)\delta l \tag{2.14}$$

and

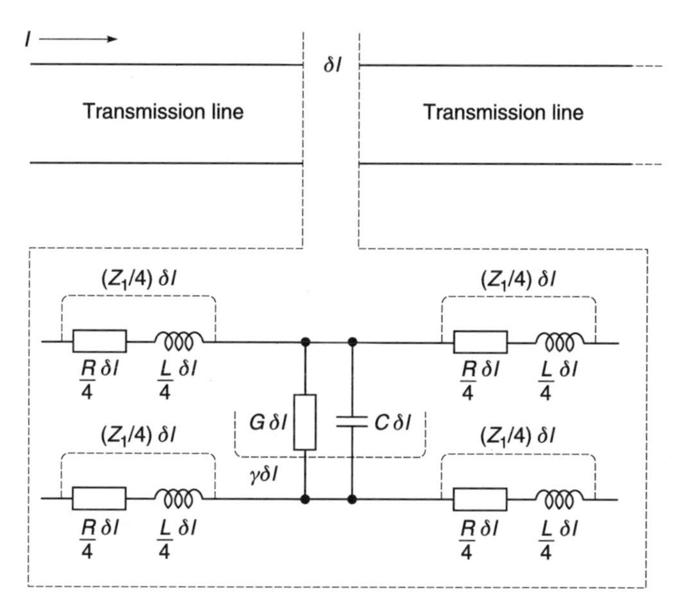

Fig. 2.8 Expanded view of a short section of transmission line

$$Z\delta l = \frac{1}{Y\delta l} = \frac{1}{(G + j\omega C)\delta l} \tag{2.15}$$

2.5.2 Derivation of line impedance

The input impedance Z_{in} of the short section δl when terminated by a matched line is given by[8]

$$Z_{in} = \frac{Z_1}{4}\delta l + \frac{Z\delta l\left(\frac{Z_1}{4}\delta l + Z_0 + \frac{Z_1}{4}\delta l\right)}{Z\delta l + \frac{Z_1}{4}\delta l + Z_0 + \frac{Z_1}{4}\delta l} + \frac{Z_1}{4}\delta l$$

Since the line is terminated by another line, $Z_{in} = Z_0$

$$Z_{in} = Z_0 = \frac{Z_1}{2}\delta l + \frac{Z\delta l\left(\frac{Z_1}{2}\delta l + Z_0\right)}{Z\delta l + \frac{Z_1}{2}\delta l + Z_0}$$

[8] The Z_0 term in the centre fraction on the right-hand side of the equation is present because the short section of line (δl) is terminated by an additional line which presents an input impedance of Z_0.

Cross-multiplying, we get

$$Z_0\left(Z\delta l + \frac{Z_1}{2}\delta l + Z_0\right) = \frac{Z_1}{2}\delta l\left(Z\delta l + \frac{Z_1}{2}\delta l + Z_0\right) + Z\delta l\left(\frac{Z_1}{2}\delta l + Z_0\right)$$

Simplifying

$$Z_0 Z\delta l + \frac{Z_0 Z_1}{2}\delta l + Z_0^2 = \frac{ZZ_1}{2}\delta l^2 + \frac{Z_1^2}{4}\delta l^2 + \frac{Z_0 Z_1}{2}\delta l + \frac{ZZ_1}{2}\delta l^2 + ZZ_0\delta l$$

and

$$Z_0^2 = ZZ_1\delta l^2 + \frac{Z_1^2}{4}\delta l^2 \tag{2.16}$$

Substituting for Z and Z_1, we get

$$Z_0^2 = \frac{(R + j\omega L)\delta l}{(G + j\omega C)\delta l} + \left(\frac{R}{4} + \frac{j\omega l}{4}\right)^2 \delta l^2$$

In the limit when $\delta l \to 0$, and taking the positive square root term

$$Z_0 = \sqrt{\frac{R + j\omega L}{G + j\omega C}} \tag{2.17}$$

If you examine the expression for Z_0 a bit more closely, you will see that there are two regions where Z_0 tends to be resistive and constant. The first region occurs at very low frequencies when $R \gg j\omega L$ and $G \gg j\omega C$. This results in

$$Z_0 \approx \sqrt{\frac{R}{G}} \tag{2.18}$$

The second region occurs at very high frequencies when $j\omega L \gg R$ and $j\omega C \gg G$. This results in

$$Z_0 \approx \sqrt{\frac{L}{C}} \tag{2.19}$$

The second region is also known as the frequency region where a transmission line is said to be 'lossless' because there are 'no' dissipative elements in the line.

Equation 2.19 is also useful because it explains why inductive loading, putting small lumped element inductors in series with lines, is used to produce a more constant impedance for the line. The frequency regions of operation described by Equations (2.18) and (2.19) are important because under these conditions, line impedance tends to remain frequency independent and a state known as 'distortionless transmission' exists. The distortionless condition is very useful for pulse waveform/digital transmissions because in these regions, frequency dispersion and waveform distortion tend to be minimal.

These statements can also be verified by the following practical example.

Example 2.4

A transmission line has the following primary constants: $R = 23\ \Omega\ \text{km}^{-1}$, $G = 4\ \text{mS km}^{-1}$, $L = 125\ \mu\text{H km}^{-1}$ and $C = 48\ \text{nF km}^{-1}$. Calculate the characteristic impedance, Z_0, of the line at a frequency of (a) 100 Hz, (b) 500 Hz, (c) 15 kHz, (d) 5 MHz and (e) 10 MHz.

Given: $R = 23\ \Omega\ \text{km}^{-1}$, $G = 4\ \text{mS km}^{-1}$, $L = 125\ \mu\text{H km}^{-1}$ and $C = 48\ \text{nF km}^{-1}$.
Required: Z_0 at (a) 100 Hz, (b) 500 Hz, (c) 15 kHz, (d) 5 MHz and (e) 10 MHz.

Solution. Use Equation 2.17 in the calculations that follow.

(a) At 100 Hz

$$R + j\omega L = (23 + j0.08)\ \Omega\ \text{km}^{-1},\ G + j\omega C = (4 + j0.030)\ \text{mS km}^{-1}$$

Hence

$$Z_0 = \sqrt{\frac{23 + j0.08}{(4 + j0.030) \times 10^{-3}}} = \underline{75.83\Omega/-2.06 \times 10^{-3}}\ \text{rad}$$

(b) At 500 Hz

$$R + j\omega L = (23 + j0.39)\ \Omega\ \text{km}^{-1},\ G + j\omega C = (4 + j0.15)\ \text{mS km}^{-1}$$

Hence

$$Z_0 = \sqrt{\frac{23 + j0.39}{(4 + j0.15) \times 10^{-3}}} = \underline{75.81\Omega/-10.30 \times 10^{-3}}\ \text{rad}$$

(c) At 15 kHz

$$R + j\omega L = (23 + j11.78)\ \Omega\ \text{km}^{-1},\ G + j\omega C = 4 + j4.52\ \text{mS km}^{-1}$$

Hence

$$Z_0 = \sqrt{\frac{23 + j11.78}{(4 + j4.52) \times 10^{-3}}} = \underline{65.42\Omega/-0.19}\ \text{rad}$$

(d) At 5 MHz

$$R + j\omega L = (23 + j3926.99)\ \Omega\ \text{km}^{-1},\ G + j\omega C = (4 + j1508)\ \text{mS km}^{-1}$$

Hence

$$Z_0 = \sqrt{\frac{23 + j3926.99}{(4 + j1508) \times 10^{-3}}} = \underline{50.03\Omega/-0.00}\ \text{rad}$$

(e) At 10 MHz

$$R + j\omega L = (23 + j7853.98)\ \Omega\ \text{km}^{-1},\ G + j\omega C = (4 + j3016)\ \text{mS km}^{-1}$$

Hence

$$Z_0 = \sqrt{\frac{23 + j7853.98}{(4 + j3016) \times 10^{-3}}} = \underline{50.03\Omega/-0.00}\ \text{rad}$$

Conclusions from Example 2.4. At low frequencies, i.e. 100–500 Hz, the line impedance Z_0 tends to remain at about 75 Ω with very little phase shift over a wide frequency range. For most purposes, it is resistive and constant in this region. See cases (a) and (b). At high frequencies, i.e. 5–10 MHz, the line impedance Z_0 tends to remain constant at about 50 Ω with little phase shift over a wide frequency range. For most purposes, it is resistive and constant in this region. See cases (d) and (e). In between the above regions, the line impedance Z_0 varies with frequency and tends to be reactive. See case (c). For radio work, we tend to use transmission lines in the 'lossless' condition (Equation 2.19) and this helps considerably in the matching of line impedances.

2.6 Characteristic impedance (Z_0) by measurement

Occasions often arise when the primary constants of a line are unknown yet it is necessary to find the characteristic impedance (Z_0). In this case, Z_0 can be obtained by measuring the short- and open-circuit impedance of the line. In Figures 2.9 and 2.10 as in Figure 2.8 let:

R represent the resistance per metre (ohms/metre)
L represent the inductance per metre (henry/metre)
G represent the conductance per metre (siemen/metre)
C represent the capacitance per metre (farad/metre)

It follows that for a short length δl, we would obtain $R\delta l$, $L\delta l$, $G\delta l$ and $C\delta l$ respectively.

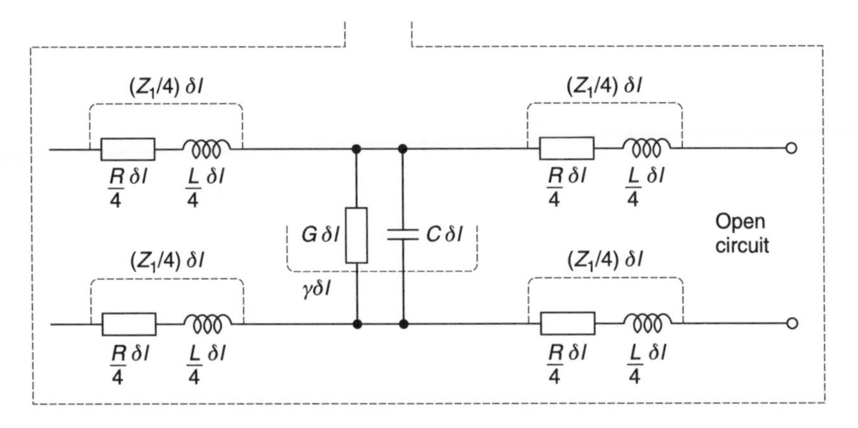

Fig. 2.9 Open-circuit equivalent of a short length of transmission line

2.6.1 Open-circuit measurement (Z_{oc})

Hence defining $Z\delta l$ as $1/Y\delta l$ we have

$$\begin{aligned} Z_{oc} &= \frac{Z_1}{4}\delta l + Z\delta l + \frac{Z_1}{4}\delta l \\ &= \frac{Z_1}{2}\delta l + Z\delta l \end{aligned} \tag{2.20}$$

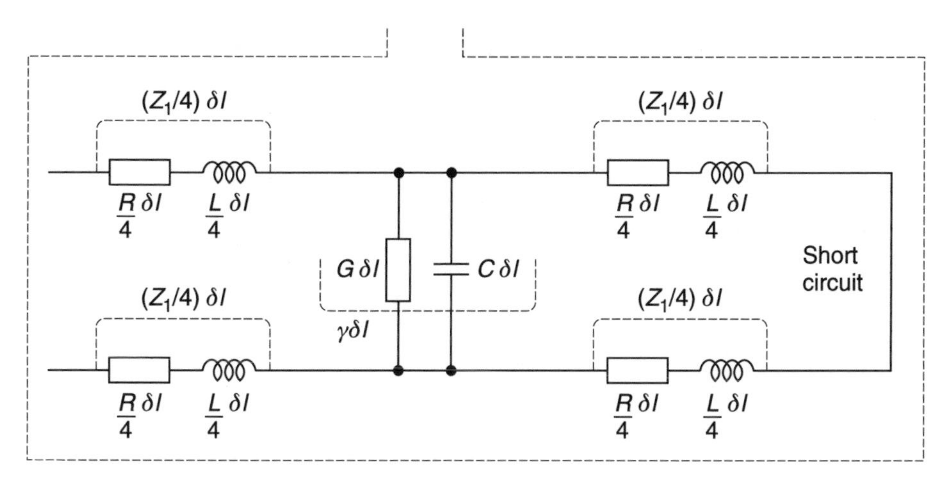

Fig. 2.10 Short-circuit equivalent of a short length of transmission line

2.6.2 Short-circuit measurement (Z_{sc})

The short-circuit impedance is

$$Z_{sc} = \frac{Z_1}{4}\delta l + \frac{Z\delta l\left(\frac{Z_1}{4}\delta l + \frac{Z_1}{4}\delta l\right)}{Z\delta l + \left(\frac{Z_1}{4}\delta l + \frac{Z_1}{4}\delta l\right)} + \frac{Z_1}{4}\delta l$$

$$= \frac{Z_1}{2}\delta l + \frac{\frac{ZZ_1}{2}\delta l^2}{Z\delta l + \frac{Z_1}{2}\delta l}$$

$$= \frac{\frac{Z_1}{2}\delta l\left(Z\delta l + \frac{Z_1}{2}\delta l\right) + \frac{ZZ_1}{2}\delta l^2}{Z\delta l + \frac{Z_1}{2}\delta l}$$

$$= \frac{ZZ_1\delta l^2 + \frac{Z_1^2}{4}\delta l^2}{Z\delta l + \frac{Z_1}{2}\delta l}$$

Using Equation 2.16 to substitute for the numerator and Equation 2.20 to substitute for the denominator, we have

$$Z_{sc} = \frac{Z_0^2}{Z_{oc}}$$

Hence

$$Z_0^2 = Z_{sc} Z_{oc}$$

or

$$Z_0 = \sqrt{Z_{sc} Z_{oc}} \tag{2.21}$$

Example 2.5

The following measurements have been made on a line at 1.6 MHz where $Z_{oc} = 900\ \Omega\ \underline{/-30^\circ}$ and $Z_{sc} = 400\ \Omega\ \underline{/-10^\circ}$. What is the characteristic impedance (Z_0) of the line at 1.6 MHz?

Given: $f = 1.6$ MHz, $Z_{oc} = 900\ \Omega\ \underline{/-30^\circ}$, $Z_{sc} = 400\ \Omega\ \underline{/-10^\circ}$.
Required: Z_0 at 1.6 MHz.

Solution. Using Equation 2.21

$$\begin{aligned} Z_0 &= \sqrt{Z_{sc} Z_{oc}} \\ &= \sqrt{900\ \Omega \underline{/-30^\circ} \times 400\ \Omega \underline{/-10^\circ}} \\ &= 600\ \Omega \underline{/-20^\circ} \end{aligned}$$

2.7 Typical commercial cable impedances

Manufacturers tend to make cables with the following characteristic impedances (Z_0). These are:

- 50 Ω – This type of cable finds favour in measurement systems and most radio instruments are matched for this impedance. It is also used extensively in amateur radio links.

  Most cable manufacturers make more than one type of 50 Ω line. For example, you can buy 50 Ω rigid lines (solid outer connector), 50 Ω low loss lines (helical and air dielectrics), 50 Ω high frequency lines for use up to 50 GHz with minimal loss. The reason for this is that different uses require different types of lines. Remember that in Equation 2.1 repeated here for convenience $Z_0 = 138 \log_{10}(D/d)/\sqrt{\varepsilon}$ and the dimension of the variables can be changed to produce the desired impedance.
- 75 Ω – This type of cable is favoured by the television industry because it provides a close match to the impedance (73.13 Ω) of a dipole aerial. Most TV aerials are designed for this impedance and it is almost certain that the cable that joins your TV set to the external aerial will have this impedance. The comments relating to the different types of 50 Ω lines also apply to 75 Ω lines.

- 140 Ω – This type of cable is used extensively by the telephone industry. The comments relating to the different types of 50 Ω lines also apply to 140 Ω lines.
- 300 Ω – This type of cable is favoured by both the radio and television industry because it provides a close match for the impedance (292.5 Ω) of a very popular antenna (folded dipole antenna) which is used extensively for VHF-FM reception. The comments relating to the different types of 50 Ω lines also apply to 300 Ω lines.
- 600 Ω – This type of cable is used extensively by the telephone industry and many of their instruments are matched to this impedance. The comments relating to the different types of 50 Ω lines also apply to 600 Ω lines.

2.8 Signal propagation on transmission lines

2.8.1 Pulse propagation on an infinitely long or matched transmission line

We are now going to use some of the ideas introduced in the previous sections, particularly Section 2.2.2, to describe qualitatively the propagation of signals along an infinitely long transmission line. In this description we will only make two assumptions:

- the transmission line is perfectly uniform, that is its electrical properties are identical all along its length;
- the line extends infinitely in one direction or is perfectly terminated.

To keep the explanation simple, we will initially only consider the propagation of a single electrical pulse along the line[9] shown in Figure 2.11. At the beginning of the line (top left hand corner) a voltage source (V_s) produces the single pulse shown in Figure 2.11. The waveforms shown at various planes (plane 1, plane 2, plane 3) on the line illustrate three of the main properties of signal propagation along a transmission line:

- **propagation delay** – the pulse appears at each successive point on the line later than at the preceding point;
- **attenuation** – the peak value of the pulse is attenuated progressively;
- **waveform distortion and frequency dispersion** – its shape differs from its original shape at successive points.

2.8.2 Propagation delay

The pulse appears later and later at successive points on the line because it takes time to travel over any distance, i.e. there is a propagation delay. As the line is uniform throughout

[9] The behaviour of a pulse travelling along an infinitely long transmission line is very similar to the example you were given in Section 2.2.2 concerning sound travelling down an infinitely long tunnel except that this time instead of voice sounds, consider the sound to originate from a single drum beat or pulse. You will no doubt remember from earlier work that a pulse is a waveform which is made up from a fundamental sinusoid and its harmonics combined together in a precise amplitude, phase and time relationship.

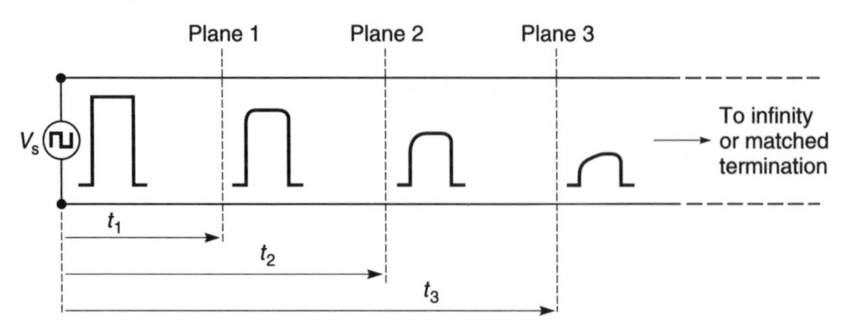

Fig. 2.11 Pulse propagation in a transmission line

its length, the amount of delay at any point is proportional to distance between that point and the source of the pulse. These time delays are shown as t_1, t_2, and t_3 in Figure 2.11. Another way of describing this is to say that the pulse propagates along the line with a uniform velocity.

2.8.3 Attenuation

The amplitude of the pulse is attenuated as it propagates down the line because of resistive losses in the wires. The amount of attenuation per unit length is uniform throughout the line because the line cross-section is uniform throughout the line length. Uniform attenuation means that the fractional reduction in pulse amplitude is the same on any line section of a given length. This is more easily understood by referring to Figure 2.11, where the pulse amplitude at plane 1 has been reduced by a factor of 0.8. At plane 2, which is twice as far from the source as plane 1, the pulse height has been reduced by a further factor of 0.8, i.e. a total of 0.8^2 or 0.64 of its original amplitude. At plane 3, which is three times as far from the source as plane 1, the reduction is 0.8^3 or 0.512 of the original amplitude.

More generally, at a distance equal to l times the distance from the source to plane 1, the height is reduced by $(0.8)^l$. Because l is the exponent in this expression this type of amplitude variation is called exponential. It can also be expressed in the form $(e^{\alpha})^l$ or $e^{\alpha l}$, where e^{α} represents the loss per unit length and is 0.8 in this particular example. In fact α is the natural logarithm of the amplitude reduction per unit length. Its unit is called the neper and loss (dB) = 8.686 nepers.[10]

Example 2.6

A transmission has a loss of two nepers per kilometre. What is the loss in dB for a length of 10 kilometres?

Given: Attenuation constant (α) = 2 nepers per km.
Required: Loss in dB for a length of 10 km.

[10] This is because dB = 20 Log (Ratio) = 20 Log (e^{α}) = $20 \times \alpha \times$ Log (e) = $20 \times \alpha \times 0.4343 = 8.686\alpha$.

Solution. If 1 km represents a loss of 2 nepers, then 10 km = $10 \times 2 = 20$ nepers. Therefore

$$\begin{aligned} \text{loss} &= 8.686 \times 20 \\ &= 173.72 \text{ dB} \end{aligned}$$

2.9 Waveform distortion and frequency dispersion

2.9.1 Amplitude distortion

The waveform of the pulse in Figure 2.11 alters as it travels along the line. This shape alteration is caused by the line constants (inductance, capacitance, resistance and conductance of the line) affecting each sinusoidal component of the waveform in a different manner. The high frequency components, which predominate on the edges of the pulse waveform, suffer greater attenuation because of increased reactive effects; the lower frequency components, which predominate on the flat portion of the waveform, suffer less attenuation. The variation of attenuation with frequency is described by the frequency response of the line.

2.9.2 Frequency distortion

In addition to attenuation, there are also time constants associated with the line components (inductance, capacitance, resistance and conductance). These cause high frequency components to travel at a different velocity from low frequency components. The variation of velocity with frequency is called the frequency dispersion of the line.

2.9.3 Phase and group velocities

As a pulse consists of sinusoidal components of different frequencies, each component will therefore be altered differently. Distinction must be made between the velocities of the sinusoidal components which are called phase velocities, $\boldsymbol{u}_{\mathbf{p}}$. The phase velocity ($\beta$) is defined as the change in radians over a wavelength and since there is a phase change of 2π radians in every wavelength, it follows that $\beta = 2\pi$ radians/wavelength (λ) or

$$\beta = \frac{2\pi}{\lambda} \tag{2.22}$$

The velocity of the complete waveform is called the group velocity, $\boldsymbol{u}_{\mathbf{g}}$. The apparent velocity of the pulse in Figure 2.11 is called its group velocity.

It is important to realise that if the line velocity and line attenuation of all the component sinusoids which make up a pulse waveform are not identical then deterioration in pulse waveform shape will occur. Pulse distortion is particularly critical in high speed data transmission where a series of distorted pulses can easily merge into one another and cause pulse detection errors.

If distortion occurs and if it is desired to know how and why a particular waveform has changed its shape, it will be necessary to examine the propagation of the constituent sinusoids of the waveform itself and to instigate methods, such as frequency and phase equalisation, to ensure minimal waveform change during signal propagation through the line.

2.10 Transmission lines of finite length

2.10.1 Introduction

In Section 2.8.1, we discussed waveforms travelling down infinitely long lines. In practice, infinitely long lines do not exist but finite lines can be made to behave like infinitely long lines if they are terminated with the characteristic impedance of the line.[11]

2.10.2 Matched and unmatched lines

A transmission line which is terminated by its own characteristic impedance, Z_0, is said to be **matched** or **properly terminated**. A line which is terminated in any impedance other than Z_0 is said to be **unmatched** or **improperly terminated**. To prevent reflections it is usual for a transmission line to be properly terminated and so it is a common condition for a transmission line to behave electrically as though it was of infinite length.

If a transmission line is to be used for signals with a wide range of frequency components, it may be difficult to terminate it properly. In general, the characteristic impedance of a transmission line will vary with frequency and if the matching load fails to match the line at all frequencies, then the line will not be properly terminated and reflections will occur.[12]

In practice, it is usual to properly terminate both ends of a transmission line, i.e. both at the sending end and the receiving end; otherwise any signal reflected from the receiving end and travelling back towards the sending end will be re-reflected again down the line to cause further reflections. The sending end can be properly terminated either by using a source generator with an impedance equal to the characteristic impedance of the line or by using a matching network to make a source generator present a matched impedance to the transmission line.

2.11 Reflection transmission coefficients and VSWR

2.11.1 Introduction

Reflection coefficients are based on concepts introduced in your childhood. Consider the case when you throw a ball at a vertical stone wall. The ball with its incident power will

[11] This argument is similar to the case mentioned in Section 2.2.2 where it was shown that if our finite tunnel was terminated with material with the same properties as an infinitely long tunnel which absorbed all the incident energy then it would also behave like an infinitely long tunnel.

[12] You see this reflection effect as multiple images on your television screen when the TV input signal is not properly terminated by the TV system. TV engineers call this effect 'ghosting'.

travel towards the wall, hit the wall which will absorb some of its incident power and then the remaining power (reflected power) will cause the ball to bounce back.

The ratio (reflected power)/(incident power) is called the reflection coefficient. The reflection coefficient is frequently represented by the Greek letter gamma (Γ). In mathematical terms, we have

$$\Gamma = \frac{\text{reflected power}}{\text{incident power}}$$

This simple equation is very useful for the following reasons.

- Its value is independent of incident power because if you double incident power, reflected power will also double.[13] If you like, you can say that Γ is normalised to its incident power.
- It gives you a measure of the hardness (**impedance**) of the wall to incident power. For example if the wall is made of stone, it is likely that very little incident power will be absorbed and most of the incident power will be returned to you as reflected power. You will get a high reflection coefficient ($\Gamma \rightarrow 1$). If the wall is made of wood, it is likely that the wood would bend a bit (less resistance), absorb more incident energy and return a weaker bounce. You will get a lower reflection coefficient ($\Gamma < 1$). Similarly if the wall was made of straw, it is more than likely that most of the incident energy would be absorbed and there would be little rebounce or reflected energy ($\Gamma \rightarrow 0$). Finally if the wall was made of air, the ball will simply go through the air wall and carry all its incident power with it ($\Gamma = 0$). There will be no reflected energy because the incident energy would simply be expended in carrying the ball further. Note in this case that the transmission medium resistance is air, and it is the same as the air wall resistance which is the load and we simply say that the load is matched to the transmission medium.
- By measuring the angle of the re-bounce relative to the incident direction, it is possible to tell whether the wall is vertical and facing the thrower or whether it is at an angle (phase) to the face-on position. Hence we can determine the direction of the wall.
- The path through which the ball travels is called the **transmission path**.
- Last but not least, you need not even physically touch the wall to find out some of its characteristics. In other words, measurement is *indirect*. This is useful in the measurement of transistors where the elements cannot be directly touched. It is also very useful when you want to measure the impedance of an aerial on top of a high transmitting tower when your measuring equipment is at ground level. The justification for this statement will be proved in Section 2.13.

2.11.2 Voltage reflection coefficient[14] (Γ_v) in transmission lines

The same principles described above can also be applied to electrical energy. This is best explained by Figure 2.12 where we have a signal generator with a source impedance, Z_s, sending electrical waves through a transmission line whose impedance is Z_0, into a load impedance, Z_L.

[13] This of course assumes that the hardness of your wall is independent of the incident power impinging on it.

[14] Some authors use different symbols for voltage reflection coefficient. Some use Γ_v, while others use ρ_v. In this book, where possible, we will use Γ_v for components and ρ_v for systems.

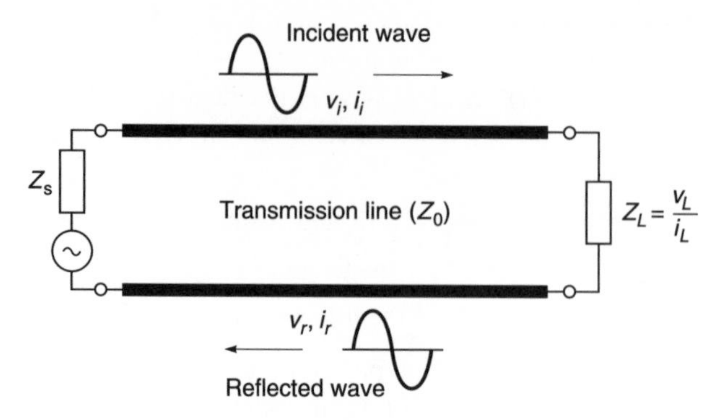

Fig. 2.12 Incident and reflected waves on a transmission line

If the load impedance (Z_L) is exactly equal to Z_0, the incident wave is totally absorbed in the load and there is no reflected wave. If Z_L differs from Z_0, some of the incident wave is not absorbed in the load and is reflected back towards the source. If the source impedance (Z_s) is equal to Z_0, the reflected wave from the load will be absorbed in the source and no further reflections will occur. If Z_s is not equal to Z_0, a portion of the reflected wave from the load is re-reflected from the source back toward the load and the entire process repeats itself until all the energy is dissipated. The degree of mis-match between Z_0 and Z_L or Z_s determines the amount of the incident wave that is reflected.

By definition

$$\text{voltage reflection coefficient} = \frac{v_{\text{reflected}}}{v_{\text{incident}}} = \Gamma_v \angle \theta \tag{2.23}$$

Also

$$\text{current reflection coefficient} = \frac{i_{\text{reflected}}}{i_{\text{incident}}} = \Gamma_\iota \angle \theta \tag{2.24}$$

From inspection of the circuit of Figure 2.12

$$Z_0 = \frac{v_i}{i_i} \tag{2.25}$$

and

$$Z_0 = \frac{v_r}{i_r} \tag{2.26}$$

The minus sign in Equation 2.27 occurs because we use the mathematical convention that current flows to the right are positive, therefore current flows to the left are negative.

$$Z_L = \frac{v_L}{I_L}$$

$$= \frac{v_i + v_r}{i_i - i_r} = \frac{v_i + v_r}{v_i/Z_0 - v_r/Z_0} = Z_0 \frac{v_i(1+\Gamma_v)}{v_i(1-\Gamma_v)} \tag{2.27}$$

Sorting out terms in respect of Γ_v

$$Z_L = Z_0 \frac{(1 + \Gamma_v)}{(1 - \Gamma_v)}$$

or

$$\Gamma_v = \frac{(Z_L - Z_0)}{(Z_L + Z_0)} \tag{2.28}$$

Returning to Equation 2.24 and recalling Equation 2.23

$$\Gamma_i = \frac{i_r}{i_i} = \frac{-v_r/Z_0}{v_i/Z_0} = -\Gamma_v \tag{2.29}$$

As the match between the characteristic impedance of the transmission line Z_0 and the terminating impedance Z_L improves, the reflected wave becomes smaller. Therefore, using Equation 2.28, the reflection coefficient decreases. When a perfect match exists, there is no reflected wave and the reflection coefficient is zero. If the load Z_L on the other hand is an open or short circuit, none of the incident power can be absorbed in the load and all of it will be reflected back toward the source. In this case, the reflection coefficient is equal to 1, or a *perfect* mismatch. Thus the *normal* range of values for the *magnitude* of the reflection coefficient is between zero and unity.

Example 2.7

Calculate the voltage reflection coefficient for the case where $Z_L = (80 - j10)\ \Omega$ and $Z_0 = 50\ \Omega$.

Given: $Z_L = (80 - j10)$, $Z_0 = 50\Omega$
Required: Γ_v

Solution. Using Equation 2.28

$$\Gamma_v = \frac{Z_L - Z_0}{Z_L + Z_0} = \frac{80 - j10 - 50}{80 - j10 + 50} = \frac{30 - j10}{130 - j10}$$

$$= \frac{31.62 \angle -18.43°}{130.38 \angle -4.40°} = 0.24 \angle -14.03°$$

Example 2.8

Calculate the voltage reflection coefficients at the terminating end of a transmission line with a characteristic impedance of 50 Ω when it is terminated by (a) a 50 Ω termination, (b) an open-circuit termination, (c) a short-circuit termination and (d) a 75 Ω termination.

Given: $Z_0 = 50\ \Omega$, Z_L = (a) $50\ \Omega$, (b) open-circuit = ∞, (c) short-circuit = $0\ \Omega$, (d) = $75\ \Omega$.
Required: Γ_v for (a), (b), (c), (d).

Solution. Use Equation 2.28.

(a) with $Z_L = 50\ \Omega$

$$\Gamma_v = \frac{Z_L - Z_0}{Z_L + Z_0} = \frac{50-50}{50+50} = 0\underline{/0°}$$

(b) with Z_L = open circuit = $\infty\ \Omega$

$$\Gamma_v = \frac{Z_L - Z_0}{Z_L + Z_0} = \frac{\infty-50}{\infty+50} = 1\underline{/0°}$$

(c) with Z_L = short circuit = $0\ \Omega$

$$\Gamma_v = \frac{Z_L - Z_0}{Z_L + Z_0} = \frac{0-50}{0+50} = -1\underline{/0°} \text{ or } 1\underline{/180°}$$

(d) with $Z_L = 75\ \Omega$

$$\Gamma_v = \frac{Z_L - Z_0}{Z_L + Z_0} = \frac{75-50}{75+50} = 0.2\underline{/0°}$$

Example 2.8 is instructive because it shows the following.

- If you want to transfer an incident voltage wave with no reflections then the terminating load (Z_L) must match the characteristic impedance (Z_0) exactly. See case (a). This is the desired condition for efficient transfer of power through a transmission line.
- Maximum in-phase voltage reflection occurs with an open circuit and maximum anti-phase voltage reflection occurs with a short circuit. See cases (b) and (c). This is because there is no voltage across a short circuit and therefore the reflected wave must cancel the incident wave.
- Intermediate values of terminating impedances produce intermediate values of reflection coefficients. See case (d).

2.11.3 Return loss

Incident power (P_{inc}) and reflected power (P_{ref}) can be related by using the magnitude of the voltage reflection coefficient (Γ). Since $\Gamma = v_{ref}/v_{inc}$, it follows that

$$\frac{P_{ref}}{P_{inc}} = \frac{v_{ref}^2/R_{load}}{V_{inc}^2/R_{load}} = \Gamma^2 \tag{2.30}$$

The return loss gives the amount of power reflected from a load and is calculated from:

$$\text{return loss (dB)} = -10 \log \Gamma^2 = -20 \log \Gamma \tag{2.31}$$

2.11.4 Mismatched loss

The amount of power transmitted to the load (P_L) is determined from

$$P_L = P_{inc} - P_{ref} = P_{inc}(1 - \Gamma^2) \tag{2.32}$$

The fraction of the incident power not reaching the load because of mismatches and reflections is

$$\frac{P_{load}}{P_{incident}} = \frac{P_L}{P_{inc}} = 1 - \Gamma^2 \tag{2.33}$$

Hence the mismatch loss (or reflection loss) is calculated from

$$\text{ML(dB)} = -10 \log (1 - \Gamma^2) \tag{2.34}$$

2.11.5 Transmission coefficient

The transmission coefficient (τ_v) is defined as the ratio of the load voltage (v_L) to the incident voltage (v_{inc}) but $v_L = v_{inc} + v_{ref}$. Hence

$$\tau_v = \frac{v_L}{v_{inc}} = \frac{v_{inc} + v_{ref}}{v_{inc}} = 1 + \Gamma_v \tag{2.35}$$

If we now use Equation 2.28 to substitute for Γ_v, we obtain

$$\tau_v = 1 + \Gamma_v = 1 + \frac{Z_L - Z_0}{Z_L + Z_0} = \frac{2Z_L}{Z_L + Z_0} \tag{2.36}$$

Sometimes Equation 2.36 is normalised to Z_0 and when Z_L/Z_0 is defined as z, we obtain

$$\tau_v = \frac{2z}{z + 1} \tag{2.36a}$$

Equation 2.36a is the form you frequently find in some articles.

2.11.6 Voltage standing wave ratio (VSWR)

Cases often arise when the terminating impedance for a transmission line is not strictly within the control of the designer. Consider a typical case where a transmitter designed for operating into a 50 Ω transmission line is made to feed an antenna with a nominal impedance of 50 Ω. In the ideal world, apart from a little loss in the transmission line, all the energy produced by the transmitter will be passed on to the antenna. In the practical world, an exact antenna match to the transmission line is seldom achieved and most antenna manufacturers are honest enough to admit the discrepancy and they use a term called the **voltage standing wave ratio**[15] to indicate the degree of mismatch.

[15] This term is based on the Standing Wave Pattern principle which was introduced in Section 2.2.2 where you walked along a tunnel which produced echoes while your friend whistled at a constant amplitude and pitch. In the tunnel case, the loudest (maximum intensity) sound occurred where the incident and reflected wave added, while the weakest sound (minimum intensity) occurred where the incident and reflected sound opposed each other.

VSWR is useful because

- it is relatively easy to measure – it is based on modulus values rather than phasor quantities which enables simple diode detectors to be used for measurement purposes;
- it indicates the degree of mismatch in a termination;
- it is related to the modulus of the reflection coefficient (shown later).

Voltage standing wave ratio is defined as

$$\text{VSWR} = \frac{|V_{\text{max}}|}{|V_{\text{min}}|} = \frac{|V_{\text{inc}}| + |V_{\text{ref}}|}{|V_{\text{inc}}| - |V_{\text{ref}}|} \tag{2.37}$$

A VSWR of |1| represents the best possible match.[16] Any VSWR greater than |1| indicates a mismatch and a large VSWR indicates a greater mismatch than a smaller VSWR. Typical Figures of VSWRs for good practical terminations range from 1.02 to 1.1.

Example 2.9

In Figure 2.12, the incident voltage measured along the transmission line is 100 V and the reflected voltage measured on the same line is 10 V. What is its VSWR?

Soloution. Using Equation 2.37

$$\text{VSWR} = \frac{|V_{\text{inc}}| + |V_{\text{ref}}|}{|V_{\text{inc}}| - |V_{\text{ref}}|} = \frac{|100| + |10|}{|100| - |10|} = 1.22$$

2.11.7 VSWR and reflection coefficient (Γ_v)

VSWR is related to the voltage reflection coefficient by:

$$\text{VSWR} = \frac{|V_{\text{inc}}| + |V_{\text{ref}}|}{|V_{\text{inc}}| - |V_{\text{ref}}|} = \frac{1 + \left|\frac{V_{\text{ref}}}{V_{\text{inc}}}\right|}{1 - \left|\frac{V_{\text{ref}}}{V_{\text{inc}}}\right|} = \frac{1 + |\Gamma_v|}{1 - |\Gamma_v|} \tag{2.38}$$

or

$$|\Gamma_v| = \frac{\text{VSWR} - 1}{\text{VSWR} + 1} \tag{2.38a}$$

Example 2.10

What is the VSWR of a transmission system if its reflection coefficient $|\Gamma_v|$ is 0.1?

[16] In a properly terminated line, there are no reflections. $V_{\text{ref}} = 0$ and substituting this value into Equation 2.37 gives

$$\text{VSWR} = \frac{|V_{\text{inc}}| + |0|}{|V_{\text{inc}}| - |0|} = |1|$$

Given: $|\Gamma_v| = 0.1$
Required: VSWR

Solution. Using Equation 2.38

$$\text{VSWR} = \frac{1+|\Gamma_v|}{1-|\Gamma_v|} = \frac{1+|0.1|}{1-|0.1|} = 1.22$$

Perfect match occurs when VSWR = |1|. This is the optimum condition and examination of Equation 2.38 shows that this occurs when $|\Gamma_v| = 0$. With this condition, there is no reflection, optimum power is transferred to the load and there are no standing wave patterns on the line to cause excessive insulation breakdown or electrical discharges to surrounding conductors and there are no 'hot spots' or excessive currents on sections of the conductors.

Example 2.11

A manufacturer quotes a maximum VSWR of 1.07 for a resistive load when it is used to terminate a 50 Ω transmission line. Calculate the reflected power as a percentage of the incident power.

Given: VSWR = 1.07, $Z_0 = 50\ \Omega$
Required: Reflected power as a percentage of incident power

Solution. Using Equation 2.38a

$$|\Gamma_v| = \frac{\text{VSWR} - 1}{\text{VSWR} + 1} = 0.034$$

Since power is proportional to V^2

$$\begin{aligned} P_{ref} &= (0.034)^2 \times P_{inc} \\ &= 0.001 \times P_{inc} \\ &= 0.1\% \text{ of } P_{inc} \end{aligned}$$

From the answer to Example 2.11, you should now realise that:

- a load with an SWR of 1.07 is a good terminating load;
- there are hardly any reflections when a transmission line is terminated with such a load;
- the transmission line is likely to behave like an infinitely long line.

2.11.8 Summary of Section 2.11

If a transmission line is not properly terminated, reflections will occur in a line. These reflections can aid or oppose the incident wave. In high voltage lines, it is possible for the aiding voltages to cause line insulation breakdown. In high current lines, it is possible at high current points for aiding currents to overheat or even destroy the metallic conductors.

The voltage reflection coefficient (Γ_v) can be calculated by

$$\Gamma_v = \frac{\text{reflected voltage wave}}{\text{incident voltage wave}} = \frac{Z_L - Z_0}{Z_L + Z_0}$$

Manufacturers tend to use VSWRs when quoting the impedances associated with their equipment. A VSWR of |1| is the optimum condition and indicates that a perfect match is possible and that there will be no reflections when the equipment is matched to a perfect transmission line. VSWR can be calculated from the reflection coefficients by

$$\text{VSWR} = \left|\frac{V_{\max}}{V_{\min}}\right| = \frac{|V_{\text{inc}}| + |V_{\text{ref}}|}{|V_{\text{inc}}| - |V_{\text{ref}}|} = \frac{1 + |\Gamma_{\text{v}}|}{1 - |\Gamma_{\text{v}}|}$$

The return loss is a way of specifying the power reflected from a load and is equal to $-10 \log \Gamma^2$. The mismatch loss or reflection loss specifies the fraction of incident power not reaching the load and is equal to $-10 \log (1 - \Gamma^2)$.

2.12 Propagation constant (γ) of transmission lines

2.12.1 Introduction

In Section 2.8, we saw that signals on transmission lines suffer attenuation, phase or time delay, and often frequency distortion. In this section, we will show the relationships between these properties and the primary constants (R, G, L and C) of a transmission line.

2.12.2 The propagation constant (γ) in terms of the primary constants

To find the **propagation constant** (γ) we start with the same equivalent circuit (Figure 2.8) used for the derivation of Z_0. It is re-drawn in Figure 2.13 with the voltage and current phasors indicated.

The propagation constant, as defined, relates V_2 and V_1 by

$$\frac{V_2}{V_1} = e^{-\gamma\delta l} \tag{2.39}$$

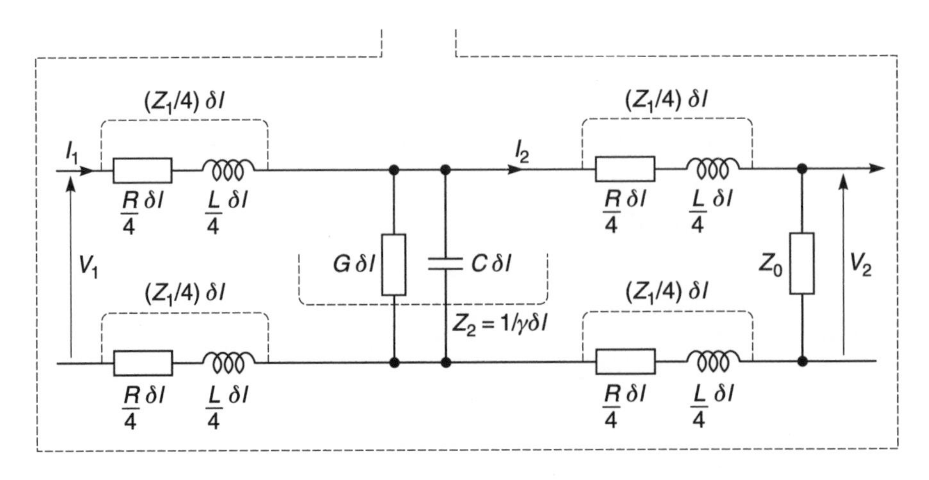

Fig. 2.13 Equivalent circuit of a very short length of line

where δl is still the short length of line referred to in Figure 2.8. It is easier to find γ using the current phasors rather than the voltage phasors; so, using $I_1 = V_1/Z_0$ and $I_2 = V_2/Z_0$

$$\frac{I_2}{I_1} = e^{-\gamma \delta l} \tag{2.40}$$

or alternatively

$$\frac{I_1}{I_2} = e^{\gamma \delta l} \tag{2.40a}$$

The current I_1 splits into two parts: I_2 and a part going through Z_2. By the current divider rule, the split is

$$I_2 = \frac{Z_2}{Z_2 + Z_1/2 + Z_0} I_1$$

giving

$$\frac{I_1}{I_2} = 1 + \frac{Z_1}{2Z_2} + \frac{Z_0}{Z_2}$$

Substituting the definitions for Z_1 and Z_2 and the formula for Z_0 derived above gives

$$\frac{I_1}{I_2} = 1 + \frac{1}{2}(R + j\omega L)(G + j\omega L)(\delta l)^2 + \sqrt{(R + j\omega L)(G + j\omega L)}\delta l$$

$$= 1 + \sqrt{(R + j\omega L)(G + j\omega L)}\delta l + \frac{1}{2}(R + j\omega L)(G + j\omega L)(\delta l)^2$$

Also $I_1/I_2 = e^{\gamma \delta l}$. To use these two expressions for I_1/I_2 to find γ, we must first expand $e^{\gamma \delta l}$ into a Taylor series. Since

$$e^x = 1 + x + \frac{x^2}{2} + \ldots$$

we can write $e^{\gamma \delta l}$ as

$$e^{\gamma \delta l} = 1 + \gamma \delta l + \frac{\gamma^2 (\delta l)^2}{2} + \ldots$$

Equating the two expressions for I_1/I_2 gives

$$1 + \gamma \delta l + \gamma^2 (\delta l)^2/2 = 1 + \sqrt{(R + j\omega L)(G + j\omega L)}\delta l + \frac{1}{2}(R + j\omega L)(G + j\omega L)(\delta l)^2$$

Subtracting 1 from each side and dividing by δl gives

$$\gamma \delta l + \gamma^2 (\delta l)^2/2 = \sqrt{(R + j\omega L)(G + j\omega L)}\delta l + \frac{1}{2}(R + j\omega L)(G + j\omega L)(\delta l)^2$$

and as δl approaches zero

$$\gamma = \sqrt{(R + j\omega L)(G + j\omega L)} \tag{2.41}$$

Since γ is complex consisting of a real term α for amplitude and β for phase, we can also write,

$$\gamma = \alpha + j\beta = \sqrt{(R + j\omega L)(G + j\omega L)} \tag{2.42}$$

If the expression for γ (Equation 2.42) is examined more closely, it can be seen that there are two regions where γ tends to be resistive and constant. The first region occurs at very low frequencies when $R \gg j\omega L$ and $G \gg j\omega C$. This results in

$$\gamma \approx \sqrt{(R)(G)} \tag{2.43}$$

In this region γ is a real number which does not depend on ω. Since the real part of γ is α, the attenuation index, there is no amplitude distortion in the very low frequency range. The second region occurs at very high frequencies when $j\omega L \gg R$ and $j\omega C \gg G$. This results in

$$\gamma \approx j\omega\sqrt{(L)(C)} \tag{2.44}$$

In this region γ is purely imaginary and is proportional to ω. Since the imaginary part of γ is β, the phase index, it means that there is no dispersion (because β is proportional to ω) in the high frequency range. The region is very useful for pulse waveform/digital transmissions because in it frequency dispersion and waveform distortion tend to be minimal.

Equation 2.44 is also useful because it explains why inductive loading, putting small lumped element inductors in series with lines, is sometimes used to reduce dispersion in lines.

Example 2.12

A transmission line has the following primary constants: $R = 23\ \Omega\ \text{km}^{-1}$, $G = 4\ \text{mS km}^{-1}$, $L = 125\ \mu\text{H km}^{-1}$ and $C = 48\ \text{nF km}^{-1}$. Calculate the propagation constant γ of the line, and the characteristic impedance Z_0 of the line at a frequency of (a) 100 Hz, (b) 500 Hz, (c) 15 kHz, (d) 5 MHz and (e) 10 MHz.

Solution. The characteristic impedance Z_0 will not be calculated here because it has already been carried out in Example 2.4. However, the results will be copied to allow easy comparison with the propagation results calculated here for the discussion that follows after this answer. Equation 2.41 will be used to calculate the propagation constant γ, and Equation 2.42 will be used to derive the attenuation constant α and the phase constant β in all the calculations that follow.

(a) At 100 Hz

$$R + j\omega L = (23 + j(2\pi \times 100 \times 125\ \mu\text{H})) = (23 + j0.08)\ \Omega\ \text{km}^{-1}$$

and

$$G + j\omega C = (4\ \text{mS} + j(2\pi \times 100 \times 48\ \text{nF})) = (4 + j0.030)\ \text{mS km}^{-1}$$

Hence

$$\gamma = \sqrt{(23 + j0.08)(4 + j0.030) \times 10^{-3})} = 0.30\underline{/0.01}$$
$$= (0.30 \text{ nepers} + 1.66 \times 10^{-3} \text{ rad}) \text{ km}^{-1}$$

and

$$Z_0 = \sqrt{\frac{23 + j0.08}{(4 + j0.030) \times 10^{-3}}} = 75.83\Omega\underline{/-2.06 \times 10^{-3}} \text{ rad}$$

(b) At 500 Hz

$$R + j\omega L = (23 + j(2\pi \times 500 \times 125\ \mu\text{H})) = (23 + j0.39)\ \Omega \text{ km}^{-1}$$

and

$$G + j\omega C = (4 \text{ mS} + j(2\pi \times 500 \times 48 \text{ nF})) = (4 + j0.15) \text{ mS km}^{-1}$$

Hence

$$\gamma = \sqrt{(23 + j0.39)(4 + j0.15) \times 10^{-3}} = 0.30\underline{/0.03} \text{ rad}$$
$$= (0.30 \text{ nepers} + 8.31 \times 10^{-3} \text{ rad}) \text{ km}^{-1}$$

and

$$Z_0 = \sqrt{\frac{23 + j0.39}{(4 + j0.15) \times 10^{-3}}} = 75.82\Omega\underline{/-10.30 \times 10^{-3}} \text{ rad}$$

(c) At 15 kHz

$$R + j\omega L = (23 + j(2\pi \times 15 \times 10^3 \times 125\ \mu\text{H})) = (23 + j11.78)\ \Omega \text{ km}^{-1}$$

and

$$G + j\omega C = (4 \text{ mS} + j(2\pi \times 15 \times 10^3 \times 48 \text{ nF})) = (4 + j4.52 \times 10^{-3}) \text{ mS km}^{-1}$$

Hence

$$\gamma = \sqrt{(23 + j11.78)(4 + j4.52) \times 10^{-3}} = 0.40\underline{/0.66} \text{ rad}$$
$$= (0.31 \text{ nepers} + 242 \times 10^{-3} \text{ rad}) \text{ km}^{-1}$$

and

$$Z_0 = \sqrt{\frac{23 + j11.78}{(4 + j4.52) \times 10^{-3}}} = 65.42\Omega\underline{/-0.19} \text{ rad}$$

(d) At 5 MHz

$$R + j\omega L = (23 + j(2\pi \times 5 \times 10^6 \times 125\ \mu\text{H})) = (23 + j3926.99)\ \Omega \text{ km}^{-1}$$

and

$$G + j\omega C = (4 \text{ mS} + j(2\pi \times 5 \times 10^6 \times 48 \text{ nF})) = (4 + j1508) \text{ mS km}^{-1}$$

Hence

$$\gamma = \sqrt{(23+3926.99)(4+j1508)\times 10^{-3}} = 76.95\underline{/1.567}\text{ rad}$$
$$= (0.33\text{ nepers} + 76.95\text{ rad})\text{ km}^{-1}$$

and

$$Z_0 = \sqrt{\frac{23+j3926.99}{(4+j1508)\times 10^{-3}}} = 50.03\Omega\underline{/-0.00}\text{ rad}$$

(e) At 10 MHz

$$R + j\omega L = (23 + j(2\pi \times 10 \times 10^6 \times 125\ \mu\text{H}) = (23 + j7853.98)\ \Omega\ \text{km}^{-1}$$

and

$$G + j\omega C = (4\text{ mS} + j(2\pi \times 10 \times 10^6 \times 48\text{ nF})) = (4 + j3016)\text{ mS km}^{-1}$$

Hence

$$\gamma = \sqrt{(23+j7853.98)(4+j3016)\times 10^{-3}} = 153.91\underline{/1.569}\text{ rad}$$
$$= (0.33\text{ nepers} + j153.9\text{ rad})\text{ km}^{-1}$$

$$Z_0 = \sqrt{\frac{23+j7853.98}{(4+j3016)\times 10^{-3}}} = 50.03\Omega\underline{/-0.00}\text{ rad}$$

Conclusions from Example 2.12. In the frequency range 100–500 Hz, the attenuation constant α tends to remain at about 0.30 nepers per km and the phase propagation β increases linearly with frequency. See cases (a) and (b). If you now compare this set of results with the same cases from Example 2.4, you will see that in this frequency range, Z_0 and α tend to remain constant and β tends to vary linearly with frequency.

What this means is that if you transmit a rectangular pulse or digital signals in this frequency range, you will find that it will pass through the transmission line attenuated but with its shape virtually unchanged. The reason for the waveform shape not changing is because the Fourier amplitude and phase relationships have not been changed.

In the frequency range 5–10 MHz, the attenuation constant α tends to remain at 0.33 nepers per km and the phase propagation β also increases linearly with frequency. See cases (d) and (e). If you now compare the above set of cases from Example 2.12 with an identical set from Example 2.4, you will see that within these frequency ranges, Z_0 and α tend to remain constant and β tends to vary linearly with frequency. Therefore the same argument in the foregoing paragraphs applies to this frequency range. This is also known as the 'distortionless' range of the transmission line.

In the intermediate frequency range of operation (see case (c) of Examples 2.4 and 2.12), both the propagation constant α and β, and the characteristic impedance of the line Z_0 vary. Fourier amplitude and phase relations are not maintained as waveforms are transmitted along the line and waveform distortion is the result.

2.12.3 Summary of propagation properties of transmission lines

There are two frequency regions where signals can be passed through transmission lines with minimum distortion; a low frequency region and a high frequency region. The low frequency region occurs when $R \gg \omega L$, and $G \gg \omega C$. The high frequency region occurs when $\omega L \gg R$ and $\omega C \gg G$. The high frequency region is also sometimes called the 'loss-less' region of transmission.

At both these high and low frequency regions of operation, the simplified expressions for Z_0 (Equations 2.18 and 2.19) and γ (Equations 2.43 and 2.44) show that there is little distortion and that the transmission line can be more easily terminated by a matched resistor.

Good cables which operate up to 50 GHz are available. They are relatively costly because of the necessary physical tolerances required in their manufacture.

2.13 Transmission lines as electrical components

Transmission lines can be made to behave like electrical components, such as resistors, inductors, capacitors, tuned circuits and transformers. These components are usually made by careful choice of transmission line characteristic impedance (Z_0), line length (l) and termination (Z_L). The properties of these components can be calculated by using well known expressions for calculating the input impedance of a transmission line.

2.13.1 Impedance relations in transmission lines

We shall now recall some transmission line properties which you learnt in Sections 2.11.2 and 2.12.2 to show you how the input impedance varies along the line and how transmission lines can be manipulated to produce capacitors, inductors, resistors and tuned circuits. These are Equations 2.28 and 2.29 which are repeated below for convenience:

$$\Gamma_v = \frac{Z_L - Z_0}{Z_L + Z_0} \tag{2.28}$$

and

$$\Gamma_i = -\Gamma_v \tag{2.29}$$

In previous derivations, voltages and currents references have been taken from the input end of the line. Sometimes, it is more convenient to take voltage and current references from the terminating or load end of the line. This is shown in Figure 2.14.

From the definition of line attenuation and for a distance l from the load, we have

$$v_i = V_{iL}\, e^{+\gamma l} \tag{2.45}$$

and

$$v_r = V_{rL}\, e^{-\gamma l} \tag{2.46}$$

and using the definition for voltage reflection coefficient Γ_v

$$\Gamma_{vl} = \frac{V_{rl}}{V_{il}} = \frac{V_{rL}\, e^{-\gamma l}}{V_{rL}\, e^{+\gamma l}} = \frac{V_{rL}}{V_{iL}} = \Gamma_L\, e^{-2\gamma l} \tag{2.47}$$

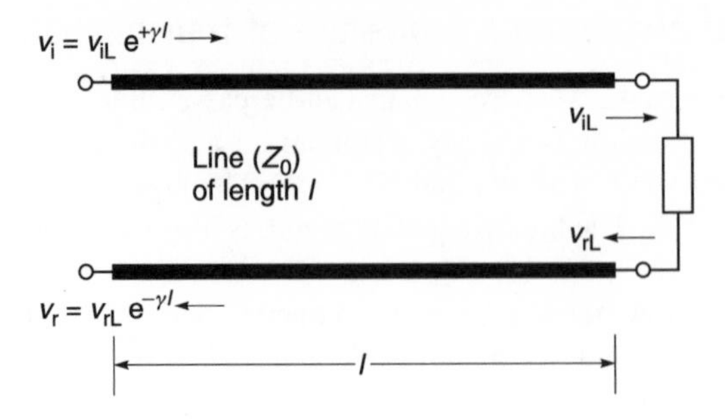

Fig. 2.14 Line voltages reference to the load end

where

l = line length equal to distance l
Γ_v = voltage reflection coefficient at load
Γ_{vl} = voltage reflection coefficient at distance l from load
γ = propagation constant = $(\alpha + j\beta)$ nepers/m

At any point on a transmission line of distance l from the load

$$v_l = v_i + v_r = v_i + v_i \Gamma_v \, e^{-2\gamma l} \tag{2.48}$$

and

$$i_l = i_i + i_r = i_i + i_i \Gamma_i \, e^{-2\gamma l} \tag{2.49}$$

Dividing Equation 2.48 by Equation 2.49 and using Equations 2.28 and 2.29

$$\frac{v_l}{i_l} = \frac{v_i + v_i \Gamma_v \, e^{-2\gamma l}}{i_i - i_i \Gamma_v \, e^{-2\gamma l}}$$

$$= \frac{v_i (1 + \Gamma_v \, e^{-2\gamma l})}{i_i (1 - \Gamma_v \, e^{-2\gamma l})} \tag{2.50}$$

Defining Z_l as impedance at point l, and Z_0 as the line characteristic impedance, Equation 2.50 becomes

$$Z_l = Z_0 \left[\frac{1 + \Gamma_v e^{-2\gamma l}}{1 - \Gamma_v e^{-2\gamma l}} \right] \tag{2.51}$$

Substituting equation 2.28 in equation 2.51 results in

$$Z_l = Z_0 \left[\frac{1 + \dfrac{Z_L - Z_0}{Z_L + Z_0} e^{-2\gamma l}}{1 - \dfrac{Z_L - Z_0}{Z_L + Z_0} e^{-2\gamma l}} \right]$$

Multiplying out and simplifying

$$Z_l = Z_0 \left[\frac{Z_L + Z_0 + (Z_L - Z_0)\, e^{-2\gamma l}}{Z_L + Z_0 - (Z_L - Z_0)\, e^{-2\gamma l}} \right]$$

Sorting out Z_0 and Z_L gives

$$Z_l = Z_0 \left[\frac{Z_L(1 + e^{-2\gamma l}) + Z_0(1 - e^{-2\gamma l})}{Z_L(1 - e^{-2\gamma l}) + Z_0(1 + e^{-2\gamma l})} \right]$$

Multiplying all bracketed terms by ($e^{\gamma l}/2$) results in

$$Z_l = Z_0 \left[\frac{Z_0 \dfrac{e^{\gamma l} - e^{-\gamma l}}{2} + Z_L \dfrac{e^{\gamma l} + e^{-\gamma l}}{2}}{Z_0 \dfrac{e^{\gamma l} + e^{-\gamma l}}{2} + Z_L \dfrac{e^{\gamma l} - e^{-\gamma l}}{2}} \right]$$

Bear in mind that by definition

$$\sinh \gamma l = \frac{e^{\gamma l} - e^{-\gamma l}}{2} \quad \text{and} \quad \cosh \gamma l = \frac{e^{\gamma l} + e^{-\gamma l}}{2}$$

Substituting for $\sinh \gamma l$ and $\cosh \gamma l$ in the above equation results in

$$Z_l = Z_0 \left[\frac{Z_0 \sinh \gamma l + Z_L \cosh \gamma l}{Z_0 \cosh \gamma l + Z_L \sinh \gamma l} \right] \tag{2.52}$$

Equation 2.52 is a very important equation because it enables us to investigate the properties of a transmission line. If the total length of the line is l then Z_l becomes the input impedance of the line. Hence, Equation 2.52 becomes

$$Z_{in} = Z_0 \left[\frac{Z_0 \sinh \gamma l + Z_L \cosh \gamma l}{Z_0 \cosh \gamma l + Z_L \sinh \gamma l} \right] \tag{2.53}$$

2.13.2 Input impedance of low loss transmission lines

From Equation 2.42, we know that $\gamma = \alpha + j\beta$. When $\alpha << \beta$, $\gamma = j\beta$.

From Equation 2.22, we know that $\beta = 2\pi/\lambda$. From mathematical tables[17] we know that $\sin(j\beta) = j \sin \beta$ and $\cos(j\beta) = \cos \beta$. If we now substitute the above facts into Equation 2.53, we will get Equation 2.54. The input impedance of a low loss transmission line is given by the expression

[17] You can also check this for yourself if you take the series for $\sin x$ and $\cos x$ and substitute $j\beta$ in place of x.

$$Z_{in} = Z_0 \left[\frac{jZ_0 \sin \frac{2\pi l}{\lambda} + Z_L \cos \frac{2\pi l}{\lambda}}{jZ_L \sin \frac{2\pi l}{\lambda} + Z_0 \cos \frac{2\pi l}{\lambda}} \right] \qquad (2.54)$$

where

Z_{in} = input impedance (ohms)
Z_0 = characteristic impedance of line (ohms)
Z_L = termination load on line (ohms)
λ = electrical wavelength at the operating frequency
l = transmission line length

2.13.3 Reactances using transmission lines

A transmission line can be made to behave like a reactance by making the terminating load a short circuit ($Z_L = 0$). In this case, Equation 2.54 becomes

$$Z_{in} = Z_0 \left[\frac{jZ_0 \sin \frac{2\pi l}{\lambda} + 0}{0 + Z_0 \cos \frac{2\pi l}{\lambda}} \right] = Z_0 \left[\frac{j \sin \frac{2\pi l}{\lambda}}{\cos \frac{2\pi l}{\lambda}} \right] = jZ_0 \tan \frac{2\pi l}{\lambda} \qquad (2.55)$$

When $l < \lambda/4$

$$Z_{in} = jZ_0 \tan \frac{2\pi l}{\lambda} \qquad (2.55a)$$

and is inductive.

When $\lambda/4 < l > \lambda/2$

$$Z_{in} = -jZ_0 \tan \frac{2\pi l}{\lambda} \qquad (2.55b)$$

and is capacitive.

Equation 2.55 follows a tangent curve and like any tangent curve it will yield positive and negative values. Therefore Equation 2.55 can be used to calculate inductive and capacitive reactances. Adjustment of Z_0 and line length, l, will no doubt set the required values.

Example 2.13

A 377 Ω transmission line is terminated by a short circuit at one end. Its electrical length is $\lambda/7$. Calculate its input impedance at the other end.

Solution. Using Equation 2.55a

$$Z_{in} = jZ_0 \tan \frac{2\pi l}{\lambda} = j377 \tan \left[\frac{2\pi}{\lambda} \frac{\lambda}{7} \right] = j377 \times 1.254 = j472.8\ \Omega$$

Similar reactive effects can also be produced by using an open-circuited load[18] and applying it to Equation 2.54 to produce inductive and capacitive reactances:

$$Z_{in} = -jZ_0 \cot \frac{2\pi l}{\lambda} \tag{2.56}$$

Equation 2.56 follows a cotangent curve and will therefore also produce positive and negative impedances. Adjustment of Z_0 and line length will set the required reactance.

Example 2.14

A 75 Ω line is left unterminated with an open circuit at one end. Its electrical length is $\lambda/5$. Calculate its input impedance at the other end.

Solution. Using Equation 2.56

$$Z_{in} = -jZ_0 \cot \frac{2\pi l}{\lambda} = -j75 \cot \frac{2\pi}{\lambda} \frac{\lambda}{5} = -j75 \times 0.325 = -j24.4\ \Omega$$

2.13.4 Transmission lines as transformers

An interesting case arises when $l = \lambda/2$. In this case Equation 2.54 becomes

$$Z_{in} = Z_0 \left[\frac{jZ_0 \sin \pi + Z_L \cos \pi}{jZ_L \sin \pi + Z_0 \cos \pi} \right] = Z_L$$

What this means is that the transmission line acts as a 1:1 transformer which is very useful for transferring the electrical loading effect of a termination which cannot be placed in a particular physical position. For example, a resistor dissipating a lot of heat adjacent to a transistor can cause the latter to malfunction. With a 1:1 transformer, the resistor can be physically moved away from the transistor without upsetting electrical operating conditions.

Another interesting case arises when $l = \lambda/4$. In this case, Equation 2.54 becomes

$$Z_{in} = Z_0 \left[\frac{jZ_0 \sin \frac{\pi}{2} + Z_L \cos \frac{\pi}{2}}{Z_L j\sin \frac{\pi}{2} + Z_0 \cos \frac{\pi}{2}} \right] = \frac{Z_0^2}{Z_L}$$

[18] Purists might argue that an open circuit does not exist at radio frequencies because any unterminated TX line has stray capacitance associated with an open circuit. We will ignore this stray capacitance temporarily because, for our frequencies of operation, its reactance is extremely high.

Therefore the transmission line behaves like a transformer where

$$Z_{in} = \frac{Z_0^{\ 2}}{Z_L} \tag{2.57}$$

At first glance Equation 2.57 may not seem to be very useful but if you refer back to Figure 2.7, you will see that the characteristic impedance (Z_0) of microstrip transmission lines can be easily changed by changing its width (w); therefore impedance matching is a very practical proposition.

Example 2.15

A transmission line has a characteristic impedance (Z_0) of 90 Ω. Its electrical length is $\lambda/4$ and it is terminated by a load impedance (Z_L) of 20 Ω. Calculate the input impedance (Z_{in}) presented by the line.

Given: $Z_0 = 90\ \Omega$, $Z_L = 20\ \Omega$, $l = \lambda/4$
Required: Z_{in}

Solution. Using Equation 2.57

$$Z_{in} = (90)^2/20 = 405\ \Omega$$

2.14 Transmission line couplers

Transmission lines can be arranged in special configurations to divide an input signal at the input port into two separate signals at the output ports. Such an arrangement is often called a signal splitter. Since the splitter is bi-directional, the same arrangement can also be used to combine two separate signals into one. These splitter/combiners are often called couplers.

The advantages of couplers are that they are very efficient (low loss), provide good matching on all ports, and offer reasonable isolation between the output ports so that one port does not interfere with the other. The greatest disadvantage of these couplers is their large physical size when used in their distributed forms.

2.14.1 The branch-line coupler

The basic configuration of the branch-line coupler is shown in Figure 2.15. It consists of transmission lines, each having a length of $\lambda/4$. Two opposite facing-lines have equal

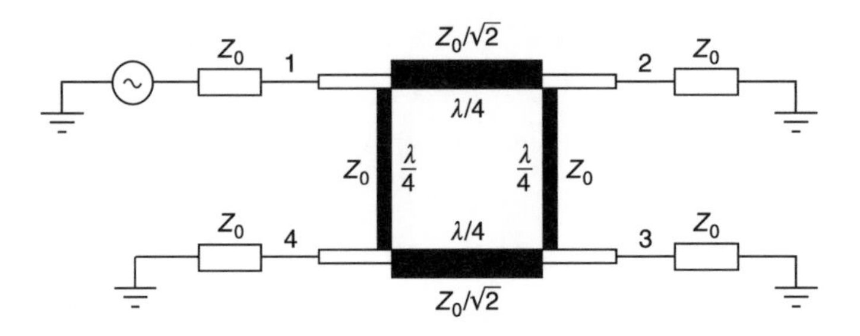

Fig. 2.15 A branch-line coupler

impedances, Z_0, and the remaining opposite facing-lines have an impedance of $Z_0/\sqrt{2}$. In the case of a 50 Ω coupler, $Z_0 = 50\ \Omega$ and $Z_0/\sqrt{2} = 35.355\ \Omega$.

Principle of operation

All ports are terminated in Z_0. Signal is applied to port 1 and it is divided equally between ports 2 and 3. There is no output signal from port 4 because the path from port 1 to 4 is $\lambda/4$ while the path from port 1 to 4 via ports 2 and 3 is $3\lambda/4$. The path difference is $\lambda/2$; hence the signals cancel each other at port 4. The net result is that the signal at port 4 is zero and it can be considered as a virtual earth point. With this virtual earth point Figure 2.15 becomes Figure 2.16(a). We know from transmission line theory (Equation 2.57) that for a $\lambda/4$ length, $Z_{in} = Z_0^2/Z_l$. In other words, a short circuit at port 4 appears as open circuits at port 1 and 3. This result is shown in Figure 2.16(b). If we now transform the impedance at port 3 to port 2, we get

$$Z_{(transformed)} = Z_0^2/Z_0 = Z_0$$

Hence we obtained the transformed Z_0 in parallel with the Z_0 termination of port 2. This situation is shown in Figure 2.16(c). We now need to transform the $Z_0/2$ termination at port 2 to port 1. At port 1

$$Z_l = \frac{(Z_0/\sqrt{2})^2}{(Z_0/2)} = \frac{Z_0^2}{2} \times \frac{2}{Z_0} = Z_0$$

This condition is shown in Figure 2.16(d).

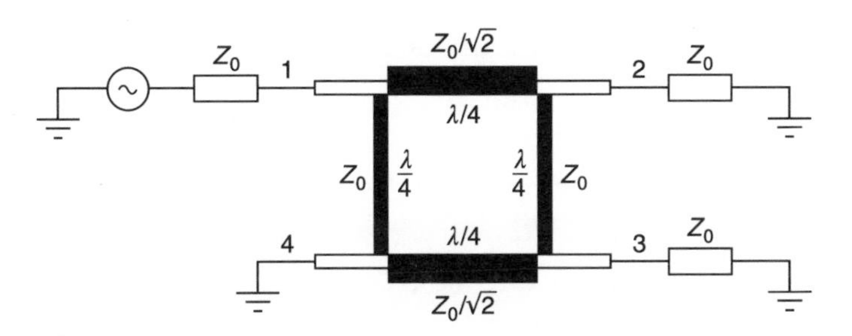

Fig. 2.16(a) Virtual short circuit at port 4

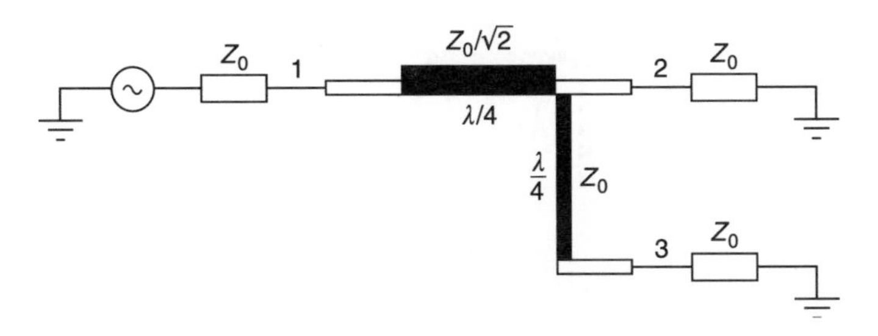

Fig. 2.16(b) Effect of virtual short circuit at port 4

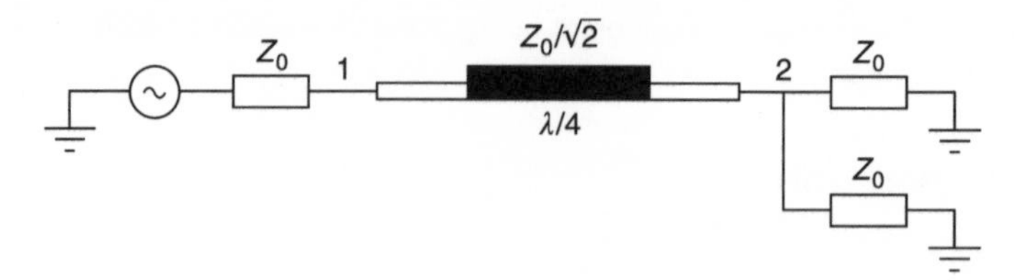

Fig. 2.16(c) Effect of port 3 transferred to port 2

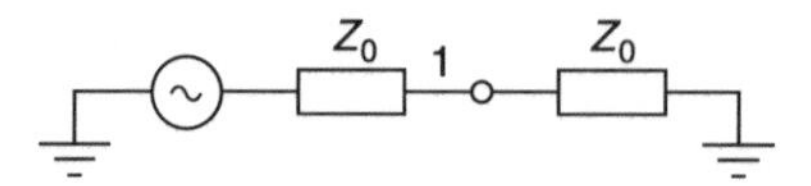

Fig. 2.16(d) Effect of port 2 transferred to port 1

We conclude that the hybrid 3 dB coupler provides a good match to its source impedance Z_0. Assuming lossless lines, it divides its signal equally at ports 2 and 3. Since signal travels over $\lambda/4$ to port 2, there is a phase delay of 90° at port 2 and another 90° phase delay from port 2 to port 3. Thus the signal arriving at port 3 suffers a delay of 180° from the signal at port 1.

The response of such a coupler designed for 5 GHz is shown in Figure 2.17. As you can see for yourself, the signal path from port 1 to port 2 (S_{21}) is 3 dB down at 5 GHz. This is also true of the signal response to port 3 (S_{31}). The signal attenuation to port 4 is theoretically infinite but this value is outside the range of the graph.

A similar analysis will show that power entering at port 2 will be distributed between ports 1 and 4 and not at port 3. A similar analysis for port 3 will show power dividing between ports 1 and 4 but not at port 2. The net result of this analysis shows that signals

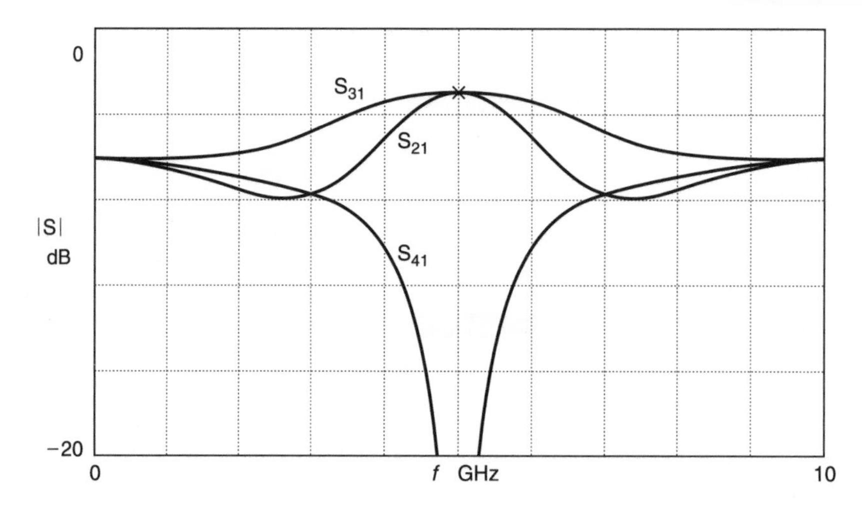

Fig. 2.17 Unadjusted response of the quadrate 3 dB coupler: S_{21} = signal attenuation path from port 1 to port 2; S_{31} = signal attenuation path from port 1 to port 3; S_{41} = signal attenuation path from port 1 to port 4

into port 2 and 3 are isolated from each other. This is a very useful feature in mixer and amplifier designs.

The advantage of the quadrature coupler is easy construction but the disadvantage of this coupler is its narrow operational bandwidth because perfect match is only obtained at the design frequency where each line is exactly $\lambda/4$ long. At other frequencies, each line length is no longer $\lambda/4$ and signal attenuation increases while signal isolation decreases between the relevant ports.

Finally, you may well ask 'if port 4 is at virtual earth, why is it necessary to have a matched resistor at port 4?' The reason is that signal balance is not perfect and the resistor helps to absorb unbalanced signals and minimises reflections.

2.14.2 The ring coupler

Ring forms of couplers have been known for many years in waveguide, coaxial and stripline configurations. The basic design requirements are similar to that of the quadrature coupler except that curved lines are used instead of straight lines. One such coupler is shown in Figure 2.18. The principle of operation of ring couplers is similar to that of branch-line or quadrature couplers.

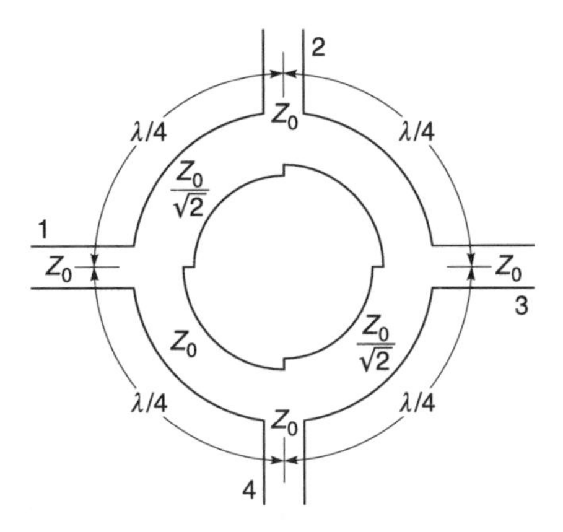

Fig. 2.18 The 3 dB ring-form branch-line directional coupler

2.14.3 The 'rat-race' coupler

A sketch of a 'rat-race' coupler is shown in Figure 2.19. The mean circumference of the ring is 1.5λ. This coupler is easy to construct and provides good performance for narrow band frequencies. The characteristic impedance of the coupler ring is $Z_0\sqrt{2}\ \Omega$ which in the case of $Z_0 = 50\ \Omega$ is a circular $70.7\ \Omega$ transmission line which is 1.5λ in circumference. The four Z_0 ports are connected to the ring in such a manner that ports 2 to 3, ports 3 to 1 and ports 1 to 4 are each separated by $\lambda/4$. Port 4 to port 2 is separated by 0.75λ. The operation of this device is illustrated in Figure 2.20.

If a signal is injected at port 1, the voltage appearing at port 2 is zero, since the path lengths differ by 0.5λ; thus port 2 can be treated as a virtual ground. Hence the transmission-line portions of the ring between ports 2 and 3, and ports 2 and 4, act as short-circuited stubs connected across the loads presented at ports 3 and 4. For centre frequency

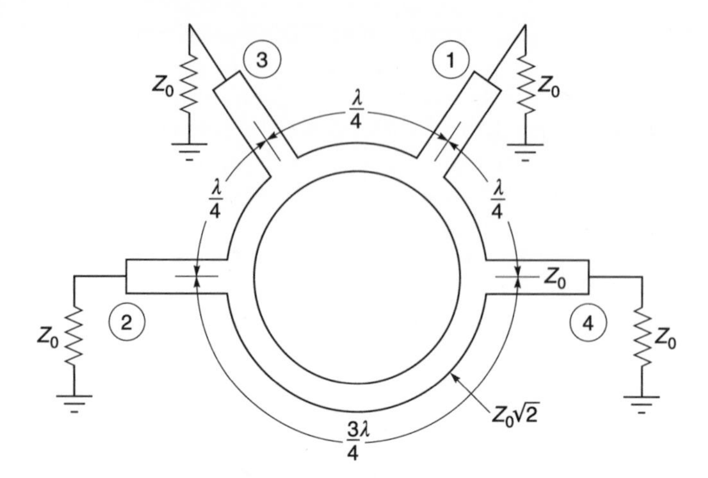

Fig. 2.19 The 'rat-race' or 'hybrid ring' coupler

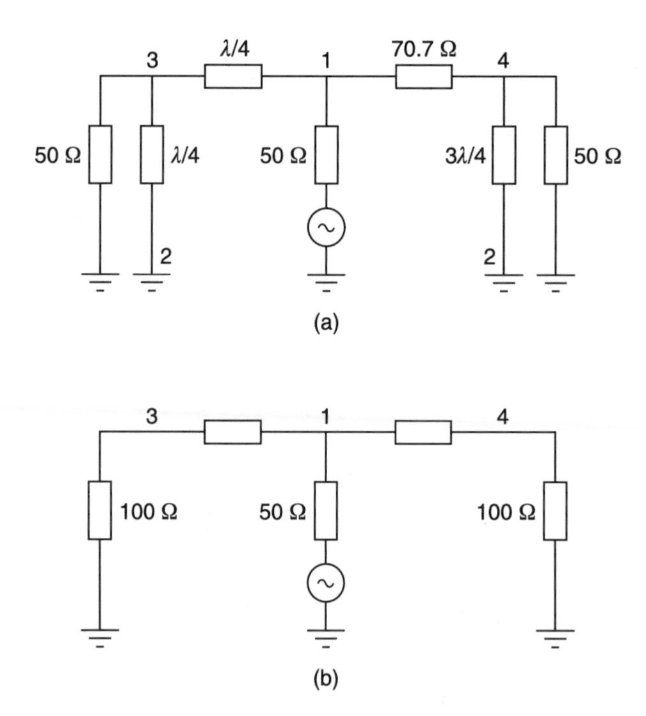

Fig. 2.20 (a) Equivalent circuit of ring hybrid with port 1 as input and ports 2 and 4 as outputs (transmission-line model with port 3 as virtual ground); (b) equivalent circuit at centre frequency

operation, these stubs appear as open circuits. Similarly, the transmission line lengths between ports 3 and 1, and ports 4 and 1, transform the 50 Ω load impedances at ports 3 and 4 to 100 Ω (2 Z_0) at port 1. When combined at port 1, these transformed impedances produce the 50 Ω impedance at port 1. See Figure 2.20(a) and (b).

A similar analysis can be applied at each port, showing that the hybrid exhibits a matched impedance of 50 Ω or Z_0 at all nodes. It should be noted that when port 1 is driven, the outputs at ports 3 and 4 are equal and in phase, while ideally there is no signal at port 2. However, when port 2 is driven, the output signals appearing at ports 3 and 4 are equal but exactly out of phase. Also there is no signal at port 1. Hence ports 1 and 2 are isolated. This is very useful especially in signal mixing circuits because it enables two slightly different frequencies, for example f_1 at port 1 and f_2 at port 2, to be applied to a balanced mixer whose diodes may be connected to ports 3 and 4 without coupling between the sources at port 1 and port 3. It also helps to combine the inputs or outputs of two amplifiers without mutual interference. The unfortunate thing about the ring is that it is a relatively narrow-band device.

2.15 Summary

Chapter 2 has provided you with a thorough basic knowledge of transmission lines and their properties which you will find very useful in circuit and system design of radio and microwave systems. You have been introduced to many properties of transmission lines in this chapter.

In Section 2.3, you were introduced to some of the more frequently used types of transmission lines. These included waveguides, coplanar waveguides, coaxial lines, microstrip and strip lines, slot lines, twin lines and finally coupled microstrip lines.

In Section 2.4, you were shown how the characteristic impedance of the coaxial, twin line and microstrip line can be calculated from its physical parameters. The information demonstrated what properties you should look for if the characteristic impedance of a line does not behave as expected.

Sections 2.5 and 2.6 demonstrated how the characteristic impedance can be calculated and measured from primary constants of the line. Section 2.7 mentions some of the more common impedances associated with commercial transmission lines but it also brought to your attention that there are many types of lines with the same impedance.

Section 2.8 explained how propagation delay, attenuation and frequency dispersion affect waveforms as they travel along transmission lines. This was followed by more discussions on the effects of these properties in Section 2.9.

In Section 2.10, we introduced the concepts of matched and unmatched lines. This was followed by a thorough discussion of reflection coefficients and voltage standing wave ratios in Section 2.11.

Section 2.12 dealt with the propagation properties of lines and showed how these can be derived from the primary constants of the line. The section also showed how optimum transmission can be achieved.

Section 2.13 showed how transmission lines can be used as transformers, impedance matching devices, inductive and capacitive reactances which in turn can be used to produce filters. Microstrip lines are particularly useful for making filters at the higher frequencies because of their versatility in allowing characteristic impedance changes to be made easily. There is also a greater tendency to use transmission lines as the tuning elements.

Section 2.14 showed you how transmission lines can be connected to act as signal couplers.

Finally, you will come across these examples again when we introduce the software program PUFF in Chapter 4 and carry out some examples to further clarify your theory and the use of the program.

3

Smith charts and scattering parameters

3.1 Introduction

3.1.1 Aims

The aims of this part are to familiarise you with two fundamental necessities in radio engineering; **Smith charts** and **scattering parameters**. We start with Smith charts because they are very useful for amplifier design, gain circles, noise circles, matching network design, impedance and admittance determination and finding reflection coefficients and voltage standing wave ratios.

Scattering parameters are important because many basic items such as filters, hybrid transformers, matching networks, transistors, amplifiers, gain blocks and monolithic microwave integrated circuits (MMIC) are used and described by manufacturers in terms of two port networks. You also need *s*-parameters for circuit and system design.

In this part, Sections 3.2 to 3.4 have been devoted to a description of the Smith chart. Section 3.5 covers its theory and Sections 3.6 to 3.9 provide examples on Smith chart applications. Section 3.10 deals with the fundamentals of scattering parameters and Section 3.11 gives examples of its applications.

3.1.2 Objectives

After reading the section on Smith charts, you should be able to:

- evaluate impedance and admittance networks;
- design impedance and admittance networks;
- design matching networks.

After reading the section on scattering parameters, you should be able to:

- understand scattering parameters;
- use two port scattering parameters efficiently;
- design two port scattering networks;
- evaluate two port scattering networks;
- calculate the frequency response of networks;
- calculate the gain of two port networks or an amplifier;
- calculate the input impedance of two port networks or an amplifier;
- manipulate two port data into other types of parameter data.

3.2 Smith charts

3.2.1 Introduction

The Smith chart is intended to provide a graphical method for displaying information, for impedance matching using lumped and distributed elements and is particularly useful in solving transmission line problems. It avoids the tedious calculations involved in applying the expressions obtained and proved in Part 2. It is an alternative check to your calculations and provides a picture of circuit behaviour to help you visualise what happens in a

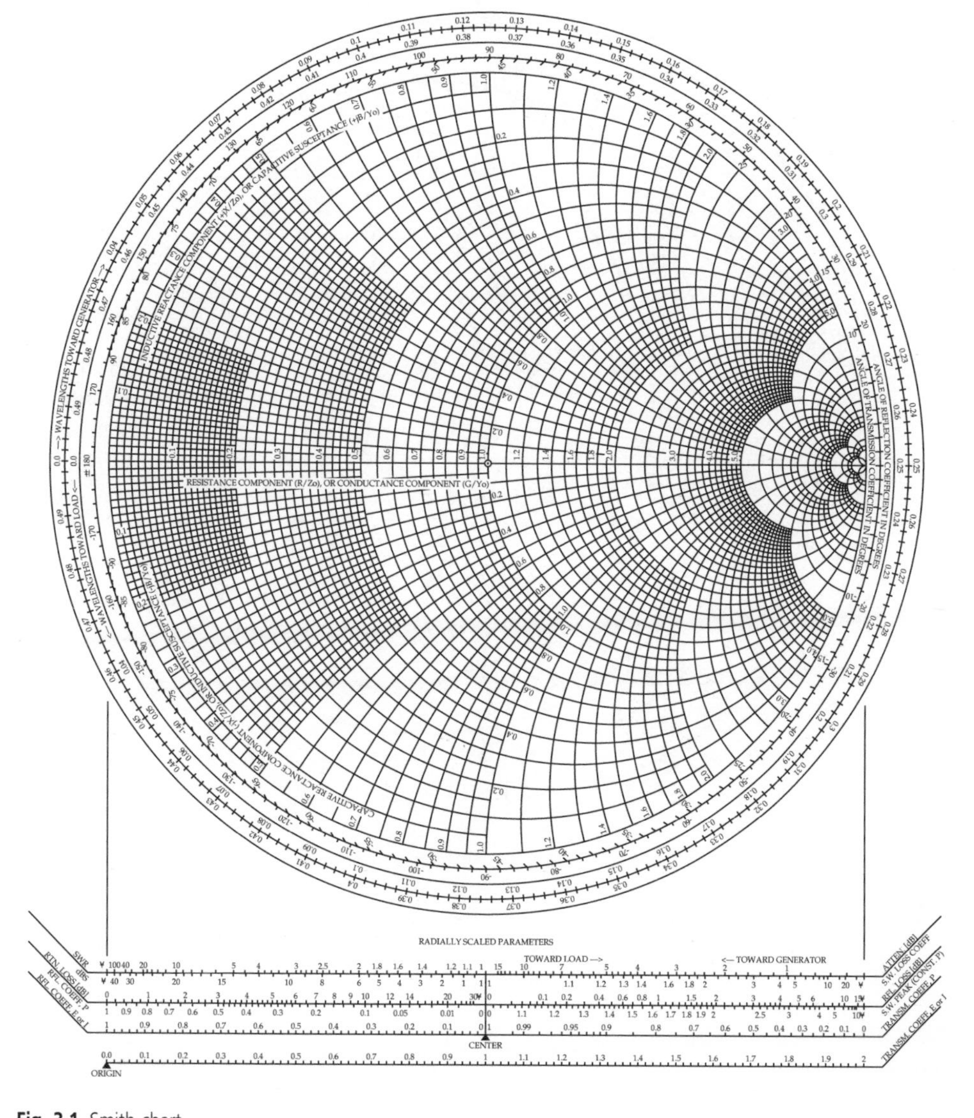

Fig. 3.1 Smith chart

circuit. The Smith chart[1] was devised by P.H. Smith in the late 1930s and an improved version was published in 1944. It is shown in Figure 3.1.

The Smith chart is based on two sets of circles which cut each other at right-angles. One set (Figure 3.2) represents the ratio R/Z_0, where R is the resistive component of the line impedance $Z_x = R + jX$. Z_0 is usually taken as the characteristic impedance of a transmission line. Sometimes Z_0 is just chosen to be a number that will provide a convenient display on the Smith chart. The other set of circles (Figure 3.3) represents the ratio jX/Z_0,

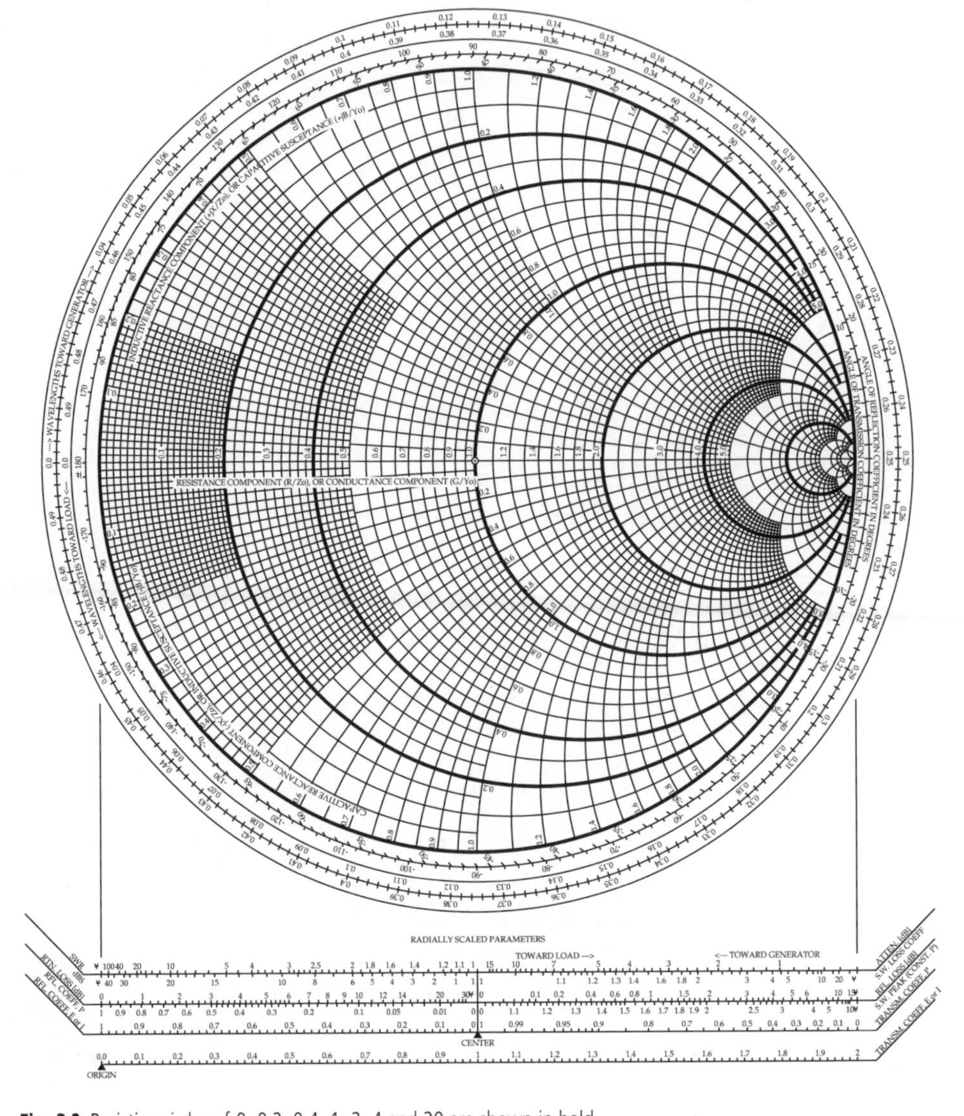

Fig. 3.2 Resistive circles of 0, 0.2, 0.4, 1, 2, 4 and 20 are shown in bold

[1] The Smith chart is a copyright of Analog Instruments Co., P.O. Box 808, New Providence, NJ07974, USA.

where X is the reactive component of the line impedance $Z_x = R + jX$. Z_0 is usually taken as the characteristic impedance of a transmission line. Sometimes Z_0 is just chosen to be a number that will provide a convenient display on the Smith chart. In both cases of normalisation, the same number must be used for both resistance and reactance normalisation. The circles have been designed so that conditions on a lossless line with a given VSWR can be found by drawing a circle with its centre at the centre of the chart.

Figures 3.1, 3.2 and 3.3 are full detailed versions of the Smith chart. When electronic versions of the chart are used, such detail tends to clutter a small computer screen. In such cases, it is best to show an outline Smith chart and to provide the actual coordinates in complex numbers. This is the system which has been used by the software program PUFF,

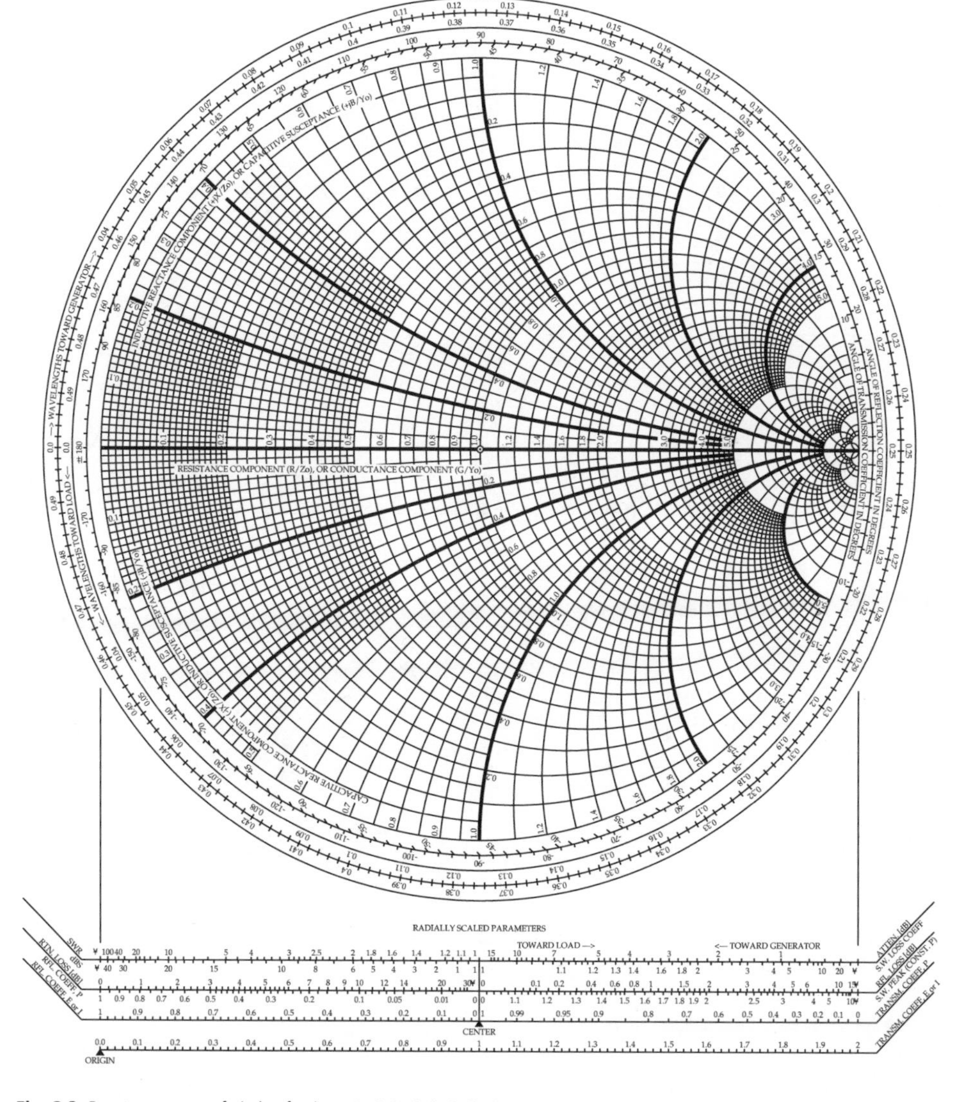

Fig. 3.3 Reactance arcs of circles for j = ±0, 0.2, 0.4, 1, 2, 4

provided with this book. There is also another Smith chart program called MIMP[2] (Motorola Impedance Matching Program).

3.2.2 Plotting impedance values

Any point on the Smith chart represents a **series** combination of resistance and reactance of the form $Z = R + jX$. Thus, to locate the impedance $Z = 1 + j1$, you would find the $R = 1$ constant resistance circle and follow it until it crosses the $X = 1$ constant reactance circle. The junction of these two circles would then represent the needed impedance value. This particular point, A shown in Figure 3.4, is located in the upper half of the chart because X is a positive reactance or inductive reactance. On the other hand, the point B which is $1 - j1$ is

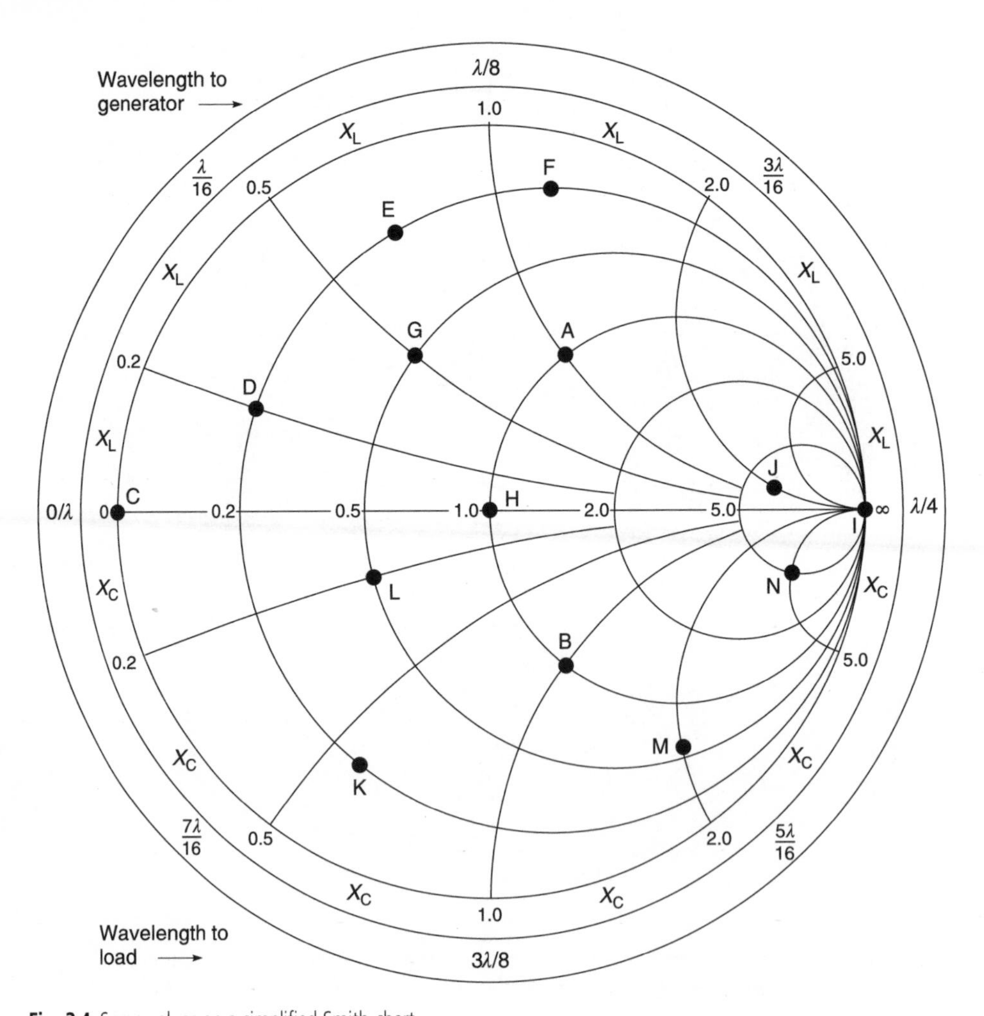

Fig. 3.4 Some values on a simplified Smith chart

[2] At the time of writing, this program can be obtained free from some authorised Motorola agents. This program provides active variable component matching facilities.

located in the *lower* half of the chart because, in this instance, X is a negative quantity and represents capacitive reactance. Thus, the junction of the $R = 1$ constant resistance circle and the $X = -1$ constant reactance circle defines that point. In general, then, to find any *series* impedance of the form $R + jX$ on a Smith chart, you simply find the junction of the R = constant and X = constant circles.

In order to give you a clearer picture of impedance values on the Smith chart, we plot additional impedance values in Figure 3.4. These values are shown in Table 3.1.

Table 3.1 Impedance values for points plotted in Figure 3.4

A = 1 + j1	B = 1 – j1	C = 0 + j0
D = 0.2 + j0.2	E = 0.2 + j0.7	F = 0.2 + j1.2
G = 0.5 + j0.5	H = 1 + j0	I = ∞ + j0
J = 6 + j2	K = 0.2 – j0.6	L = 0.5 – j0.2
M = 0.6 – j2	N = 5 – j5	

Try to plot these values on Figure 3.4 and check if you get the correct values.

In some cases, you will not find the circles that you want; when this happens, you will have to interpolate between the two nearest values that are shown. Hence, plotting impedances on the Smith chart produces a plotting error. However, the error introduced is relatively small and is negligible for practical work.

If you try to plot an impedance of $Z = 20 + j20\ \Omega$, you will not be able to do it accurately because the $R = 20$ and $X = 20\ \Omega$ circles would be (if they were drawn) on the extreme right edge of the chart – very close to infinity. In order to facilitate the plotting of larger impedances, **normalisation** is used. That is, each impedance to be plotted is divided by a convenient number that will place the new normalised impedance near the centre of the chart where increased accuracy in plotting is obtained. For the preceding example, where $Z = 20 + j20\ \Omega$, it would be convenient to divide Z by 100, which yields the value $Z = 0.2 + j0.2$. This is very easily found on the chart.

The important thing to realise is that if normalisation is carried out for one impedance then all impedances plotted on that chart must *be* divided by the same number in the normalisation process. Otherwise, you will not be able to use the chart. Last but not least, when you have finished with your chart manipulations, you must then **re-normalise** (multiply by the same number used previously) to get your true values.

Example 3.1

Plot the points (0.7 – j0.2), (0.7 + j0.3), (0.3 – j0.5) and (0.3 + j0.3) on the Smith chart in Figure 3.4.

Solution. The above points are all shown on the Smith chart in Figure 3.5. Check your plotting points in Figure 3.4 against those in Figure 3.5.

3.2.3 *Q* of points on a Smith chart

Since the quality factor Q is defined as reactance/resistance, it follows that every point on the Smith chart has a value of Q associated with it. For example, the plotted points of Example 3.1 are shown in Table 3.2.

Table 3.2 Q values of the points in Example 3.1

Resistance (R)	Reactance (X)	$Q = \lvert X \rvert / R$
0.7	−0.2	0.286
0.7	+0.3	0.429
0.3	−0.5	1.667
0.3	+0.3	1.000

At the moment, it is not clear as to what can be done with these Q values but you will understand their validity later when we investigate broadband matching techniques in Section 3.8.2.

3.2.4 Impedance manipulation on the chart

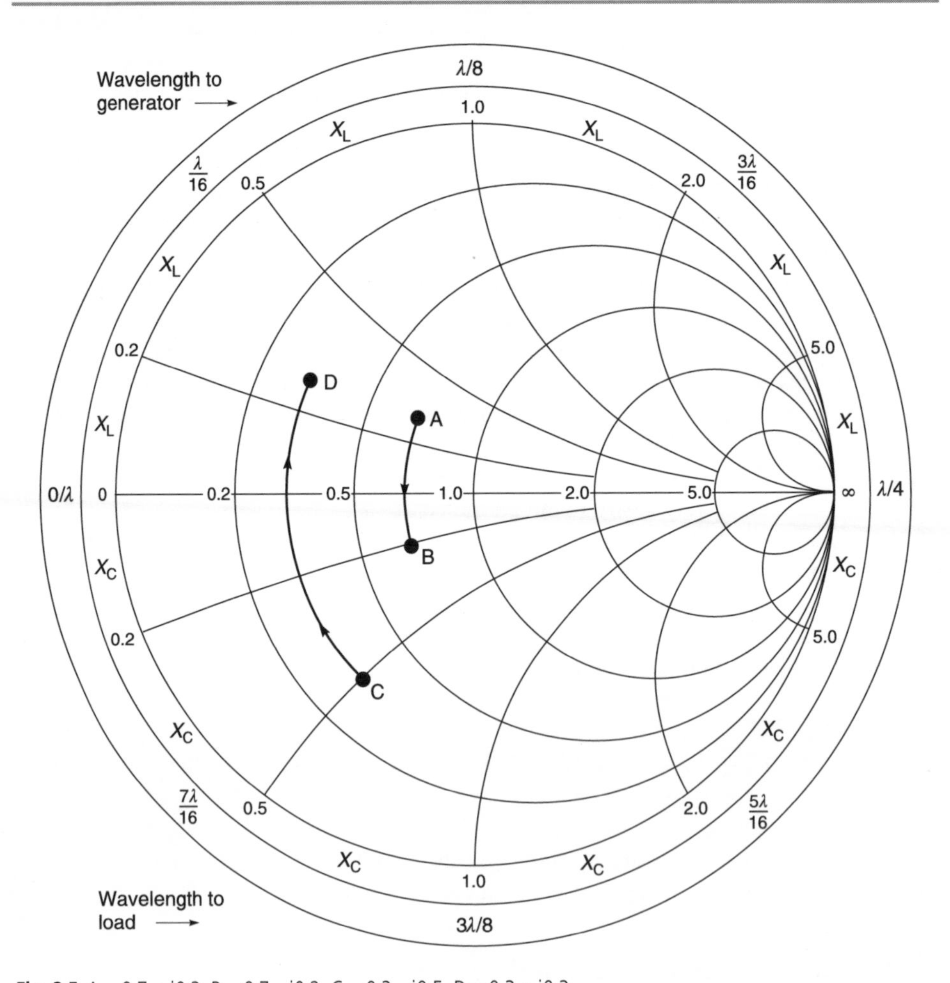

Fig. 3.5 A = 0.7 + j0.3, B = 0.7 – j0.2, C = 0.3 – j0.5, D = 0.3 + j0.3

Figure 3.5 indicates graphically what happens when a series capacitive reactance of –j0.5 Ω is added to an impedance of $Z = (0.7 + j0.3)\ \Omega$. Mathematically, the result is

$$Z = (0.7 + j0.3 - j0.5)\ \Omega = (0.7 - j0.2)\ \Omega$$

which represents a series *RC* quantity. Graphically, we have plotted (0.7 + j0.3) as point A in Figure 3.5. You then read the reactance scale on the periphery of the chart and move *anti-clockwise* along the $R = 0.7\ \Omega$ constant resistance circle for a distance of $X = -j0.5\ \Omega$. This is the plotted impedance point of $Z = (0.7 - j0.2)\ \Omega$, shown as point B in Figure 3.5.

Adding a series inductance to a plotted impedance value simply causes a move *clockwise* along a constant resistance circle to the new impedance value. Consider the case in Figure 3.5 where a series inductance j0.8 Ω is added to an impedance of (0.3 – j0.5) Ω. Mathematically the result is

$$Z = (0.3 - j0.5 + j0.8)\ \Omega = (0.3 + j0.3)\ \Omega$$

Graphically, we have plotted point (0.3 – j0.5) in Figure 3.5 as point C then moved along the 0.3 resistance circle and added j0.8 to that point to arrive at point D.

In general the addition of a series capacitor to a plotted impedance moves that impedance *counter-clockwise* along a constant resistance circle for a distance that is equal to the reactance of the capacitor. The addition of a series inductor to a plotted impedance moves that impedance *clockwise* along a constant resistance circle for a distance that is equal to the reactance of the inductor.

3.2.5 Conversion of impedance to admittance

The Smith chart, described so far as a family of impedance coordinates, can easily be used to convert any impedance (Z) to an admittance (Y), and vice-versa. In mathematical terms, an admittance is simply the inverse of an impedance, or

$$Y = \frac{1}{Z} \tag{3.1}$$

where, the admittance (Y) contains both a real and an imaginary part, similar to the impedance (Z). Thus

$$Y = G \pm jB \tag{3.2}$$

where

G = conductance in Siemens (S)
B = susceptance in Siemens (S)

To find the inverse of a series impedance of the form $Z = R + jX$ mathematically, you would simply use Equation 3.1 and perform the necessary calculation. But, how can you use the Smith chart to perform the calculation for you without the need for a calculator? The easiest way of describing the use of the chart in performing this function is to first work a problem out mathematically, and then plot the results on the chart to see how the two functions are related. Take, for example, the series impedance $Z = (1 + j1)\ \Omega$. The inverse of Z is

$$Y = \frac{1}{1 + j1\ \Omega} = \frac{1}{1.414\ \Omega\ \angle 45^\circ} = 0.7071\ \text{S}\ \angle -45^\circ = (0.5 - j0.5)\ \text{S}$$

If we plot the points (1 + jl) and (0.5 – j0.5) on the Smith chart, we can easily see the graphical relationship between the two. This construction is shown in Figure 3.6. Note that

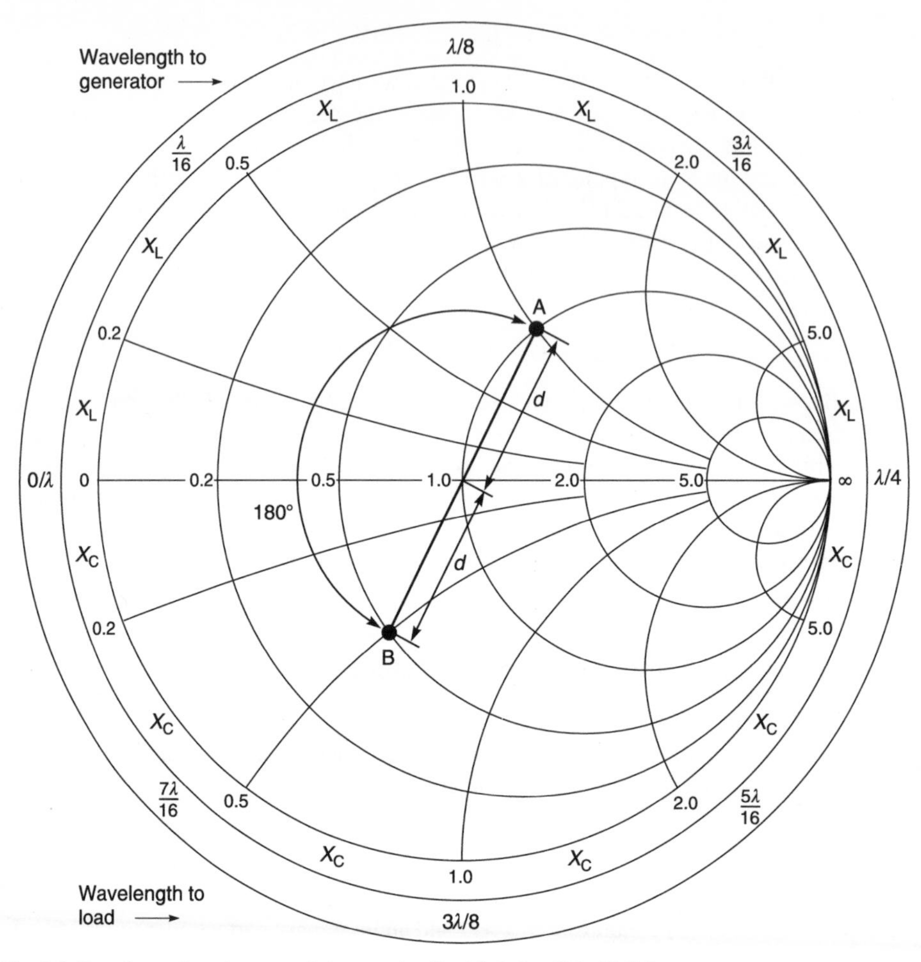

Fig. 3.6 Changing an impedance to admittance: A = (1 + j1) Ω, B = (0.5 – j0.5) S

the two points are located at exactly the same distance (d) from the centre of the chart but in opposite directions (180°) from each other. Indeed, the same relationship holds true for *any* impedance and its inverse. Therefore, without the aid of a calculator, you can find the reciprocal of an impedance or an admittance by simply plotting the point on the chart, measuring the distance (d) from the centre of the chart to that point, and then plotting the measured result the same distance from the centre but in the opposite direction (180°) from the original point. This is a very simple construction technique that can be done in seconds.

Example 3.2

Use the Smith chart in Figure 3.6 to find the admittance of the impedance (0.8 – j1.6).

Given: $Z = (0.8 - j1.6)$
Required: Admittance value Y

Solution. The admittance value is located at the point (0.25 + j0.5). You can verify this yourself by entering the point (0.8 – j1.6) in Figure 3.6. Measure the distance from your point to the chart centre. Call this distance *d*. Draw a line of length 2*d*, from your point through the centre of the chart. Read off the coordinates at the end of this line. You should now get (0.25 + j0.5).

3.3 The immittance Smith chart

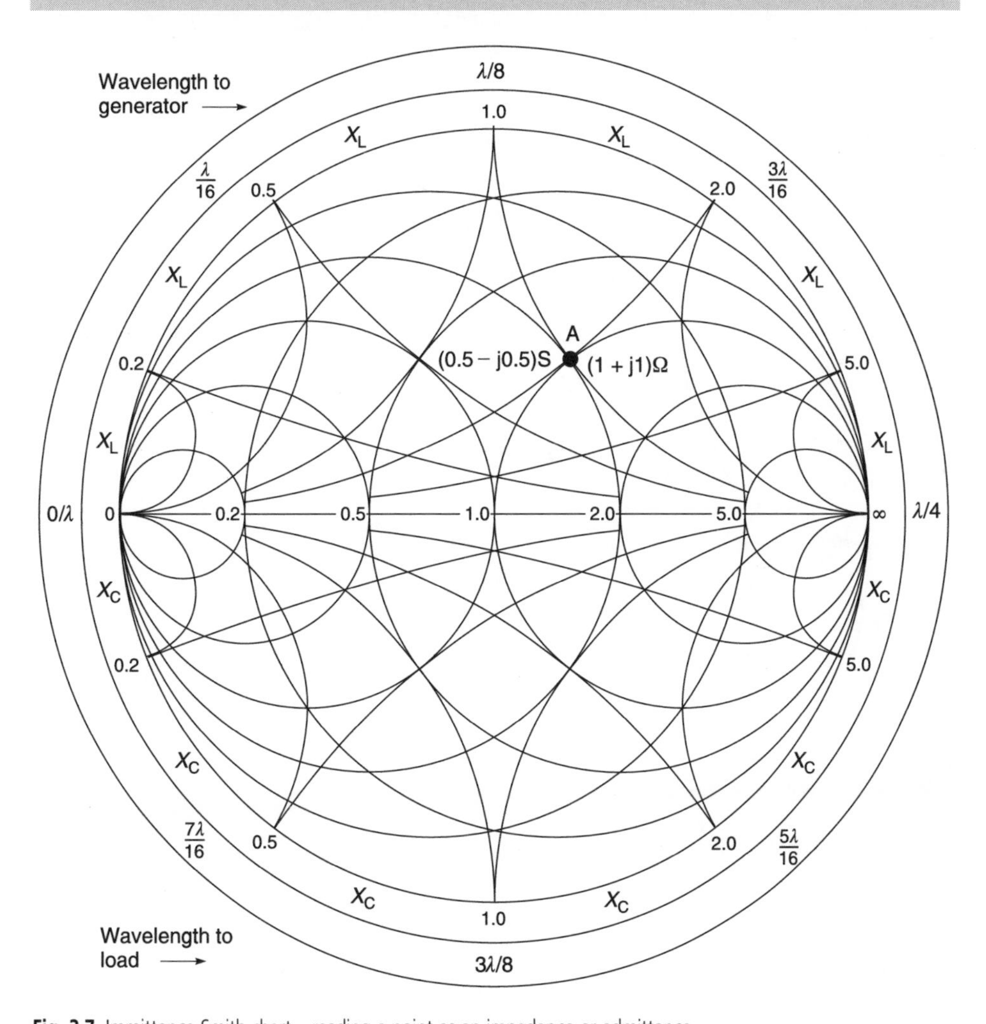

Fig. 3.7 Immittance Smith chart – reading a point as an impedance or admittance

Alternatively, we can use another Smith chart, rotate it by 180°, and overlay it on top of a conventional Smith chart. Such an arrangement is shown in Figure 3.7. The chart that you see in Figure 3.7 is one which I have prepared for you. Detailed charts are also obtainable commercially as Smith chart[3] Form ZY-01-N. With these charts, we can plot

[3] Smith chart Form ZY-01-N is a copyright of Analog Instruments Co, P.O. Box 808, New Providence, NJ 07974, USA.

the coordinates (1 + j1) directly on the impedance chart and read its admittance equivalent (0.5 – j0.5) on the rotated admittance chart directly. Another approach that we could take (if we are working solely with admittances) is to just rotate the chart itself 180° and manipulate values on the chart as admittances. This will be shown more clearly in the next section.

3.4 Admittance manipulation on the chart

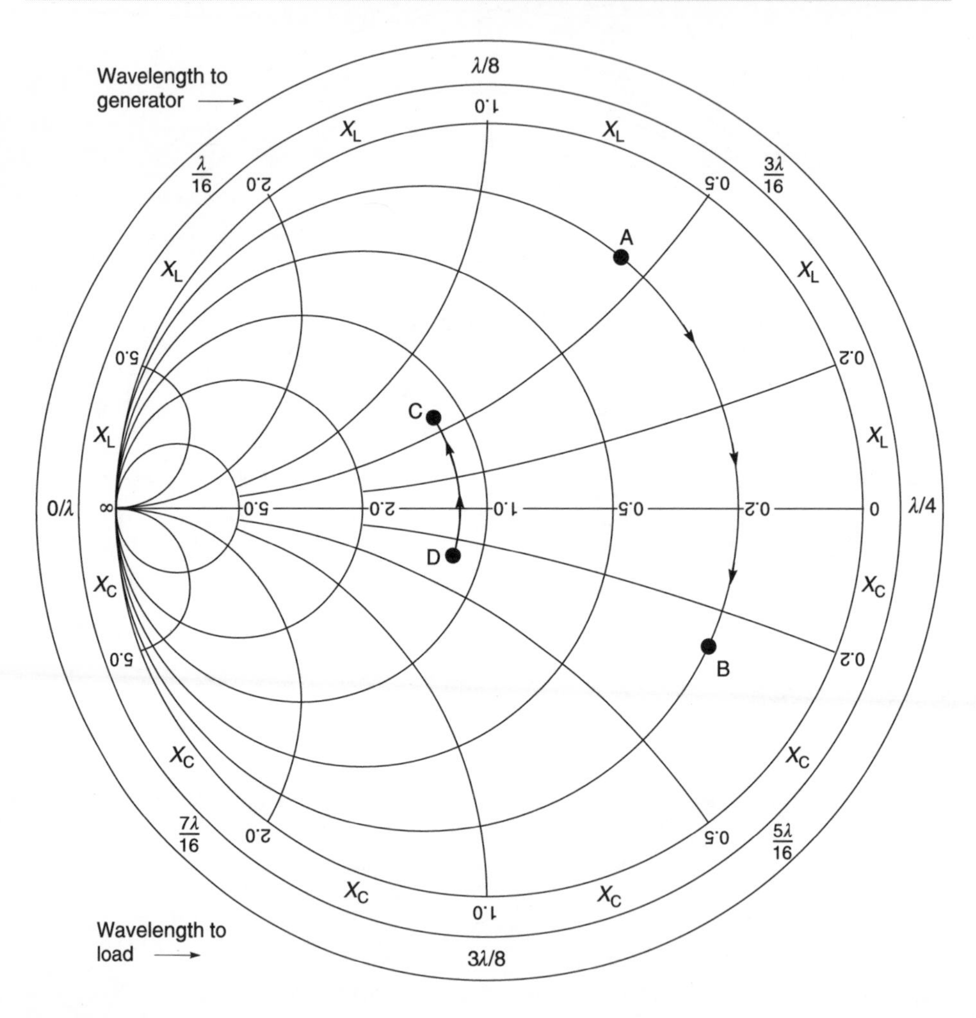

Fig. 3.8 Impedance chart used as an admittance chart: A = (0.2 – j0.6) S, B = (0.2 + j0.3) S, C = (1.2 + j0.4) S, D = (1.2 – j0.6) S

In this section, we want to present a visual indication of what happens when a **shunt** element is added to an **admittance**. The addition of a shunt capacitor is shown in Figure 3.8. For an example we will choose an admittance of Y = (0.2 – j0.6) S and add a shunt capacitor with a susceptance (reciprocal of reactance) of +j0.9 S. Mathematically, we know that parallel susceptances are simply added together to find the equivalent susceptance. When this is done, the result becomes:

$$Y = (0.2 - j0.6)\ S + j0.9\ S = (0.2 + j0.3)\ S$$

If this point is plotted on the admittance chart, we quickly recognise that all we have done is to move along a constant conductance circle (G) *clockwise* a distance of jB = 0.9 S. In other words, the real part of the admittance has not changed, only the imaginary part.

Similarly, if we had a point (1.2 + j0.4) S and added an inductive susceptance of (–j10) S to it, we would get (1.2 + j0.4) S – j10 S = (1.2 – j0.6) S. This is also shown in Figure 3.8. Hence, adding a shunt inductance to an admittance moves the point along a constant conductance circle *counter-clockwise* a distance (–jB) equal to the value of its susceptance, –j10 S, as shown in Figure 3.8.

3.5 Smith chart theory and applications

This section deals with the derivation of the resistance (R) and reactance (X) circles of the Smith chart (see Figure 3.9). If you are prepared to accept the Smith chart and are not interested in the theory of the chart, then you may ignore all of Section 3.5.1 because it will not stop you from using the chart.

3.5.1 Derivations of circles

Our definitions are:

(i) the normalised load impedance is

$$z = \frac{Z_L}{Z_0} = \frac{R + jX}{Z_0} = r + jx \tag{3.3}$$

(ii) the reflection coefficient is

$$\Gamma = p + jq \tag{3.4}$$

Here p (for in-phase) is the real part of Γ and q (for quadrature) is the imaginary part. From Part 2, we use the definition for $\Gamma = (Z_L - Z_0)/(Z_L + Z_0)$ and changing into normalised values by using Equation 3.3, we obtain

$$\Gamma = \frac{z - 1}{z + 1}$$

and by transposing

$$z = \frac{1 + \Gamma}{1 - \Gamma}$$

Substituting z from Equation 3.3 and Γ from Equation 3.4 gives

$$r + jx = \frac{1 + p + jq}{1 - p - jq}$$

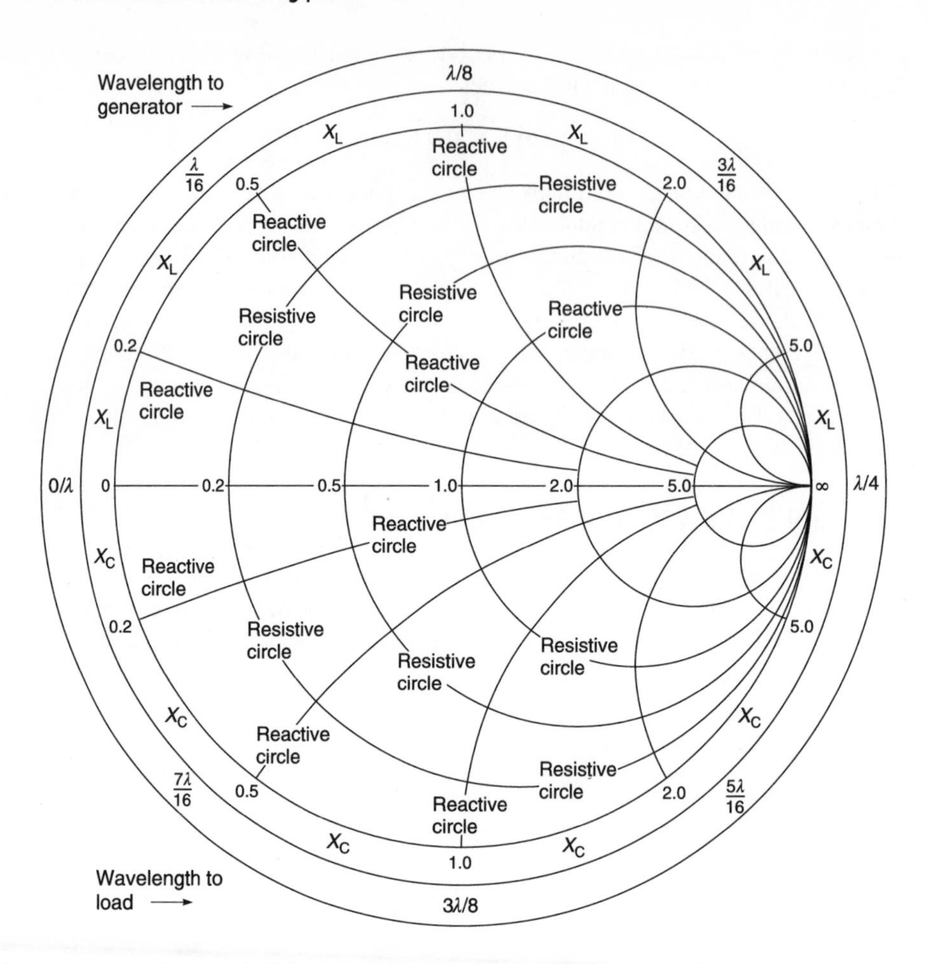

Fig. 3.9 Resistive and reactive circles in the Smith chart

To rationalise the denominator, multiply through by $[(1 - p) + jq]$. This gives

$$r + jx = \frac{[(1 + p) + jq][(1 - p) + jq]}{[(1 - p) - jq][(1 - p) + jq]}$$

$$= \frac{(1 - p^2 - q^2) + j^2 q}{(1 - p)^2 + q^2}$$

Equating real parts

$$r = \frac{(1 - p^2 - q^2)}{(1 - p)^2 + q^2} \tag{3.5}$$

Equating imaginary parts

$$x = \frac{2q}{(1 - p)^2 + q^2} \tag{3.6}$$

First, we derive the equation of an r circle. From Equation 3.5

$$r(1 - 2p + p^2 + q^2) = (1 - p^2 - q^2)$$

and

$$(r + 1)p^2 - 2pr + (r + 1)q^2 = 1 - r$$

Dividing throughout by $(r + 1)$

$$\left(p^2 - \frac{2pr}{r+1}\right) + q^2 = \frac{1-r}{r+1} \times \frac{(1+r)}{(1+r)} = \frac{1-r^2}{(r+1)^2}$$

To complete the square in p, add $r^2/(r + 1)^2$ to both sides. This gives

$$\left(p^2 - \frac{2pr}{r+1} + \frac{r^2}{(r+1)^2}\right) + q^2 = \frac{1-r^2}{(r+1)^2} + \frac{r^2}{(r+1)^2}$$

and

$$\left(p - \frac{r}{r+1}\right)^2 + q^2 = \left(\frac{1}{r+1}\right)^2 \tag{3.7}$$

This is the equation of a circle, with centre at $[p = r/(r+1), q = 0]$ and radius $1/(r + 1)$. For any given value of r, the resultant is called a constant-r circle, or just an r circle.

Next, we derive an equation of an x circle. From Equation 3.6

$$x(1 - 2p + p^2 + q^2) = 2q$$

Dividing throughout by x

$$(p^2 - 2p + 1) + \left(q^2 - \frac{2q}{x}\right) = 0$$

To complete the square, we add $1/x^2$ to both sides:

$$(p^2 - 2p + 1) + \left(q^2 - \frac{2q}{x} + \frac{1}{x^2}\right) = \frac{1}{x^2}$$

and

$$(p-1)^2 + \left(q - \frac{1}{x}\right)^2 = \left(\frac{1}{x}\right)^2 \tag{3.8}$$

This is the equation of a circle, with centre at $[p = 1, q = 1/x]$ and radius, $1/x$. For any given value of x, the resultant circle is called a constant-x circle, or just an x circle.

3.5.2 Smith chart applications

As the proof and theory of the Smith chart has already been explained in Section 3.5.1, we will now concentrate on using the chart. We will commence by using the chart to find reflection coefficients and impedances of networks, then progress on to using the chart for matching with $\lambda/4$ transformers and tuning stubs.

3.6 Reflection coefficients and impedance networks

3.6.1 Reflection coefficients

The Smith chart can be used to find the reflection coefficient at any point, in modulus-and-angle form. If we plot the point (0.8 – j1.6) denoted by point A on the Smith chart of Figure 3.10 and extend the line OA to B, we will find by measurement that the angle BOC is about –55.5°. You will not see this angle scale in Figure 3.10 because our condensed Smith chart does not have an angle scale.[4]

The modulus of the reflection coefficient can be found from Equation 2.38 which states that

$$\text{VSWR} = \frac{1+|\Gamma_v|}{1-|\Gamma_v|}$$

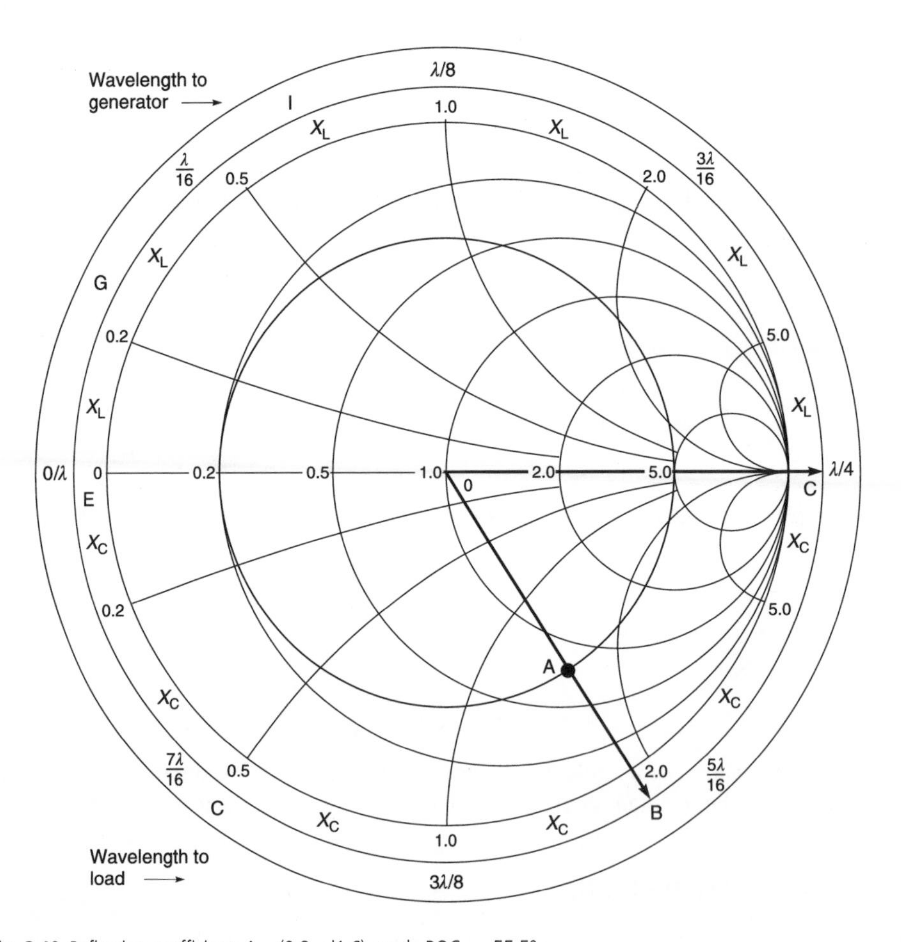

Fig. 3.10 Reflection coefficient: A = (0.8 – j1.6), angle BOC = –55.5°

[4] The full Smith chart also carries an angle scale. See Figure 3.1.

So, rearranging

$$|\Gamma_v| = \frac{\text{VSWR} - 1}{\text{VSWR} + 1}$$

In our example, VSWR[5] ≈ 5, so $|\Gamma_v| \approx (5 - 1)/(5 + 1) = 0.667$ and hence the reflection coefficient = $0.667\underline{/-55.5°}$.

Some Smith charts such as Figure 3.1 do have 'radially scaled parameters' scales, and it is possible to obtain the reflection coefficient magnitude directly by simply measuring the radius of the VSWR circle and reading off the same distance on the voltage reflection coefficient scale. Alternatively, use Equation 2.38a to calculate it.

Example 3.3

In Figure 3.11, the VSWR circle has a radius of 0.667. An impedance is shown on the Smith chart as point B which is (0.25 – j0.5). What is its reflection coefficient?

Given: VSWR circle radius = 0.667.
Impedance at point B = (0.25 – j0.5).
Required: Voltage reflection coefficient Γ.

Solution. The answer is $0.667\underline{/-124°}$. This is shown as point C in Figure 3.11.

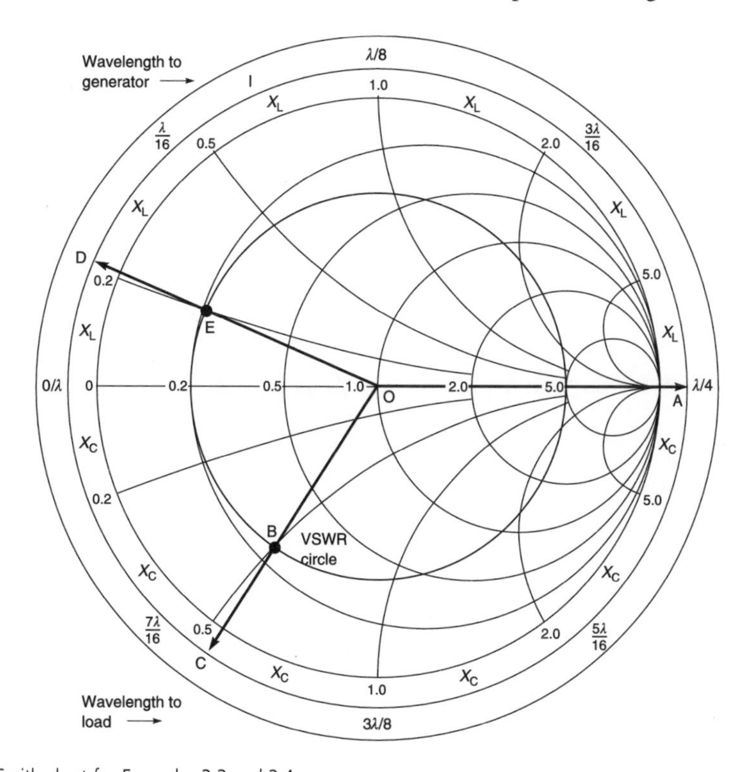

Fig. 3.11 Smith chart for Examples 3.3 and 3.4

[5] VSWR is obtained by completing the circle enclosing the point A (Figure 3.10). It is then read off the intersection between the circle and the real axis and in this case = 5. Proof of this will be given when we derive Equation 3.14.

Example 3.4

In Figure 3.11, the VSWR circle has a radius of 0.667. If the angle AOD is +156°, what impedance does the point E represent on the Smith chart?

Given: VSWR circle radius = 0.667
AOD = +156°.
Required: Impedance at point E.

Solution. The answer is approximately 0.21 + j0.21. This is shown as point E in Figure 3.11.

3.6.2 Impedance of multi-element circuits

The impedance and/or admittance of multi-element networks can be found on the Smith chart without any calculations.

Example 3.5

What is (a) the impedance and (b) the reflection coefficient looking into the network shown in Figure 3.12?

Given: Network shown in Figure 3.12.
Required: (a) Input impedance, (b) reflection coefficient.

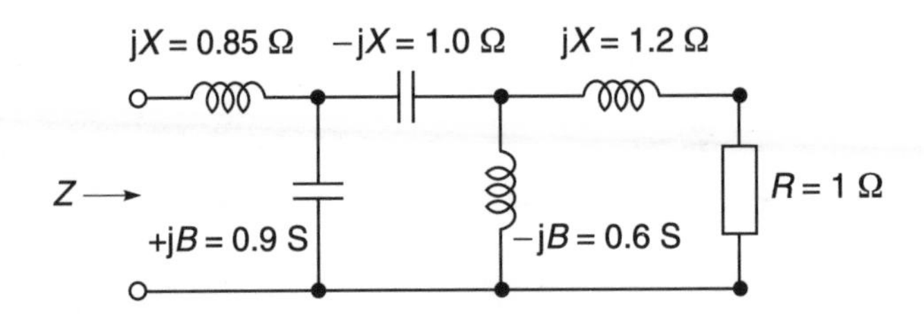

Fig. 3.12 Network to be analysed

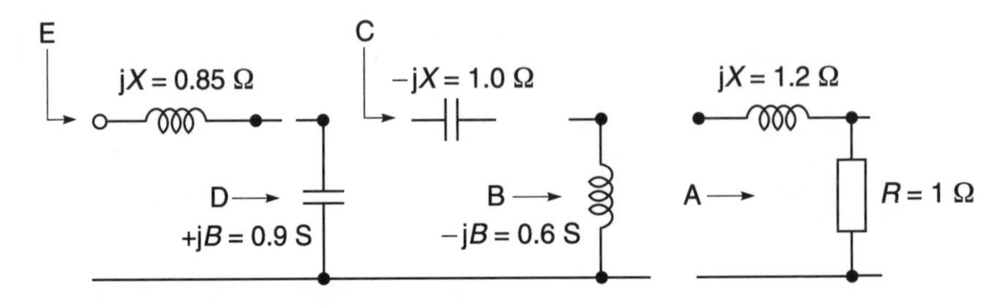

Fig. 3.13 Circuit of Figure 3.12 dis-assembled for analysis

Solution. The problem is easily handled on a Smith chart and not a single calculation needs to be performed. The solution is shown by using Figure 3.13.

(a) To find the impedance, proceed as follows.

(1) Separate the circuit down into individual branches as shown in Figure 3.13. Plot the series branch where $Z = (1 + j1.2)\ \Omega$. This is point A in Figures 3.13 and 3.14.

(2) Add each component back into the circuit – one at a time. The following rule is particularly important. *Every time you add an impedance, use the impedance part of the chart. Every time you add an admittance, use the admittance part of the chart.* If you observe the above rule, you will have no difficulty following the construction order below:

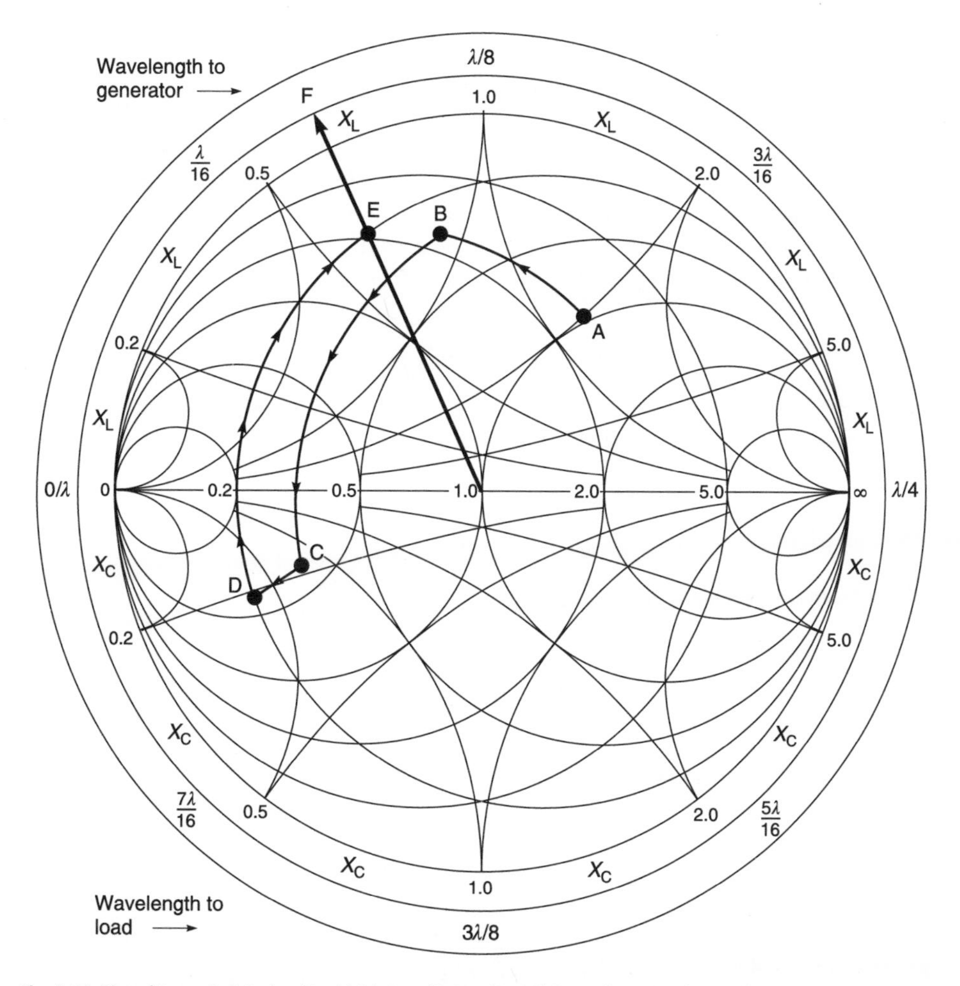

Fig. 3.14 Plot of Example 3.5: A = $(1 + j1.2)\ \Omega$ or $(0.41 - j0.492)$ S, B = $(0.3 + j0.8)\ \Omega$ or $(0.41 - j1.092)$ S, C = $(0.3 - j0.2)\ \Omega$ or $(2.3326 + j1.522)$ S, D = $(0.206 - j0.215)\ \Omega$ or $(2.326 + j2.422)$ S, E = $(0.206 + j0.635)\ \Omega$ or $(0.462 - j1.425)$ S, F = $0.746 \angle 113.58°$

Arc AB = shunt L = $-jB = -0.6$ S
Arc BC = series C = $-jX = -1.0\ \Omega$
Arc CD = shunt L = $+jB = +0.9$ S
Arc DE = series C = $+jX = +0.85\ \Omega$

(3) The impedance value (point E) can then be read directly from Figure 3.14. It is $Z = (0.2 + j0.63)$.

(b) To find the reflection coefficient, proceed as follows.
(1) Draw a line from the centre of the chart (Figure 3.14) through the point E to cut the chart periphery at point F. Measure the distance from the chart centre O to point E and transfer this distance to the reflection coefficient scale (if you have one) to obtain its value.[6] This value is 0.74.
(2) Read the angle of intersection of the line OE and the periphery. This angle is 114°.
(3) Hence the value of the reflection coefficient is $0.74\ \angle 114°$.

3.7 Impedance of distributed circuits

In the previous example, we showed you how the Smith chart can be used for lumped circuit elements. In Sections 3.7 and 3.8, we will show you how the Smith chart can also be used with distributed circuit elements like transmission lines.

For ease of verifying Smith chart results, some of the transmission line expressions derived earlier will now be repeated without proof. These are:

$$Z_l = Z_0 \left[\frac{1 + \Gamma_v e^{-2\gamma l}}{1 - \Gamma_v e^{-2\gamma l}} \right] \tag{3.9}$$

For a lossless line, with $\gamma = j\beta$, this becomes

$$Z_l = Z_0 \left[\frac{1 + \Gamma_v e^{-j2\beta l}}{1 - \Gamma_v e^{-j2\beta l}} \right] \tag{3.10}$$

Equation 3.9 has been shown to be

$$Z_{in} = Z_0 \left[\frac{Z_0 \sinh \gamma l + Z_L \cosh \gamma l}{Z_0 \cosh \gamma l + Z_L \sinh \gamma l} \right] \tag{3.11}$$

Equation 3.11 can also be written in the form

$$\frac{Z_{in}}{Z_0} = \left[\frac{Z_0 \sinh \gamma l + Z_L \cosh \gamma l}{Z_0 \cosh \gamma l + Z_L \sinh \gamma l} \right] \tag{3.12}$$

It is also possible to divide each term in Equation 3.12 by ($Z_0 \cosh \gamma l$) and, remembering that for a lossless line that $\tanh \gamma l = j \tan \beta l$, we get

[6] The unfortunate part of this condensed Smith chart is that we do not have a voltage reflection coefficient scale. So please accept my answer for now until we use PUFF. Alternatively, use Equation 2.38a to calculate it.

$$\frac{Z_{in}}{Z_0} = \left[\frac{j\tan\beta l + Z_L/Z_0}{j(Z_L/Z_0)\tan\beta l + 1}\right] \tag{3.13}$$

Note Equations 3.12 and 3.13 have been normalised with respect to Z_0. Although stated earlier, we will re-iterate that normalisation is used on Smith charts because it enables a larger range of values to be covered with greater accuracy on the same chart. However, if we divide a value by Z_0 (normalised) before entering it on the Smith chart, it stands to reason that we must then multiply the Smith chart result by Z_0 (re-normalised) to gets its true value.

Nowadays, the above mathematical calculations can be easily programmed into a computer, and you may well question the necessity of the Smith chart for line calculations. However, as you will see shortly the Smith chart is more than just a tool for line calculations; it is invaluable for gaining an insight into line conditions for matching purposes, parameter presentation, constant gain circles, stability and instability circles, Qs of elements, standing-wave ratios, and voltage maxima and minima positions.

The input and output impedances of transistors are usually complex, with a significant reactive component. The Smith chart is particularly useful for matching purposes when a transmission line is terminated by the input of a transistor amplifier, or when the line is driven from an amplifier. Matching is necessary because it avoids reflections on the lines and ensures maximum power transfer from source to load.

3.7.1 Finding line impedances

Smith charts can be used to find line impedances as demonstrated by the following examples.

Example 3.6

A transmission line with a characteristic impedance $Z_0 = 50\ \Omega$ is terminated with a load impedance of $Z_L = (40 - j80)\ \Omega$. What is its input impedance when the line is (a) 0.096λ, (b) 0.173λ and (c) 0.206λ?

Given: $Z_0 = 50\ \Omega$, $Z_L = (40 - j80)\ \Omega$.
Required: The input impedance Z_{in} of the terminated line when (a) the line is 0.096λ, (b) 0.173λ and (c) 0.206λ.

Solution. First, the load impedance is expressed in normalised form as

$$\frac{Z_L}{Z_0} = \frac{(40 - j80)\Omega}{50\Omega} = 0.8 - j1.6$$

This value is plotted on the chart as point A in Figure 3.15. Note that this point is the intersection of the arcs of two circles. The first is that which cuts the horizontal axis labelled 'resistance component (R/Z_0)' at the value 0.8. Because the reactive component is negative, the second circle is that which cuts the circular axis on the periphery labelled 'capacitance reactance component ($-jX/Z_0$)' at the value 1.6. A line is now drawn from the centre of the Smith chart (point 1 + j0) through the point 0.8 – j1.6 and projected to the periphery of the circle to cut the 'wavelengths towards generator' circle at 0.327λ. This is shown as point B.

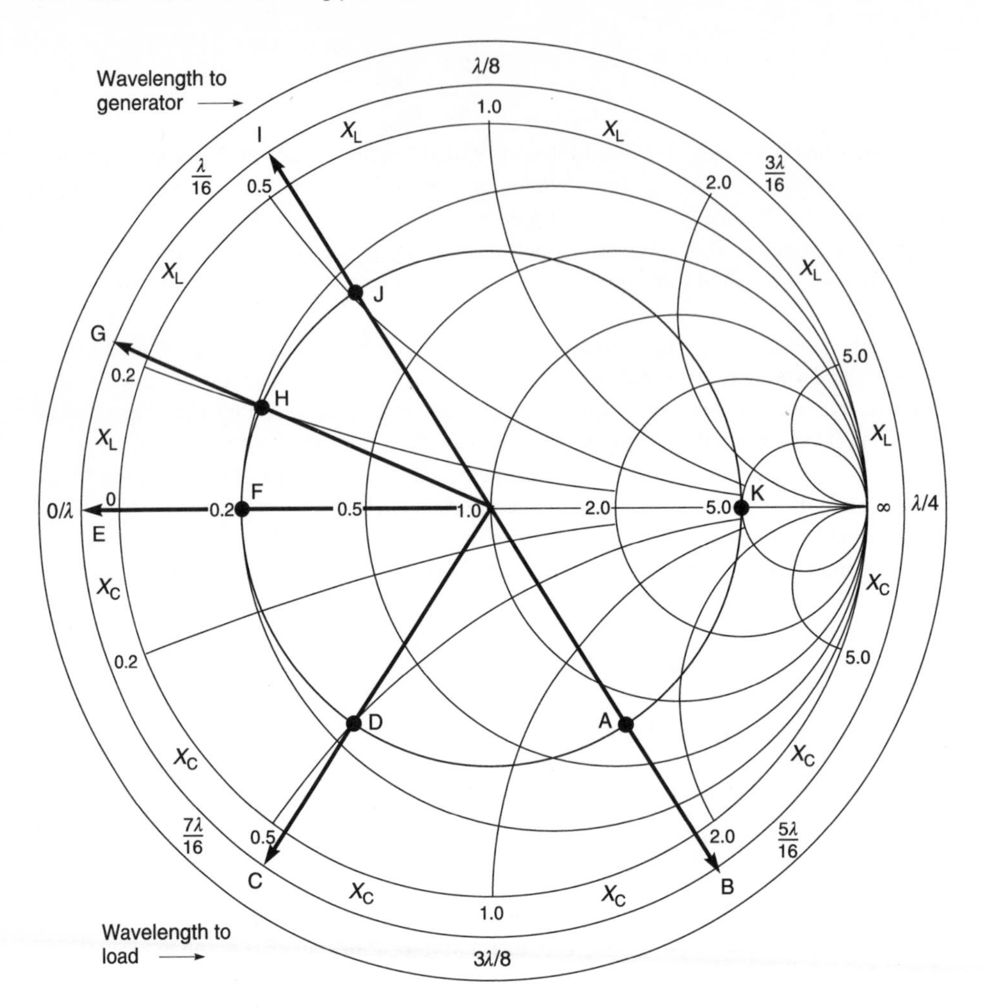

Fig. 3.15 Finding input impedance of a line: A = 0.8 – j1.6, B = 0.327λ, C = 0.423λ, D = 0.25 – j0.5 E = 0.5λ, or 0λ, F = 0.2 + j0, G = 0.033λ, H = 0.2 + j02, I = 0.077λ, J = 0.25 – j0.5, K = 5 + j0

Next, a circle is drawn with its centre at the centre of the chart (point 1 + j0), passing through the load impedance value (point A) in Figure 3.15. This circle represents all possible line impedances along the transmission line and we will call it our impedance circle.[7]

(a) We wish to know Z_{in} when the TX line is 0.096λ long. For this, we start at the point B, 0.327λ on the 'wavelengths towards generator' circle and move 0.096λ in the direction of the generator (i.e. away from the load). Since 0.327λ + 0.096λ = 0.423λ, the point we want is 0.423λ on the 'wavelengths towards generator' circle. I have denoted this point as point C. From C, draw a straight line to the Smith chart centre. Where this line cuts our impedance circle, read out the value at the intersection. I read it as (0.25 – j0.5) and have marked it as point D. To get the actual value, I must re-normalise the chart value by 50 Ω and get (0.25 – j0.5) × 50 Ω = (12.5 – j25) Ω.

[7] This circle is also called the voltage standing wave ratio (VSWR) circle.

(b) For the case where the line length is 0.173λ, I must move 0.173λ on the 'towards the generator' scale from the same starting point of 0.327λ, i.e. point B. So I must get to the point $(0.327 + 0.173)\lambda$ or 0.5λ on the 'wavelengths towards generator' scale. This is shown as point E in Figure 3.15. Joining point E with the Smith chart centre shows that I cut our original impedance circle at the point F, where the value is (0.2 + j0). To get the true value, I must re-normalise, (0.2 + j0) × 50 Ω = (10 + j0) Ω.

Note: It is possible to get a pure resistance from a complex load by choosing the right length of transmission line. This is very important as you will see later when we do designs on matching complex transistor loads to pure resistances.

(c) For the case where the line length is 0.206λ, I must move 0.206λ on the 'towards the generator' scale from the same starting point of 0.327λ, i.e. point B. This scale only goes up to 0.5λ and then restarts again. So I must get to 0.5λ which is a movement of $(0.5 - 0.327)\lambda$ or 0.173λ then add another $(0.206 - 0.173)\lambda$ or 0.033λ to complete the full movement of 0.206λ to arrive at point G. Joining point G with the Smith chart centre shows that I cut our original impedance circle at the point H, where the value is (0.2 + j0.2). To get the true value, I must re-normalise, (0.2 + j0.2) × 50 Ω = (10 + j10) Ω.

If you wish, you may confirm these three values (12.5 – j25) Ω, 10 Ω and (10 + j10) Ω by calculations using Equation 3.11 but bear in mind that you will have to change the wavelength[8] into radians before applying the equation.

Example 3.7

Show that, when $Z_L = R_L + j0$ on a lossless line, the VSWR equals R_L/Z_0. (Hint: First find the reflection coefficient of the load.)

Solution. Using Equation 2.28

$$\Gamma_v = \frac{Z_L - Z_0}{Z_L + Z_0}$$

When $Z_L = R_L$, this becomes

$$\Gamma_v = \frac{R_L - Z_0}{R_L + Z_0}$$

Here, both R_L and Z_0 are purely resistive, so Γ_v is a real number. Using Equation 2.38

$$\text{VSWR} = \frac{1 + |\Gamma_v|}{1 - |\Gamma_v|} = \frac{1 + \Gamma_v}{1 - \Gamma_v}$$

because Γ_v is real. So

[8] One wavelength = 2π radians.

$$\begin{aligned}\text{VSWR} &= \frac{1+(R_L - Z_0)/(R_L + Z_0)}{1-(R_L - Z_0)/(R_L + Z_0)} \\ &= \frac{(R_L + Z_0)+(R_L - Z_0)}{(R_L + Z_0)-(R_L - Z_0)} \\ &= \frac{2R_L}{2Z_0}\end{aligned}$$

Hence

$$\text{VSWR} = \frac{R_L}{Z_0} \tag{3.14}$$

It has been shown in Example 3.7 that when $Z_L = R_L$, and Z_0 is resistive, the VSWR is given simply by VSWR = R_L/Z_0. In this case $R_L/Z_0 = 5.0$ (point K in Figure 3.15), so the VSWR is 5.0. Thus the point where the circle cuts the right-hand horizontal axis gives the VSWR on the line.

Looking at successive points around the standing-wave circle drawn through the load impedance is equivalent to looking at successive points along a lossless line on which the VSWR equals that of the circle. The successive values of input line impedances at points D, F, H, K around the circle correspond to line impedances at successive points along the line. The distance along the line is directly proportional to the angle around the standing-wave circle. One complete revolution takes us from a voltage minimum at the point F in Figure 3.15, where $Z_{in} = 0.2\ Z_0$, to the point opposite this on the circle with a voltage maximum where $Z_{in} = 5\ Z_0$ at point K, and back to the first minimum. Since standing-wave minima repeat every half wavelength, one complete revolution corresponds to $\lambda/2$. The peripheral scales marked 'wavelengths towards generator' and 'wavelengths towards load' are calibrated accordingly.

Figure 3.15 shows that, for our example, the position (point F) of the first voltage minimum is at 0.173λ back from the load, clockwise around the scale 'wavelengths toward generator'. The first maximum (point K) is shown at $[(0.5 - 0.327) + 0.25]\lambda$ or 0.423λ from the load, clockwise around the 'wavelengths towards generator' scale. The distance between voltage maxima and minima is 0.25λ as you should expect from transmission line theory.

Example 3.8

Use the Smith chart in Figure 3.15 to find the line impedance at a point one quarter wavelength from a load of $(40 - j80)\ \Omega$.

Given: $Z_L = (40 - j80)\Omega$, $l = \lambda/4$.
Required: Z @ $\lambda/4$ from Z_L.

Solution. Moving around the chart away from the load and towards the generator (that is clockwise) through 0.25λ (that is 180°) brings us to the normalised impedance $(0.25 + j0.5)$, point J in Figure 3.15. Since the chart was normalised to 50 Ω, to get the true value we must multiply $(0.25 + j0.5) \times 50 = (12.5 + j25)\ \Omega$.

3.8 Impedance matching

When a load such as a transistor input, with a substantial reactive component, is driven from a transmission line, it is necessary to match the load to the line to avoid reflections and to transfer the most power from source to load. The reactive component of the load can be tuned out, and the resistive component matched to the line, using a matching network of inductors and capacitors. However, at UHF and microwave frequencies, lumped inductors and capacitors can be very lossy, and much higher Q values can be obtained by using additional sections of transmission line instead. There are two common techniques used, the quarter-wavelength transformer and the stub tuner. These are described below.

3.8.1 Impedance matching using a $\lambda/4$ transformer

This is shown by Example 3.9 which first uses a line length to convert a complex load to a resistive load and then uses a 1/4 line transformer to transform the resistive load to match the desired source impedance.

Example 3.9

Figure 3.15 shows how a quarter-wavelength section of line can be used to match a load, such as the input of a transistor, to a line. Suppose the load has the normalised value of the previous example, that is (0.8 – j1.6), point A in Figure 3.15. First, the Smith chart is used to choose a length l of line which, when connected to the load, will have a purely resistive input impedance. In this example, the length could be either 0.173λ, with a normalised input impedance of 0.2 (point F in Figure 3.15) or 0.432λ with a normalised input impedance of 5.0 (point K in Figure 3.15). Suppose we choose the higher value, with $l = 0.432\lambda$ and $Z/Z_0 = 5.0$.

Next, we calculate the characteristic impedance of the quarter-wave section. It is required to match Z_{in} to the input line which is Z_0, so its input impedance Z_{in} must equal Z_0.

The $\lambda/4$ transformer. The equation for a quarter-wave transformer has been derived in Chapter 2 as Equation 2.57. To distinguish the main transmission line impedance (Z_0) from the $\lambda/4$ transformer line impedance, we shall denote the latter as Z_{0t}. Therefore

$$Z_{in} = \frac{Z_{0t}^2}{Z_L} \quad \text{or} \quad Z_{0t} = \sqrt{Z_{in} Z_L}$$

Since we wish to make Z_{in} of the $\lambda/4$ transformer match Z_0, we therefore get

$$Z_{0t} = \sqrt{Z_{in} Z_L} = \sqrt{Z_0 Z_L}$$

Normalising, this becomes

$$\frac{Z_{0t}}{Z_0} = \frac{\sqrt{Z_0 Z_L}}{Z_0} = \sqrt{\frac{Z_L}{Z_0}}$$

In the example, $Z_L/Z_0 = 5.0$. Therefore

$$\frac{Z_{0t}}{Z_0} = \sqrt{5.0} \approx 2.24$$

Therefore, the $\lambda/4$ transformer characteristic impedance, $Z_{0t} = 2.24\ Z_0$. With $Z_0 = 50\ \Omega$

$$Z_{0t} = 2.24 \times 50\ \Omega \approx 112\ \Omega$$

In microstrip, an impedance of 112 Ω is possible but the microstrip itself is beginning to get narrow and fabrication accuracy of the 112 Ω line might be more difficult.

Example 3.10

Find the characteristic impedance Z_{0t} of a $\lambda/4$ transformer required for the case when $Z_L/Z_0 = 0.2$, and $Z_0 = 50\ \Omega$.

Solution. Again using Equation 2.57

$$Z_{0t}/Z_0 = \sqrt{Z_L/Z_0} = \sqrt{0.2} = 0.447$$

So, with $Z_0 = 50\ \Omega$

$$Z_{0t} = 0.447 \times 50 \approx 22\ \Omega$$

The answer to Example 3.7 shows that the required $\lambda/4$ line's characteristic impedance turns out to be low. In microstrip, a low line impedance means that the microstrip itself is wider and in some cases this might be an easier fabrication process.

In general, a higher value of Z_{0t} results if a $\lambda/4$ transformer is coupled to a standing-wave voltage maxima on the matching line, and a lower value of Z_{0t} results if a $\lambda/4$ transformer is coupled to a standing-wave voltage minima on the matching line. The choice as to where the $\lambda/4$ transformer is coupled will depend on circumstances such as physical size, dielectric constants of microstrip line, etc.

3.8.2 Broadband matching using $\lambda/4$ transformers

Broadband matching is used when we want to achieve the best compromise match between source and load across a given bandwidth. Compromise is necessary because perfect matching at all individual frequencies is not feasible. The term 'broadband' implies that impedance compensation should be achieved over frequency ranges larger than 50% of the central frequency.

Distributed elements, due to their fixed geometric characteristics, are usually poor performers when broadband performance is required. There are, nevertheless, distributed networks that exhibit better broadband performance than others. For example, the $\lambda/4$ line transformer of Figure 3.16(a) allows matching over a small frequency band, while two $\lambda/4$ lines in cascade (Figure 3.16(b)) provide a greater matching bandwidth and, of course, the

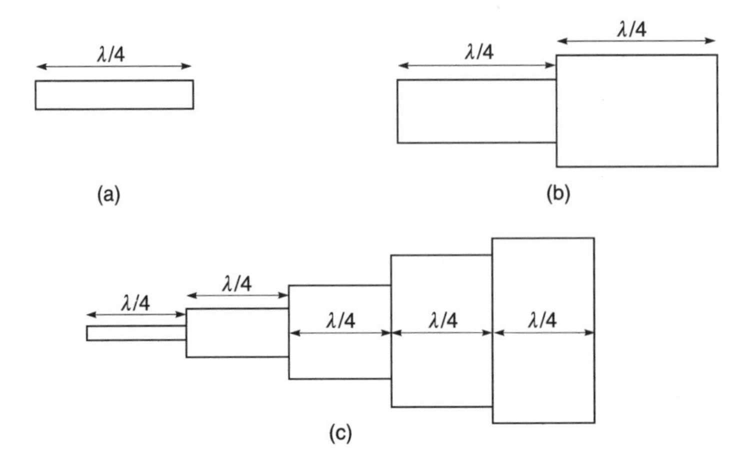

Fig. 3.16 Layout of line transformers: (a) one $\lambda/4$; (b) two $\lambda/4$ lines in cascade; (c) five $\lambda/4$ lines cascaded

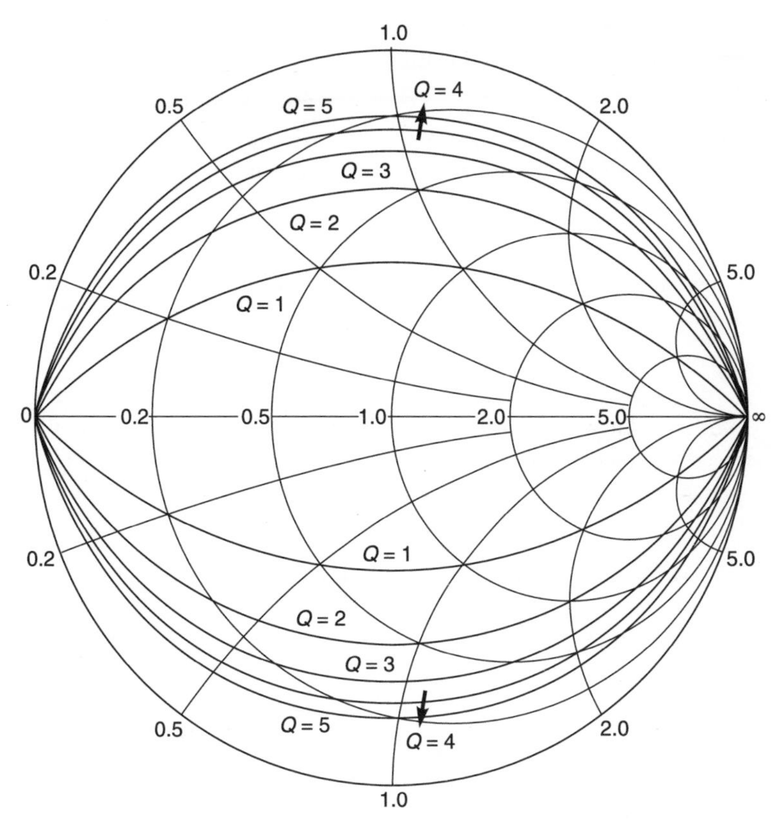

Fig. 3.17 Constant-Q arcs on the Smith chart

five $\lambda/4$ cascaded transformers of Figure 3.16(c) would provide an even greater bandwidth. In general cascading more quarter-wave transformers provides greater bandwidth. However, you should be aware that $\lambda/4$ line transformers can only be used in the GHz range because below these frequencies $\lambda/4$ lengths can be very long. For example at 30 MHz in air, $\lambda/4 = 2.5$ m. When quarter-wave length lines are required at low frequencies, the lumped circuit equivalent of a quarter-wave line length (often a π circuit) is used.

In the next example, we will investigate the difference in bandwidth obtained between using a single $\lambda/4$ transformer and a transformer produced by two $\lambda/4$ lines in cascade. However, before commencing, it is worth investigating how the Smith chart can be used to help design. In Section 3.2.3, we showed that any point on the Smith chart has a Q value associated with it. The locus of impedances on the chart with the same Q is an arc that crosses the open-circuit and short-circuit loads. Several Q arcs are shown in Figure 3.17. The Q arcs in the Smith chart can be used to provide the limits within which the matching network should remain in order to provide a larger operational bandwidth. Remembering that $Q = f_{\text{centre}}/f_{\text{bandwidth}}$, it follows that for a given centre frequency a wider bandwidth requires a lower Q.

Example 3.11

A source impedance Z_S of $(50 + j0)$ is to be matched to a load of $(100 + j0)$ over a frequency range of 600–1400 MHz. Match the source and load by using (a) one quarter-wave transformer and (b) two quarter-wave transformers. (c) Sketch a graph of the reflection coefficient against frequency.

Given: $Z_S = (50 + j0)\ \Omega$, $Z_L = (100 + j0)\ \Omega$, bandwidth = 600–1400 MHz.
Required: (a) Matching network using one $\lambda/4$ transformer, (b) matching network using two $\lambda/4$ transformers, (c) a sketch of their network reflection coefficient against bandwidth.

Solution

(a) Use one quarter-wave transformer as in the circuit of Figure 3.16(a). We start by using Equation 2.57 which is $Z_{in} = Z_0^2/Z_L$ which yields

$$Z_0 = \sqrt{Z_{in} Z_L} = \sqrt{(50 + j0)(100 + j0)} = 70.71\Omega$$

(b) Use two quarter-wave transformers as in Figure 3.18. Note in this case, I have called the characteristic impedance of the first $\lambda/4$ line from the load, Z_{0t1} and the characteristic impedance of the second $\lambda/4$ line from the load, Z_{0t2}.

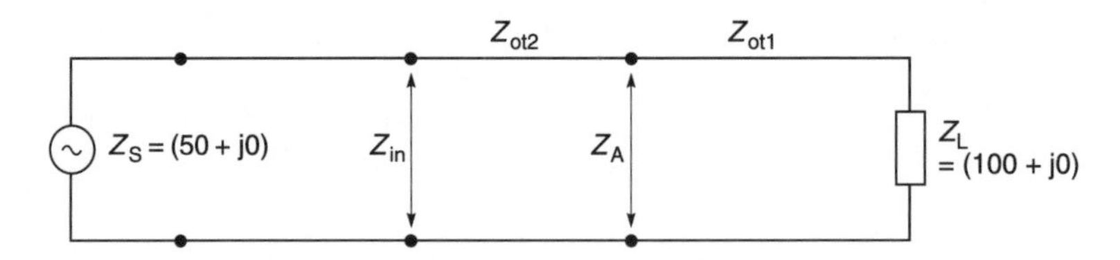

Fig. 3.18 Matching with two quarter-wave transformers

From Equation 2.57

$$Z_A = (Z_{0t1})^2/Z_L$$

Again using Equation 2.57, and substituting for Z_A

$$Z_{in} = (Z_{0t2})^2/Z_A = (Z_{0t2})^2/(Z_{0t1})^2/Z_L$$

Sorting out, we get

$$\frac{Z_{in}}{Z_L} = \frac{(Z_{0t2})^2}{(Z_{0t1})^2} \quad \text{or} \quad \frac{Z_{0t1}}{Z_{0t2}} = \sqrt{\frac{Z_L}{Z_{in}}}$$

Bearing in mind that Z_{in} must match the Z_s and substituting in values

$$\frac{Z_{0t1}}{Z_{0t2}} = \sqrt{\frac{100 + j0}{50 + j0}} = 1.414$$

If I choose a value of 60 Ω for Z_{0t2} then $Z_{0t1} = 1.414 \times 60 = 84.85\ \Omega$

(c) If the reflection coefficients of the network 1 and network 2 are plotted against frequency, you will get Figure 3.19. I have also included Table 3.3 to give you some idea of the difference between the networks. From both the table and the graph, you should note that the two $\lambda/4$ network has lowered the reflection coefficient by approximately 6 dB at 600 MHz and 1400 MHz. There has also been a reduction of about 12.8 dB at 800 MHz and 1200 MHz.

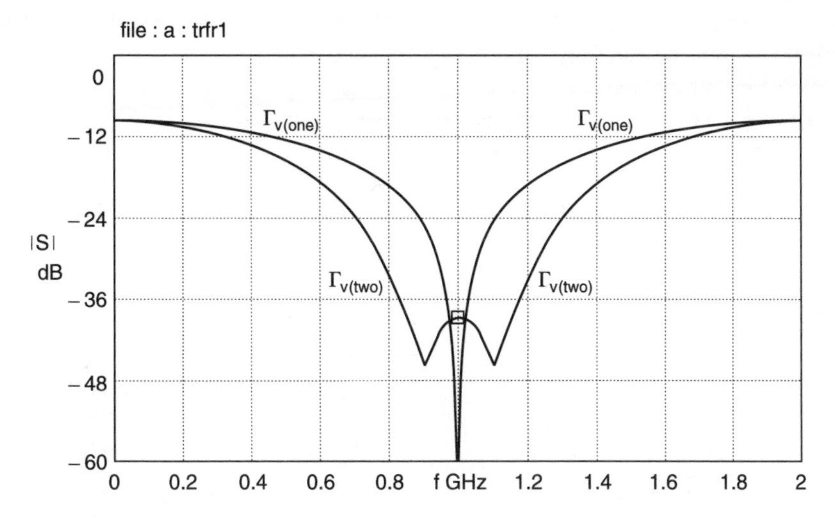

Fig. 3.19 Matching with $\lambda/4$ line transformers: using one $\lambda/4$ transformer; using two $\lambda/4$ transformers

Table 3.3 Reflection coefficient (dB) against frequency (GHz)

(GHz)	0.6	0.8	1.0	1.2	1.4
One $\lambda/4$ TX	–13.83	–19.28	> –60	–19.28	–13.83
Two $\lambda/4$ TXs	–18.81	–32.09	–38.69	–32.09	–18.81

Note: The Smith chart graphics and calculations to obtain this graph are quite long; to save work, I have used the PUFF software supplied with this book.

3.8.3 Impedance matching using a stub tuner

An alternative method to impedance matching by $\lambda/4$ transformers is shown in Figure 3.20 where a transmission line stub (stub matching) is tapped into the main transmission line R_0 at a distance l_1 from the load Z_L to provide a good match between Z_L and a source

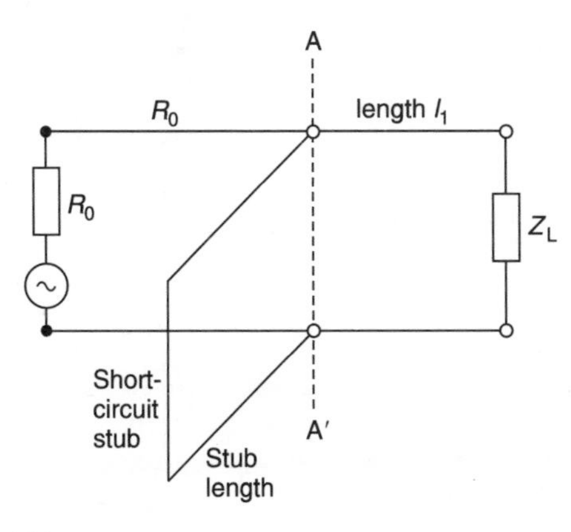

Fig. 3.20 Principle of stub matching

generator, R_0. The process appears simple enough but the line length l_1 and the stub length are **critical** and must be carefully controlled for a good match.

There are several methods of calculating these line lengths and we will give you two methods at this stage. These are explained and illustrated in Tables 3.4 and 3.5 . Examples of how these methods are used in calculating line lengths are given in Example 3.11 which follows after the explanation.

Example 3.12 is an example of how stub matching is carried out on a Smith chart.

Table 3.4 (Matching method 1)

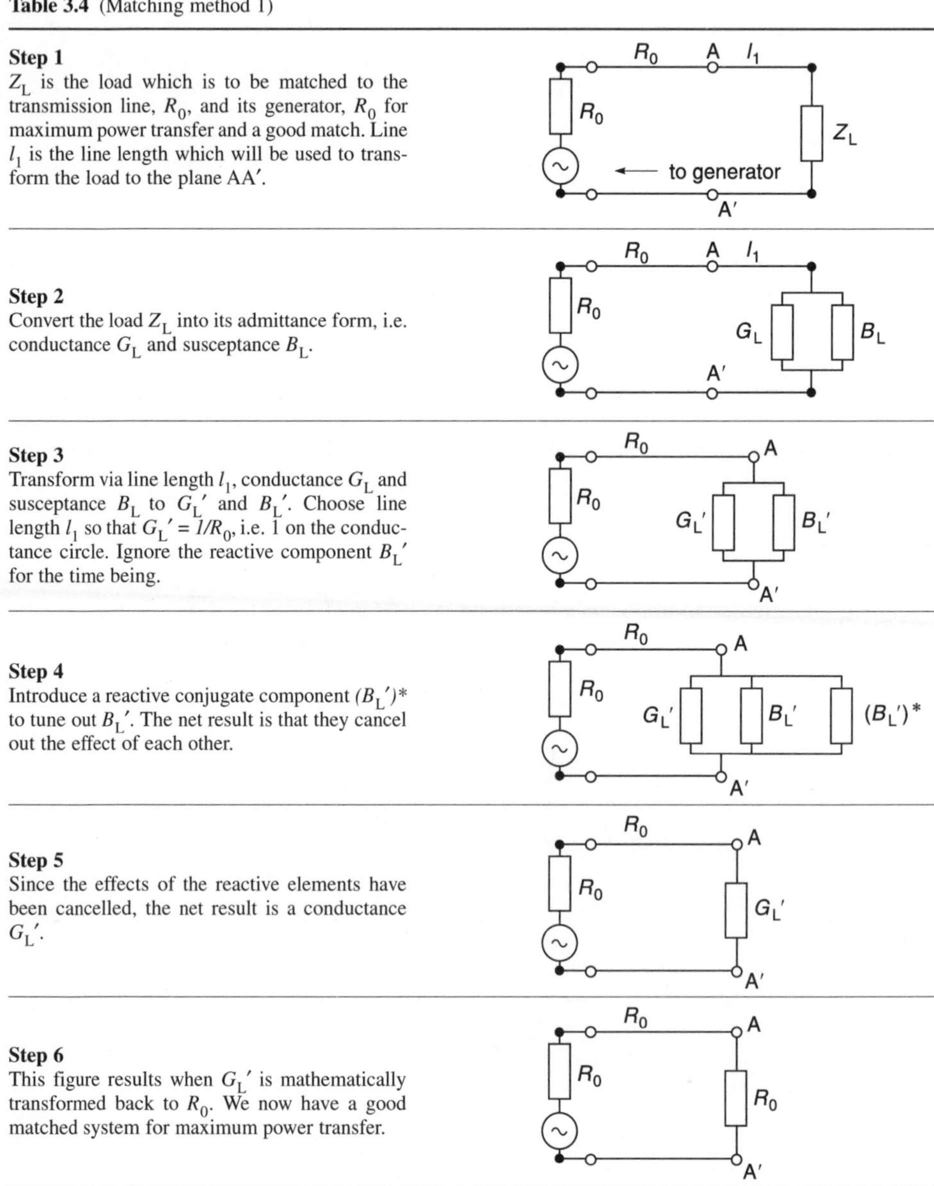

Step 1 Z_L is the load which is to be matched to the transmission line, R_0, and its generator, R_0 for maximum power transfer and a good match. Line l_1 is the line length which will be used to transform the load to the plane AA′.	
Step 2 Convert the load Z_L into its admittance form, i.e. conductance G_L and susceptance B_L.	
Step 3 Transform via line length l_1, conductance G_L and susceptance B_L to G_L' and B_L'. Choose line length l_1 so that $G_L' = 1/R_0$, i.e. 1 on the conductance circle. Ignore the reactive component B_L' for the time being.	
Step 4 Introduce a reactive conjugate component $(B_L')^*$ to tune out B_L'. The net result is that they cancel out the effect of each other.	
Step 5 Since the effects of the reactive elements have been cancelled, the net result is a conductance G_L'.	
Step 6 This figure results when G_L' is mathematically transformed back to R_0. We now have a good matched system for maximum power transfer.	

Table 3.5 (Matching method 2)

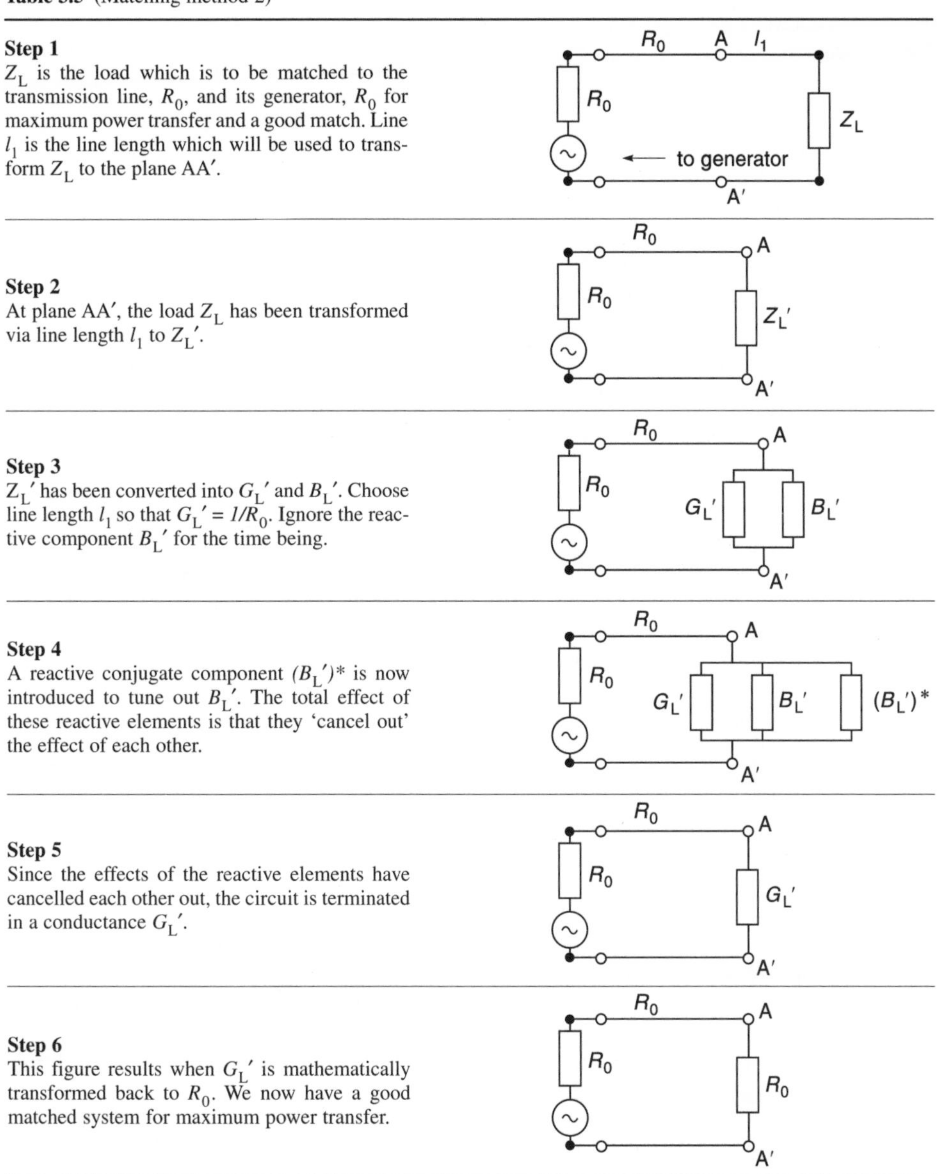

Step 1	Z_L is the load which is to be matched to the transmission line, R_0, and its generator, R_0 for maximum power transfer and a good match. Line l_1 is the line length which will be used to transform Z_L to the plane AA′.
Step 2	At plane AA′, the load Z_L has been transformed via line length l_1 to Z_L'.
Step 3	Z_L' has been converted into G_L' and B_L'. Choose line length l_1 so that $G_L' = 1/R_0$. Ignore the reactive component B_L' for the time being.
Step 4	A reactive conjugate component $(B_L')^*$ is now introduced to tune out B_L'. The total effect of these reactive elements is that they 'cancel out' the effect of each other.
Step 5	Since the effects of the reactive elements have cancelled each other out, the circuit is terminated in a conductance G_L'.
Step 6	This figure results when G_L' is mathematically transformed back to R_0. We now have a good matched system for maximum power transfer.

Example 3.12

A series impedance load (40 – j80) Ω is to be matched to a generator source of 50 Ω via a 50 Ω transmission line. Design a single stub matching system to provide this result.

Given: $Z_L = (40 - j80)\ \Omega$, $Z_g = 50\ \Omega$, $Z_0 = 50\ \Omega$.
Required: A single stub matching system.

Solution. Two methods will be given. Method 1 is depicted in Figure 3.21 and Method 2 is shown in Figure 3.22.

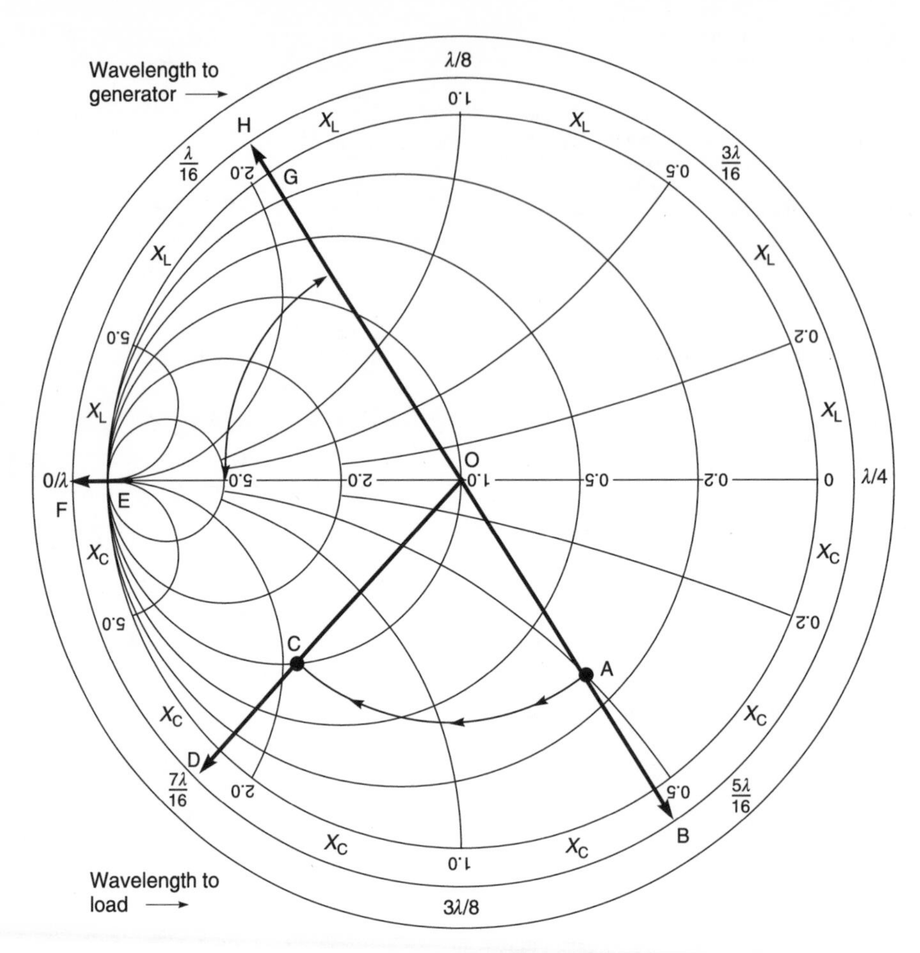

Fig. 3.21 Using matching method 1: A = (0.25 + j0.50) Ω, B = 0.077072λ, C = (1 + j1.8027) S, D = 0.183377λ, E = (∞ + j ∞) S, G = (0 – j1.8027) S, arc BD = 0.106 305λ, arc FH = 0.080 606λ

Method 1 based on Table 3.4. To simplify this example, I will use an ordinary Smith chart in its admittance form.

1 Normalise the load impedance with respect to 50 Ω. This gives (40 – j80)/50 = 0.8 – j1.6. (See step 1 of Table 3.4.)
2 Convert the normalised impedance into its admittance form by calculation. This gives 1/(0.8 – j1.6) = 0.25 + j0.5. (See step 2 of Table 3.4). The value 0.25 + j0.5 is shown as point A in Figure 3.21. Draw a constant SWR circle, using the centre of the Smith chart (point O) and a radius equal to OA.
3 Project the line from the centre of the Smith chart (point O), through point A to point B. Read point B on the 'wavelengths towards generator' scale. This is 0.077λ.
4 Transform point A along the SWR circle in the 'towards generator' direction until you obtain a conductance of 1 or unity. (Step 3 of Table 3.4.) This is shown as point C in Figure 3.21. At point C, the admittance is (1 + j1.8) S. This tells you that the

conductance of the circle is matched but that you must get rid of a susceptance of j1.8.

5 Project the line OC to point D. Read this value on the wavelength towards generator scale. It is 0.183λ. Subtract this value from the value at point B, i.e. $(0.183 - 0.077)\lambda = 0.106\lambda$. This is the length of the line l_1.
6 The unwanted susceptance of +j1.8 obtained in step 4 above must be cancelled out. For this, we must introduce a susceptance of –j1.8 to tune out the unwanted susceptance of +j1.8. (See step 4 of Table 3.4.) Bear in mind that a short circuit (0 Ω) has a conductance of ∞ Ω. We start at point E and increase the stub length until we obtain a susceptance of –j1.8. This is point G in Figure 3.21.
7 Project line OG to H and line OE to F. Measure the wavelength distance on the 'towards generator scale' between the points F and H to obtain the length of the stub. In this case, it is $0.331\lambda - 0.250\lambda = 0.081\lambda$. Hence the length of the short-circuited tuning stub to produce a susceptance of – j1.8 to cancel out the unwanted +j1.8 is 0.081λ. (See steps 5 and 6 of Table 3.4.) The generator, line and load are now all matched to each other.

Results using method 1

- The position of the stub l_1 from the load is 0.106λ.
- The length of the short-circuited stub line is 0.081λ.

These results are obtained from the chart.

Just to convince you that the graphical method is correct, I have also calculated out all the values to several decimal points by using a spreadsheet. These numbers are given in the caption in Figure 3.21; however, you should be aware that in practice, accuracy beyond two decimal points is seldom necessary. In other words, the accuracy of the Smith chart is sufficient for practical design.

Method 2 based on Table 3.5. To simplify this example, we will use Smith chart type ZY-01-N which was first introduced to you in Figure 3.7. This chart is used because it affords easy conversion from impedance to admittance plots and vice-versa. If you have forgotten how to use this chart refer back to Figure 3.7.

1 Normalising Z_L with respect to 50 Ω gives $(40 - j80)/50 = 0.8 - j1.6$. Since this value is impedance, the solid coordinate (impedance) lines are used to locate the point in Figure 3.22. This is plotted as point A in the figure. (See step 1 of Table 3.5.) Extend the line OA to the outer periphery (point B). Note the reading on the 'wavelengths towards generator scale'. This is 0.327λ and is denoted by point B.
2 Draw a constant SWR circle with its centre at the Smith chart centre (point O) and a radius equal to the distance from the chart centre to point A.
3 Move clockwise (towards the generator) along the constant SWR circle until you come to the unity conductance circle (admittance coordinates) at point C. The reason why point C is chosen is because at the unity conductance circle, the conductance element is matched to the system. (See step 2 of Table 3.5.) Draw a line from the Smith chart centre (point O) through point C to the same 'wavelengths scale'. This line cuts the circle (point D) at 0.433λ. Subtracting the wavelength value of D from C $(0.433–0.327)\lambda = 0.106\lambda$. This distance is designated as 'length l_1' in Figure 3.20. It tells you that the stub must be placed at a point 0.106λ from the load.

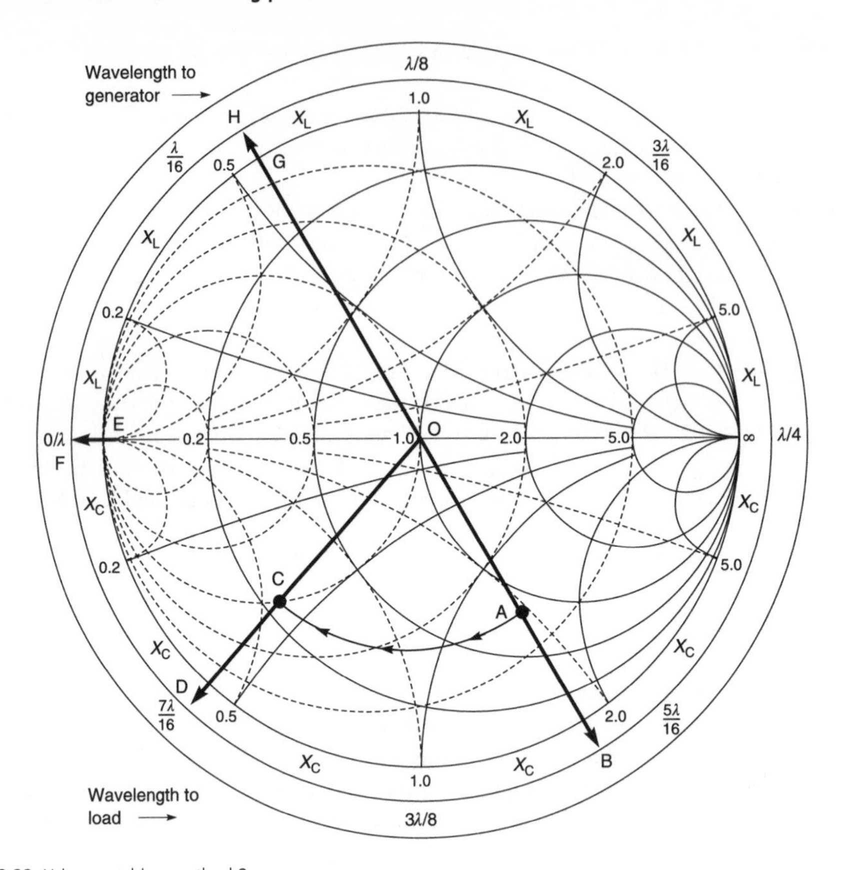

Fig. 3.22 Using matching method 2

4 Return to point C and read its admittance value which is (1 + j1.8). The conductance value is 1 and it is telling you that at this point the transformed conductive element is already matched to Y_o. (See step 3 of Table 3.5.) However, at point C, we also have a susceptance of +j1.8. We want only a conductance element and do not want any susceptance and will nullify the unwanted susceptance effect by tuning it out with –j1.8 from the stub line. (See steps 4, 5 and 6 of Table 3.5.)

5 The stub line used in Figure 3.20 is a short-circuit line which means that $Z_L = 0 + j0$ and that its admittance load is (∞ – j∞). This is shown as point E on the admittance scale in Figure 3.22. The extended line OE also cuts the 'wavelengths towards generator' circle at 0.00λ (point F). We now have to move clockwise (towards the generator away from the short-circuit load) until we generate a susceptance of –j1.8. This is shown as point G in Figure 3.22. The extended line OH from the Smith chart centre through the –j1.8 point is 0.081λ. Therefore the stub length is (0.081 – 0.00)λ = 0.081λ. The generator, line and load are now all matched to each other.

Results using method 2

- The position of the stub from the load is 0.106λ.
- The length of the short-circuited stub line is 0.081λ.

Summing up. Within the limits of graphical accuracy, both methods produce the same results.

- The position of the stub from the load is 0.106λ.
- The length of the short-circuited stub line is 0.081λ.

Example 3.13

A transistor amplifier has an input resistance of 100 Ω shunted by a capacitance of 5 pF. Find (a) the position of short-circuit stub on the line required to match the amplifier input to a 50 Ω line at 1 GHz and (b) its length.

Given: Transistor input impedance = 100 Ω shunted by 5 pF, frequency = 1 GHz.
Required: Matching circuit to a 50 Ω line (a) determine length of short-circuit stub, (b) determine its position from the load.

Solution

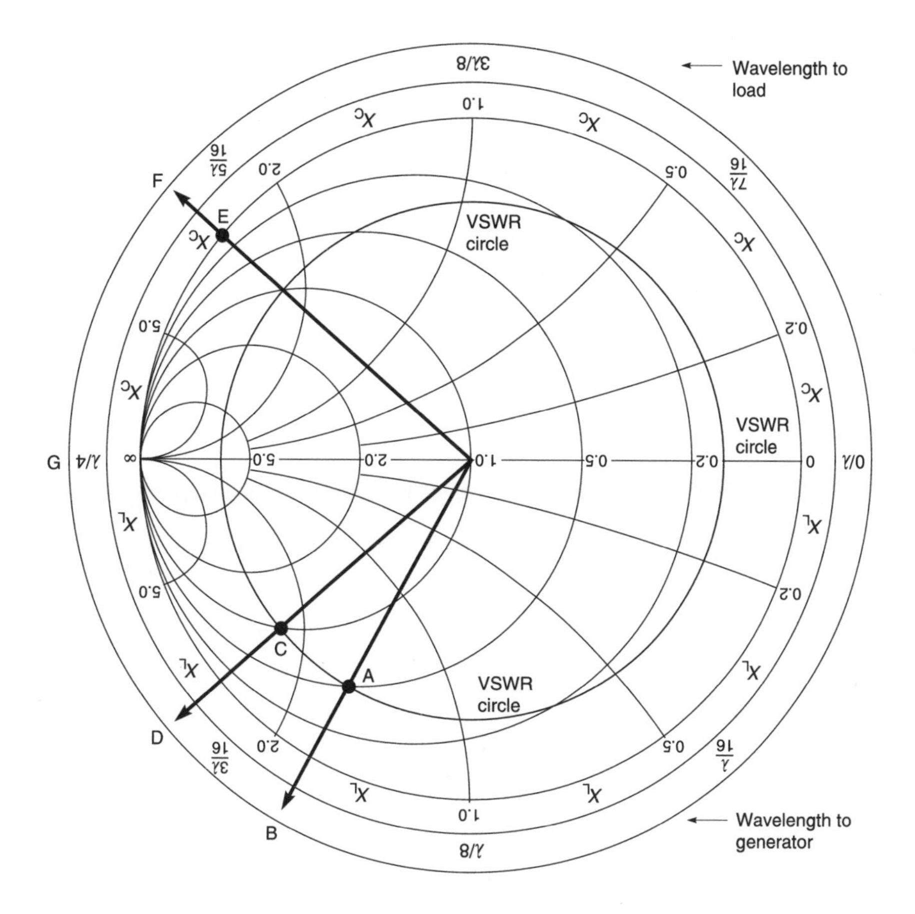

Fig. 3.23 Matching of transistor input impedance: A (0.5 + j1.57), B (0.165λ), C (1 + j2.3), D (0.193λ), E (–j2.3), F (0.315λ), G (0.25λ)

(a)
$$Y_o = 1/50\ \Omega = 20\ \text{mS}$$

and

$$Y_L = G_L + j\omega C_L = 1/100\ \Omega + j2\pi \times 1\ \text{GHz} \times 5\ \text{pF}$$
$$= 10\ \text{mS} + j31.4\ \text{mS}$$

Hence

$$Y_L/Y_o = 0.5 + j1.57$$

This is plotted in Figure 3.23 as point A. The radius through this cuts the 'wavelengths towards generator' scale at 0.165λ (point B). The VSWR circle cuts the $G/Y_o = 1$ circle at $1 + j2.3$ (point C), which corresponds to 0.193λ (point D) toward the generator. So the stub connection point should be at $(0.193\lambda - 0.165)\lambda = 0.028\lambda$ from the transistor input.

(b) The required normalised stub susceptance is $-j2.3$. This is plotted as point E. The radius through this cuts the 'wavelengths towards generator' scale at 0.315λ (point F). The short-circuit stub length should be $0.315\lambda - 0.25\lambda = 0.065\lambda$.

Summing up

- The stub connection point is 0.028λ from the transistor input.
- The short-circuit stub length is 0.065λ at the connection point.

The program PUFF issued with this book has facilities for single stub matching.

3.8.4 Impedance matching using multiple stubs

In single stub matching (Figure 3.20) the distance from the load to the stub and the length of the stub must be accurately controlled. In some situations, for example an antenna mounted on a tower, you cannot easily control the distance from antenna to stub, therefore we add one or more stubs to provide matching. One arrangement of double stub matching is shown in Figure 3.24.

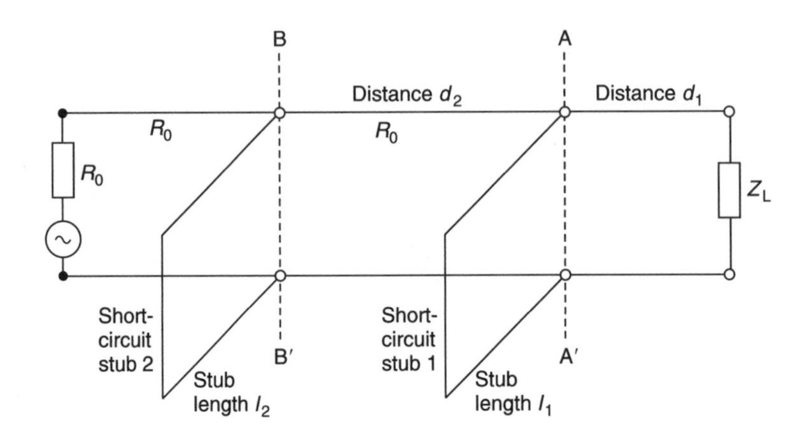

Fig. 3.24 Double stub matching network

In double stub impedance matching, two stubs are shunted at fixed positions across the main transmission line. Each stub may be either short-circuited or open-circuited. Its lengths are given by l_1 and l_2 respectively. The distance, d_2, between the stubs is usually fixed at 1/8, 3/8, or 5/8 of a wavelength, whereas the position of the nearest stub from the load, d_1, is determined by the distance from the load. Explanation is best given by an example.

Example 3.14

A system similar to that shown in Figure 3.24 has a load $Z_L = (50 + j50)\ \Omega$ which is to be matched to a transmission line and source system with a characteristic impedance of 50 Ω. The distance, d_1, between the load and the first stub is 0.2λ at the operating frequency. The distance, d_2, between the two stubs is 0.125λ at the operating frequency. Use a Smith chart to estimate the lengths of l_1 and l_2 of the stubs.

Solution. This problem will be solved by normalising all values in the question by 50 Ω. Hence (50 + j50) Ω and 50 Ω will become (1 + j1) n and 1 n respectively; the 'n' is used to denote normalised values. Values will then be plotted on the immittance Smith chart of Figure 3.25 and the chart result will be re-normalised to produce the correct answer.

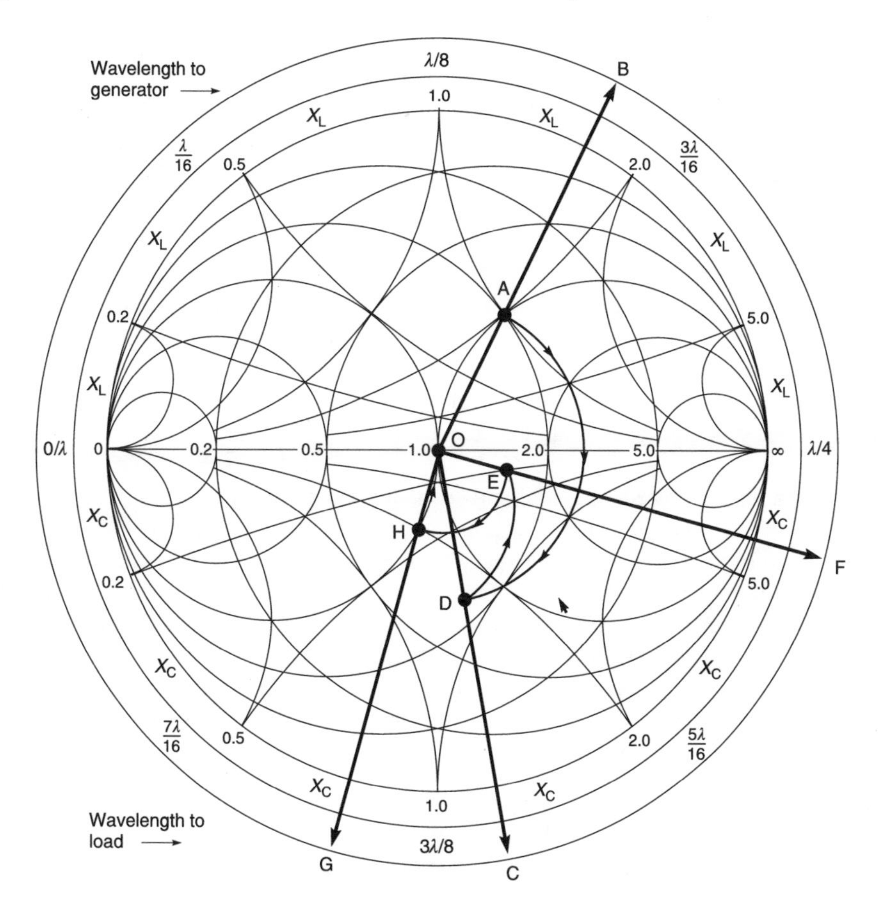

Fig. 3.25 Double stub matching: A = (1 + j1) n, Arc BC = 0.2λ, D = (0.759 – j0.838) n, E = (1.653 – j0.224) n, Arc FG = 0.125λ, H = (0.781 – j0.421) n, O = (1.008 – j0.001) n

Before starting, you should realise that distance d_1 and d_2 are out of your control; d_1 is fixed by the system structure, d_2 is fixed after you have selected your double tuning stub device which is 0.125λ in this case. You can only vary the susceptance of the stubs. Therefore, you vary stub 1 to a convenient point E in Figure 3.25 so that when that value is moved distance d_2 (0.125λ), the new point will be on the unit conductance circle where you can use stub 2 to vary the susceptance value until it reaches the impedance (1 + j0) n.

1 The load is plotted at (1 + j1) using the impedance coordinates. This is shown as A in Figure 3.25. Project OA until it cuts the wavelengths to generator scale at point B which reads 0.161λ. Move along this scale for 0.2λ. This is denoted by the arc BC. Point C is 0.361λ. Draw a line from C to O.
2 An arc of a constant $|\Gamma|$ circle with radius OA is drawn clockwise from the load point A to D which cuts the line OC. Point D is the value of the transferred load through 0.2λ. From point D, the first susceptance stub moves the transferred load to E. Point E was found experimentally by altering the length of stub 1. Extend line OE to F on the periphery. Move the arc along a periphery distance of 0.125λ to G. Arc FG represents the 0.125λ distance between the two stubs. Join O to G.
3 An arc of a constant $|\Gamma|$ circle with radius OE is drawn clockwise until it cuts the line OG at H. Point H is the transferred load from E after moving through 0.125λ. Ideally, point H should be on the unit conductance circle which means that the resistive element is matched and that stub 2 can now be used to move point H to point O which is the desired point (1 + j0).
4 The normalised values of all the points are given in the annotation for Figure 3.25. From these values, we can now calculate the susceptance which each stub must provide.
   As before, for stub 1, we need (–j0.1 at E) – (–j0.65 at D) = +j0.55. The length l_1 of the stub is found by plotting its load impedance at the point S and following round the 'wavelengths towards generator' scale to the point where the 0.55 susceptance circle cuts the perimeter, at about 0.17λ. (The calculated value is 0.167λ.)
   For stub 2, we need (j0 at O) – (–j0.55 at H) = +j0.55. The length l_2 of this stub is found as for stub 1 yielding, again, a value of about 0.17λ. (The calculated value is 0.172λ.)

Finally when you are faced with trial and error methods such as selecting point E in the above example, it is much easier if you have a dynamic impedance matching computer program. One such program is the Motorola Impedance Matching Program often called MIMP. This program allows you to alter values and see results instantaneously. MIMP has been written by Dan Moline and at the time of writing this book, Motorola generally issues a copy of it free of charge to bona fide engineers.

The program PUFF issued with this book has facilities for checking multiple stub matching. In fact, Example 3.14 is repeated electronically in Secton 4.13.4.

3.9 Summary of Smith charts

The Smith chart is a phasor diagram of the reflection coefficient, Γ, on which constant-r and constant-x circles are drawn, where r and x are the normalised values of the series resistive and reactive parts of the load impedance. The horizontal and vertical axes of the chart are the real and imaginary axes of the reflection coefficient, but they are not labelled as such.

Any circle centred on the Smith chart centre is a constant-$|\Gamma|$ circle and a constant VSWR circle, too.

A load impedance, or the impedance looking into a line towards the load, is represented by the intersection of an r circle and an x circle.

If a series lumped reactance is added to the load, the r circle through the load impedance point is followed and the added normalised reactance is represented by the increase or decrease in the corresponding value of the x circle crossed.

If a series line is added at any point then a constant-$|\Gamma|$ circle is followed, clockwise 'towards the generator' through an angle on the chart corresponding to its length in wavelengths.

The admittance chart is a version of the impedance Smith chart rotated through 180°. The r and x circles become g and b circles and their intersections represent admittances.

The immittance chart is a combination of both the impedance chart and the admittance chart.

If a lumped susceptance is shunted across the load, the g circle through the load admittance point is followed and the added normalised susceptance is represented by the increase or decrease in the corresponding value of the b circle crossed.

If a short-circuited shunt line (stub) is shunted across the line at any point, then the g circle through the point is followed, through a susceptance change corresponding to the stub length in wavelengths. For lengths less than a quarter wavelength, the short-circuit stub appears capacitive and rotation is clockwise round the g circle. For lengths up to three-quarters of a wavelength, the stub appears inductive and rotation is anticlockwise.

Open-circuit stubs have the opposite susceptance, with rotation in the opposite direction around the g circle.

Double stubs are useful when loads are variable. Usually the stub spacing is kept fixed, but the stub lengths are adjustable to achieve matching.

3.10 Scattering parameters (*s*-parameters)

3.10.1 Introduction

Voltages and currents are difficult to measure in microwave structures because they are distributed values and vary with their position in microwave structures. In fact, the widely spread current in a waveguide is virtually impossible to measure directly.

Waves are more easily measured in microwave networks. One method of describing the behaviour of a two port network is in terms of incident and reflected waves. This is shown in Figure 3.26. This method is known as **scattering parameters** or usually denoted as *s*-parameters. The *s*-parameter approach avoids many voltage and current problems particularly in the measurement of transistors where short- and open-circuit terminations can cause transistor instability and in some cases failure. In many cases, measurement is carried out in *s*-parameters using an automated computer corrected network analyser. This method is fast and accurate and the results obtained are then mathematically converted into the requisite *z*, *h*, *y* and *ABCD* parameters. The converted information can be trusted because the accuracy of the original measured data is high.

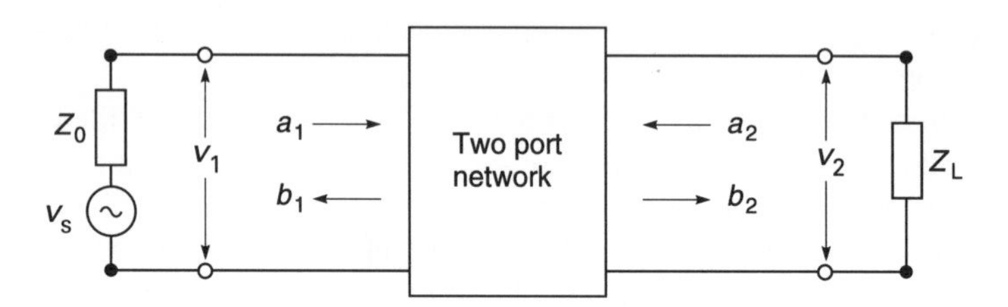

Fig. 3.26 Two port scattering network with source and load

3.10.2 Overall view of scattering parameters

Figure 3.26 represents a scattering parameter two-port network driven from a source with impedance Z_0, and driving a load of impedance Z_L. In the figure, a_1 and a_2 represent incident voltage waves; and b_1 and b_2 represent reflected voltage waves. These four waves are related by the following equations where s_{11}, s_{12}, s_{21} and s_{22} are the 'scattering', or *s*-parameters:

$$b_1 = s_{11}a_1 + s_{12}a_2 \tag{3.15}$$

and

$$b_2 = s_{21}a_1 + s_{12}a_2 \tag{3.16}$$

Equations 3.15 and 3.16 are also written in matrix form as

$$\begin{bmatrix} b_1 \\ b_2 \end{bmatrix} = \begin{bmatrix} s_{11} & s_{12} \\ s_{21} & s_{22} \end{bmatrix} \begin{bmatrix} a_1 \\ a_2 \end{bmatrix} \tag{3.17}$$

When scattering parameters are to be measured, the applied source is a generator which has the source impedance Z_0 equal to the system characteristic impedance and this generator is connected to the system by a line of characteristic impedance Z_0, as in Figure 3.27. The load is purely resistive, with impedance Z_0, and connected by a line of impedance Z_0. So the source seen by the two port's input is Z_0, and the load seen by its output is also Z_0. In this case, there is no power reflected from the load, so $a_2 = 0$.

From Equation 3.15, if $a_2 = 0$, then $b_1 = s_{11}a_1$. So s_{11} can be defined as

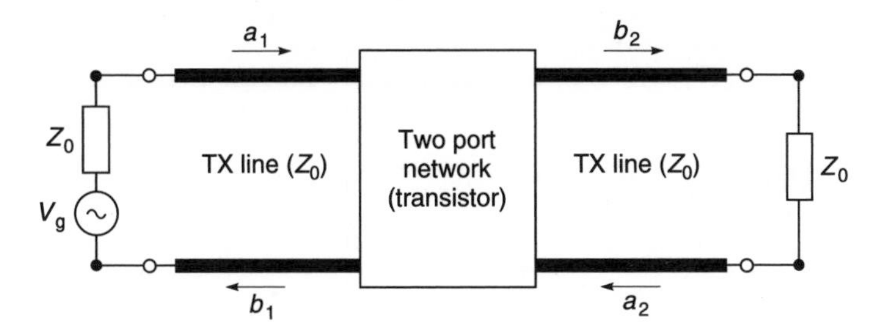

Fig. 3.27 Measurement of s-parameters

$$s_{11} = \left.\frac{b_1}{a_1}\right|_{a_2=0} \tag{3.18}$$

and s_{11} is the reflection coefficient at the input port (port 1) of the network.

From Equation 3.16, with $a_2 = 0$, $b_2 = s_{21}a_1$. So s_{21} can be defined as

$$s_{21} = \left.\frac{b_2}{a_1}\right|_{a_2=0} \tag{3.19}$$

Since this is the ratio of the output wave voltage to the incident wave voltage, $|s_{21}|^2$ is the insertion power gain of the network.

The other two s-parameters, s_{12} and s_{22}, are found by inter-changing the electrical connections to the two ports, so that port 2 is driven from the source, and port 1 is loaded by Z_0. Now $a_1 = 0$, and

$$s_{12} = \left.\frac{b_1}{a_2}\right|_{a_1=0} \tag{3.20}$$

and

$$s_{22} = \left.\frac{b_2}{a_2}\right|_{a_1=0} \tag{3.21}$$

$|s_{12}|^2$ is the reverse insertion power gain, and s_{22} is the output port reflection coefficient.

It should be clear, but two points are worth stressing.

- The scattering parameters are defined, and measured, relative to a fixed system impedance Z_0. In practice, the chosen value is nearly always 50 Ω resistive.
- The scattering parameters are complex quantities, representing ratios of phasors at a defined plane at each port.

It is now necessary to define symbols a_1, a_2, b_1 and b_2 in terms of voltages and currents.

3.10.3 Incident and reflected waves in scattering parameters

It will ease understanding if the explanation of incident and reflected waves is taken in two parts. We will begin with the 'ideal' situation where there is complete match within the

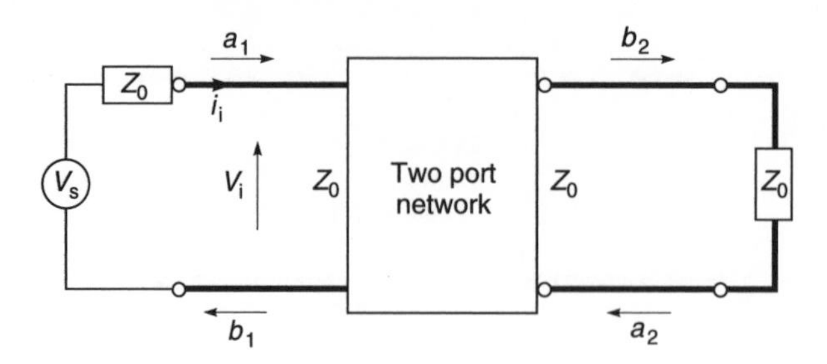

Fig. 3.28 'Ideal' two-port network

system, i.e. where the source generator, connecting lines, two-port network and load impedance all have characteristic impedances of Z_0. Then, we will proceed with the real life practical situation where the two-port network does not match the measuring system.

In Figure 3.28, we consider the ideal situation where a generator (v_s) with an internal impedance Z_0 feeds a transmission line whose impedance is Z_0 which in turn feeds a two-port network whose input and output impedances are Z_0. The output from the two-port network is then fed through another transmission line of Z_0 to a termination load where $Z_L = Z_0$. In other words because everything in the system matches, we have no reflected power, therefore the incident voltage (v_i) represents the input voltage to the network and the incident current (i_i) represents the current flowing into the network.

Now consider the practical case (Figure 3.29) where the-two port network is not matched to the same system. Due to the mismatch, we will now have reflected power. This reflected power will produce a reflected voltage v_r and a reflected current i_r. If we now defined v_1 as being the sum of the incident and reflected voltages and i_1 as the difference of the incident and reflected currents, we have

$$v_1 = v_i + v_r \tag{3.22}$$

and

$$i_1 = i_i - i_r \tag{3.23}$$

By the definition of impedances, we have

$$Z_0 = \frac{v_i}{i_i} = \frac{v_r}{i_r} \tag{3.24}$$

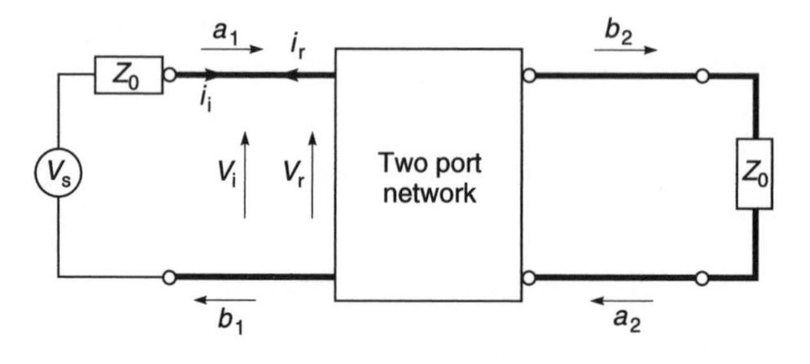

Fig. 3.29 Practical two-port network

Equation 3.24 can be re-written to yield

$$i_i = \frac{v_i}{Z_0} \tag{3.25}$$

and

$$i_r = \frac{v_r}{Z_0} \tag{3.25a}$$

Substituting Equations 3.25 and 3.25a into Equation 3.23 gives

$$i_1 = \frac{v_i}{Z_0} - \frac{v_r}{Z_0}$$

Hence

$$Z_0 i_1 = v_i - v_r \tag{3.26}$$

Adding Equations 3.22 and 3.26 yields

$$2v_i = v_1 + Z_0 i_1$$

Hence

$$v_i = \frac{1}{2}[v_1 + Z_0 i_1] \tag{3.27}$$

Subtracting Equation 3.26 from Equation 3.22 yields

$$2v_r = v_1 - Z_0 i_1$$

Hence

$$v_r = \frac{1}{2}[v_1 - Z_0 i_1] \tag{3.28}$$

The incident wave v_i is defined as the square root of the incident power. Therefore,

$$a_1 = \sqrt{\frac{v_i^2}{Z_0}} = \frac{v_i}{\sqrt{Z_0}} \tag{3.29}$$

Using Equation 3.27 to substitute for v_i in Equation 3.29 and dividing by $\sqrt{Z_0}$, we get

$$a_1 = \frac{v_i}{\sqrt{Z_0}} = \frac{1}{2}\left[\frac{v_1}{\sqrt{Z_0}} + \sqrt{Z_0}\ i_1\right] \tag{3.30}$$

Similarly, the reflected wave v_r is defined as the square root of the reflected power. Therefore

$$b_1 = \sqrt{\frac{v_r^2}{Z_0}} = \frac{v_r}{\sqrt{Z_0}} \tag{3.31}$$

Using Equation 3.28 to substitute for v_r in Equation 3.31 and dividing by $\sqrt{Z_0}$, we get

$$b_1 = \frac{v_r}{\sqrt{Z_0}} = \frac{1}{2}\left[\frac{v_1}{\sqrt{Z_0}} - \sqrt{Z_0}\ i_1\right] \tag{3.32}$$

Again using similar arguments, it can be shown that

$$a_2 = \frac{1}{2}\left[\frac{v_2}{\sqrt{Z_0}} + \sqrt{Z_0}\ i_2\right] \tag{3.33}$$

and

$$b_2 = \frac{1}{2}\left[\frac{v_2}{\sqrt{Z_0}} - \sqrt{Z_0}\ i_2\right] \tag{3.34}$$

Thus, we have now evaluated a_1, a_2, b_1 and b_2 in terms of incident voltages and currents and the characteristic impedance of the measuring system.

3.10.4 *S*-parameters in terms of impedances

From Equations 3.18, 3.27, 3.28, 3.30 and 3.32 we write

$$s_{11} = \left.\frac{b_1}{a_1}\right|_{a_2=0} = \frac{\frac{1}{2}[v_1 - Z_0 i_1]}{\frac{1}{2}[v_1 + Z_0 i_1]} = \frac{\frac{i_1}{2}\left[\frac{v_1}{i_1} - Z_0\right]}{\frac{i_1}{2}\left[\frac{v_1}{i_1} + Z_0\right]}$$

and since v_1/i_1 = input impedance at port 1 of the two-port network which we will call Z_1, we have

$$s_{11} = \left.\frac{b_1}{a_1}\right|_{a_2=0} = \frac{Z_1 - Z_0}{Z_1 + Z_0} \tag{3.35}$$

Note that Z_1 is really the load for the signal generator in this case; in some cases, it is common to write Z_L instead of Z_1 which makes Equation 3.35 identical to the transmission line reflection coefficient (Γ_1) so that we have

$$s_{11} = \left.\frac{b_1}{a_1}\right|_{a_2=0} = \frac{Z_L - Z_0}{Z_L + Z_0} = \Gamma_1 \tag{3.36}$$

Equation 3.36 also confirms what we have shown in Figure 3.3 that when the input impedance of the two-port network = Z_0, the reflection coefficient is zero and that there is no reflected wave.

Using the same process as above, it is possible to show that

$$s_{22} = \left.\frac{b_2}{a_2}\right|_{a_1=0} = \frac{Z_2 - Z_0}{Z_2 + Z_0} = \Gamma_2 \tag{3.37}$$

where Z_2 is the driving impedance of output port (port 2) of the two-port network.

3.10.5 Conversion between *s*-parameters and *y*-parameters

Most radio frequency measurements are now carried out using automated computer controlled network analysers with error correction. The measurements are then converted from *s*-parameters to other types of parameters such as transmission parameters (ABCD), hybrid *h*-parameters, and admittance *y*-parameters. We provide you with Table 3.6 to enable conversion between *s*- and *y*-parameters.

Table 3.6 Conversion between scattering *s*-parameters and *y*-parameters

$s_{11} = \frac{(1-y_{11})(1-y_{22})+y_{12}y_{21}}{(1+y_{11})(1+y_{22})-y_{12}y_{21}}$ †	$y_{11} = \left(\frac{(1+s_{22})(1-s_{11})+s_{12}s_{21}}{(1+s_{11})(1+s_{22})-s_{12}s_{21}}\right)\frac{1}{Z_0}$ *
$s_{12} = \frac{-2y_{12}}{(1+y_{11})(1+y_{22})-y_{12}y_{21}}$ †	$y_{12} = \left(\frac{-2s_{12}}{(1+s_{11})(1+s_{22})-s_{12}s_{21}}\right)\frac{1}{Z_0}$ *
$s_{21} = \frac{-2y_{21}}{(1+y_{11})(1+y_{22})-y_{12}y_{21}}$ †	$y_{21} = \left(\frac{-2s_{21}}{(1+s_{11})(1+s_{22})-s_{12}s_{21}}\right)\frac{1}{Z_0}$ *
$s_{22} = \frac{(1+y_{11})(1-y_{22})+y_{12}y_{21}}{(1+y_{11})(1+y_{22})-y_{12}y_{21}}$ †	$y_{22} = \left(\frac{(1+s_{11})(1-s_{22})+s_{12}s_{21}}{(1+s_{11})(1+s_{22})-s_{12}s_{21}}\right)\frac{1}{Z_0}$ *

* where Z_0 = the characteristic impedance of the transmission lines used in the scattering parameter system, usually 50 Ω.
† Notice that when you are converting from admittance (*Y*) to *s*-parameters, (left hand column of Table 3.6), each individual *Y* parameter must first be multiplied by Z_0 before being substituted into the equations.

3.11 Applied examples of *s*-parameters in two-port networks

Most people who have not encountered *s*-parameters earlier tend to find *s*-parameter topics a bit abstract because they have only been used to tangible voltages, currents and lumped circuitry. In order to encourage familiarity with this topic, we offer some examples.

3.11.1 Use of *s*-parameters for series elements

Example 3.15

Calculate the *s*-parameters for the two-port network shown in Figure 3.30 for the case where $Z_0 = 50\ \Omega$.

Given: Network of Figure 3.30 with $Z_0 = Z_L = 50\ \Omega$.
Required: *s*-parameters.

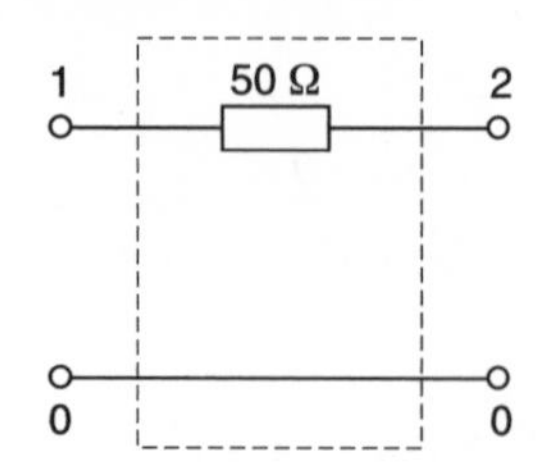

Fig. 3.30 Resistive network

Solutions

s_{11}: Terminate the output in Z_0 and determine Γ_1 at the input. See Figure 3.31(a).
By inspection:

$$Z_1 = 50\ \Omega + 50\ \Omega = 100\ \Omega$$

From Equation 3.36

$$s_{11} = \Gamma_1 = \frac{Z_1 - Z_0}{Z_1 + Z_0} = \frac{100 - 50}{100 + 50} = \frac{1}{3}$$

or

$$s_{11} = 0.333 \angle 0° \quad \text{or} \quad -9.551 \text{ dB} \angle 0°$$

s_{22}: Terminate the input with Z_0 and determine Γ_2 at the output. See Figure 3.31(b).
By inspection:

$$Z_2 = 50\ \Omega + 50\ \Omega = 100\ \Omega$$

From Equation 3.37

$$s_{22} = \Gamma_2 = \frac{Z_2 - Z_0}{Z_2 + Z_0} = \frac{100 - 50}{100 + 50} = \frac{1}{3}$$

or

$$s_{22} = 0.333 \angle 0° \text{ or } -9.551 \text{ dB} \angle 0°$$

Note: s_{11} and s_{22} are identical. This is what you would expect because the network is symmetrical.

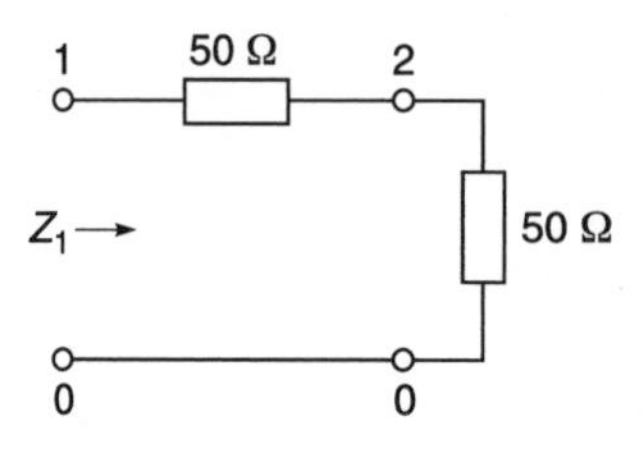

Fig. 3.31(a) Calculating s_{11}

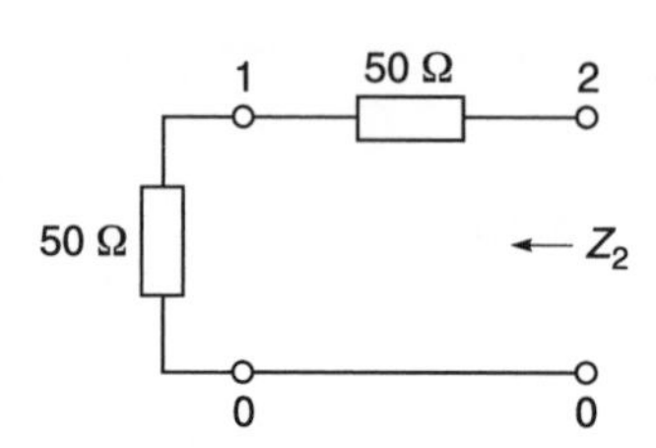

Fig. 3.31(b) Calculating s_{22}

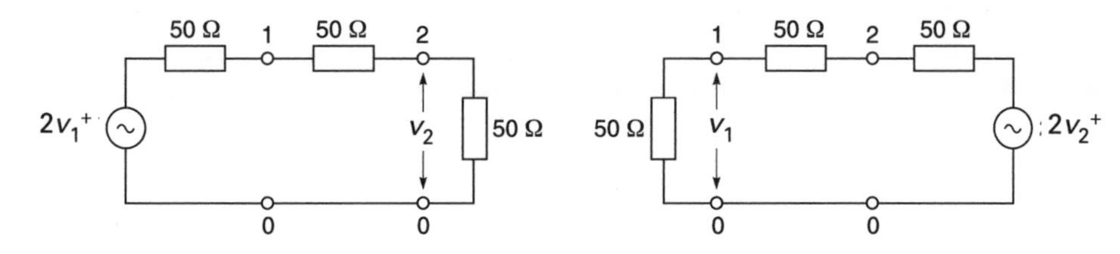

Fig. 3.31(c) Calculating s_{21}

Fig. 3.31(d) Calculating s_{12}

s_{21}: Drive port 1 with the 50 Ω generator and open-circuit voltage of $2V_1^+$. The multiplication factor of 2 is used for mathematical convenience as it ensures that the voltage incident on the matched load will be V_1^+. The superscript + sign following the voltage is meant to indicate that the voltage is **incident** on a particular port. Calculate voltage V_2. See Figure 3.31(c).

$$s_{21} = \frac{V_0}{V_1^+} = \frac{V_2}{V_1^+}$$

By inspection:

$$V_{2+} = \frac{50}{50 + (50 + 50)}(2V_1^+) = \frac{50}{150}(2V_1^+) = \frac{2}{3}V_1^+$$

Hence

$$s_{21} = \frac{V_0}{V_1^+} = \frac{V_2}{V_1^+} = \frac{2V_1^+}{3V_1^+} = \frac{2}{3}$$

or

$$s_{21} = 0.667 \angle 0° \text{ or } -3.517 \text{ dB} \angle 0°$$

s_{12}: See Figure 3.31(d). By inspection:

$$V_1 = \frac{50}{50 + (50 + 50)}(2V_2^+) = \frac{2}{3}V_2^+$$

$$s_{12} = \frac{V_1}{V_2^+} = \frac{2V_2^+}{3V_2^+} = \frac{2}{3}$$

$$s_{12} = 0.667\angle 0° \quad \text{or} \quad -3.517 \text{ dB} \angle 0°$$

Note: s_{12} and s_{21} are the same because the network is symmetrical.

Summing up

For *s*-parameters:

$$s_{11} = 0.333 \angle 0° \qquad s_{12} = 0.667 \angle 0°$$
$$s_{21} = 0.667 \angle 0° \qquad s_{22} = 0.333 \angle 0°$$

or in matrix notation

$$[s] = \begin{bmatrix} 0.333 \angle 0° & 0.667 \angle 0° \\ 0.667 \angle 0° & 0.333 \angle 0° \end{bmatrix}$$

Later on, you will see that this symmetry in a network often leads to considerable simplification in manipulating networks.

3.11.2 Use of *s*-parameters for shunt elements

Example 3.16

Calculate the *s*-parameters for the two port network shown in Figure 3.32 for the case where $Z_0 = 50\ \Omega$.

Given: Network of Figure 3.32 with $Z_0 = Z_L = 50\ \Omega$.
Required: *s*-parameters.

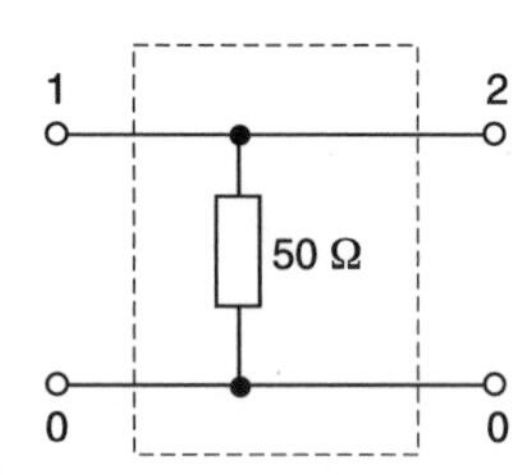

Fig. 3.32 Resistive network

Solutions

s_{11}: Terminate the output in Z_0 and determine Γ_1 at the input. See Figure 3.33(a). By inspection:

$$Z_1 = \frac{(50 \times 50)\Omega}{(50 + 50)\Omega} = 25\,\Omega$$

From Equation 3.36

$$S_{11} = \Gamma_1 = \frac{Z_1 - Z_0}{Z_1 + Z_0} = \frac{25 - 50}{25 + 50} = \frac{-1}{3}$$

or

$$s_{11} = 0.333 \angle 180° \text{ or } -9.551 \text{ dB} \angle 180°$$

s_{22}: Terminate the input with Z_0 and determine Γ_2 at the output. See Figure 3.33(b). By inspection:

$$Z_2 = \frac{(50 \times 50)\ \Omega}{(50 + 50)\ \Omega} = 25\ \Omega$$

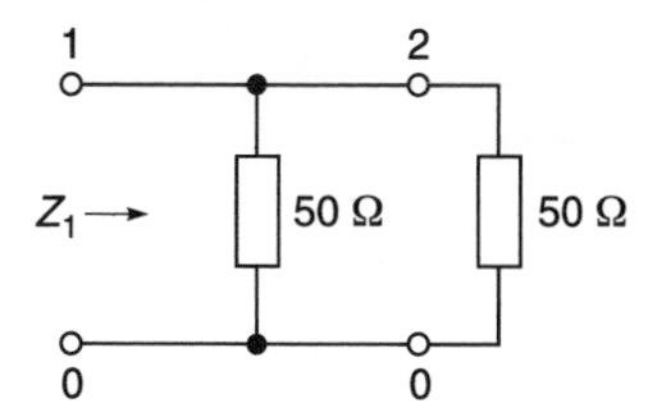

Fig. 3.33(a) Calculating s_{11}

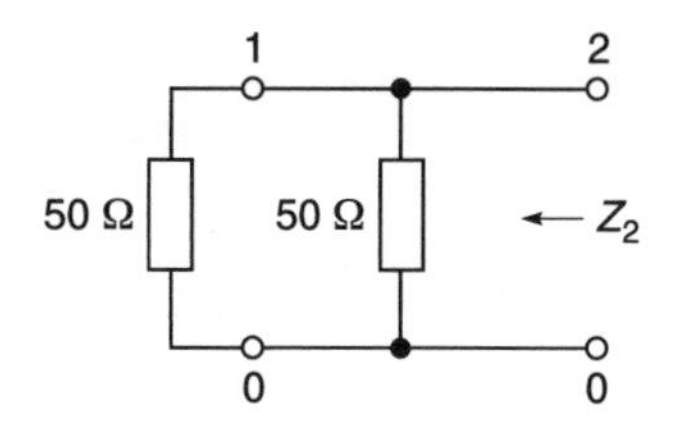

Fig. 3.33(b) Calculating s_{22}

From Equation 3.37

$$s_{22} = \rho_2 = \frac{Z_1 - Z_0}{Z_1 + Z_0} = \frac{25 - 50}{25 + 50} = \frac{-1}{3}$$

or

$$s_{22} = 0.333 \angle 180° \text{ or } -9.551 \text{ dB} \angle 180°$$

Note: s_{11} and s_{22} are identical. This is what you would expect because the network is symmetrical.

s_{21}: Drive port 1 with the 50 Ω generator and open-circuit voltage of $2V_1^+$. The multiplication factor of 2 is used for mathematical convenience as it ensures that the voltage incident on the matched load will be V_1^+. The superscript + sign following the voltage is meant to indicate that the voltage is **incident** on a particular port. Calculate voltage V_2. See Figure 3.33(c).

$$s_{21} = \frac{V_0}{V_1^+} = \frac{V_2}{V_1^+}$$

By inspection and bearing in mind that two 50 Ω resistors in parallel = 25 Ω

$$V_2 = \frac{25}{50 + 25}(2V_1^+) = \frac{25}{75}(2V_1^+) = \frac{2}{3}V_1^+$$

Hence

$$s_{21} = \frac{2V_1^+}{3V_1^+} = \frac{2}{3}$$

or

$$s_{21} = 0.667 \angle 0° \quad \text{or} \quad -3.517 \text{ dB} \angle 0°$$

s_{12}: See Figure 3.33(d). By inspection, and bearing in mind that two 50 Ω resistors in parallel = 25 Ω

$$V_1 = \frac{25}{50 + 25}(2V_2^+) = \frac{2}{3}V_2^+$$

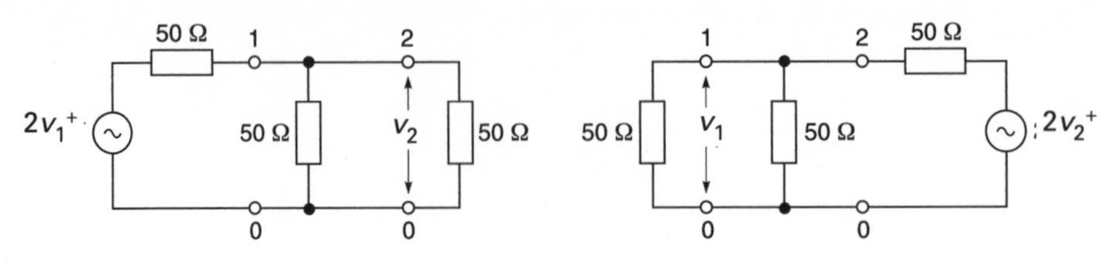

Fig. 3.33(c) Calculating s_{21}

Fig. 3.33(d) Calculating s_{12}

$$s_{12} = \frac{V_1}{V_2^+} = \frac{2V_2^+}{3V_2^+} = \frac{2}{3}$$

or

$$s_{12} = 0.667\angle 0° \quad \text{or} \quad -3.517 \text{ dB} \angle 0°$$

Note: s_{12} and s_{22} are the same because the network is symmetrical.

Summing up
For s-parameters:

$$s_{11} = 0.333 \angle 180° \qquad s_{12} = 0.667 \angle 0°$$
$$s_{21} = 0.667 \angle 0° \qquad s_{22} = 0.333 \angle 180°$$

3.11.3 Use of *s*-parameters for series and shunt elements

Example 3.17

(a) Calculate the s-parameters for the two-port network shown in Figure 3.34 for the case where $Z_0 = 50\ \Omega$.
(b) Find the return loss at the input with $Z_L = Z_0$.
(c) Determine the insertion loss for the network when the generator and the termination are both 50 Ω.

Given: Network of Figure 3.34 with $Z_0 = Z_L$.
Required: (a) s-parameters, (b) return loss, (c) insertion loss.

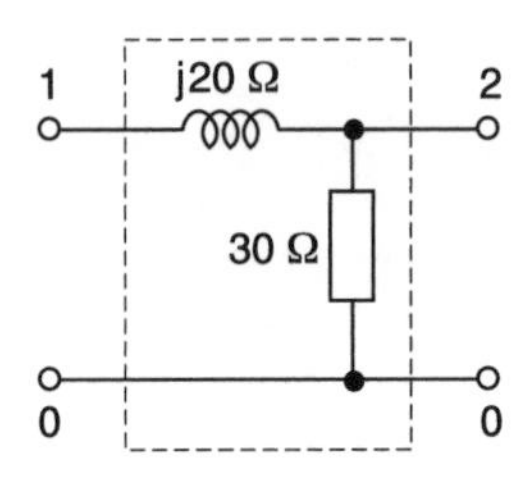

Fig. 3.34 Complex network

Solutions

(a) s_{11}: Terminate the output in Z_0 and determine ρ at the input. See Figure 3.35(a). By inspection the combined value of the 30 Ω and 50 Ω resistors is:

$$(30 \times 50)/(30 + 50) = 18.75\ \Omega$$

Hence

$$Z_1 = 18.75\ \Omega + j20\ \Omega$$

From Equation 3.36

$$s_{11} = \rho_1 = \frac{Z_1 - Z_0}{Z_1 + Z_0} = \frac{(18.75 + j20) - 50}{(18.75 + j20) + 50}$$
$$= \frac{-31.25 + j20}{68.75 + j20} = \frac{37.10\angle 147.38^\circ}{71.60\angle 16.22^\circ}$$

or

$$s_{11} = 0.518 \angle 131.16^\circ \quad \text{or} \quad -5.713 \text{ dB} \angle 131.16^\circ$$

s_{22}: Terminate the input with Z_0 and determine ρ at the output. See Figure 3.35(b). By inspection:

$$Z_2 = \frac{(50 + j20)(30)}{(50 + j20 + 30)} = \frac{1500 + j600}{80 + j20}$$
$$= \frac{1615.549\angle 21.801^\circ}{82.462\angle 14.036^\circ} = 19.591 \angle 7.765^\circ \quad \text{or} \quad 19.412 + j2.647$$

From Equation 3.37

$$s_{22} = \rho_2 = \frac{Z_2 - Z_0}{Z_2 + Z_0} = \frac{(19.412 + j2.647) - 50}{(19.412 + j2.647) + 50} = \frac{-30.588 + j2.647}{69.412 + j2.647}$$
$$= \frac{30.702 \angle 175.054^\circ}{69.462 \angle 2.184^\circ} = 0.442 \angle 172.870^\circ$$

or

$$s_{22} = 0.442 \angle 172.87^\circ \quad \text{or} \quad -7.092 \text{ dB} \angle 172.87^\circ$$

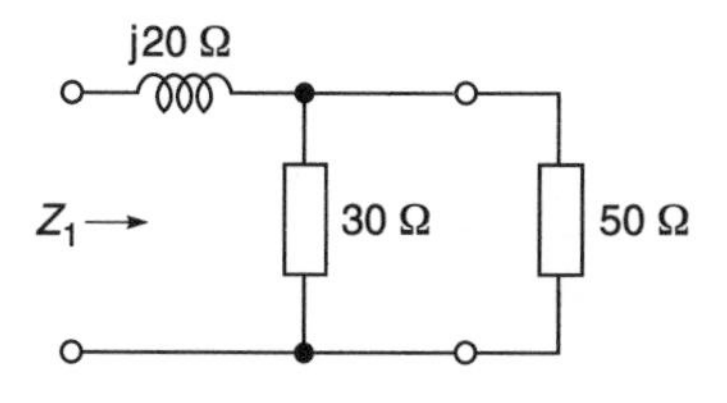

Fig. 3.35(a) Calculating s_{11}

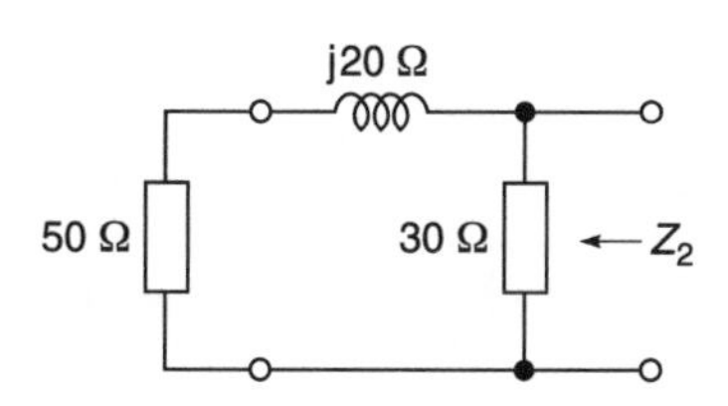

Fig. 3.35(b) Calculating s_{22}

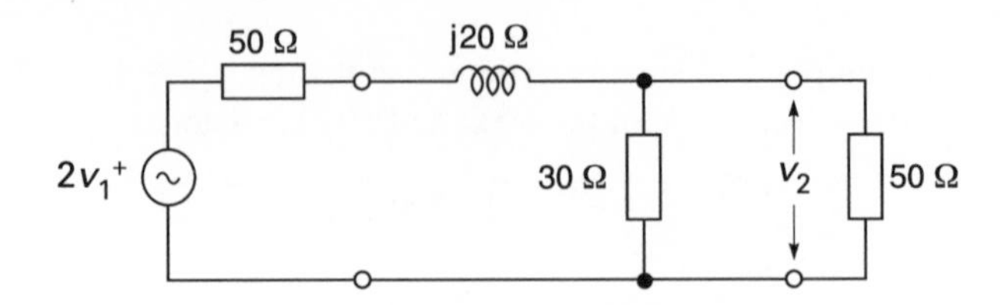

Fig. 3.35(c) Calculating s_{21}

s_{21}: Drive port 1 with the 50 Ω generator and open-circuit voltage of $2V_1^+$. The multiplication factor of 2 is used for mathematical convenience as it ensures that the voltage incident on the matched load will be V_1^+. The superscript + sign following the voltage is meant to indicate that the voltage is incident on a particular port. Measure voltage across V_2. See Figure 3.35(c).

$$s_{21} = \frac{V_0}{V_1^+} = \frac{V_2}{V_1^+}$$

By inspection:

$$V_2 = \frac{30 // 50}{30 // 50 + (50 + \text{j}20)}(2V_1^+) = \frac{18.75}{68.75 + \text{j}20}(2V_1^+)$$

$$= \frac{(18.75)(2V_1^+)}{71.60 \angle 16.22°} = (0.524 \angle -16.22°)V_1^+$$

Hence

$$s_{21} = \frac{(0.524 \angle -16.22°)V_1^+}{V_1^+}$$

or

$$s_{21} = 0.524 \angle -16.22° \quad \text{or} \quad -5.613 \text{ dB} \angle -16.22°$$

s_{12}: The Thévenin equivalent of the generator to the right of the plane 'c c' is obtained first. See Figures 3.35(d) and 3.35(e).

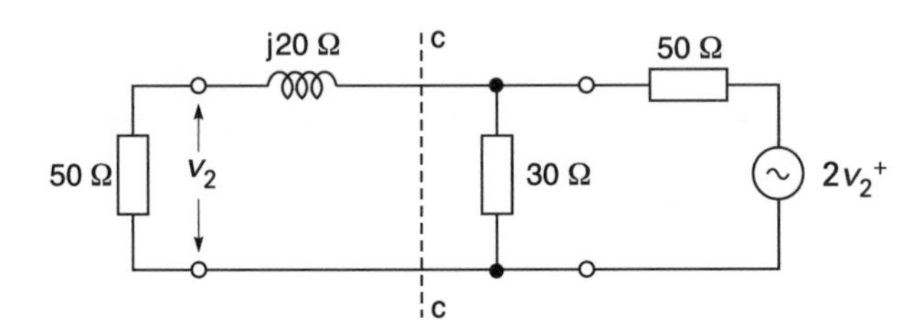

Fig. 3.35(d) Circuit before applying Thévenin's theorem

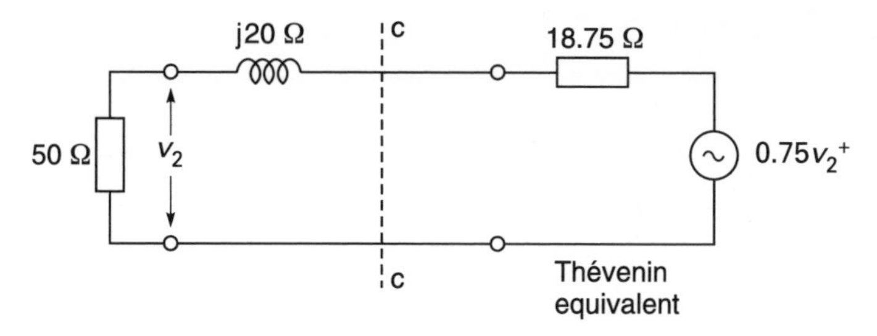

Fig. 3.35(e) Circuit after applying Thévenin's theorem

The Thévenin equivalent generator voltage is

$$\frac{30}{30+50}(2V_2^+) = 0.75V_2^+$$

The Thévenin equivalent internal resistance is

$$\frac{30 \times 50}{30+50} = 18.75\,\Omega$$

By inspection:

$$V_1 = \frac{50}{(18.75+50+\text{j}20)}(0.75V_2^+) = \frac{37.5}{71.60\angle 16.22°}V_2^+$$
$$= 0.524\angle -16.22°$$

$$s_{12} = \frac{V_1}{V_2^+} = \frac{(0.524\angle -16.22°)V_2^+}{V_2^+} = 0.524\angle -16.22°$$

or

$$s_{12} = 0.524\angle -16.22° \quad \text{or} \quad -5.613\text{ dB}\angle -16.22°$$

To sum up for *s*-parameters

(a) $s_{11} = 0.518\angle 131.16°$ $\quad s_{12} = 0.524\angle -16.22°$
$s_{21} = 0.524\angle -16.22°$ $\quad s_{22} = 0.442\angle 172.87°$

(b) From part (a), $\Gamma\angle\theta = 0.518\angle 131.16°$. Therefore

$$\text{return loss (dB)} = -20\log_{10}|0.518| = -20 \times (-0.286) = 5.713\text{ dB}$$

(c) The forward power gain of the network will be $|s_{21}|^2$.

$$|s_{21}|^2 = (0.524)^2 = 0.275$$

This represents a loss of $-10\log_{10} 0.275 = 5.61$ dB.

3.11.4 Use of *s*-parameters for active elements

Example 3.18

A 50 Ω microwave integrated circuit (MIC) amplifier has the following *s*-parameters:

$$s_{11} = 0.12 \angle -10° \qquad s_{12} = 0.002 \angle -78°$$
$$s_{21} = 9.8 \angle 160° \qquad s_{22} = 0.01 \angle -15°$$

Calculate: (a) input VSWR, (b) return loss, (c) forward insertion power gain and (d) reverse insertion power loss.

Given: $s_{11} = 0.12 \angle -10°$ $\quad s_{12} = 0.002 \angle -78°$
$s_{21} = 9.8 \angle 160°$ $\quad s_{22} = 0.01 \angle -15°$

Required: (a) Input VSWR, (b) return loss, (c) forward insertion power gain, (d) reverse insertion power loss.

Solution

(a) From Equation 2.38

$$\text{VSWR} = \frac{1+|\Gamma|}{1-|\Gamma|} = \frac{1+|s_{11}|}{1-|s_{11}|} = \frac{1+0.12}{1-0.12}$$
$$= 1.27$$

(b) Return loss (dB) = $-20 \log_{10} 0.12 = 18.42$ dB
(c) Forward insertion gain = $|s_{21}|^2 = (9.8)^2 = 96.04$ or
Forward insertion gain = $10 \log_{10} (9.8)^2$ dB = 19.83 dB
(d) Reverse insertion gain = $|s_{12}|^2 = (0.002)^2 = 4 \times 10^{-6}$ or
Reverse insertion gain = $10 \log_{10} (0.002)^2$ dB = –53.98 dB

The amplifier is virtually unilateral with (53.98 – 19.83) or 34.15 dB of output to input isolation.

3.12 Summary of scattering parameters

Section 3.10 has been devoted to the understanding of two-port scattering networks. Section 3.11 has been devoted to the use of two-port scattering networks. You should now be able to manipulate two-port networks skilfully and have the ability to change two-port parameters given in one parameter set to another parameter set.

An excellent understanding of scattering parameters is vitally important in microwave engineering because most data given by manufacturers are in terms of these parameters. In fact, you will find it difficult to proceed without a knowledge of *s*-parameters. This is the reason why we have provided you with several examples of *s*-parameter applications. The examples will be repeated using a software program called PUFF which has been supplied to you with this book. The purpose of these software exercises is to reinforce the concepts you have learnt and also to convince you that what we have been doing is correct.

Do not be unduly perturbed if you initially found *s*-parameters difficult to understand.

Understanding of s-parameters is slightly more difficult because they deal with waves, which is a very different concept from the steady-state voltages and currents which we have used in the past.

Finally, the information you have acquired is very important because most information in radio and microwave engineering is given in terms of scattering and admittance parameters. We have devoted particular attention to s-parameters because, later on, when you start analysing microwave components, filters, amplifiers, oscillators and measurements, you will be confronted with scattering parameters again and again. This is the reason why we have provided you with many examples on the use of scattering parameters.

4

PUFF software

4.1 Introduction

4.1.1 Aims

The aims of this chapter are threefold: (i) to help you install software program PUFF 2.1 on your computer; (ii) to use PUFF (software program supplied with this book) to verify the examples which you had worked with in the earlier chapters; and (iii) to give you confidence in using the software.

The software program that we introduce here is known as PUFF Version 2.1 It is a radio and microwave design and layout computer program developed by the California Institute of Technology (CalTech) Pasadena, USA. The program has been licensed to the publishers for use with this book. The conditions of the licence are that you use it for private study and experimental designs, and that copies of the program must not be made available to the general public and on the Internet, World Wide Web, etc. For ease of understanding, in all the discussions that follow, we will refer to PUFF Version 2.1 as PUFF.

Note: When installed on a computer with a colour monitor, PUFF displays a default colour screen. In the descriptions that follow, it is not feasible to display colour pictures, and every effort has been made to annotate the graphs. However, if you still have difficulty, run PUFF on your computer using the PUFF examples supplied on the PUFF disk. In fact, set up each example and check the results for yourself. It will give you confidence in using PUFF.

To install your software and to be able to manipulate the program, read Sections 4.2 to 4.10.

4.1.2 Objectives

The objectives of Part 4 are to teach you how to use PUFF for the following topics:

- amplifier designs
- calculating *s*-parameters for circuits
- calculating *s*-parameters for distributed components
- calculating *s*-parameters for lumped components
- circuit layout

- coupled circuits
- filter frequency response and matching
- line transformer matching and frequency response
- stub line matching and frequency response
- transistor matching and frequency response

4.2 CalTech's PUFF Version 2.1

CalTech's PUFF is a computer program that allows you to design a circuit using pre-selected lumped components (*R*, *L*, *C* and transistors) and/or distributed components (microstrip lines and striplines variants). The components may be arranged to form circuits. The layout of the circuit can be computer magnified, printed to provide a template for printed circuit layout and construction. The program provides facilities for calculating the scattering parameters (*s*-parameters) of the designed circuit layout with respect to frequency. The results can be read directly, plotted in Cartesian coordinates (*X*-*Y* plots) and in Smith chart impedance or admittance form. The Smith chart can also be used for matching purposes and for oscillator design.

4.3 Installation of PUFF

4.3.1 Introduction

The following notes are written with the express objective of explaining how to install PUFF on a personal computer. CalTech's PUFF is supplied on a compact disk. The program is designed to work on all PC, IBM PC/XT/AT and compatible computers. The minimum hardware requirements for the computer are a CD drive, a hard disk, 640K RAM and a 80286 or higher processor. If you do not have a CD drive, then ask a friend to copy all the PUFF folder files to a high density 3.5″ disk for transfer to your hard disk. The processor should have its matching coprocessor; if no coprocessor is present PUFF will run in its floating point mode and operate less rapidly. MS-DOS(R) versions 3.0 or higher should be used.

All printed outputs are directed to the parallel port LPT1. Printing graphic screens requires an Epson or a Laser compatible printer. PUFF provides six printer drivers for 'screen dumps'.

4.3.2 Installation

In the instructions which follow, I will use these conventions.

- **Bold** letters for what you have to type.
- Plus signs to indicate that two or more keys must be pressed together. For example, **ctrl + f** means that you keep the **ctrl** key pressed down while you type **f**. Similarly **alt + shift + g** means that you keep the **alt** and **shift** keys pressed down while you type **g**.
- The output from the screen is shown in *italics*.
- To simplify explanation I shall assume that your hard disk is c: and that you are installing from drive a:. If your hard disk drive is not c: then substitute the drive letter of your hard

disk drive whenever you see c:. If you are not installing from drive a:, then substitute the drive letter of your drive whenever you see a:.

4.3.3 Installing PUFF

This program is relatively small. It is not compressed. It can be run directly from the supplied disk initially. We do not recommend that you do this because if you damage your original disk, then you will not have a 'back-up' program. We recommend that you run PUFF from a directory on your hard disk. You install it simply by making a directory called PUFF on your hard disk and copy all programs including the sub-directory VGA_eps from your diskette to your PUFF directory. Details are given in the following sections.

Windows installation

1 Insert your PUFF disk in drive a:.
2 Start Windows in the usual manner and ensure that the *Program Manager* is displayed.
3 From *Program Manager*, click on the *main menu* and select *File Manager*.
4 From the *File* menu, select *create a directory*. In the Create Directory box type **c:puff.** Press the **RETURN** key.
5 Keep in the *File Manager* window. Click on disk *a* on the menu bars to display all the files on disk a:.
6 Select all items on disk a: and copy them all to the new directory PUFF which you have just created. You select *copy* from the *File* menu in *File Manager*. Press the **RETURN** key. When presented with the Copy box, type **c:puff**. Press the **RETURN** key.
7 All PUFF folder files from disk a: should be copied in your directory c:PUFF.
8 Installation is now complete.

DOS installation

1 Insert the PUFF disk in drive a:.
2 Create a directory PUFF using the DOS MKDIR command. At the DOS prompt *C:\>*, type **MKDIR PUFF** and press the **RETURN** key.
3 To copy all the files from the PUFF disk in drive a: into the directory C :\PUFF. type **COPY A: *.* C:\PUFF*.*** and press the **RETURN** key.
4 Check that all the files including subdirectory VGA_eps have been copied into your PUFF directory by typing **DIR C:\PUFF*.*** and press the **RETURN** key.

4.4 Running PUFF

4.4.1 Running under Windows

PUFF will run in a DOS window if accessed via File Manager in Windows 3.1 or My Computer or Windows Explorer or File Manager in Windows 95, 98 and ME. The ability of Windows to support PUFF is dependent on computer memory. If you attempt unsuccessfully the techniques below, then close down Windows and follow the DOS instructions.

Running under Windows 3.1

Double click on the c:PUFF directory in *File Manager*. Double click on PUFF.exe. PUFF should load into a DOS window.

An alternative to this is to double click on the MSDOS icon and follow the DOS instructions in Section 4.4.2.

Running under Windows 95/98/2000

My Computer

1 Double click on the *My Computer* icon on the Windows Desktop.
2 Double click on [c:].
3 Double click on the *PUFF* folder icon.
4 Double click on the PUFF program icon. PUFF should now run in a DOS window.

Windows Explorer

1 Select *Programs* from the *Start menu.*
2 Click on *Windows Explorer.*
3 Double click on the *c:PUFF* folder.
4 Double click on the *PUFF* application. PUFF should load into a DOS window.

An alternative to either of these methods is to select *Programs* from the *Start* menu, then choose the MS-DOS prompt and follow the DOS instructions.

4.4.2 DOS instructions

To run PUFF, you must first call up the directory PUFF and then type PUFF to start the program. Follow this procedure.

1 Type **CD C:\PUFF** and press the **RETURN** key. On the screen you should see *C:\PUFF>*,.
2 Type **PUFF** and press the **RETURN** key.

The program will start with an information text screen giving details of the program and your computer. This is essentially an information sheet. At the last line, it will say *Press ESC to leave the program or any other key to continue.* Type any key.

If your computer has a colour monitor then you will see Figure 4.1 in colour. Otherwise you will see it in monochrome. Throughout this block, I will use monochrome but where or when necessary, I will add annotation to enable easy identification of program parameters. The Figure 4.1 screen has been set by a template called *setup.puf* which is called up automatically when you start the program without defining a particular template. Templates are used to keep the program size small so that PUFF remains versatile and will run on personal computers. This template specifies the physical properties (size, thickness, dielectric constant, terminal connections) and the electrical parameters (dielectric constant, dielectric loss tangent, metal conductivity, etc.) of the board which will be used to construct the circuit. Frequency ranges and components are also defined by the template. Some of these properties can be changed directly within the program, others will have to be changed by modifying the template. We will show you how to change these properties later, but for now just accept the default template.

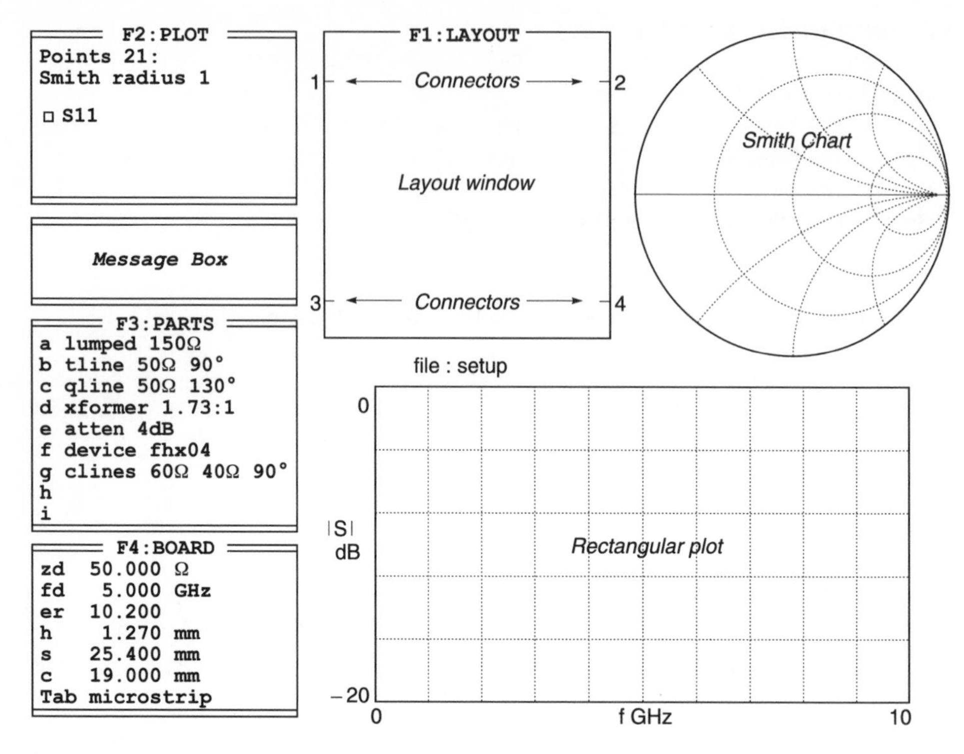

Fig. 4.1 Default screen for PUFF (words in italics have been added for easy identification)

We will now examine the elements of Figure 4.1.

1 Box **F1: LAYOUT** is a blank board on which we design our circuit. We can insert and join up many types of components on it.
2 Box **F2: PLOT** specifies the parameters which will be used for plotting the circuit. You can change these parameters and plot up to four *s*-parameters simultaneously. Within limits, you can also change the frequency and the number of plotting points.
3 **Message Box** between **F2** and **F3**. This box is an information box used for communications between the user and the program. It is normally blank when the program is started. It can also be used to yield **S11**, **S22**, **S33**, **S44** in terms of impedance or admittance.
4 Box **F3: PARTS** specifies the components which can be used in the design. At the moment, we have only shown seven components but you can specify up to 18 different components in the design. These can be resistors, capacitors, inductors, transistors, transformers, attenuators, lossless transmission lines (tlines), lossy transmission lines (qlines) and coupled lines (clines). If on your screen, you see Ü instead of Ω, then your machine has been configured differently to the way expected by the program. This will not affect your work but just make a note that Ü means Ω.
5 Box **F4: BOARD** displays some of the layout board properties. It tells you that the impedance (zd) of the connecting lines, its source impedance and load impedance are 50 Ω. It specifies the dielectric constant (er) of the board as 10.2 at 5 GHz. The

thickness (h) of the board is 1.27 mm, its size (s) is 25.4 mm square and the distance between connectors (c) is 19.00 mm. The board is configured for microstrip layout.

6 The **Smith chart** on the top right side shows you the *s*-parameters of the designed circuit. It can also be expanded and changed into admittance form.

7 The **Rectangular plot** (amplitude vs frequency graph) at the bottom of the screen plots *s*-parameters in dB against frequency. In future, I will simply call this plot the *rectangular plot.*

8 Pressing the **F10** key will give you a small *help screen.*

4.5 Examples

The use of the above properties will now be illustrated.

4.5.1 Example 4.1

Example 4.1 is relatively simple but it does help you gain confidence in using the program. We will start by constructing a 50 Ω transmission line on the layout board. This is shown in Figure 4.2. To carry out the above construction proceed as follows.

1 If necessary switch your computer on, call your PUFF directory, type **PUFF**, press the **RETURN** key and you should get Figure 4.1.

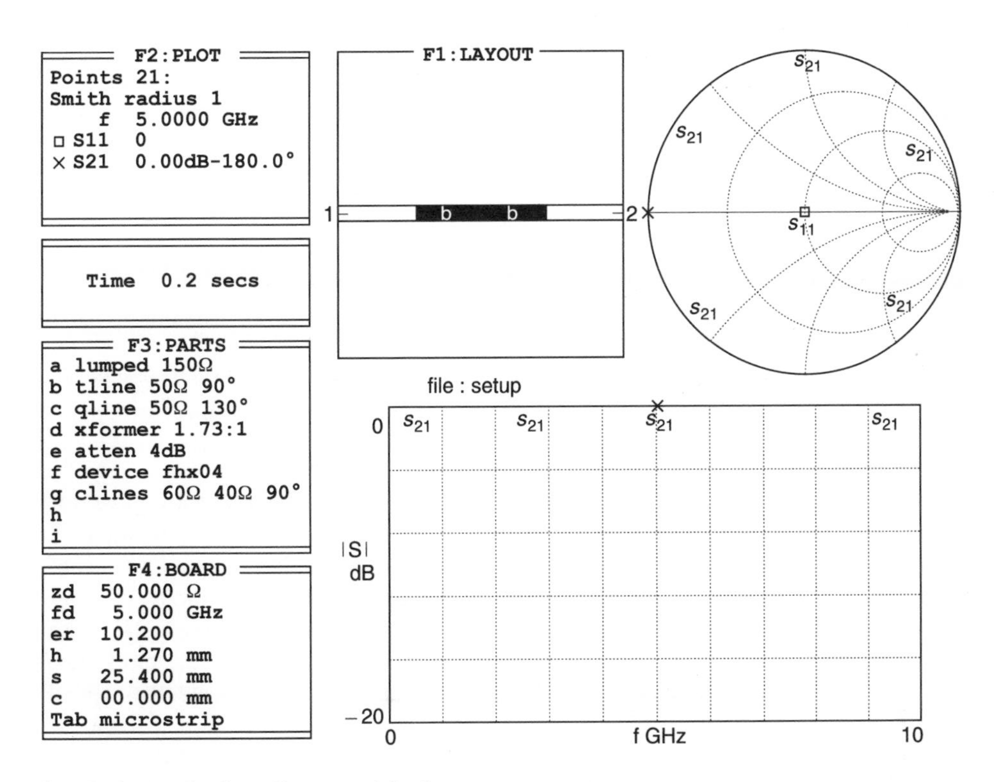

Fig. 4.2 Diagram showing a 50 Ω transmission line

2 Press the **F4** key. F4 will now be highlighted and you will be permitted to change values in the F4 box. An underlined cursor will appear on the zd line. Press the **down arrow** key (five times) until you get to the *c 19.00 mm* line. Throughout this set of instructions, keep looking at Figure 4.2 for guidance and confirmation of your actions.
3 Press the **right arrow** key until the cursor is under the *1* on the line, then type **00** (zero). Line *c* should now read *c 00.000 mm.* What you have effectively done is reduce the spacing between connectors 1 and 3, and 2 and 4 to zero. You will not see the effect as yet.
4 Press the **F1** key. F1 will now be highlighted and you will be permitted to lay components in the F1 box. An *X* will appear in the centre of the board. Note that there are now only two connecting terminals on the centre edges of the board. This is because of the action carried out in step 3. Notice also that in the F3 box, line *a* is highlighted. This means that you have selected a lumped 150 Ω resistor to be put on the board. We do not want this; we only want the 50 Ω transmission line on line *b* to be selected, so type **b**. Line *b tline 50 Ω 90°* will be highlighted.
5 Press the **left arrow** key and you will see the circuit of Figure 4.2 emerging. (If you make any mistake in carrying out these instructions, erase by retracing your step with the **shift** key pressed down. For example, if you want to erase what you have just done, press **shift+right arrow** keys. You can also erase the entire circuit by pressing **ctrl+e** keys.)
6 Type **1** and the tline will be joined to terminal 1.
7 Press **right arrow** key twice. See Figure 4.2.
8 Type **2** to join the right-hand section of line to terminal 2. Our circuit is now complete and we have put two sections of 50 transmission line *(b)* between a 50 Ω generator and load. We are now in a position to investigate its electrical properties.
9 Press the **F2** key. F2 will now be highlighted and you will be permitted to specify your measurement parameters in the F1 box.
10 Press the **down arrow** key three times. This will produce a new line *XS*.
11 Type **21** because we want to measure the parameters S11 and S21. Again refer to Figure 4.2 for guidance.
12 Type **p** to plot your parameters. You will now get Figure 4.2.
13 If you do not get this figure then repeat the above steps.
14 To save Figure 4.2, press the **F2** key. Type **Ctrl+s**. In the message box you will see *File to save?* Type **TX50** and press the **RETURN** key. Figure 4.2 is now saved under the file name *TX50*.

I will now explain to you the meaning of Figure 4.2. From the F2 box, you can see that we have measured S11 and S21 over 21 frequency points within the frequency range from 0–10 GHz. This frequency range is marked on the rectangular plot (amplitude vs frequency graph) on the bottom right-hand side of the screen. S21 is indicated on the graph and on the outer periphery of the Smith chart. (If you have a colour monitor, it is the blue line.) S11 is indicated on the centre of the Smith chart. *S*11 cannot be shown on the rectangular plot because its value is minus infinity and outside the range of the plot. PUFF reports any magnitude as small as –100 dB as zero and any magnitude greater than 100 dB as zero.

You can also obtain the values of S11 and S21 at discrete frequency points by referring to the F2 box. At the moment, it is showing that at 5 GHz, S11 = 0 and S21 = 0 dB (ratio of 1). This is expected because the reflection coefficient (S11) is zero and as tlines are considered as lossless lines in the program, the gain (S21) is also zero.

You can also check the input impedance of the line by moving the cursor to the S11 line and typing = . The actual value of the input line is shown on the Message box as $R_s = 50$ and $X_s = 0$. This tells you that the input impedance of the line is 50 Ω.

You should still be in the F2 mode. Press the **PageUp** key and watch the changes of frequency, S11, and S21 in the F2 box. In addition the symbols for S11 and X for S21 also move on the Smith chart and the rectangular plot. Press this again and watch the same movement. To lower the frequency, press the **PageDown** key and watch the same parameters noted previously. Note the **PageUp** and **PageDown** keys will only function when F2 is highlighted. The reason why you do not see drastic changes in the *S*-parameters is because from theory, we know that a lossless transmission with a characteristic impedance of 50 Ω inserted within a matched 50 system will only produce phase changes with frequency.

4.5.2 Smith chart expansion

While we are in F2 mode, it is also possible to expand the Smith chart to get a clearer view. **Press alt+s**. You will now get Figure 4.3. You can also press the **TAB** key to change the Smith chart form into an admittance display. See Figure 4.4.

When you have finished, use **alt+s** to toggle back to the normal display. Remember the

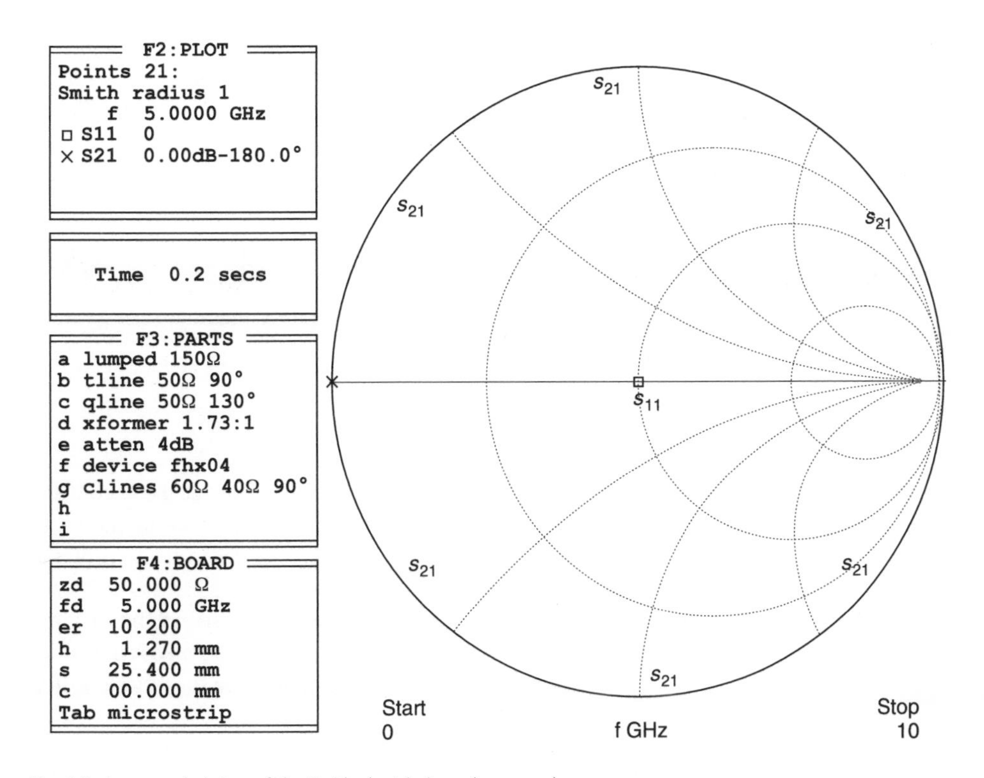

Fig. 4.3 An expanded view of the Smith chart in impedance mode

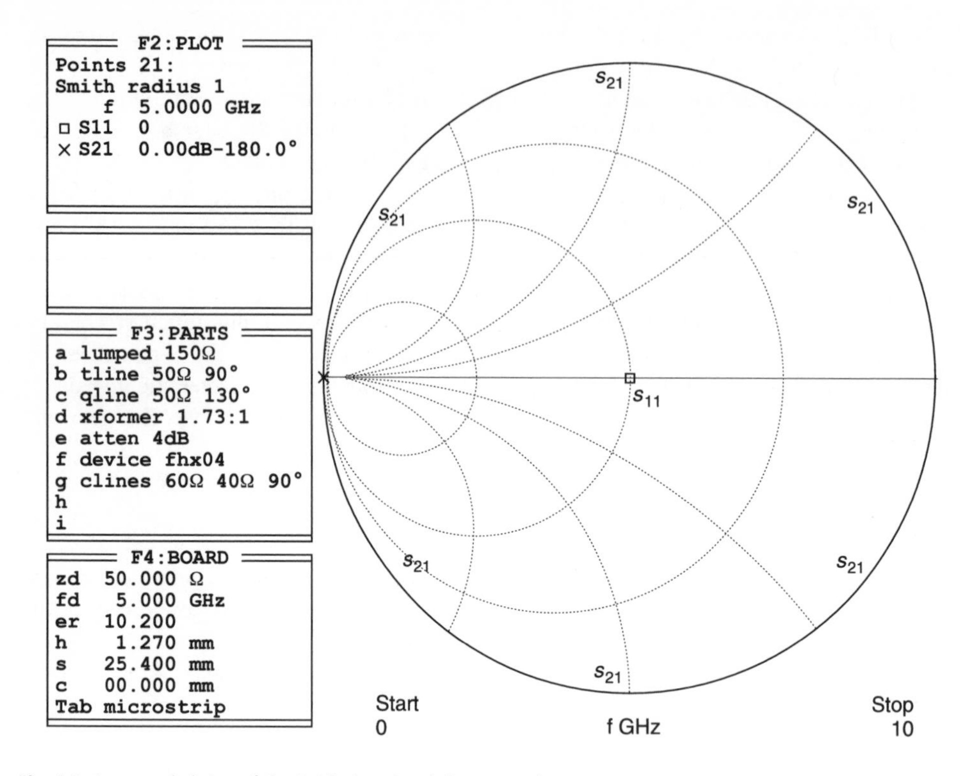

Fig. 4.4 An expanded view of the Smith chart in admittance mode

TAB and **alt+s** actions act as toggle switches and only perform these functions when the F2 box is active.

Another point that you should be aware of is that in the F2 mode, if you press the **TAB** key to change the Smith chart from impedance to admittance and if you move the cursor to S11 and type **=**, you will get the parallel values of the line rather than the series values mentioned earlier.

4.5.3 Printing and fabrication of artwork

If you have the proper printer connected, you should be able to print out the layout for photo-etching purposes to make the actual printed circuit board. The print-out shown in Figure 4.5 is five times the actual physical size of the layout. The magnification of the layout is chosen to reduced the 'jagged edges' (constant in a printer) to an insignificant width of the line. The print-out is then photographed, and reduced back to the original layout size. In the photographic reduction process, these jagged edges are also reduced by five, so that its effect on the true width of the line is less. The net result is that the line impedance is reproduced more accurately.

Note: Do not attempt to print at this time; it will be covered later.

Fig. 4.5 Print-out of microstrip line

4.5.4 Summary of Example 4.1

From Example 4.1 you have learnt how to use PUFF to:

1. change the position of the terminal connections (c in F4 mode);
2. select parts from the Parts Board (b in F3 mode);
3. layout and connect selected components to the board terminals; erase your layout circuit or components (**ctrl+e**, **shift+arrow**, in F1 mode);
4. plot and read the results of your circuit layout (S11 and S21 in F2 mode);
5. expand the *Z*- and *Y*-display of the Smith chart for more accurate readings (**alt+s** and **TAB** in F2 mode);
6. be able to obtain series and parallel values using S11 and the = sign;

7 save a file;
8 print out your layout for subsequent circuit construction.

Now try Example 4.1 on your own to see if you have remembered the procedures.

4.6 Bandpass filter

The electrical results of Example 4.1 have not been too interesting because it is commonly known that if a lossless 50 Ω line is inserted into a 50 Ω system, then little change, other than phase, takes place. However, it was deliberately chosen to produce minimum confusion in learning how to use the program and also to show that the program actually produces a well known and expected result.

4.6.1 Example 4.2

In Example 4.2, we will become a bit more adventurous and introduce some quarter-wave line short-circuited stubs across the junctions A, B, and C in the system. This case is shown in Figure 4.6. The procedure for Example 4.2 is almost identical to that for Example 4.1 except that you will have to remember (i) that to short-circuit a component to the ground plane, you must type the equal (=) sign at the end you want grounded; and (ii) to lay a component in the vertical direction, you must type either the **up arrow** key or the **down arrow** key.

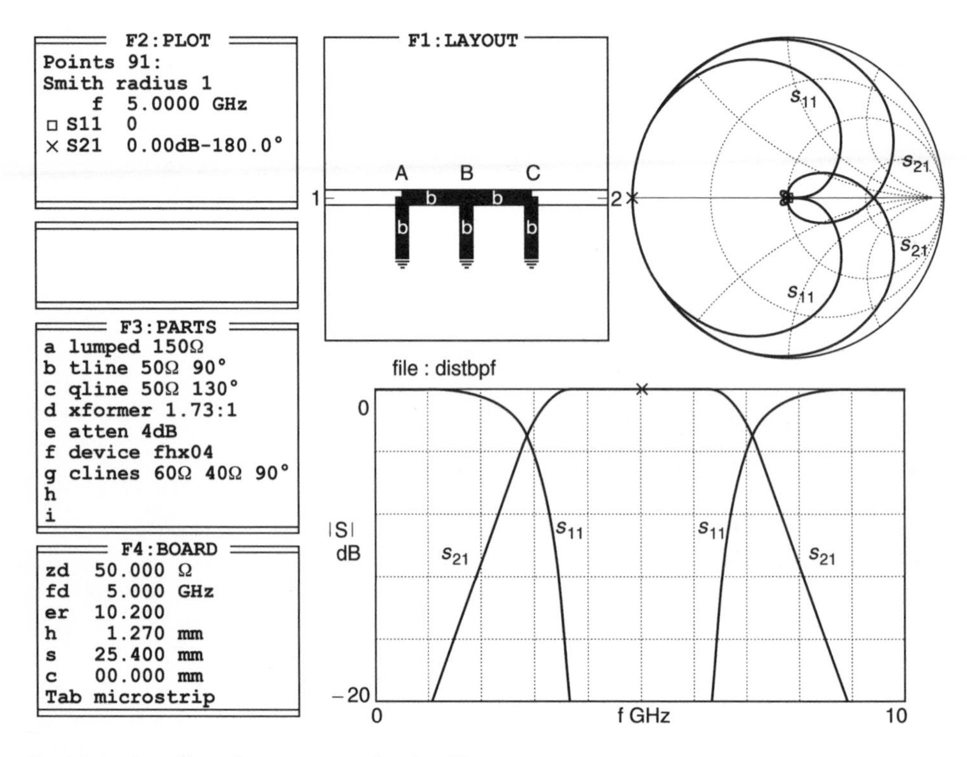

Fig. 4.6 Bandpass filter using quarter-wave lengths of line

In Figure 4.6, we have produced a bandpass filter centred at the centre frequency f_0 (5 GHz in this case) where the lines are exactly a quarter-wave long. You will no doubt remember from previous transmission line theory that the transformation of a $\lambda/4$ line is $Z_{in} = Z_0^2 / Z_{load}$. At f_0, all three short-circuited lines produce an infinite impedance across junctions A, B, and C. This means that signal transmission is unimpaired at f_0, and you will get an identical result to that of Example 4.1, i.e. the 180° phase shift and, since tlines in this program are assumed to be lossless, you will also obtain zero attenuation.

When the frequency is not f_0, then the shorted stubs do not present infinite impedances[1] at the junctions. At $f < f_0$, the shorted stubs will be inductive and will shunt signal to ground. At $f > f_0$, the shorted stubs will be capacitive and shunt signal to ground. The result is shown in Figure 4.6.

I suggest that you try to reproduce Figure 4.6 on your own but do not despair if you do not succeed because the details are given below. However, here are a few hints which might prove useful before you begin.

Hint: From F1 use the **down arrow** keys when you want to lay a component downwards, an **up arrow** key when you want to move upwards and use the equal sign **(=)** key when you want to ground a component. Now carry out the relevant procedures of Example 4.2. If you are still unable to get Figure 4.6, then carry out the instructions given below but throughout this set of instructions, keep looking at Figure 4.6 for guidance and confirmation of your actions.

1 If necessary switch on your computer. Call your PUFF directory, type PUFF, press the **RETURN** key and you should get Figure 4.1.
2 Press the **F4** key. F4 will now be highlighted and you will be permitted to change values in the F4 box. An underlined cursor will appear on the zd line. Press the **down arrow** key (five times) until you get to the *c 19.00 mm* line.
3 Press the **right arrow** key until the cursor is under the *1* on the line, type **00** (zero). Line *c* should now read *c 00.000 mm* as in Figure 4.6. What you have effectively done is reduce the space between connectors 1 and 3, and 2 and 4 to zero. You will not see the effect as yet.
4 Press the **F1** key. F1 will now be highlighted and you will be permitted to lay components in the F1 box. An *X* will appear in the centre of the board. Note that there are now only two connecting terminals on the centre edges of the board. This is because of the action carried out in step 3. Notice also that in the F3 box, line *a* is highlighted. This means that you have selected a lumped 50 Ω resistor to be put on the board. We do not want this; we only want a 50 Ω transmission line 90° long to be inserted into the system which means that we want to use part *b* for our construction.
5 Press the **F3** key. Press the **down arrow** key. This will get us to line *b*. Confirm that this line reads *b tline 50 Ω 90°*. If this is not the case, then overtype the line to correct it.
6 Press the **F1** key to return to the Layout box. An *X* will appear in the centre of the board. If line *b* of F3 is not already selected, type **b** to select the 50 Ω 90° transmission line.

[1] Remember the expression for a short-circuited transmission line $Z_{in} = jZ_0 \tan(\beta l)$. When $\beta l = 90°$, Z_{in} = infinity; when $\beta l < 90°$, Z_{in} is inductive; when $\beta l > 90°$, Z_{in} is capacitive.

Press the **left arrow** key once, then the **down arrow** key followed by an = key. Press the **up arrow** key. This will get you back to the line junction A (not shown on computer screen), press the **1** key. Your construction should now look like the left-hand side of the layout board of Figure 4.6. (If you make any mistake in carrying out these instructions, erase by retracing your step with the **shift** key pressed down. For example, if you want to erase the horizontal transmission line, press **shift+right arrow** keys. You can also erase the entire circuit by pressing **ctrl+e** keys.)

7 Press the **right arrow** key once; it will now return your cursor to the centre of the board (see junction B of Figure 4.6). Press the **down arrow** key once and press the = key. This should now give you the centre of Figure 4.6.
8 Press the **up arrow** key once to return to the centre of the board, press the **right arrow** key once to get to junction C. See Figure 4.6. Press the **down arrow** key once and press the = key. Press the **up arrow** key once; follow by typing the **2** key. You should now get the complete construction of Figure 4.6.
9 Our circuit is now complete and we have put two $\lambda/4$ sections of 50 Ω transmission line sandwiched between a 50 Ω generator and a 50 Ω load. We also have three short-circuited $\lambda/4$ lines between junctions A, B, C and ground. We are now in a position to investigate the electrical properties of the filter.
10 Press the **F2** key. F2 will now be highlighted and you will be permitted to specify your measurement parameters in the F2 box.
11 Press the **down arrow** key three times. This will produce a new line *X S*.
12 Type **21** because we want to measure the parameters S11 and S21. Again refer to Figure 4.6 for guidance.
13 Type **p** to plot your parameters. You will now get the entire picture of Figure 4.6.
14 If you do not get this figure then repeat the above steps again.
15 To save Figure 4.6, press the **F2** key. Type **ctrl+s**. In the Message box you will see *File to save?* Type **distbpf** and press the **RETURN** key. Figure 4.6 is now saved under the file name *distbpf*.

4.6.2 Printing and fabrication of artwork

If you have the proper printer connected, you should be able to print out the layout for photo-etching purposes to make the actual printed circuit board. The print-out shown in Figure 4.7 is five times the actual physical size of the layout. The magnification of the layout is chosen to reduced the 'jagged edges' (constant in a printer) to an insignificant width of the line. The print-out is then photographed and reduced back to the original layout size. In the photographic reduction process, these jagged edges are also reduced by five, so that its effect on the true width of the line is less. The net result is that the line impedance is reproduced more accurately. Note that in the print-out, there is no distinction between the width of the 50 Ω lines. Ground points are also not shown as these have to be drilled through the board. Do not attempt to print at this time; it will be covered later in the guide.

4.6.3 Summary of Example 4.2

In Example 4.2, you have:

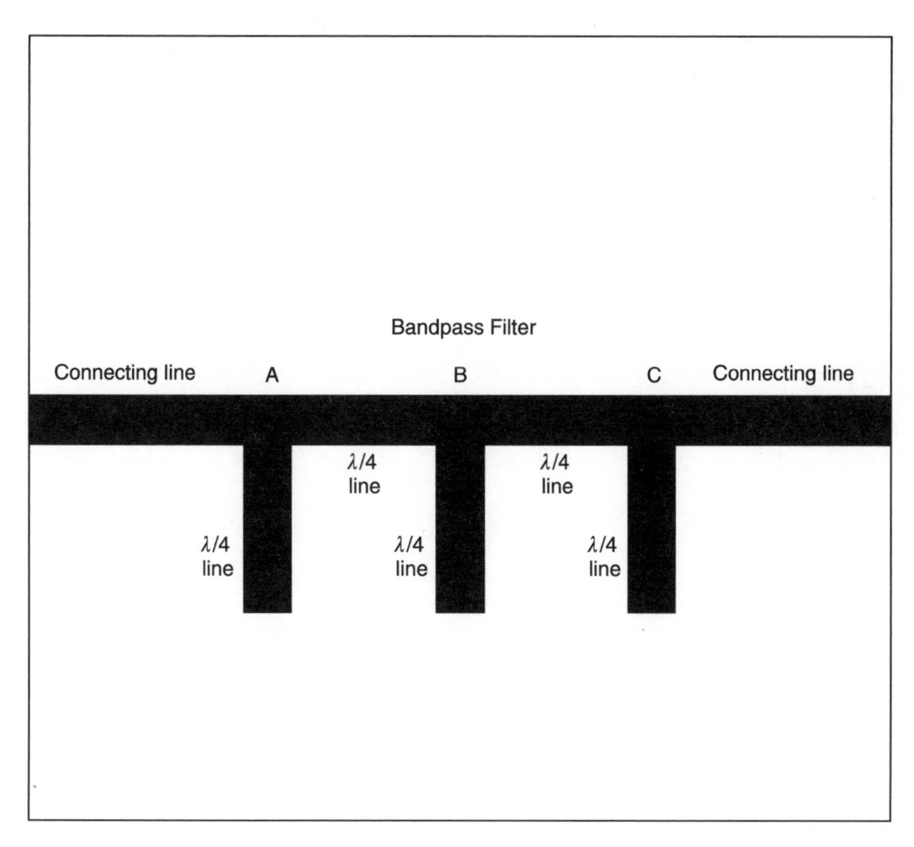

Fig. 4.7 Print-out of a bandpass filter in a 50 Ω system

1 learnt how a bandpass filter can be constructed from $\lambda/4$ lines;
2 reinforced your ideas of how to use PUFF;
3 read and interpreted the rectangular plot and Smith chart of the bandpass filter introduced in a 50 Ω system;
4 saved another file;
5 understood the artwork for Example 4.2.

Self test question 4.1

What do the S11 and S21 rectangular plots (amplitude vs frequency graph) tell you?

Answer. The S11 rectangular plot shows that a very good match exists in the passband of the filter and that poor match occurs outside the filter passband. You can check this by pressing the **F2** key to enter the plot mode, and by pressing the **PageUp** and **PageDown** keys to change the frequency to read S11 at 5 GHz, where the return loss tends toward infinity. You cannot see this on the rectangular plot because, for practical reasons, PUFF reports any magnitude as small as –100 dB as zero and any magnitude greater than 100 dB as zero. At frequencies 3 GHz and 7 GHz, S11 is only about –4 dB. The S21 plot shows the transmission plot loss as varying between 0 dB at 5 GHz to about 12 dB at 2 GHz and 8 GHz.

4.7 PUFF commands

At this stage, it is becoming increasingly difficult to remember all the commands that you have been shown. To facilitate your work, I have tabulated some commands in Tables 4.1 to 4.4.

Table 4.1

F1 box commands	Function
right arrow	to lay a previously selected component to the right
left arrow	to lay a previously selected component to the left
up arrow	to lay a previously selected component above
down arrow	to lay a previously selected component below
1	to connect a point to connector 1
2	to connect a point to connector 2
3	to connect a point to connector 3
4	to connect a point to connector 4
shift+right arrow	to erase a component inserted by the left arrow key
shift+left arrow	to erase a component inserted by the right arrow key
shift+up arrow	to erase a component inserted by the down arrow key
shift+down arrow	to erase a component inserted by the up arrow key
shift+e	to erase the entire circuit
shift+n	to move between nodes
shift + 1	to move selector to port 1
shift + 2	to move selector to port 2
shift + 3	to move selector to port 3
shift + 4	to move selector to port 4
=	to earth a point

Table 4.2

F2 box commands	Function
p	plot
ctrl+p	plot new modified parameters and keep previous plot
page up	move measurement up in frequency
page down	move measurement down in frequency
arrow to Points	retyping changes number of measurement points
arrow to Smith	retyping changes radius of Smith chart
arrow to S lines	to add additional S-parameter measurements
TAB	toggles Smith chart between Y- and Z-parameters
alt+s	to toggle an enlarged Smith chart
alt+s then TAB	toggles an enlarged Smith chart to Y- or Z-parameters
ctrl+a	prints board artwork on appropriate printer
ctrl+s	saves file
=	cursor on S_{xx} and Smith chart on impedance yields series resistance and reactance of the circuit at port $_{xx}$
=	cursor on S_{xx} and Smith chart on admittance yields parallel resistance and reactance of the circuit at port $_{xx}$

Table 4.3

F3 box commands	Function
up arrow	move up a line
down arrow	move down a line
right arrow	move a space to the right
left arrow	move a space to the left
insert key	allows the insertion of characters
alt+d	inserts the symbol for degrees
alt+o	inserts the symbol for ohm
ctrl+r	reads a file
j	inserts symbol for positive reactance
–j	inserts symbol for negative reactance
S	symbol for susceptance
mm	denotes component size in millimetres
M	megohms
+	used for series connections of components, e.g. R + jxx – jxx means resistance + inductance + capacitance in series
alt p (//)	used when you want components in parallel, e.g. R//jxx//–jxx means resistor, inductance and capacitance are in parallel
TAB	elongates F3 list to accommodate 18 different components

Table 4.4

F4 box commands	Function
zd XXX	allows change of system impedance
fd XXX	allow change of central frequency
er XXX	allows change of board dielectric constant
h XXX	changes thickness of substrate board
s XXX	changes size of substrate board
c XXX	changes distances between connectors
TAB	toggles layout between microstrip, stripline & Manhattan modes. Microstrip and stripline modes are scaled. Manhattan mode is not scaled but allows PUFF to be used for evaluation and plotting of circuits

4.8 Templates

4.8.1 Introduction

At the beginning, we told you that the reason why PUFF is a powerful but relatively small program is because it uses templates to store information. We also asked you to temporarily accept the default templates introduced earlier.

4.8.2 Setup templates

We are now in a position to show you the default template, *setup.puf*, which is automatically called up when you start up PUFF.[2]

[2] If you want to start up PUFF with a specified template called *another*, then you must specify it, when you start PUFF by typing **PUFF another**. PUFF will then start up using the template called *another*.

Default template: setup.puf

```
\b{oard} {.puf file for PUFF, version 2.1}
d    0    {display: 0 VGA or PUFF chooses, 1 EGA, 2 CGA, 3 One-color}
o    1    {artwork output format: 0 dot-matrix, 1 LaserJet, 2 HPGL file}
t    0    {type: 0 for microstrip, 1 for stripline, 2 for Manhattan}
zd   50.000   Ohms {normalizing impedance. 0<zd}
fd   5.000 GHz    {design frequency. 0<fd}
er   10.200     {dielectric constant. er>0}
h    1.270 mm    {dielectric thickness. h>0}
s    25.400 mm    {circuit-board side length. s>0}
c    00.000 mm    {connector separation. c>=0}
r    0.200 mm    {circuit resolution, r>0, use Um for micrometers}
a    0.000 mm    {artwork width correction.}
mt   0.010 mm    {metal thickness, use Um for micrometers.}
sr   0.000 Um    {metal surface roughness, use Um for micrometers.}
lt   0.0E+0000   {dielectric loss tangent.}
cd   5.8E+0007   {conductivity of metal in mhos/meter.}
p    5.000     {photographic reduction ratio. p<=203.2mm/s}
m    0.600     {mitering fraction. 0<=m<1}
\k{ey for plot window}
du 0 {upper dB-axis limit}
dl –20 {lower dB-axis limit}
fl 0 {lower frequency limit in GHZ. fl>=0}
fu 10 {upper frequency limit in GHZ. fu>fl}
pts 91 {number of points, positive integer}
sr 1  {Smith-chart radius. sr>0}
S  11  {subscripts must be 1, 2, 3, or 4}
S . . . 21
\p{arts window} {O = Ohms, D = degrees, U = micro, |=parallel}
lumped 150O
tline 50O 90D
qline 50O 130D
xformer 1.73:1
atten 4dB
device fhx04
clines 60O 40O 90D
{Blank at Part h }
{Blank at Part i }
{Blank at Part j }
{Blank at Part k }
{Blank at Part l }
{Blank at Part m }
{Blank at Part n }
{Blank at Part o }
{Blank at Part p }
{Blank at Part q }
{Blank at Part r }
```

As you can see for yourself, the template contains a lot of information. First and foremost, we shall examine the contents in the default template. Later we will create our own template.

In viewing the template, you will have to constantly refer to the first characters of a line to follow my discussion.

\b This line is for identification purposes to enable PUFF to know that it is a template.

d The number specifies your display screen. For a VGA screen use 0.

o The number specifies the type of printer which you will use to produce the artwork.[3] For a bubblejet or deskjet printer try using 1.

zd Specifies the normalising impedance of your system. Some antenna systems use 75 normalising impedance. If you use PUFF for other impedances then you must change the value accordingly.

fd Specifies the design frequency of your system. Change it if you want a different design frequency.

er The bulk dielectric constant of your substrate. Alumina and some Duroids have high dielectric constants. The dielectric constant determines the physical size of your distributed components. Lay your components only on the board with the dielectric constant which you will use in your construction.

h Dielectric thickness also affects the physical size of your distributed components. Ensure that your dielectric thickness is that which you will use in your construction.

s The square size of your board. The standard default size is 25.4 mm (1 inch square). This is a common size for experimental work but you can choose a size (within limits) to suit yourself. Remember though that your artwork will be several times larger and your printer might not be able to handle it.

c The ability to move connector spacing is important because it enables short leads to the connector.

r If the distance between components is less than the circuit resolution PUFF will connect the parts together.

a Artwork width correction. This is used when you want to alter line width from that calculated from the program.

mt The metal thickness is needed for calculating line losses.

sr Surface roughness is required for calculating line losses.

lt Dielectric loss tangent is also required for calculating line losses.

cd Metallic conductivity is also required for calculating line losses.

p The photographic reduction ratio used in printing the artwork. You can alter it for greater accuracy but bear in mind your maximum printing size.

m Mitreing function is used because when a tx line changes direction or when an open-circuited tx line ends, there is associated stray capacitance. To minimise these effects, corner joints and end joints are frequently 'thinned' or mitred.

\k Informs PUFF what limits to display on your Smith chart and rectangular plot.

du Sets your upper limit in dB (y axis).

[3] The reason why I told you not to attempt printing the artwork for the earlier examples is because I have no way of knowing the type (dot-matrix, Laser jet, HPGL file) of your printer. In my case, I have successfully used the Laserjet driver for my Hewlett-Packard bubble-jet printer. Some bubble-jet printers do not respond in this manner and you may be left with half a printed artwork, etc. However, as some of those files were saved, you can print them if you have a printer that will respond to one of the software drivers.

dl Sets your lower limit in dB (y axis).
fl Sets the lower limit in GHz (x axis).
fu Sets the upper limit in GHz (x axis).
pts Sets the number of points to which you want the circuit calculated and plotted. A larger number gives greater accuracy but takes a longer time. To make your display symmetrical about the centre frequency, you should choose an odd number of points.
sr Determines the Smith chart radius. In most cases you would want a factor of 1, but in oscillator design it is usual to use a Smith chart radius greater than 1.
S11 Determines the subscripts you want in your scattering parameter measurements. You can simultaneously measure up to four scattering parameters.
\p Informs PUFF that what follows relates to the F3 parts window and that O = ohms, D = degrees, U = micro, | = parallel.

lumped 150O = 150 Ω lumped resistor.
tline 50O 90D = 50 Ω lossless tx line 90° length at centre frequency
qline 50O 130D = 50 Ω lossy tx line 130° length at centre frequency
xformer 1.73:1 = transformer with a transformation ratio of 1.73:1
atten 4dB = attenuator with 4 dB attenuation
device fhx04 = a transistor called fhx04 whose *s*-parameters are enclosed
clines 60O 40O 90D = coupled lines, one of 60 Ω, one of 40 Ω, length = 90°
{Blank at Part h} etc. = no parts specified.

Note: It is very important that you realise that the template is written in ASCII text and can only be read in ASCII text by PUFF. Do not under any circumstances try to change the default template directly because if you mess it up, PUFF will *not* run. Follow this procedure:

1 always make a copy of the template before altering it;
2 alter the copied template with an ASCII text editor;
3 save your new template in ASCII TEXT.

Most DOS provide an ASCII text editor. For example, MSDOS supplies *edlin.com* for text editing. If you do not have one, then proceed very carefully with a word processor only if it can handle *textfiles*. Open the template in ASCII text. Make all your changes in ASCII text and save the file in ASCII text. Then and only then will PUFF read the file.

When you save a file, PUFF saves the file on a template and includes calculated answers and drawing instructions for the file. You can obtain a full version of it from your *distbpf .puf* file. I offer an abridged version here for those of you who cannot read the file.

Example 4.2 saved template (abridged)

```
\b{oard} {.puf file for PUFF, version 2.1}
d   0    {display: 0 VGA or PUFF chooses, 1 EGA, 2 CGA, 3 One-color}
o   1    {artwork output format: 0 dot-matrix, 1 LaserJet, 2 HPGL file}
t   0    {type: 0 for microstrip, 1 for stripline, 2 for Manhattan}
zd  50.000   Ohms {normalizing impedance. 0<zd}
fd  5.000 GHz    {design frequency. 0<fd}
er  10.200   {dielectric constant. er>0}
```

```
    h   1.270 mm   {dielectric thickness. h>0}
    s   25.400 mm   {circuit-board side length. s>0}
    c   19.000 mm   {connector separation. c>=0}
    r   0.200 mm   {circuit resolution, r>0, use Um for micrometers}
    a   0.000 mm   {artwork width correction.}
    mt   0.000 mm   {metal thickness, use Um for micrometers.}
    sr   0.000 Um   {metal surface roughness, use Um for micrometers.}
    lt   0.0E+0000   {dielectric loss tangent.}
    cd   5.8E+0007   {conductivity of metal in mhos/meter.}
    p   5.000   {photographic reduction ratio. p<=203.2mm/s}
    m   0.600   {mitering fraction. 0<=m<1}
    \k{ey for plot window}
    du 0 {upper dB-axis limit}
    dl –20 {lower dB-axis limit}
    fl 0   {lower frequency limit. fl>=0}
    fu 10   {upper frequency limit. fu>fl}
    pts 91   {number of points, positive integer}
    sr 1   {Smith-chart radius. sr>0}
    S   11   {subscripts must be 1, 2, 3, or 4}
    S   21
    \p{arts window} {O = Ohms, D = degrees, U = micro, |=parallel}
    lumped 150O
    tline 50O 90D
    qline 50O 130D
    xformer 1.73:1
    atten 4dB
    device fhx04
    clines 60O 40O 90D
    {Blank at Part h }
    . . . . . .(abridged)
    {Blank at Part r }
    \s{parameters}
f           S11                    S21
0.00000    1.00000     –180.0   2.4E–0013       0.0
0.11111    0.99996      177.5   8.7E–0003      87.5
0.22222    0.99985      175.0   1.8E–0002      85.0
0.33333    0.99965      172.5   2.7E–0002      82.5
0.44444    0.99936      169.9   3.6E–0002      79.9
0.55556    0.99896      167.3   4.6E–0002      77.3
. . . abridged
4.00000   2.3E–0002   154.7   0.99974     –115.3
4.11111   4.5E–0002   147.1   0.99897     –122.9
4.22222   5.9E–0002   139.6   0.99826     –130.4
4.33333   6.5E–0002   132.3   0.99790     –137.7
4.44444   6.4E–0002   125.2   0.99794     –144.8
4.55556   5.8E–0002   118.1   0.99832     –151.9
4.66667   4.7E–0002   111.0   0.99888     –159.0
```

```
4.77778   3.3E-0002  104.0  0.99944    -166.0
4.88889   1.7E-0002   97.0  0.99985    -173.0
5.00000   1.5E-0010   90.2  1.00000    -180.0
5.11111   1.7E-0002  -97.0  0.99985     173.0
5.22222   3.3E-0002 -104.0  0.99944     166.0
5.33333   4.7E-0002 -111.0  0.99888     159.0
5.44444   5.8E-0002 -118.1  0.99832     151.9
5.55556   6.4E-0002 -125.2  0.99794     144.8
5.66667   6.5E-0002 -132.3  0.99790     137.7
5.77778   5.9E-0002 -139.6  0.99826     130.4
5.88889   4.5E-0002 -147.1  0.99897     122.9
6.00000   2.3E-0002 -154.7  0.99974     115.3
. . . abridged
 9.00000  0.99588   -156.3  9.1E-0002   -66.3
 9.11111  0.99694   -159.2  7.8E-0002   -69.2
 9.22222  0.99778   -162.0  6.7E-0002   -72.0
 9.33333  0.99844   -164.7  5.6E-0002   -74.7
 9.44444  0.99896   -167.3  4.6E-0002   -77.3
 9.55556  0.99936   -169.9  3.6E-0002   -79.9
 9.66667  0.99965   -172.5  2.7E-0002   -82.5
 9.77778  0.99985   -175.0  1.8E-0002   -85.0
 9.88889  0.99996   -177.5  8.7E-0003   -87.5
10.00000  1.00000   -180.0  1.5E-0010   -89.9
   \c{ircuit} (instructions for layout)
   98   0 b
   203  2 left
   208  3 down
   61   3 =
   200  2 up
   49   2 1
   205  1 right
   208  4 down
   61   4 =
   200  1 up
   205  5 right
   208  6 down
   61   6 =
   200  5 up
   50   5 2
```

From the above two examples, you should now be able to specify a template for different displays.

Self test question 4.2

The measurement frequency range for Figure 4.1 is 0–10 GHz. If the measurement frequency range for Figure 4.1 is to be changed from 4.5 to 5.5 GHz, how would you modify the template to measure the frequency range 4.5–5.5 GHz?

Answer

1 Copy the setup template into an ASCII text editor.
2 Find the line beginning with *fl 0*. Change the value *0* to **4.5**.
3 Find the line beginning with *fu 10*. Change the value *10* to **5.5**. This will make the rectangular plot display its frequency axis as *4.5 to 5.5 GHZ*.
4 Save the file in ASCII text format. For ease of explanation, we will call the saved file *setup1*.
5 When you restart PUFF, type **PUFF setup1**. PUFF will now start using *setup1* as the default template.
6 Alternatively, you can start PUFF in the conventional manner. If necessary, press the **F3** key to enter the F3 box. Type **ctrl+r**. You will obtain the reply *File to read*. Type **setup1**.

Alternative method Press F2 key. Press 'down arrow' key until cursor is under 0. Overtype 4.5. Next, press 'down arrow' key until cursor is under 10. Overtype 5.5.

Self test question 4.3

How would you change the amplitude axis of PUFF's rectangular plot to display an amplitude of +20–40 dB?

Answer

1 Copy the setup template into an ASCII text editor.
2 Find the line beginning with *du 0*. Change the value *0* to **20**.
3 Find the line beginning with *dl –20*. Change the value *–20* to **–40**. This will make the rectangular plot display its amplitude axis as 20 to –40.
4 Save the file in ASCII text format. For ease of explanation, we will call the saved file *setup2*.
5 When you restart PUFF, type **PUFF setup2**. PUFF will now start using *setup2* as the default template.

Example 4.3

A template is required to make PUFF operate in the range 0–300 MHz and over a dynamic range of 0–40 dB. Make such a template and name it *setup300*.

Solution

1 Copy the setup template into an ASCII text editor.
2 Find the line beginning with *dl –20*. Move the cursor under the hyphen in *20*, type **–40**. This will make the rectangular plot display its amplitude axis as *0 to –40 dB*.
3 Find the line beginning with *fu 10*. Move the cursor under the *1* in *10*, type **.3** . This will make the rectangular plot display its frequency axis as *0 to 0.3 GHz*.
4 Save the file as **setup300** in ASCII text format.
5 Restart PUFF. Type **PUFF setup300**. PUFF will now start using *setup300* as the default template.

Alternative method Press F2 key. Press 'down arrow' key until cursor is under –20 position. Overtype –40. Next, press 'down arrow' key until cursor is under 10. Overtype 0.3.

4.9 Modification of transistor templates

PUFF comes with the template for a HEMT (high electron mobility transistor) called FHX04.dev. It is possible to use other transistor templates provided they are in ASCII text and provided they follow the exact layout for the FHX04.dev detailed below. Again, if you want to make your own template we suggest that you copy the transistor template below and modify it.

Template for FHX04.dev

{FHX04FA/LG Fujitsu HEMT (89/90), f=0 extrapolated; Vds=2V, Ids=10mA}

f	s11		s21		s12		s22	
0.0	1.000	0.0	4.375	180.0	0.000	0.0	0.625	0.0
1.0	0.982	–20.0	4.257	160.4	0.018	74.8	0.620	–15.2
2.0	0.952	–39.0	4.113	142.0	0.033	62.9	0.604	–28.9
3.0	0.910	–57.3	3.934	124.3	0.046	51.5	0.585	–42.4
4.0	0.863	–75.2	3.735	107.0	0.057	40.3	0.564	–55.8
5.0	0.809	–92.3	3.487	90.4	0.065	30.3	0.541	–69.2
6.0	0.760	–108.1	3.231	75.0	0.069	21.0	0.524	–82.0
7.0	0.727	–122.4	3.018	60.9	0.072	14.1	0.521	–93.6
8.0	0.701	–135.5	2.817	47.3	0.073	7.9	0.524	–104.7
9.0	0.678	–147.9	2.656	33.8	0.074	1.6	0.538	–115.4
10.0	0.653	–159.8	2.512	20.2	0.076	–4.0	0.552	–125.7
11.0	0.623	–171.1	2.367	7.1	0.076	–10.1	0.568	–136.4
12.0	0.601	178.5	2.245	–5.7	0.076	–15.9	0.587	–146.4
13.0	0.582	168.8	2.153	–18.4	0.076	–21.9	0.611	–156.2
14.0	0.564	160.2	2.065	–31.2	0.077	–28.6	0.644	–165.4
15.0	0.533	151.6	2.001	–44.5	0.079	–36.8	0.676	–174.8
16.0	0.500	142.8	1.938	–58.8	0.082	–48.5	0.707	174.2
17.0	0.461	134.3	1.884	–73.7	0.083	–61.7	0.733	163.6
18.0	0.424	126.6	1.817	–89.7	0.085	–77.9	0.758	150.9
19.0	0.385	121.7	1.708	–106.5	0.087	–97.2	0.783	139.1
20.0	0.347	119.9	1.613	–123.7	0.098	–119.9	0.793	126.6

Note that each S-parameter is denoted by an amplitude ratio and angle in degrees. For example at 1 GHz

$$S11 = 0.982 \angle -20.0^\circ \qquad S21 = 4.257 \angle 160.4^\circ$$
$$S12 = 0.018 \angle 74.8^\circ \qquad S22 = 0.620 \angle -15.2^\circ$$

Self test question 4.4

If you wanted to change the $S12$ parameter for HEMT FHX04 at 5 GHz to *0.063 31.4*, how would you do it?

Answer

1 Copy the transistor template (FHX04.dev) into an ASCII text editor.
2 Find the line beginning with *5.0*. It should read:

5.0 0.809 –92.3 3.487 90.4 0.065 30.3 0.541 –69.2

3 Change it to read:

5.0 0.809 –92.3 3.487 90.4 **0.063 31.4** *0.541 –69.2*

4 Save the file in ASCII text format. For ease of explanation, we will call the saved file *FHX04A.dev*.

You can then insert or change this part name in your chosen set-up template so that whenever you start PUFF, the device will be shown in the F3 box. Alternatively you can insert the part directly into the F3 box whenever you want to use the transistor.

4.10 Verification of some examples given in Chapters 2 and 3

4.10.1 Verification of microstrip line

We can now use PUFF to verify some of the conclusions reached in Chapters 2 and 3. In doing so, I will only quote the example number, and where appropriate its question and answer. I will then use PUFF to show that the conclusion is correct.

Example 2.3

Two microstrip lines are printed on the same dielectric substrate. One line has a wider centre strip than the other. Which line has the lower characteristic impedance? Assume that there is no coupling between the two lines.

Solution. The broader microstrip has the lower characteristic impedance. Using PUFF this is confirmed in Figure 4.8 where the broad microstrip has a characteristic impedance of 20 Ω and the narrow microstrip has a Z_0 of 90 Ω.

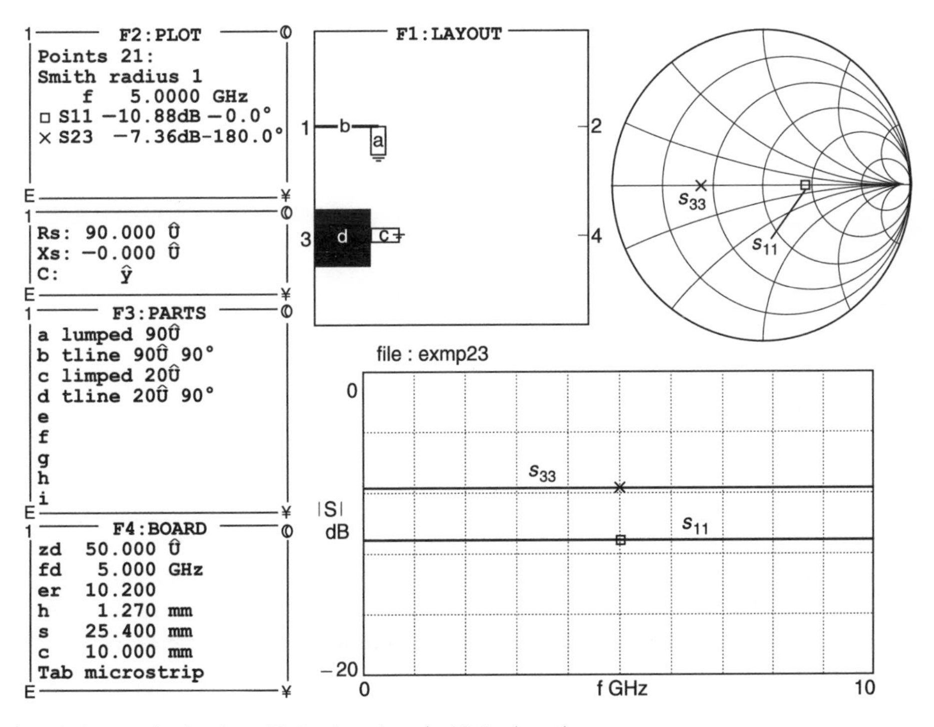

Fig. 4.8 PUFF results showing a 20 Ω microstrip and a 90 Ω microstrip

4.10.2 Verification of reflection coefficient

Example 2.7

Calculate the reflection coefficient for the case at 5 GHz where $Z_L = (80 - j10)\ \Omega$ and $Z_0 = 50\ \Omega$.

$$\Gamma = \frac{Z_L - Z_0}{Z_L + Z_0} = \frac{80 - j10 - 50}{80 - j10 + 50} = \frac{30 - j10}{130 - j10}$$

$$= \frac{31.62\angle -18.43^\circ}{130.38\angle -4.40^\circ} = 0.24\angle -14.03^\circ \quad \text{or} \quad -12.305\ \text{dB}\angle -14.03^\circ$$

Solution. Using Equation 2.24

When the above answer is written in dB, we get –12.3 dB ∠ –14.03°. Compare this answer with $S11$ in the F2 box of Figure 4.9.

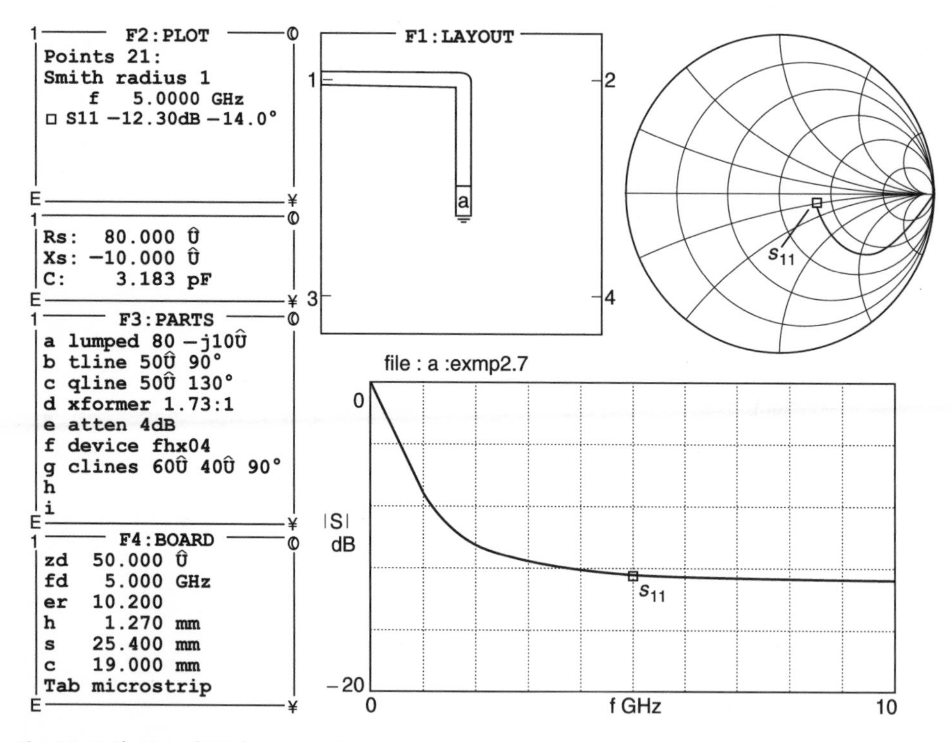

Fig. 4.9 Verification of S11 for Example 2.7

Example 2.8

Calculate the voltage reflection coefficients at the terminating end of a transmission line with a characteristic impedance of 50 Ω when it is terminated by (a) a 50 Ω termination, (b) an open-circuit termination, (c) a short-circuit termination, and (d) a 75 Ω termination.

Given: $Z_0 = 50\ \Omega$, Z_L = (a) 50 Ω, (b) open-circuit = ∞, (c) short-circuit = 0 Ω, (d) = 75 Ω.
Required: Γ_v for (a), (b), (c), (d).

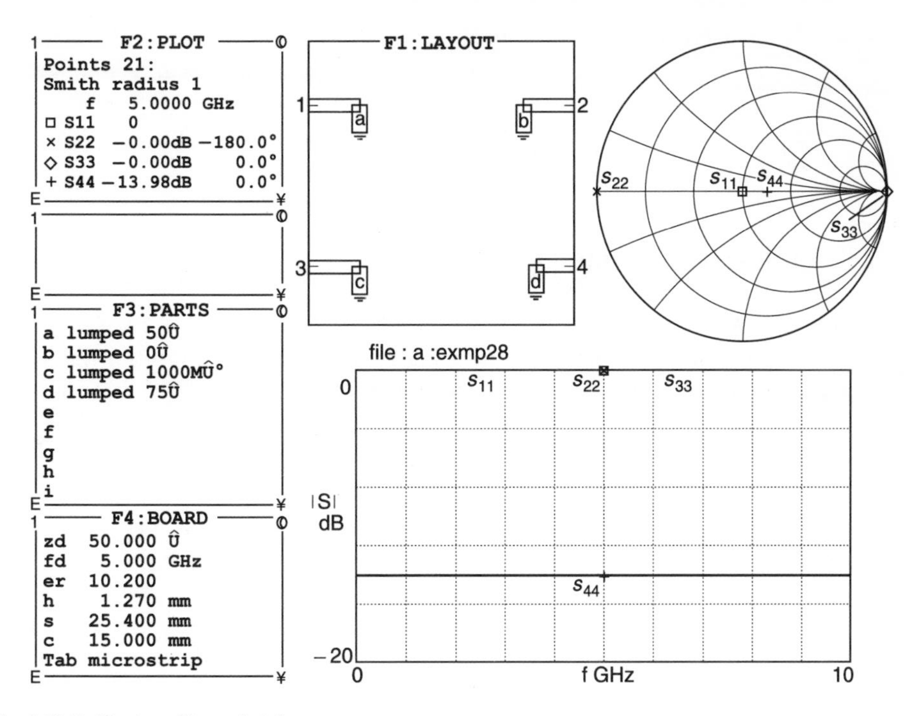

Fig. 4.10 Verification of Example 2.8

Solution. Use Equation 2.24.

(a) $Z_L = 50\underline{/0°}$

$$\Gamma_v = \frac{Z_L - Z_0}{Z_L + Z_0} = \frac{50\underline{/0°} - 50\underline{/0°}}{50\underline{/0°} + 50\underline{/0°}} = 0\underline{/0°} \;\; 0 \text{ dB}$$

(b) Z_L = open-circuit = $\infty\underline{/0°}$

$$\Gamma_v = \frac{Z_L - Z_0}{Z_L + Z_0} = \frac{\infty\underline{/0°} - 50\underline{/0°}}{\infty\underline{/0°} + 50\underline{/0°}} = 1\underline{/0°} \; = \; 0 \text{ dB} \angle 0°$$

(c) Z_L = short-circuit = $0\underline{/0°}$

$$\Gamma_v = \frac{Z_L - Z_0}{Z_L + Z_0} = \frac{0\underline{/0°} - 50\underline{/0°}}{0\underline{/0°} + 50\underline{/0°}} = -1\underline{/0°} \quad \text{or} \quad 1\underline{/180°} \; = \; 0 \text{ dB} \angle 180°$$

(d) $Z_L = 75\underline{/0°}$

$$\Gamma_v = \frac{Z_L - Z_0}{Z_L + Z_0} = \frac{75\underline{/0°} - 50\underline{/0°}}{75\underline{/0°} + 50\underline{/0°}} = 0.2\underline{/0°} = -13.98 \text{ dB} \angle 0°$$

In Figure 4.10, I have plotted case (a) as $S11$, case (b) as $S33$,case (c) as $S22$, case (d) as $S44$. For $S33$, I have used a 1000 MΩ resistor to represent a resistor of infinite ohms.

Alternatively, you could have used nothing to represent an open circuit. Most r.f. designers do not like open circuits because open circuits can pickup static charges which can destroy a circuit. It is far better to have some very high resistance so that static charges can be discharged to earth. Note how all the answers agree with the calculated ones.

4.10.3 Verification of input impedance

Example 2.13

A 377 Ω transmission line is terminated by a short circuit at one end. Its electrical length is $\lambda/7$. Calculate its input impedance at the other end.

Solution. Using Equation 2.55

$$Z_{in} = jZ_0 \tan\frac{2\pi l}{\lambda} = j377\tan\left[\frac{2\pi}{\lambda}\frac{\lambda}{7}\right] = j377 \times 1.254 = j472.8\ \Omega$$

Remembering that $\lambda/7 = 51.43°$ and using PUFF, we see from the Message box that Rs = 0 and Xs = 472.767 Ω (see Figure 4.11). This confirms the calculated value.

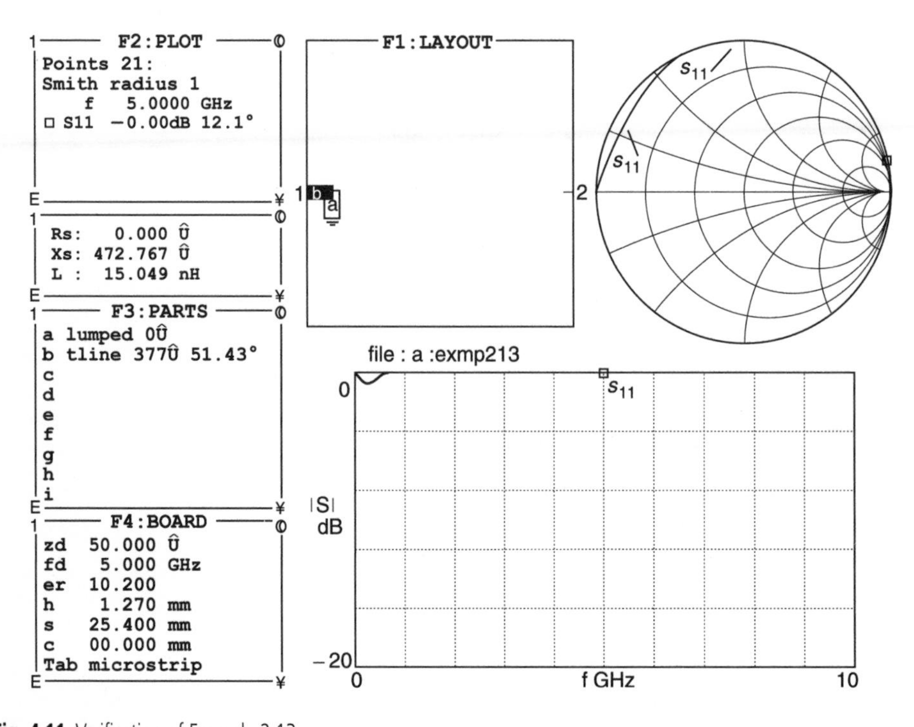

Fig. 4.11 Verification of Example 2.13

Example 2.14

A 75 Ω line is left unterminated with an open circuit at one end. Its electrical length is $\lambda/5$. Calculate its input impedance at the other end.

Solution. Using Equation 2.56

$$Z_{\text{in}} = \text{j}Z_0 \cot \frac{2\pi l}{\lambda} = -\text{j}75 \cot\left[\frac{2\pi}{\lambda}\frac{\lambda}{5}\right] = -\text{j}75 \times 0.325 = -\text{j}24.4\ \Omega$$

Bearing in mind that $\lambda/5 = 72°$ and using 1000 MΩ to simulate an open circuit, PUFF gives the answer in the Message box as –j24.369 Ω (see Figure 4.12).

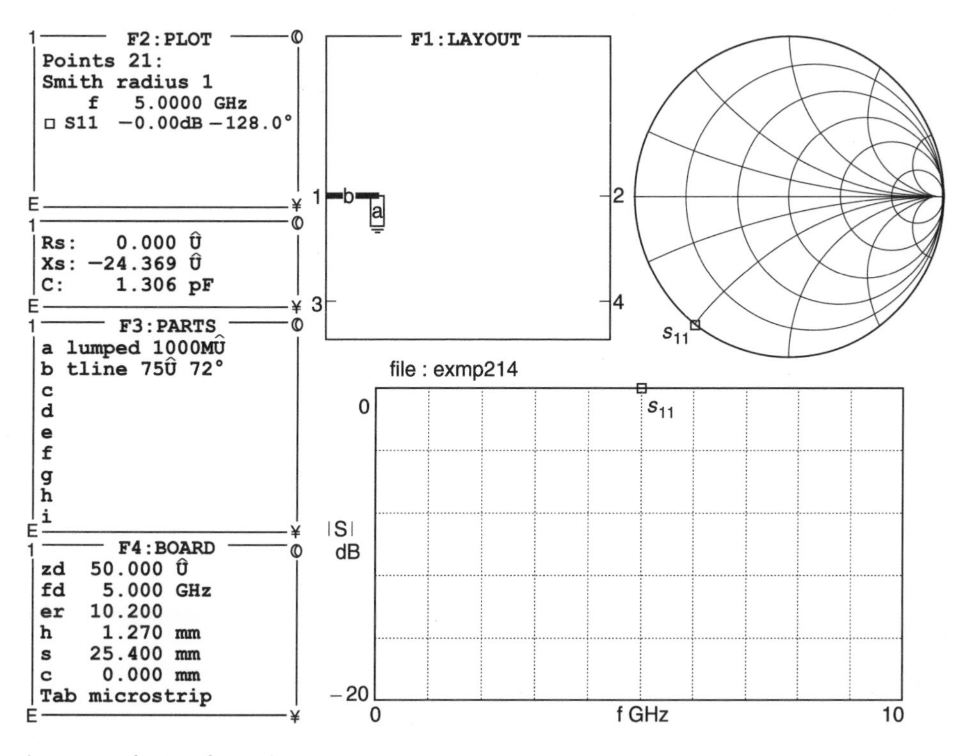

Fig. 4.12 Verification of Example 2.14

Example 2.15

A transmission line has a characteristic impedance (Z_0) of 90 Ω. Its electrical length is $\lambda/4$ and it is terminated by a load impedance (Z_L) of 20 Ω. Calculate the input impedance (Z_{in}) presented by the line.

Solution. Using Equation 2.57

$$Z_{\text{in}} = (90)^2/20 = 405\ \Omega$$

Using PUFF and reading the Message box, we get Rs = 405 Ω, Xs = 0 Ω (see Figure 4.13).

The above examples should now convince you that much of the transmission line theory covered in Chapter 2 has been proven.

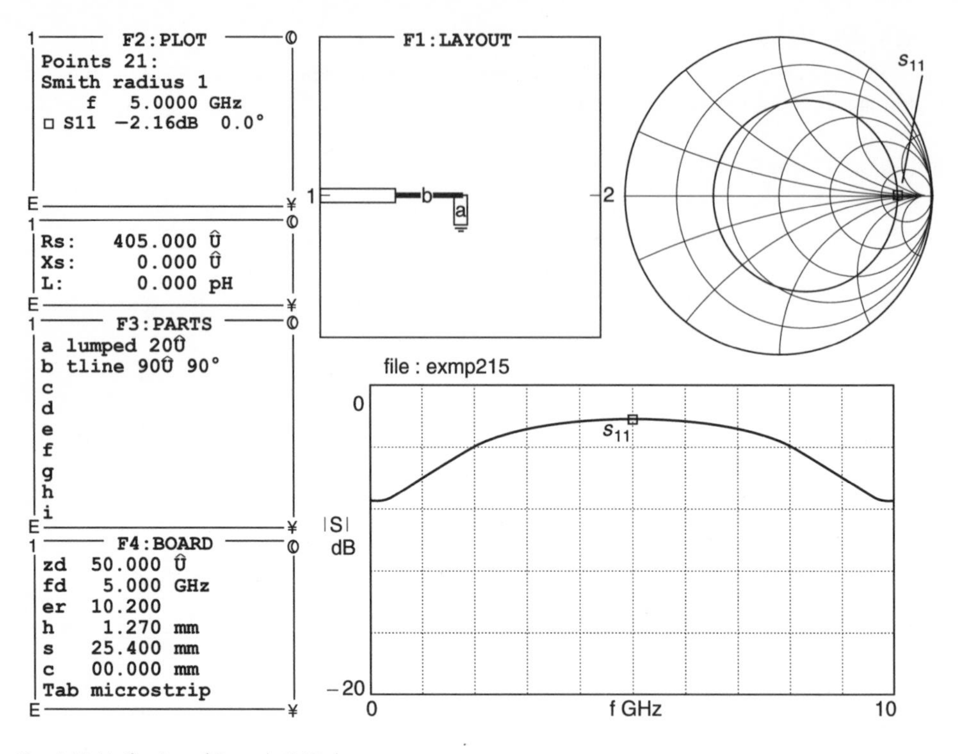

Fig. 4.13 Verification of Example 2.15

4.11 Using PUFF to evaluate couplers

In Section 2.14, we investigated the theory of two popular couplers. These are (a) the branch-line coupler and (b) the rat-race coupler.

4.11.1 Branch-line coupler

The theory for the branch-line coupler was covered in Section 2.14.1. For the branch-line coupler, I have used vertical lines with Z_0 of 50 Ω and horizontal lines with Z_0 of 35.55 Ω. In the F2 box (Figure 4.14), you can see that the match ($S11$ to a 50 Ω system) is excellent. $S21$ and $S41$ show that the input power from $S11$ is divided equally between the two ports but that there is a phase change as explained in the text. $S31$ shows you that there is very little or no transmission to port 3. All the above statements confirm the theory presented in Section 2.14.1.

4.11.2 Rat-race coupler

The theory for the rat-race coupler was covered in Section 2.14.3. Here PUFF confirms what we have discussed. In using PUFF, I have chosen Z_0 for the ring as 70.711 Ω and used a rectangle to represent the ring but all distances between the ports have been kept as before. This is shown in Figure 4.15.

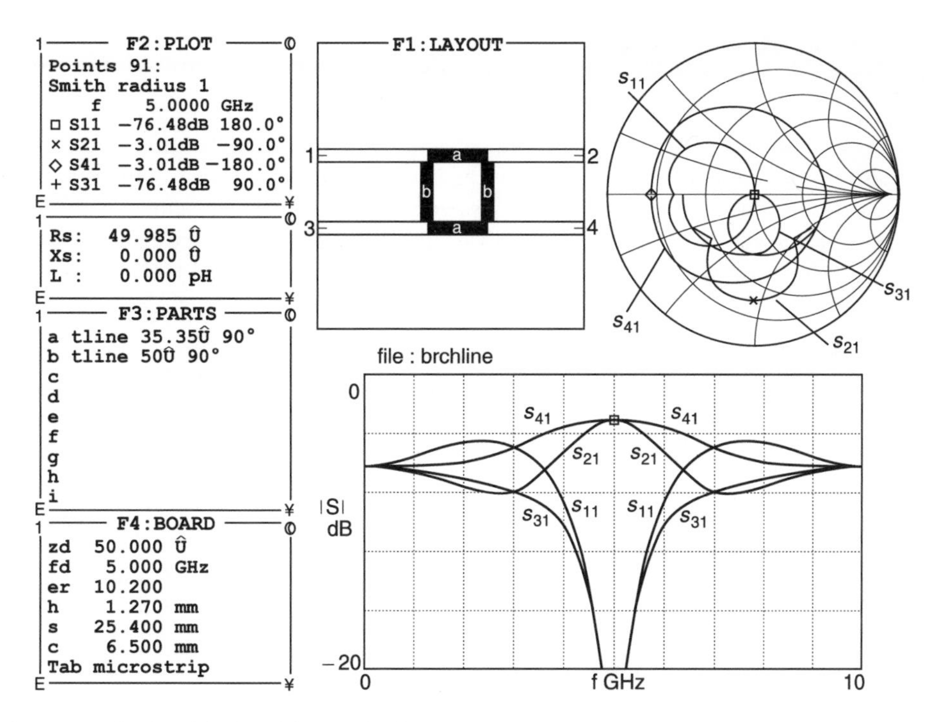

Fig. 4.14 Verification of the branch-line coupler theory

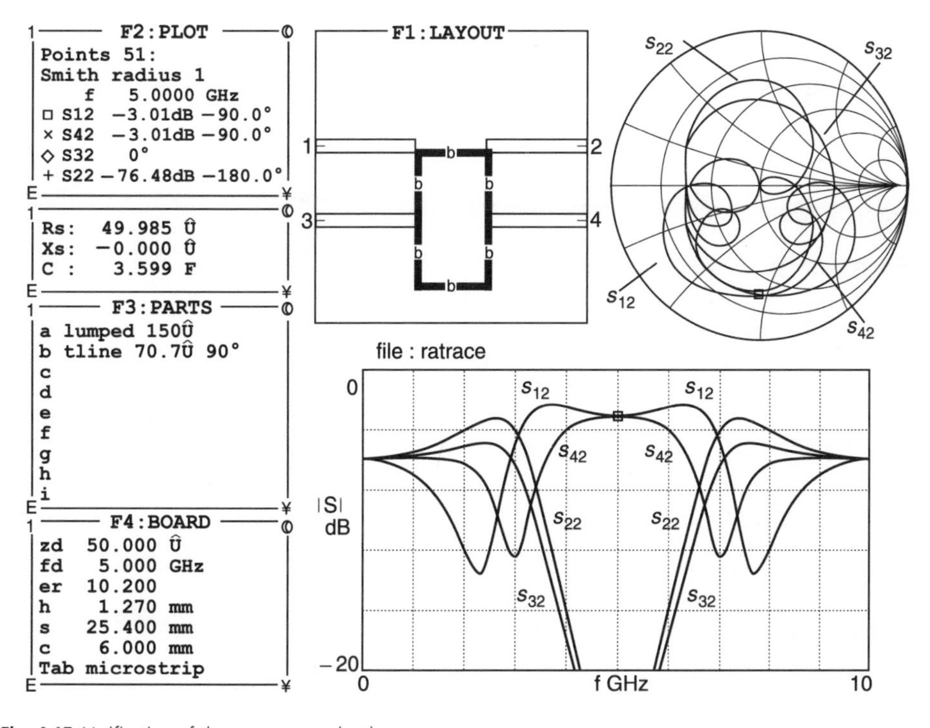

Fig. 4.15 Verification of the rat-race coupler theory

4.12 Verification of Smith chart applications

In Chapter 3, we used the Smith chart to derive admittances from impedances, calculate line input impedance, and solve matching networks. In this section, we show you how it can also be done with PUFF but note that the intermediate steps in the solutions are not given and sometimes it can be difficult to visualise what is actually happening in a circuit. We shall begin with Example 3.2. As usual we will simply use a numbered example, introduce its context and solution and then show how it can be solved with PUFF.

4.12.1 Admittance

Example 3.2

Use the Smith chart in Figure 3.6 to find the admittance of the impedance (0.8 – j1.6).

Solution. The admittance value is located at the point (0.25 + j0.5). You can verify this yourself by entering the point (0.8 – j1.6) in Figure 3.6. Measure the distance from your point to the chart centre and call this distance d. Draw a line of length $2d$ from your point through the centre of the chart. Read off the coordinates at the end of this line. You should now get (0.25 + j0.5) S.

With PUFF, we simply insert its value in the F3 box, and draw it in the F1 box. In the F2 box, press the **TAB** key to change the Smith chart into its admittance form. Move the cursor to $S11$, press **p** for plot and the equals sign (=) to read its value in the

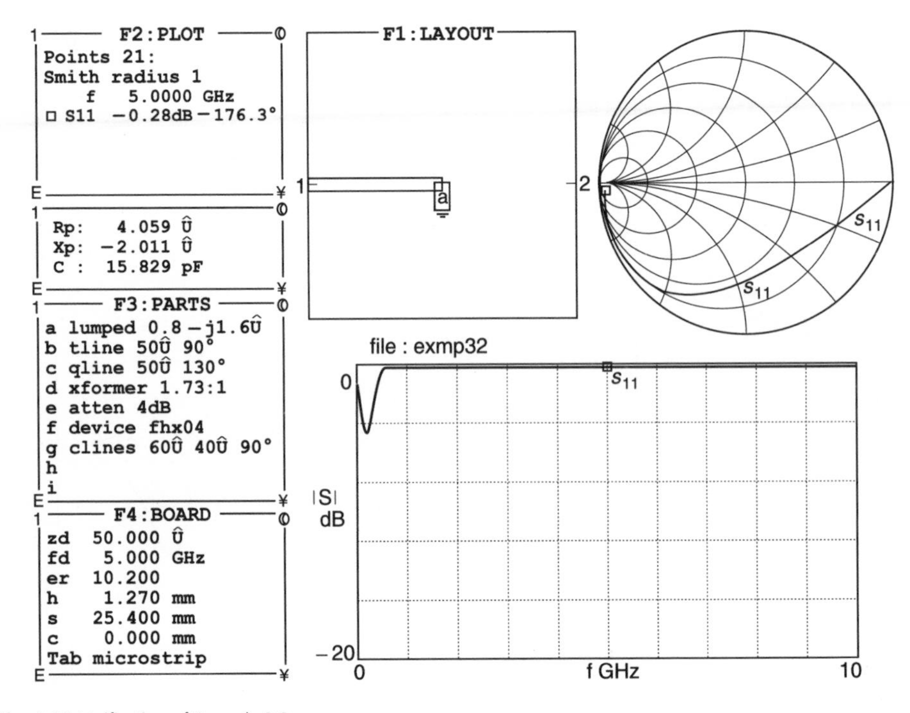

Fig. 4.16 Verification of Example 3.2

Message box. Note that Rp and Xp are given as parallel elements and its units are in ohms. However, remembering that (0.25 + j0.5) S is a combination of a conductance of 0.25 S to represent a resistor and j0.5 S to represent a capacitor, we simply take the reciprocal of each element to get the desired answer for each element. This is shown in Figure 4.16.

4.12.2 Verification of network impedances

For Example 3.5, we simply draw the network in the F1 box and seek its results in the F2 box and Message boxes.

Example 3.5

What is (a) the impedance and (b) the reflection coefficient looking into the network shown in Figure 3.12?

Solution. The solution to this problem was given in the annotation to Figure 3.14 as:

(a) impedance $Z = (0.206 + j0.635)\ \Omega$
(b) reflection coefficient $\Gamma = 0.746 \angle 113.58°$

The PUFF solutions (Figure 4.17) give:

(a) impedance $Z = (0.206 + j0.635)\ \Omega$
(b) reflection coefficient $\Gamma = -2.55$ dB $\angle 113.6°$ which is $0.746 \angle 113.6°$

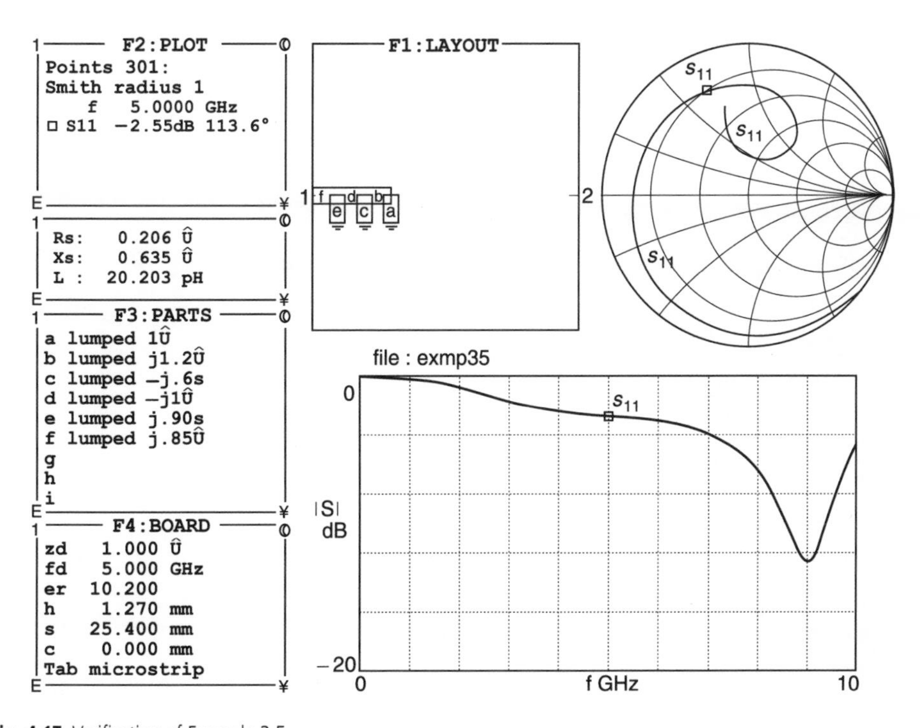

Fig. 4.17 Verification of Example 3.5

Note:

- In Figure 4.17 the drive impedance (zd) in PUFF has to be reduced to 1 Ω instead of the usual 50 Ω because the values in the circuit have been normalised.
- PUFF only gives the overall input impedance. If you had wanted intermediate values, then you would have to add one immittance at a time and read out its value.

4.12.3 Verification of input impedance of line

For this we will use Example 3.6.

Example 3.6

A transmission line with a characteristic impedance $Z_0 = 50\ \Omega$ is terminated with a load impedance of $Z_L = (40 - j80)\ \Omega$. What is its input impedance when the line is (a) 0.096λ, (b) 0.173λ, and (c) 0.206λ?

Solution. The answers calculated previously are:

(a) $(0.25 - j0.5)\ \Omega$ which after re-normalisation yields $(12.5 - j25)\ \Omega$
(b) $(0.20 - j0.0)\ \Omega$ which after re-normalisation yields $(10.0 - j0)\ \Omega$
(c) $(0.21 + j0.2)\ \Omega$ which after re-normalisation yields $(10.5 + j10)\ \Omega$

With PUFF, we obtain:

(a) $(12.489 - j24.947)\ \Omega$ – shown in Figure 4.18.
(b) $(9.897 + j0.022)\ \Omega$ – not shown in Figure 4.18.
(c) $(10.319 + j10.111)\ \Omega$ – not shown in Figure 4.18.

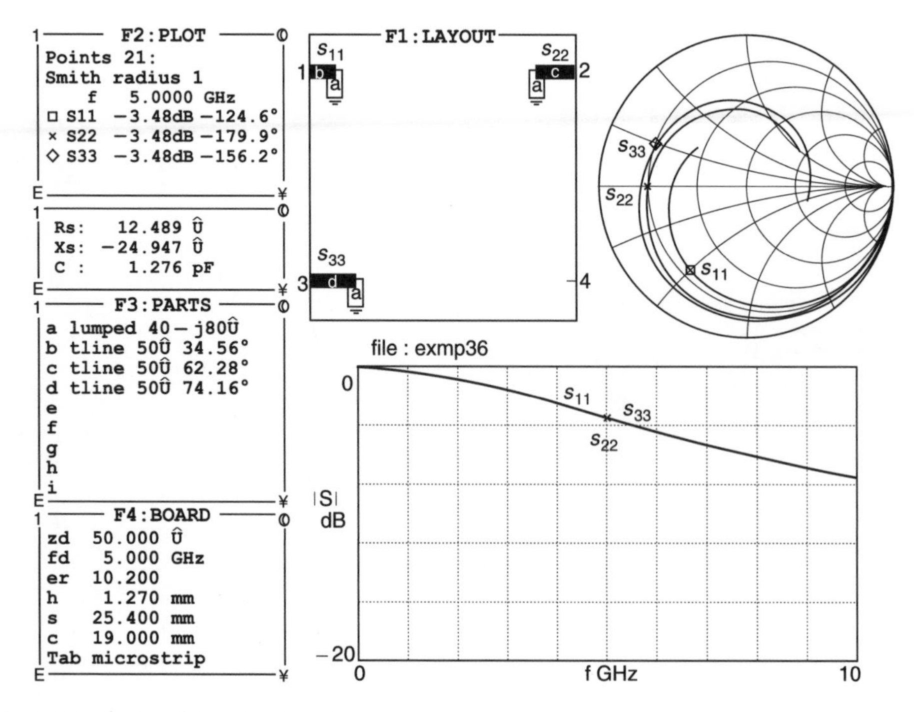

Fig. 4.18 Verification of Example 3.6 results

Items (b) and (c) were obtained from PUFF by moving the cursor in the F2 box to *S*22 and pressing the = key, and then to *S*33 and pressing the = key.

4.12.4 Verification of quarter-wave transformers

PUFF can be used to investigate the effect of quarter-wave line transformer matching. For this we will use Example 3.11.

Example 3.11

A source impedance of (50 + j0) is to be matched to a load of (100 + j0) over a frequency range of 600–1400 MHz. Match the source and load by using (a) one quarter-wave transformer, and (b) two quarter-wave transformers. Sketch a graph of the reflection coefficient against frequency.

Solution. Use Equation 2.57. For the one $\lambda/4$ transformer, we had previously calculated Z_{0t1} as 70.711 Ω. For the two $\lambda/4$ transformers, we had previously calculated Z_{0t1} as 60 Ω and Z_{0t2} as 84.85 Ω. We had also obtained the following table.

(GHz)	0.6	0.8	1.0	1.2	1.4
One $\lambda/4$ TX	– 13.83	– 19.28	> – 60	– 19.28	– 13.83
Two $\lambda/4$ TXs	18.81	– 32.09	38.69	32.09	– 18.81

Using PUFF, we show the two results in Figure 4.19.

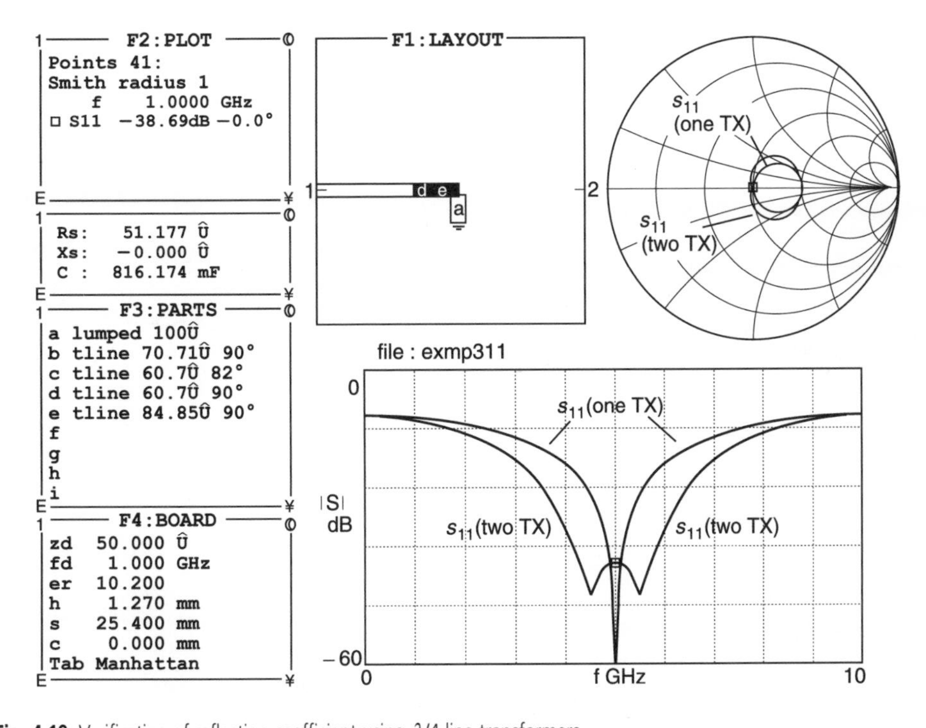

Fig. 4.19 Verification of reflection coefficient using $\lambda/4$ line transformers

4.13 Verification of stub matching

Stub matching is very important. PUFF can be used to match both passive and active networks. To give you an idea of how this is achieved, I will detail the matching of Example 3.12 which was carried out manually in Section 3.8.3.

Example 3.12

Use microstrip lines to match a series impedance of (40 – j80) Ω to 50 Ω at 1 GHz.

Solution

1 Switch on PUFF. Press the **F3** key. Type **ctrl+r**. The program will reply *File to read:?*
2 Type **match1**. Press the **RETURN** key. You will obtain Figure 4.20(a). Note that in the F3 box, we have:
   - the series impedance we wish to match, (40 – j80);
   - a transmission line (b) whose length is designated as *?50°*;
   - a transmission line (c) whose length is designated as *50°*;
   - note also that the rectangular plot x axis is marked in degrees and not frequency.

   The next step is to construct the circuit shown in Figure 4.20(b).
3 Press the **F1** key. Type **a**. Press **shift+right arrow** key seven times until the cursor is positioned to the right of the layout board.

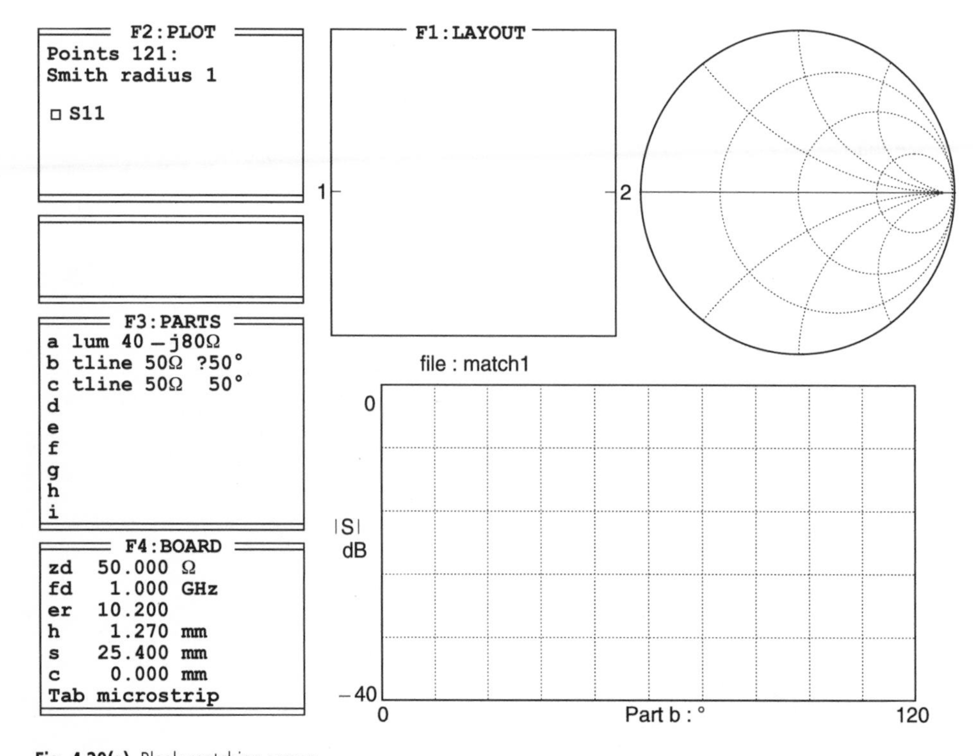

Fig. 4.20(a) Blank matching screen

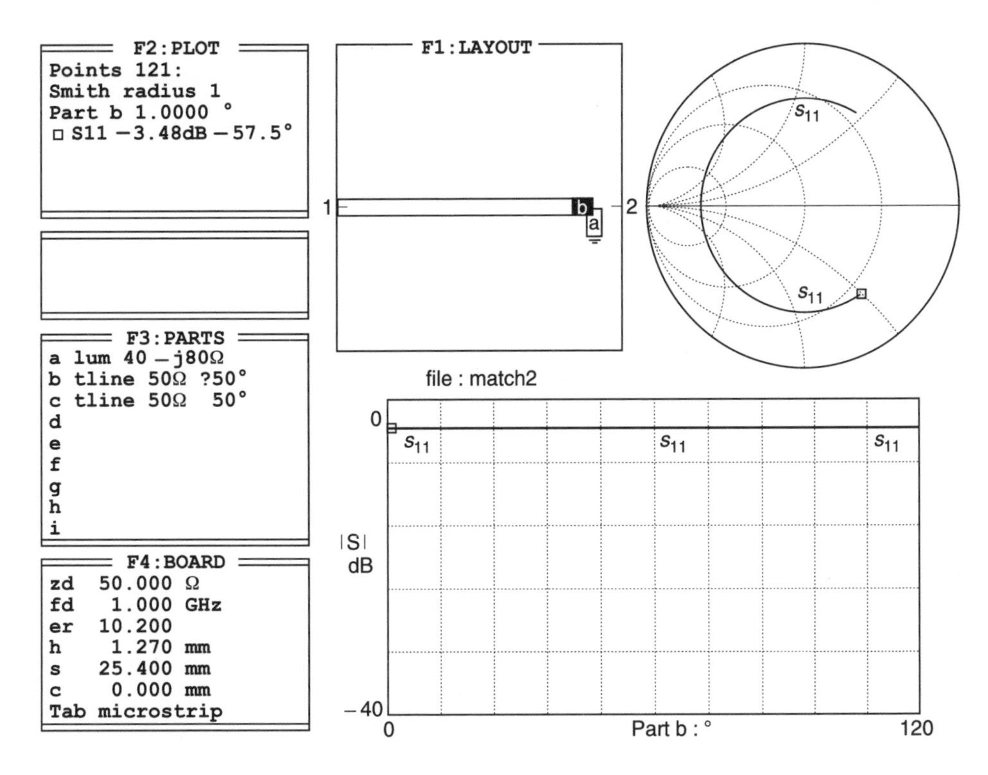

Fig. 4.20(b) Second stage of matching

4 Press the **down arrow** key. Type =. This grounds part a.

5 Press the **up arrow** key. Type **b**. Press **left arrow** key. Type **1**. The layout board is now completed.

6 Press the **F2** key. Type **p**. Press the **TAB** key to change the Smith chart coordinates to an admittance display. You will now obtain Figure 4.20(b). Note that the little square marker is on the lower right of the Smith chart.

7 Type **alt+s**. You will get an expanded view of the Smith chart similar to that of Figure 4.20(c).

8 We now want to move the marker until the square marker intersects the unity circle. Press the **page up** key several times and the marker will begin to move towards the intersection point with the unity circle. At the same time, look in the F2 box and you will see part *b* length increasing in degrees. The square marker should reach the unity circle when line *b* is approximately *38°*. At the intersection point, the conductive part of the line is matched to the input, but there is also a reactive part which must be 'tuned out'. At this stage, we do not know its value but from its chart position, we know that the reactive part is capacitive.

9 Type **alt+s** to revert back to the display of Figure 4.20(b).

10 Press the **F3** key. Use **arrow** keys to move the cursor to line *b* and overtype *?50* with **38**. You have now fixed line length *b* at *38°*. Use **arrow** keys to move the cursor to line *c* and overtype *50* with **?50**. See Figure 4.20(d) for both actions.

11 Press the **F1** key. Note how line length *b* has changed. If not already there, use **arrow** keys to move the cursor until it is at the junction of line *1* and line *b*.

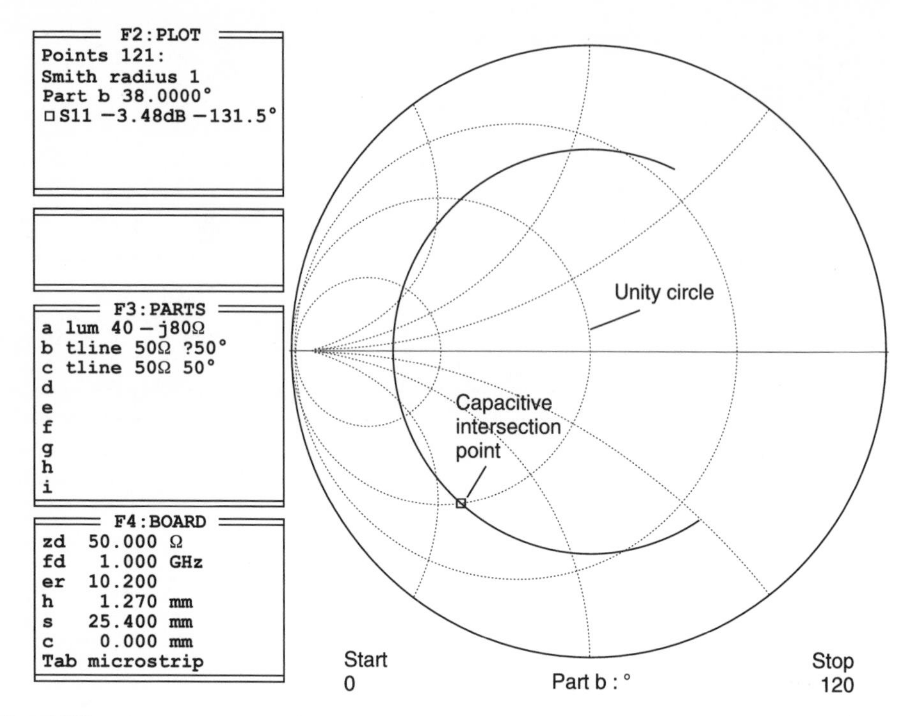

Fig. 4.20(c) Intersection point in Smith chart

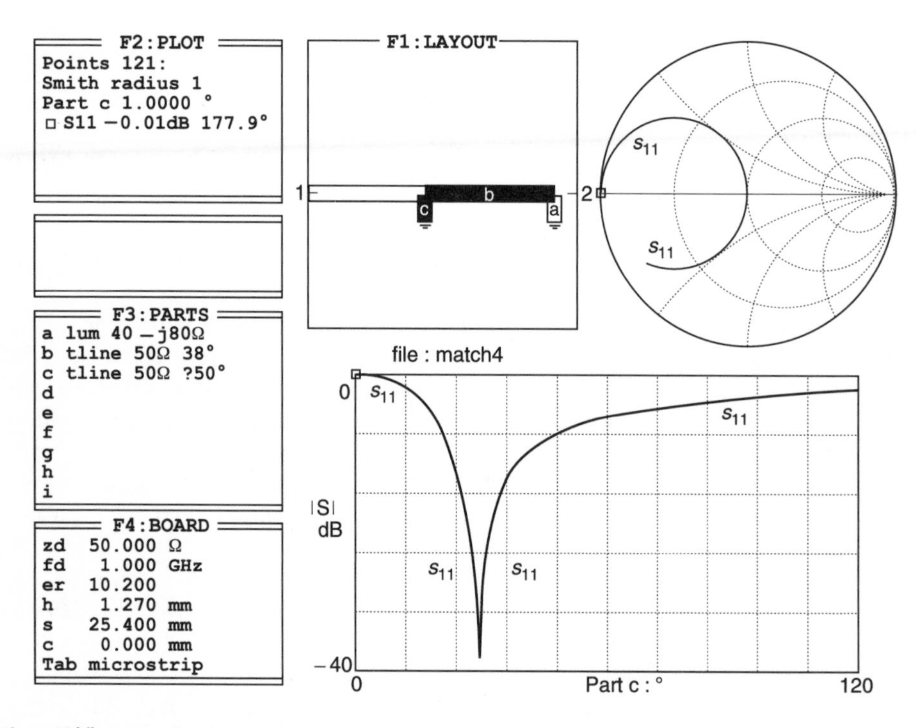

Fig. 4.20(d) Connecting the tuning stub line

12 Type **c** to select the other line. To layout line c in the position shown in Figure 4.20(d), press the **down arrow** key once. Type =.
13 Press the **F2** key. Type **p**. Press the **TAB** key. You will now get Figure 4.20(d).
14 Type **alt+s**. You will now obtain the expanded Smith chart shown in Figure 4.20(e) but with the exception that the square marker is on the left edge of the Smith chart.
15 We now want to move the square marker on the unity circle until it reaches the *match point* in the centre of the Smith chart. Press the **page up** key several times and the marker will begin to move towards the *match point* at the centre of the Smith chart. At the same time, look in the F2 box and you will see part c length decreasing in degrees.
16 The square marker should reach the unity circle *match point* when line c is approximately 29°. Note this length. At the *match point*, the reactive part will have been 'tuned out' by the stub. We can now leave the Smith chart.
17 Summing up: we now know the two line lengths required; line b is *38°* and line c is a short-circuited stub of *29°*.
18 Press the **F3** key. Type **ctrl+r**. When the program replies *File to read?* type **match6**. Press the **RETURN** key.
19 You will get Figure 4.20(f) where the previously calculated line lengths have been used. Note that the match is best (return loss ≈ 38 dB) at our design frequency of 1 GHz.

It is possible to get Figure 4.20(f) directly from Figure 4.20(d). After obtaining and entering line lengths b as 38° and c as 29°, press the F2 key. Press the down-arrow key nine times. Retype over 120 to show 2. Press p and you will obtain the file MATCH6.

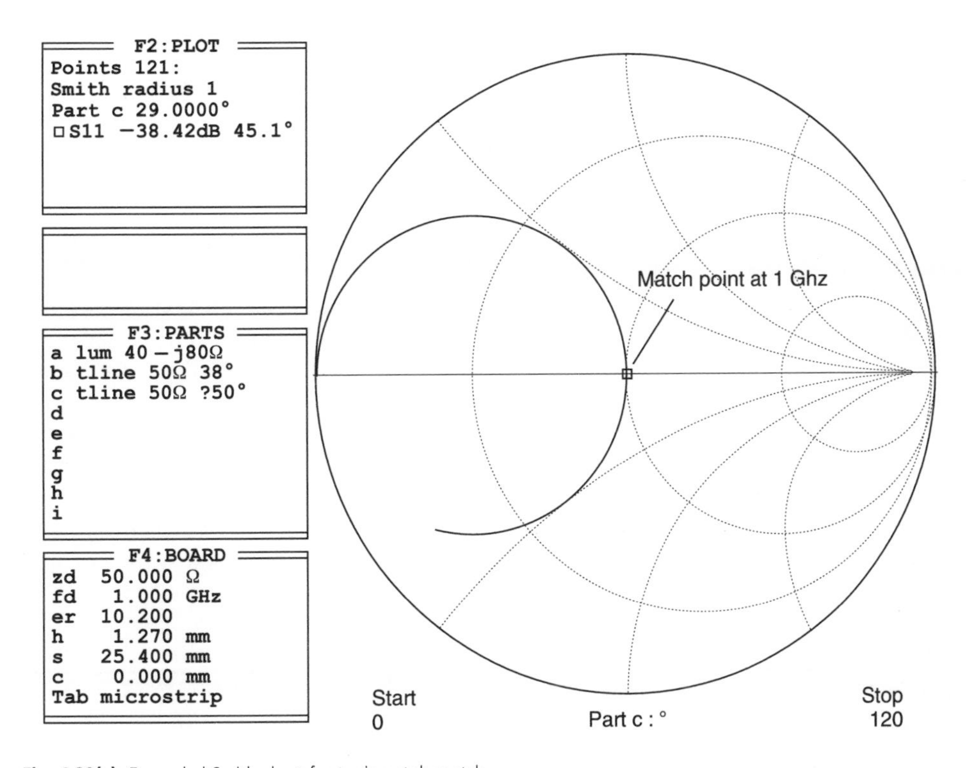

Fig. 4.20(e) Expanded Smith chart for tuning stub match

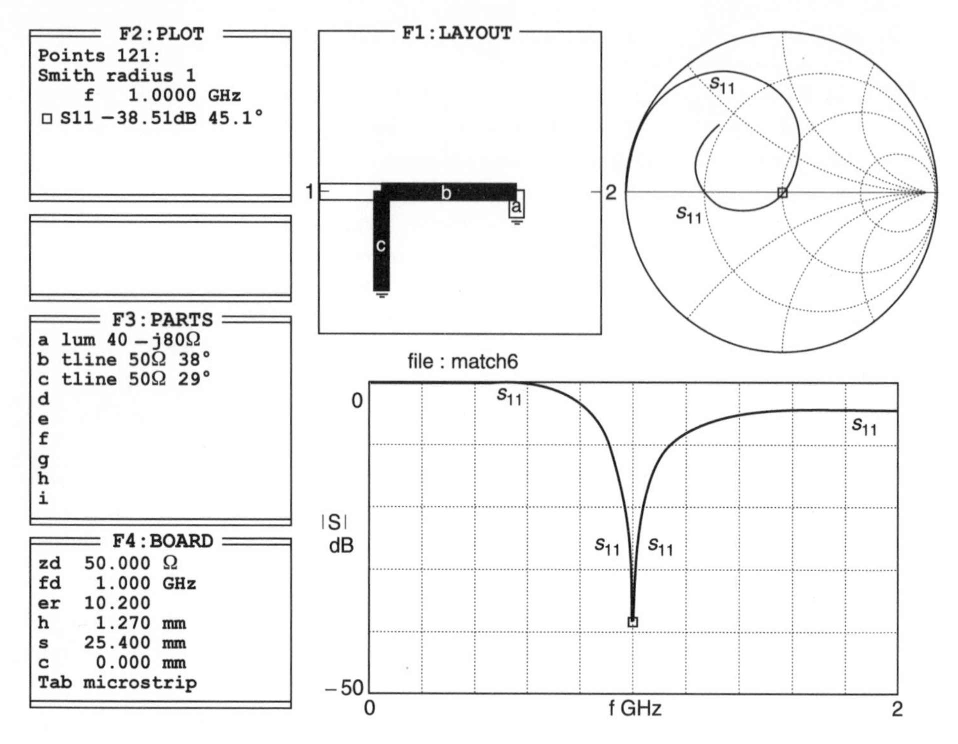

Fig. 4.20(f) Showing the effect of the matching stubs at 1 GHz

4.13.1 Summary of matching methods

I would now like to summarise the methods we have used for matching. For easy comparison remember that 1 wavelength = 360°, so 29° = 0.081λ and 38° = 0.106λ. The results for three methods are given in Table 4.5.

From the above, you will see that there is little difference whichever method you use. The direct calculations have been found through a 'goal seek' program in an Excel spreadsheet. The answers are definitely more accurate, but in practical situations we do not require such accuracy. Also do not worry unduly if you find that when you repeat the same calculations on PUFF your answers may differ slightly (≈0.1 dB). This is due to truncation errors in the program.

The graphical methods give an insight into what can be achieved more easily. For example, in microstrip, a short-circuited stub is not easy to manufacture. In this particular case, we could have increased the length of line *b* to move the matching point into the inductive part of the Smith chart and then used a capacitive stub for tuning out the inductive part of the circuit. I will not show you how this is done in this example but it is done in Example 4.4 which follows immediately.

Table 4.5

PUFF matching	Smith chart	Direct calculations
Example 3.12 (PUFF)	Example 3.12 of Chapter 3	Eqn 2.54
l_1 = 38° = 0.106λ	l_1 = 0.106λ	l_1 = 0.106 305λ
Stub = 29° = 0.081λ	Stub = 0.081λ	Stub = 0.0 806 036λ

4.13.2 Matching transistor impedances

Example 4.4 shows how the input impedance of a transistor (type fhx04) can be matched to 50 Ω at 5 GHz. In this example, I will only provide you with intermediate diagrams because the methodology is identical to that of Example 3.12.

Two methods are shown:

- a capacitive stub matching system shown in Figures 4.21(a) to (d);
- an inductive stub matching system shown in Figures 4.22(a) to (c).

Either method is suitable, but in microstrip circuits it is much easier to make an open-circuited stub than a short-circuited stub; therefore the capacitive stub matching method is preferred. However, you should compare Figure 4.21(d) and Figure 4.22(c) and note that although matching is achieved at our desired frequency of 5 GHz, there are matching differences on 'off-frequency' matching.

If you wish to try these matching networks out for yourself, Figures 4.21(a) to (d) are given as files *sweep1* to *sweep4* respectively on your disk. Figures 4.22(a) to (c) are given on your disk as files *sweep22*, *sweep33* and *sweep44* respectively. As before, I suggest that you start off with Figure 4.21(a) and build up the capacitive stub matching system. This ensures that if you run into trouble, you will have the other templates available. Similarly, start off with Figure 4.22(a) for the inductive stub matching system and build up the circuit accordingly.

Transistor matching using a capacitive stub

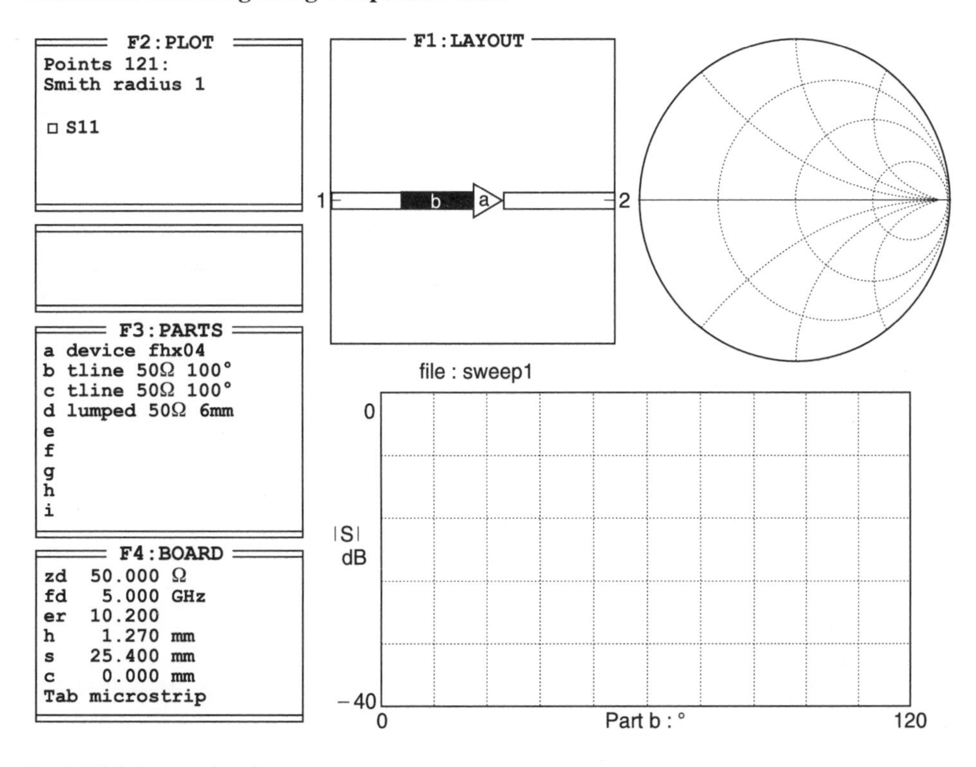

Fig. 4.21(a) First matching line

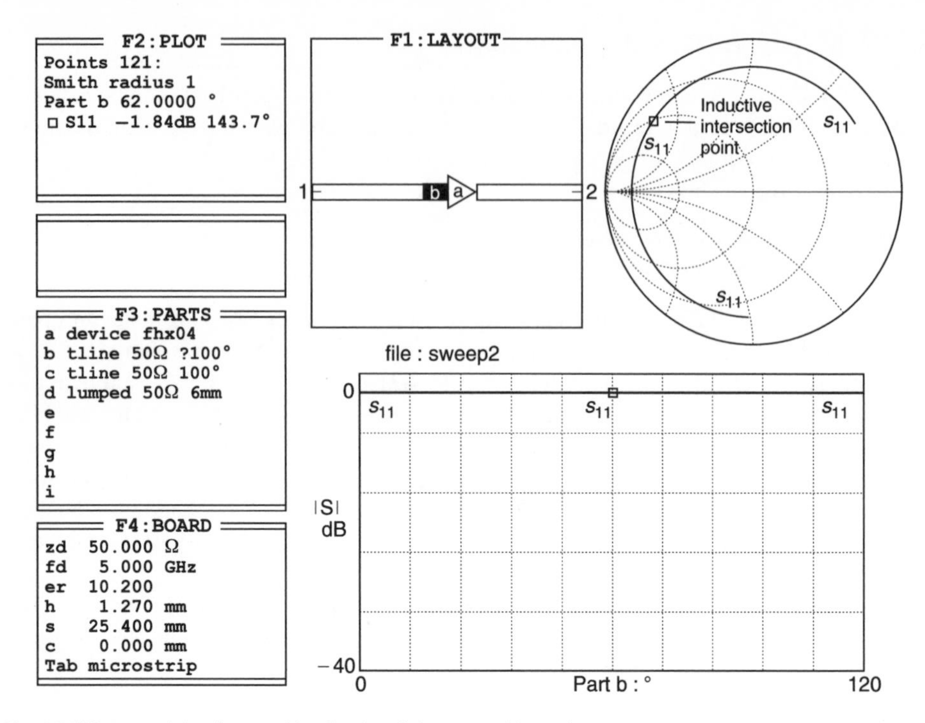

Fig. 4.21(b) Determining first matching line length for a capacitive stub

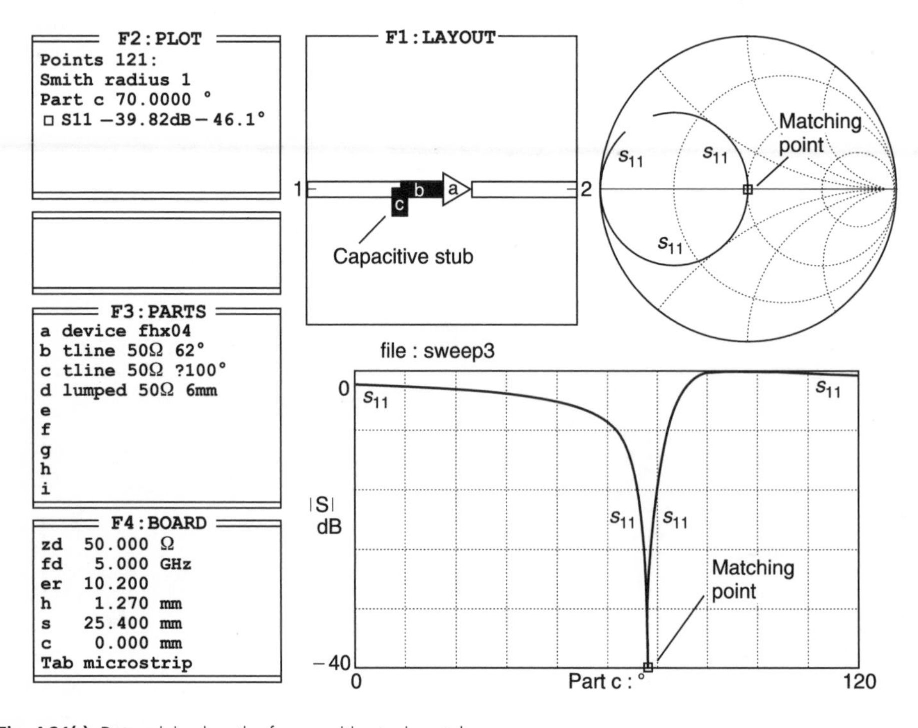

Fig. 4.21(c) Determining length of a capacitive tuning stub

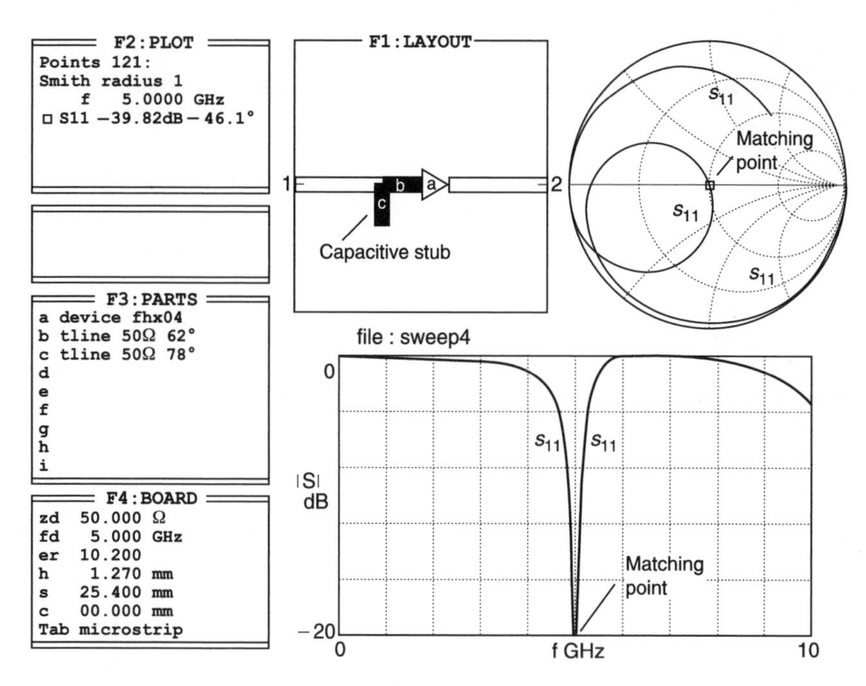

Fig. 4.21(d) Matching using a capacitive stub

Transistor matching using an inductive stub

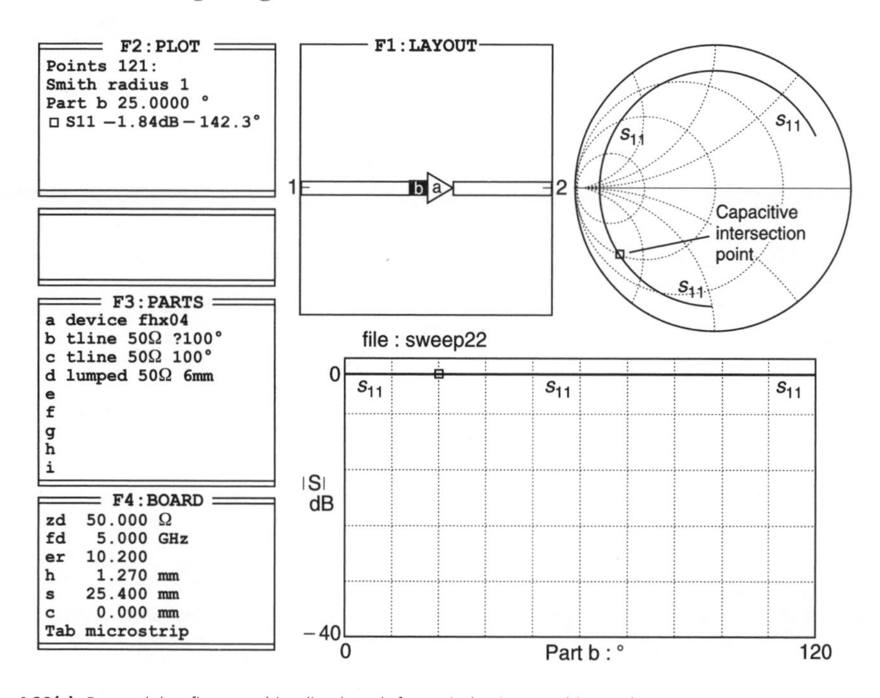

Fig. 4.22(a) Determining first matching line length for an inductive matching stub

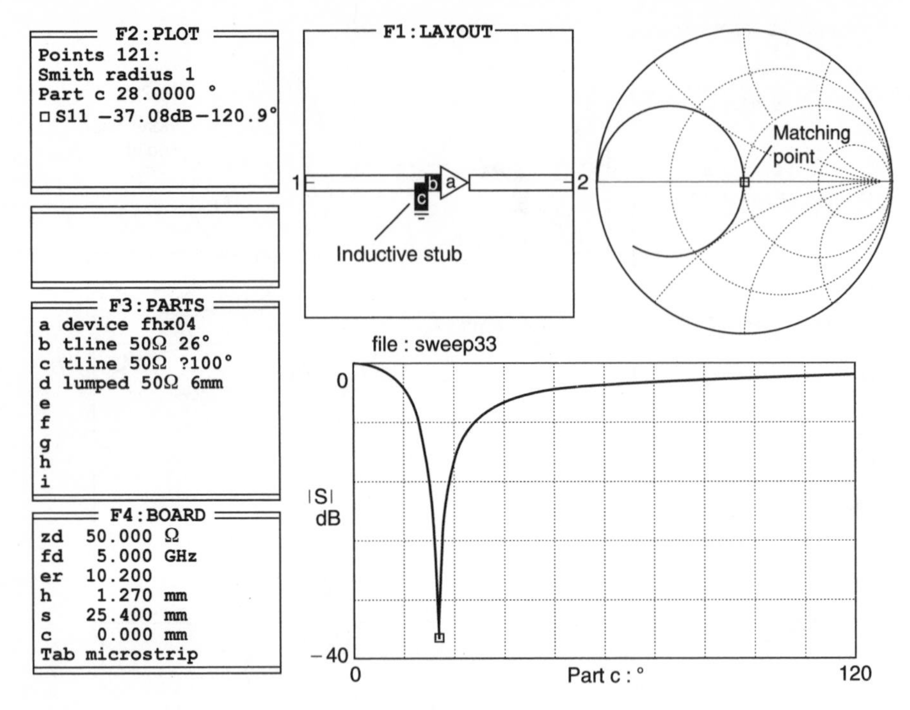

Fig. 4.22(b) Determining length of an inductive tuning stub

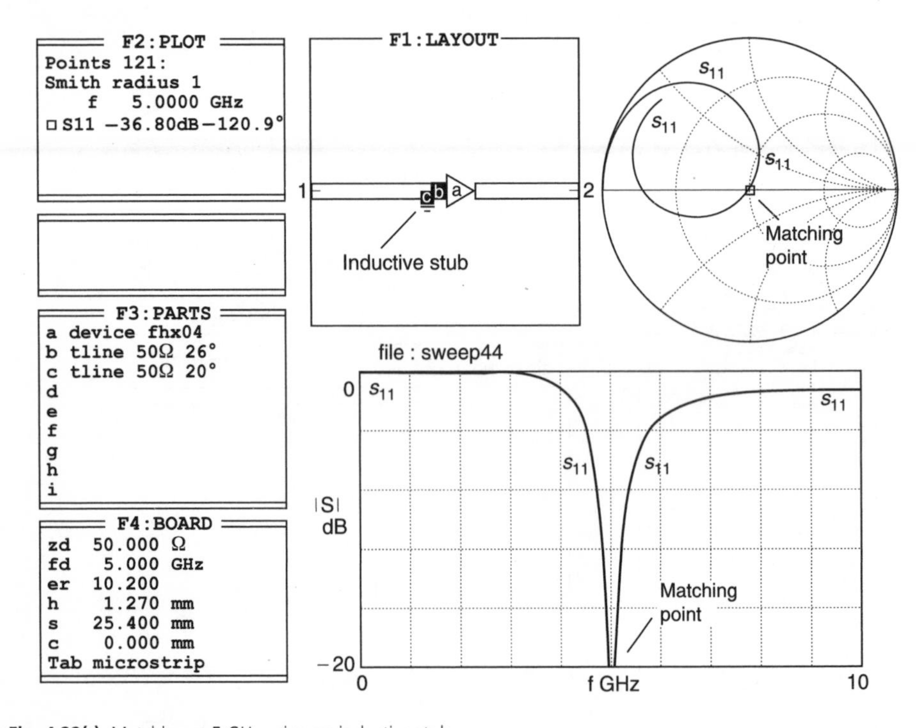

Fig. 4.22(c) Matching at 5 GHz using an inductive stub

Verification of output impedance matching

If you want to match the output impedance of any active or passive device use similar procedures to that of Example 4.4. The detailed matching procedure is again identical to Example 3.12.

4.13.3 Stub matching second example

Here is another example of stub matching. Try to see if you can carry out the stub matching for Example 3.13 which was done manually in Chapter 3.

Example 3.13

A transistor amplifier has an h.f. input resistance of 100 Ω shunted by a capacitance of 5 pF. Find the length of short-circuit stub, and its position on the line, required to match the amplifier input to a 50 Ω line at 1 GHz.

Solution

- The stub connection point is 0.028λ from the transistor input.
- The short-circuit stub length is 0.065λ at the connection point.

Example 3.13 is another example of stub matching. You already know the answer. See if you can use PUFF to match the circuit on your own. Hint: if you cannot do it, look at Figure 4.23.

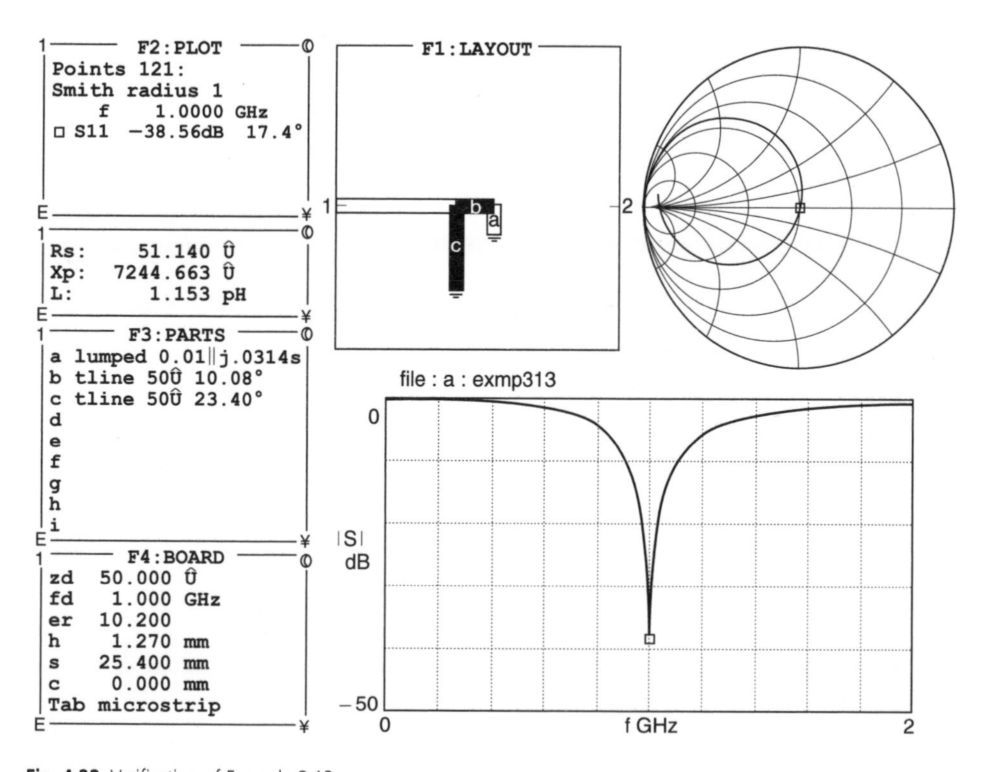

Fig. 4.23 Verification of Example 3.13

4.13.4 Double stub tuning verification

PUFF can also be used for verifying double stub matching. We will use Example 3.14 which was carried out manually in Chapter 3.

Example 3.14

A system similar to the double stub matching system shown in Figure 3.16 has a load Z_L = (50 + j50) Ω which is to be matched to a transmission line and source system with a characteristic impedance of 50 Ω. The distance, d_1, between the load and the first stub is 0.2λ (72°) at the operating frequency. The distance, d_2, between the two stubs is 0.125λ (45°) at the operating frequency. Use a Smith chart to estimate the lengths l_1 and l_2 of the stubs.

Solution. In Example 3.14, we found that:

- stub 1 required an electrical length of 0.167λ or 60.12°;
- stub 2 required an electrical length of 0.172λ or 61.92°;

The construction using PUFF is shown in Figure 4.24.

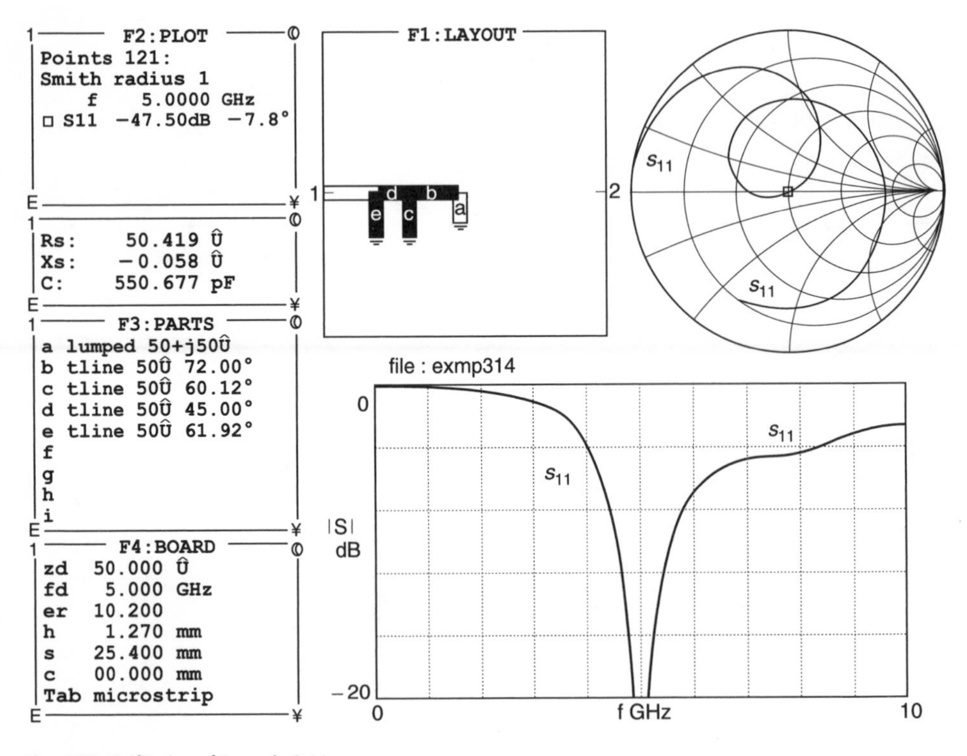

Fig. 4.24 Verification of Example 3.14

4.14 Scattering parameters

PUFF can also be used for calculating *S*-parameters. We demonstrate this with Examples 3.15, 3.16 and 3.17.

4.14.1 Series elements

Example 3.15

Calculate the S-parameters for the two-port network shown in Figure 3.30 for the case where $Z_0 = 50\ \Omega$.

Solution. Summing up for *S*-parameters:

$S11 = 0.333 \angle 0°$ or -9.551 dB $\angle 0°$ $\quad S12 = 0.667 \angle 0°$ or -3.517 dB $\angle 0°$

$S21 = 0.667 \angle 0°$ or -3.517 dB $\angle 0°$ $\quad S22 = 0.333 \angle 0°$ or -9.551 dB $\angle 0°$

This circuit is shown in Figure 4.25.

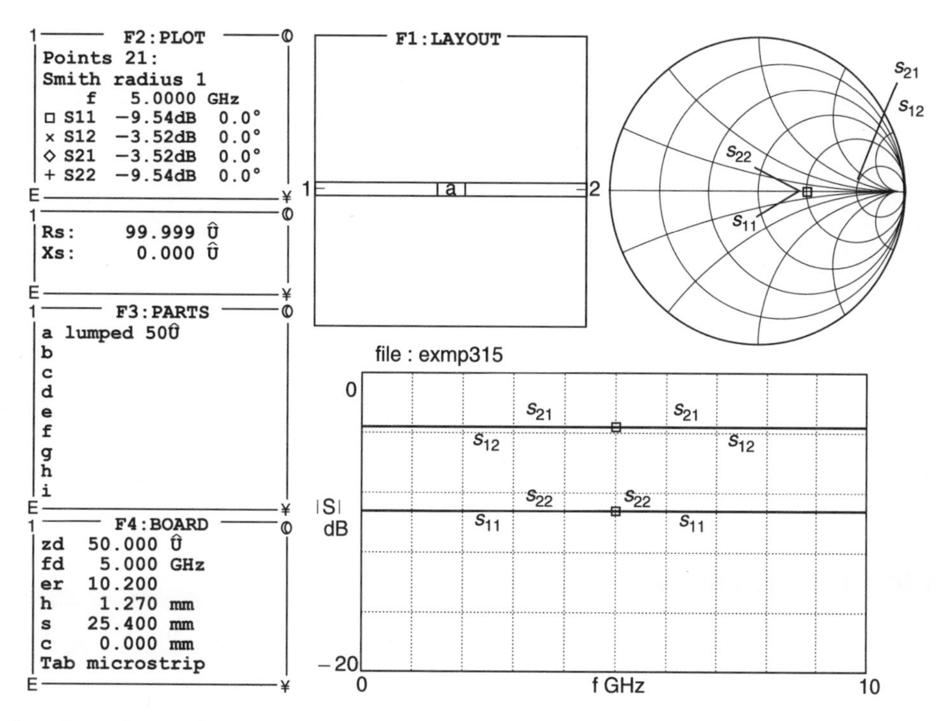

Fig. 4.25 Verification of Example 3.15

4.14.2 Shunt elements

PUFF can be used to calculate shunt elements as shown in Example 3.16.

Example 3.16

Calculate the *S*-parameters for the two port network shown in Figure 3.32 for the case where $Z_0 = 50\ \Omega$.

Solution. Summing up for *S*-parameters:

$S11 = 0.333 \angle 180°$ or -9.551 dB $\angle 180°$ $\quad S12 = 0.667 \angle 0°$ or -3.517 dB $\angle 180°$

$S21 = 0.667 \angle 0°$ or -3.517 dB $\angle 180°$ $\quad S22 = 0.333 \angle 180°$ or -9.551 dB $\angle 180°$

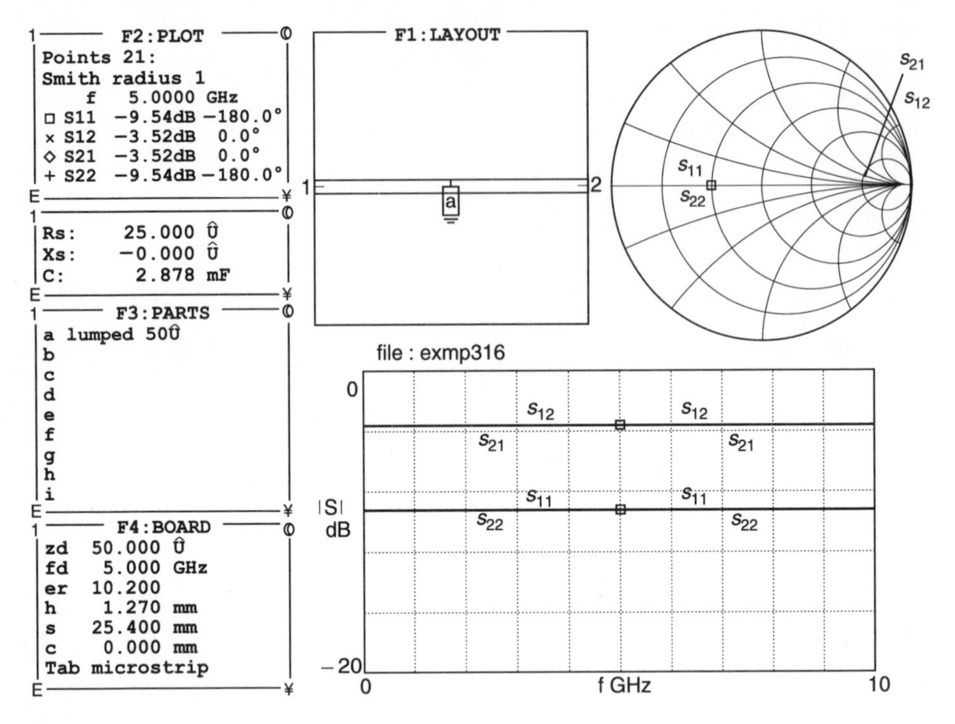

Fig. 4.26 Verification of Example 3.16

The values calculated by PUFF are shown in Figure 4.26. You can see that they are similar.

4.14.3 Ladder network

PUFF can be used for calculating ladder networks. This is shown by Example 3.17 which was carried out manually in Chapter 3.

Example 3.17

(a) Calculate the *S*-parameters for the two-port network shown in Figure 3.32 for the case where $Z_0 = 50\ \Omega$.
(b) Find the return loss at the input with $Z_L = Z_0$.
(c) Determine the insertion loss for the network when the generator and the termination are both 50 Ω.

Solution. To sum up for *S*-parameters

(a)

$$S11 = 0.518 \angle 131.16° \text{ or } -5.713 \text{ dB} \angle 131.16°$$
$$S12 = 0.524 \angle -16.22° \text{ or } -5.613 \text{ dB} \angle -16.22°$$
$$S21 = 0.524 \angle -16.22° \text{ or } -5.613 \text{ dB} \angle -16.22°$$
$$S22 = 0.442 \angle 172.87° \text{ or } -7.092 \text{ dB} \angle 131.16°$$

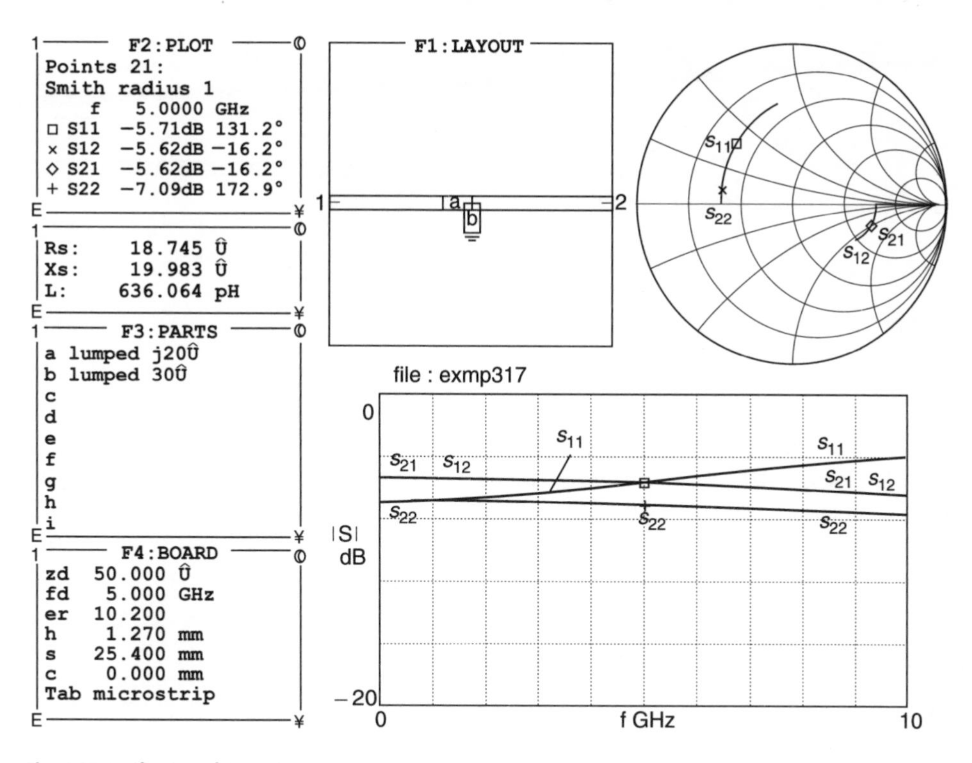

Fig. 4.27 Verification of Example 3.17

(b) From part (a), $\Gamma \angle \theta = 0.518 \angle 131.16°$. Hence

$$\text{return loss (dB)} = -20 \log_{10} |0.518| = -20 \times (-0.286) = 5.71 \text{ dB}$$

(c) The forward power gain of the network will be $|S21|^2$.

$$|S21|^2 = (0.524)^2 = 0.275$$

This represents a loss of $-10 \log_{10} 0.275 = 5.61$ dB.

Figure 4.27 gives the answers for the parameters. To derive the other two answers, namely (b) return loss and (c) forward power, you merely read off $S11$ and $S21$ and calculate to get the values.

4.15 Discontinuities: physical and electrical line lengths

This section is vitally important in the construction of your circuits. In all the problem solving given earlier, the theoretical (electrical) line length has been assumed. In other words, we have taken a physical line length of one wavelength and assumed it to be 360 electrical degrees. In the practical case, if you were to do this with a transmission line, you would find that a physical length of one wavelength is not likely to be 360 electrical

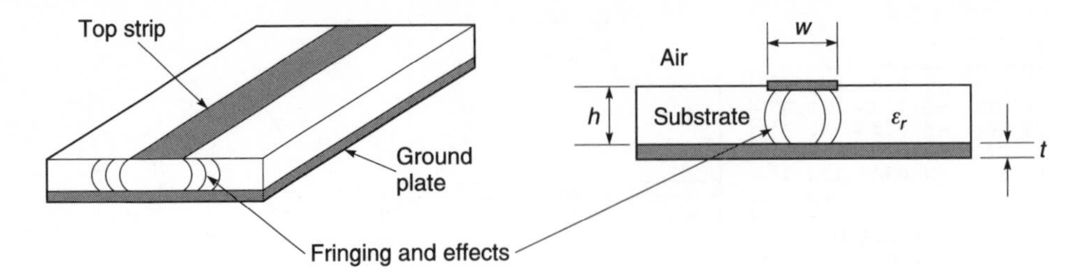

Fig. 4.28 Effect of fringing fields in transmission lines

degrees. This is due to end effects and line fringing effects, shown in Figure 4.28. The general name for these effects is discontinuities.

PUFF does not take these effects into account when drawing the artwork; therefore you must compensate for them when you use PUFF to draw the artwork. UCLA (PUFF program writers)[1] suggest that we consider four dominant discontinuities in microstrip. These are:

- excess capacitance of a corner;
- capacitive end effects for an open circuit;
- step change in width;
- length correction for the shunt arm of a tee junction.

4.15.1 Excess capacitance of a corner

When a sharp right-angle bend occurs in a circuit (Figure 4.29(a)) there will be a large reflection from the corner capacitance. PUFF mitres corners to reduce the capacitance and minimise this reflection as shown in Figure 4.29(b). You can change the value of the mitre fraction (m) set in the *setup.puf* template as 0.6. The mitre fraction (*m*) is defined as:

$$m = 1 - b\Big/\sqrt{w_1^2 + w_2^2} \tag{4.1}$$

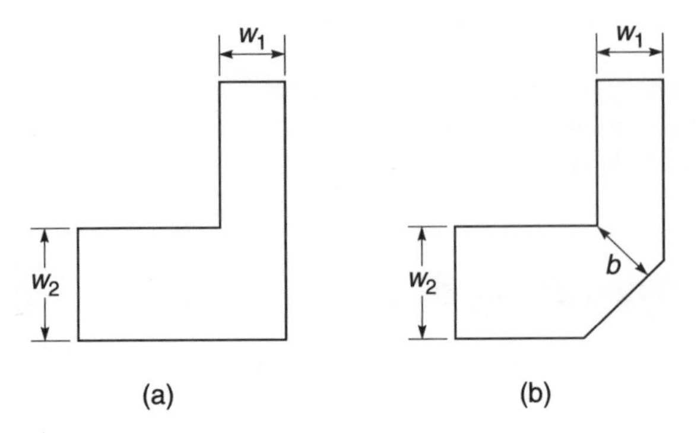

Fig. 4.29 Chamfering (mitreing) of corners

[1] This is also on the CD-ROM PUFF manual.

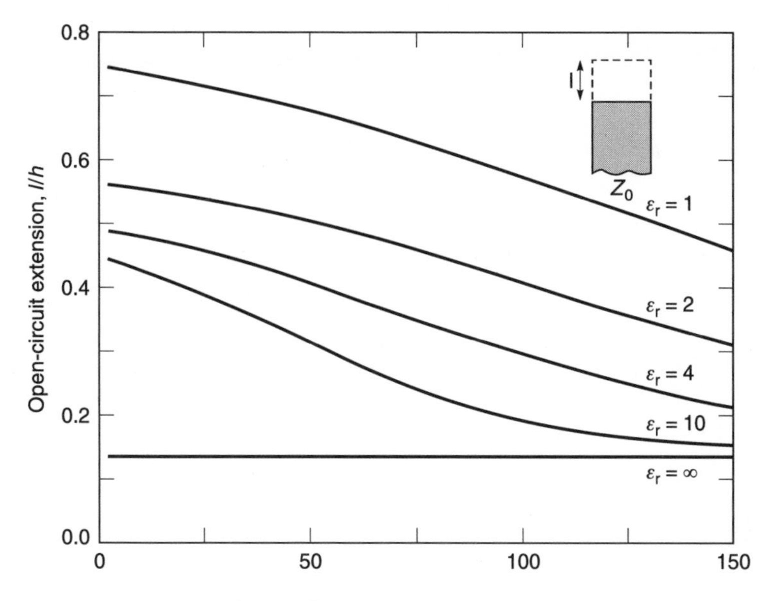

Fig. 4.30 Line length compensation for end effects

4.15.2 Capacitance end effects for an open circuit

In an open-circuit line, the electric fields extend beyond the end of the line. This excess capacitance makes the electrical length longer than the nominal length of the line, typically by a third to a half of the substrate thickness. To compensate for this effect in the artwork, a negative length correction can be added to the parts list. Hammerstad and Bekkadal[4] give an empirical formula for the length extension l in microstrip:

$$\frac{l}{h} = 0.412\left(\frac{\varepsilon_{\text{eff}} + 0.3}{\varepsilon_{\text{eff}} - 0.258}\right)\left(\frac{w/h + 0.262}{w/h + 0.813}\right) \tag{4.2}$$

where ε_{eff} is the effective dielectric constant of the through arm.

Note: In the above correction, the length l must be negative, i.e. the length l must be subtracted from the desired length in the parts list.

4.15.3 Step change in width of microstrip

A similar method may be used to compensate for a step change in width between high and low impedance lines. This is shown in Figure 4.31. The discontinuity capacitance at the end of the low impedance line will have the effect of increasing its electrical length.

[4] E.O. Hammerstad and F. Bekkadal, *A Microstrip Handbook*, ELAB Report, STF 44 A74169, N7034, University of Trondheim, Norway, 1975.

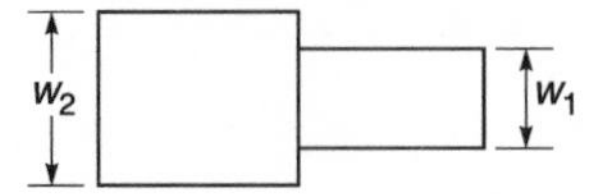

Fig. 4.31 Step change in width of microstrip

Assuming the wider low impedance line has width w_2, and the narrow high impedance line has width w_1, compensate using the expression suggested by Edwards[5]

$$\frac{l_s}{h} = 0.412\left(\frac{\varepsilon_{\text{eff}} + 0.3}{\varepsilon_{\text{eff}} - 0.258}\right)\left(\frac{w/h + 0.262}{w/h + 0.813}\right)\left[1 - \frac{w_1}{w_2}\right] \tag{4.3}$$

or

$$\frac{l_s}{h} = \frac{l}{h}\left[1 - \frac{w_1}{w_2}\right] \tag{4.3a}$$

where l_s is the step length correction for line w_2 and l/h is the value obtained from Equation 4.2 and Figure 4.30.

4.15.4 Length correction for the shunt arm of a tee-junction

In the tee-junction shown in Figure 4.32, the electrical length of the shunt arm is shortened by distance d_2. The currents effectively take a short cut, passing close to the corner. It is particularly noticeable in the branch-line coupler because there are four tee-

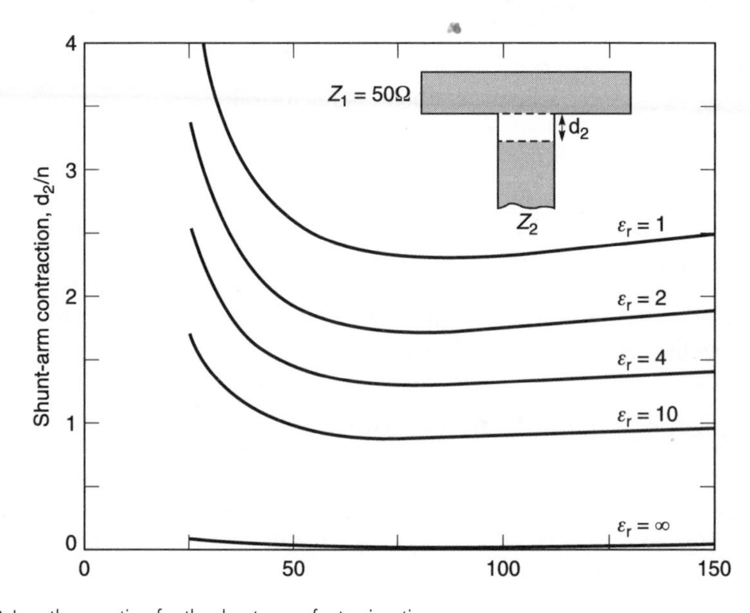

Fig. 4.32 Length correction for the shunt arm of a tee-junction

[5] T.C. Edwards, *Foundations for microstrip circuit design*, second edition, John Wiley & Sons, ISBN 0 471 93062 8, 1992.

junctions. Hammerstad and Bekkadal (see footnote 4) give an empirical formula for d_2 in microstrip:

$$\frac{d_2}{h} = \frac{120\pi}{Z_1\sqrt{\varepsilon_{\text{eff}}}}\left\{0.5 - 0.16\frac{Z_1}{Z_2}[1 - 2\ln(Z_1/Z_2)]\right\} \quad (4.4)$$

where ε_{eff} is the effective dielectric constant of the through arm. Equation 4.4 is plotted in Figure 4.32 for a 50 Ω through line. Additional help on discontinuity modelling for both microstrip and stripline can be found in a book by Gupta.[6]

4.16 Summary

By now, I am sure you will agree that your studies in Parts 2 and 3 on transmission lines, Smith charts and S-parameters are beginning to bring rewards and help you toward the goal of being a good h.f. and microwave engineer.

Sections 4.1 to 4.3 of this part have been devoted to the installation of PUFF. In Section 4.4, we covered the principles of using PUFF. Section 4.5 provided some simple examples of how to use the facilities provided by PUFF for printing and production of artwork. In Section 4.6, we designed and produced the artwork, and measured the frequency response of a bandpass filter using transmission lines.

Section 4.7 was used to collect and collate all the PUFF commands that you had learnt previously. The use, design and modification of templates for the PUFF system were discussed in Section 4.8. In Section 4.9, we learnt how to manipulate and alter transistor templates for PUFF.

In Section 4.10, we achieved our goal of verifying and checking that the work carried out in Parts 2 and 3 was valid. Fifteen examples (2.3, 2.7, 2.8, 2.13, 2.14, 2.15, 3.2, 3.5, 3.6, 3.11, 3.12, 3.13, 3.14, 3.15, 3.16 and 3.17) were entered into PUFF. Their results were compared with the examples produced manually in Parts 2 and 3. The fact that both sets of results agree should give you confidence in the use of either method.

In Section 4.11, we used PUFF to investigate the properties of the branch-line coupler and the rat-race coupler. We evaluated their transmission and matching properties.

Section 4.12 was used to show how PUFF can be used to find admittances (Example 3.2), network impedance and reflection coefficient (Example 3.5), input impedance of transmission lines (Example 3.6), quarter-wave transformers (Example 3.11), and cascading of quarter-wave transformers. The examples created manually in Part 3 all agree with the solutions provided by PUFF.

The very important technique of stub matching was detailed and demonstrated in Section 4.13. It provided details on how single stub matching can be carried out with PUFF. The PUFF answer agreed well with Example 3.12 which was previously carried out manually and also with direct calculations. See Table 4.5 in Section 4.13.1 for details. Section 4.13.2 provided details on how matching can be carried out using inductive or capacitive tuning stubs. Section 4.13.3 provided an electronic matching of Example 3.13. Double stub electronic matching of Example 3.14 was verified in Section 4.13.4.

[6] K.C. Gupta, R. Garg and R. Chadha, *Computer-aided design of microwave circuits*, Artech House, Deham, Mass. USA, ISBN 0–89006–105–X, 1981.

Section 4.14 demonstrated how PUFF can be used to calculate the *S*-parameters of series elements (Example 3.15), shunt elements (Example 3.16) and networks (Example 3.17).

The important subject of discontinuities in microstrip and how they may be compensated for in the PUFF artwork was covered in Section 4.15. Four types were discussed, and compensation methods for minimising these effects were shown.

Now that you are familiar with many passive networks and their solutions, we will be moving on to active circuits, mainly the design of amplifiers, in the following parts. However, this is not the last of PUFF because we will be using it in the design of filters and amplifiers.

Last but not least, the use of PUFF in the design and layout of circuits detailed in an article called 'Practical Circuit Design' is reproduced on the disk accompanying this book. However, I advise you to defer reading it until you have reached the end of Part 7 because many of the principles and techniques used in the article have yet to be explained.

5

Amplifier basics

5.1 Introduction

The information gained in the previous parts has now allowed us to move into the realms of amplifier design. Small signal r.f. amplifiers assume many configurations. We show two common configurations. In Figure 5.1, we show the circuit of a single stage amplifier. It consists of five main sections:

- input source with a source impedance Z_s;
- an input tuned/matching circuit comprising C_1, L_1 and C_2;
- a transistor amplifier (transistor biasing is not shown);
- an output tuned/matching circuit comprising L_2 and C_3;
- load (Z_L).

In Figure 5.2, we show the circuit of a multi-stage integrated amplifier circuit. It consists of five main sections:

- input source with a source impedance Z_s;
- a multi-stage amplifier gain block (sometimes called 'gain blok');

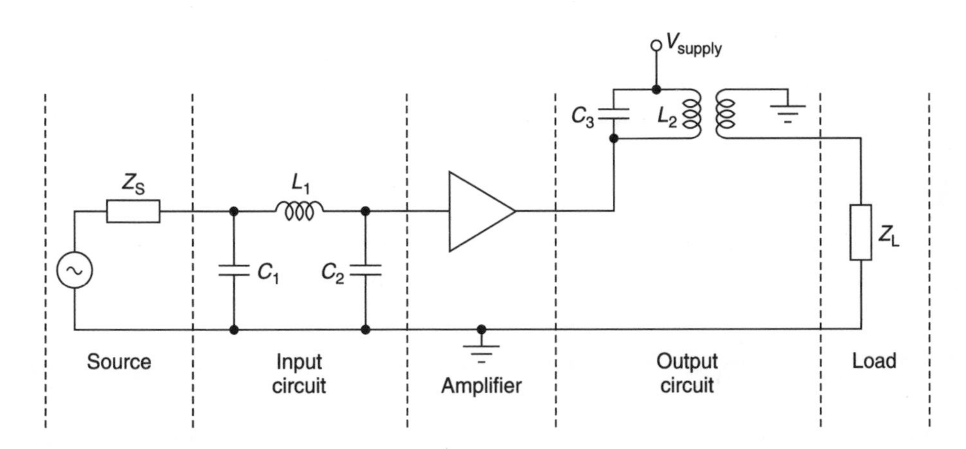

Fig. 5.1 Single stage amplifier

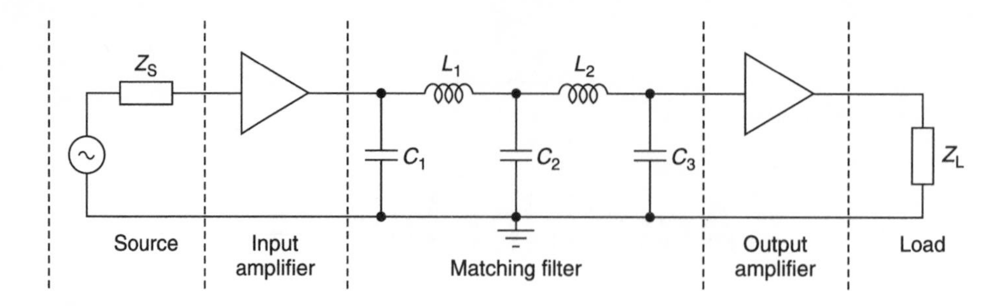

Fig. 5.2 Multi-stage amplifier

- a multi-tuned/matching filter circuit comprising C_1, L_1, C_2, L_2 and C_3;
- a multi-stage amplifier gain block (sometimes called 'gain blok');
- load (Z_L).

From the above figures, it is clear that to design a circuit, we must understand:

- tuned circuits
- filters
- matching techniques
- amplifier parameters
- gain block parameters

Much material will be devoted to matching circuits in this chapter because after selection of a transistor or gain block for a particular design, there is not much you can do within the active device other than present efficient ways in which energy can be coupled in and out of the device. This in turn calls for efficient matching circuits for the intended purposes.

In this chapter, we will investigate tuned circuits, filters and impedance matching techniques. This will enable us to deal with transistors, and semiconductor devices in the next chapter.

5.1.1 Aims

The aims of this chapter are to introduce you to the passive elements and/or devices which are used in conjunction with active devices (transistors, etc.) to design complete circuits.

5.1.2 Objectives

The objectives of this chapter are to show how passive, discrete and distributed elements can be used in the design of tuned circuits, filters and impedance matching networks.

5.2 Tuned circuits

As these equations are readily available in any elementary circuit theory book, we shall simply state the equations associated on single series and parallel tuned circuits.

5.2.1 Series circuits

The series C, L and R circuit is shown in Figure 5.3. The fundamental equations relating to the series circuit are:

$$Z = R + j\omega L + 1/(j\omega C) \tag{5.1}$$

$$\omega_o = 1/\sqrt{LC} \tag{5.2}$$

$$\frac{v_r}{v_s} = \frac{R}{R + j(\omega L - 1/\omega C)} \tag{5.3}$$

$$Q = \omega_o L/R \quad \text{or} \quad 1/(\omega_o CR) \tag{5.4}$$

$$Q = \omega_o/(\omega_2 - \omega_1) \tag{5.5}$$

where

Z = input impedance with R (ohms), L (Henries), C (Farads)
ω_o = resonant frequency in radians per second
v_r = voltage across resistor R
v_s = open-circuit source voltage
Q = quality factor
ω_2 = upper frequency (rads/sec) where the response has fallen by 3 dB
ω_1 = lower frequency (rads/sec) where the response has fallen by 3 dB
$\omega_2 - \omega_1$ = 3 dB bandwidth of the circuit

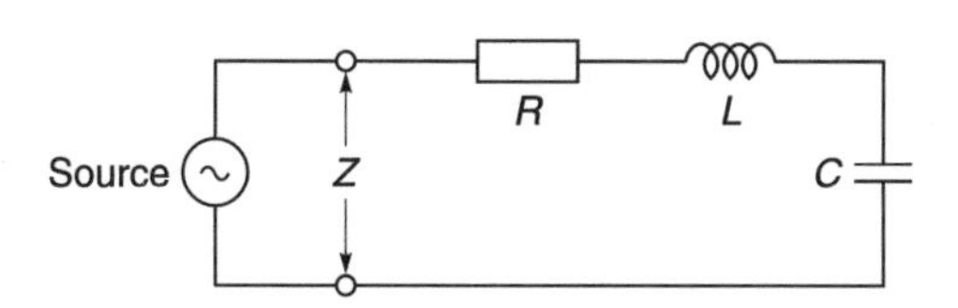

Fig. 5.3 Series circuit

Example 5.1

A series CLR circuit has $R = 3\ \Omega$, $L = 20$ nH and a resonant frequency (f_0) of 500 MHz. Estimate (a) its impedance at resonance, (b) the value of the capacitance needed for resonance at 500 MHz, (c) Q of the circuit at resonance, and (d) the 3 dB bandwidth.

Given: $R = 3\ \Omega$, $L = 20$ nH and $f_0 = 500$ MHz.
Required: (a) Impedance at resonance, (b) value of series capacitance for resonance at 500 MHz, (c) Q of the circuit at resonance and (d) the 3 dB bandwidth of the circuit.

Solution

(a) Using Equation 5.1

$$Z = R + j\omega L + (1/(j\omega C) = 3 + j(X_L) - j(X_C)$$

Since $X_L = X_C$ at resonance

$$Z = 3\ \Omega$$

(b) Using Equation 5.2

$$\omega_o = 1/\sqrt{LC} = 2\pi \times 500\ \text{MHz} = (20\ \text{nH} \times C\ \text{pF})^{-0.5}$$

Transposing

$$C = 5.066\ \text{pF}$$

(c) Using Equation 5.4

$$Q = \omega_o L/R \quad \text{or} \quad 1/(\omega_o CR) = (2\pi \times 500\ \text{MHz} \times 20\ \text{nH})/3 = 62.832/3 = 20.944$$

(d) Using Equation 5.5

$$Q = \omega_o/(\omega_2 - \omega_1)$$

Therefore 20.944 = 500 MHz/bandwidth MHz. Hence

$$3\ \text{dB bandwidth} \approx 23.873\ \text{MHz}$$

Using PUFF

To find C, we invoke the simple optimiser in PUFF which is called the component sweep. Instead of sweeping with frequency, a circuit's scattering parameters may be swept with respect to a changing component parameter. This feature is invoked by placing a question mark (?) in front of the parameter to be swept in the appropriate position of a part description in the F3 box. This is shown in Figure 5.4 where a question mark (?) has been placed in front

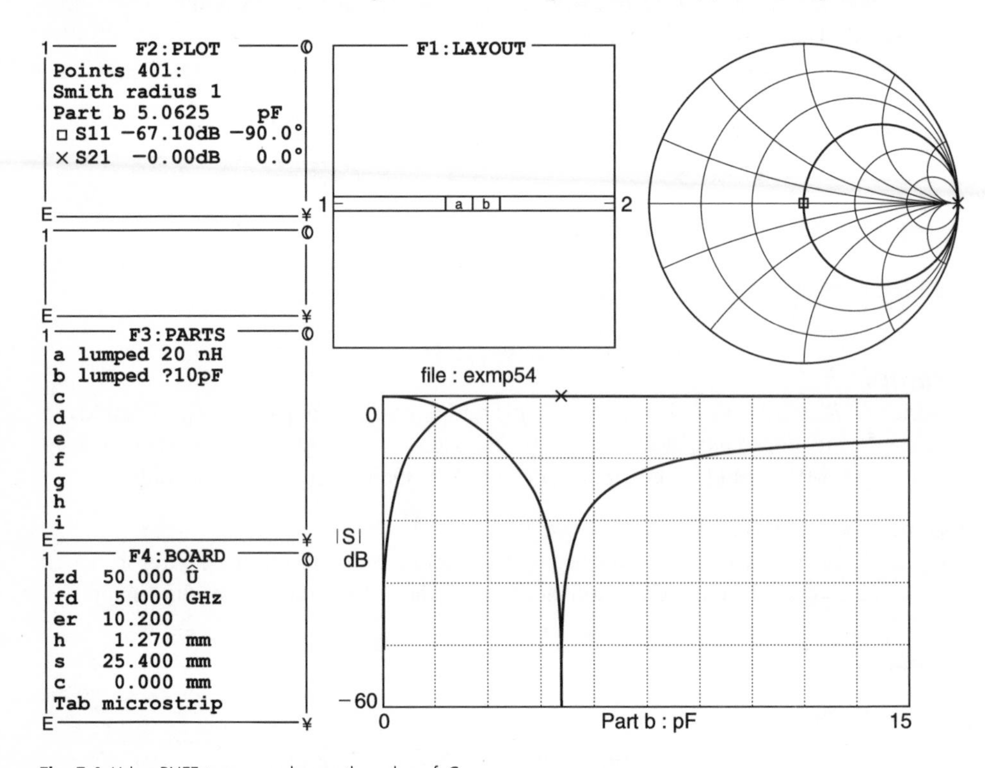

Fig. 5.4 Using PUFF to sweep-change the value of C

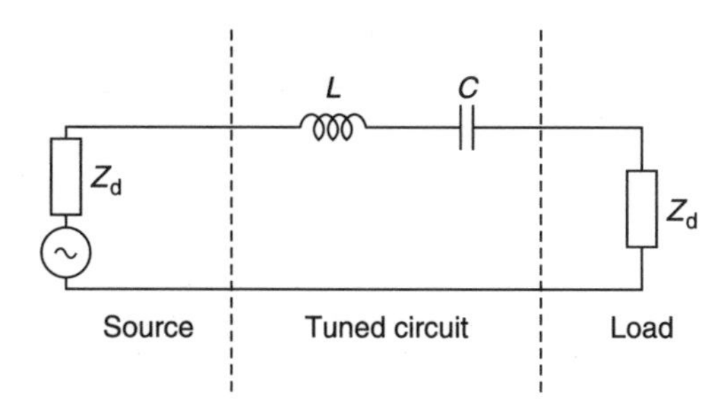

Fig. 5.5 Equivalent circuit used by PUFF

of the 10 pF in the F3 box.[1] PUFF now knows that we wish to sweep-change the value of C until we get resonance at the f_d frequency which has been set to 0.5 GHz or 500 MHz in the F4 box.

If you now press the **F2** key and press the **p** key, you will get the plot shown in Figure 5.4. Note that in the third sentence in the F2 box, we have the *Part b 5.0625 pF.* You will not get this value initially, because the first displayed value is not in resonance; however, if while in the p plot mode, you press the **page up** and/or **page down** keys, you will find the value of *Part b* changing. You will also see the *X* mark on the rectangular plot move simultaneously. Keep pressing the **page up** or **page down** keys, until $S11$ shown in the F2 box is reading the largest negative number, in this case, $S11 \approx -67$ dB. Now *Part b* will show 5.06 pF approximately. This is the value of C which will resonate with the 20 nH to produce resonance at 500 MHz.

The reason why I have asked you to use the $S11$ indicator rather than the $S21$ indicator is because it is easier to locate the minimum point. This is best explained by Figure 5.5 where I have shown the equivalent circuit used by PUFF. You know that you only get perfect match ($S11 = 0$) when Z_d (source) is terminated by Z_d (load). This will only occur when $j(\omega L - 1/\omega C) = 0$. Having found the value of C as 5.06 pF, we can now plot the circuit conventionally and obtain the response of Figure 5.6. To obtain better scaling in Figure 5.6, I have copied and modified the PUFF set-up template and changed the frequency range from 400 MHz to 600 MHz, f_d to 500 MHz and Z_d to 1.5 Ω so that the total resistance in the circuit is 1.5 Ω + 1.5 Ω (see Figure 5.5), i.e. 3 Ω.

If we now press the **F2** key and **p**, we will get the response of the Q curve and by using the **page up** and **page down** keys, we can observe that the upper –3 dB point occurs at approximately 512.5 MHz. The low –3 dB point occurs at 488.5 MHz. Hence the 3 dB bandwidth is (512.5 – 488.5) MHz = 24 MHz. To sum up:

Item	**Calculation**	**PUFF**
Value of capacitance	5.066 pF	5.0625 pF
3 dB bandwidth	23.873 MHz	24 MHz

You may well ask whether doing it by PUFF is worth the effort when you can obtain good results by calculation. It is worth it, because PUFF gives you a picture of the circuit

[1] Swept lumped components are restricted to single resistors, capacitors or inductors. A description such as *lumped ?1+5j–5j Ω* is not allowed since it is a series *CLR* circuit. In addition the parallel sign || cannot be used in the lumped specification. The unit and prefix given in the part description (following the ?) is inherited by the component sweep.

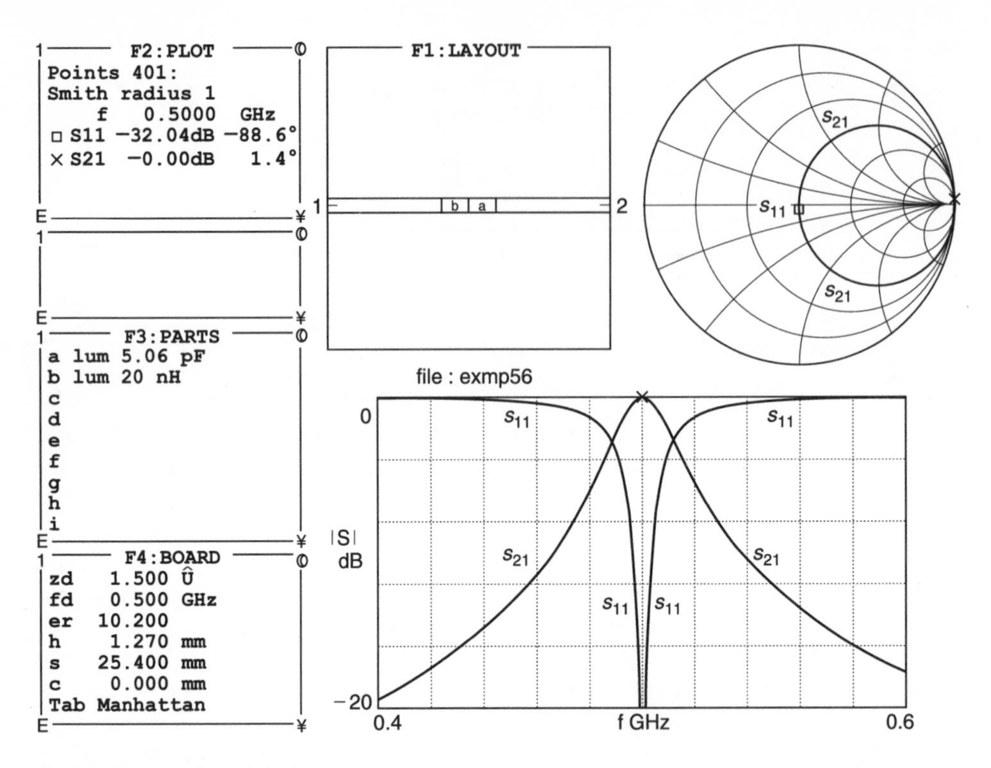

Fig. 5.6 Response of tuned circuit

response. It tells you the amount of rejection (attenuation) for frequencies outside the resonance frequency. It also shows you how the Q must be modified to get the results you want. Last, but not least, using PUFF gives you confidence for more complicated circuits later in the book.

5.2.2 Parallel circuits

Similar results can also be obtained either by calculation or by using PUFF for parallel circuits which are normally used for the load impedance of amplifiers. The fundamental equations relating to the parallel circuit are:

$$Z = 1/Y \tag{5.6}$$

$$Y = G + \mathrm{j}\omega C - \mathrm{j}(1/\omega L) \tag{5.7}$$

$$\omega_o = 1/\sqrt{LC} \tag{5.8}$$

$$\frac{i_r}{i_s} = \frac{G}{G + \mathrm{j}\omega C - \mathrm{j}1/\omega L} \tag{5.9}$$

$$Q = R/\omega_o L \quad \text{or} \quad \omega_o CR \tag{5.10}$$

$$Q = \omega_o/(\omega_2 - \omega_1) \tag{5.11}$$

where

Z = input impedance with R (ohms), L (henries), C (farads)
Y = input admittance with G (Siemens), L (henries), C (farads)
ω_o = resonant frequency in radians per second
i_r = current across conductance G
i_s = total current through admittance
Q = quality factor
ω_2 = upper frequency (rads/sec) where the response has fallen by 3 dB
ω_1 = lower frequency (rads/sec) where the response has fallen by 3 dB
$\omega_2 - \omega_1$ = 3 dB bandwidth of the circuit

Example 5.2

A parallel circuit (Figure 5.7) consists of an inductor of 20 nH, a capacitance of 5.06 pF and a resistance across the tuned circuit of 2.5 kΩ. It is driven from a current source. Plot its frequency response from 400 MHz to 600 MHz.

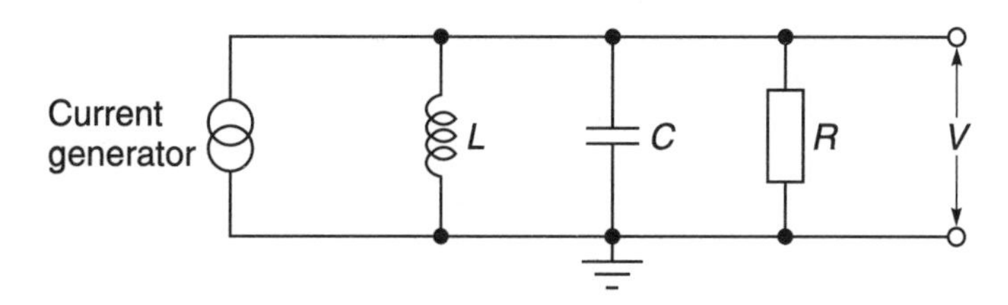

Fig. 5.7 Parallel tuned circuit

Solution. Using PUFF, we obtain Figure 5.8. In the F3 box, we have used a symbol | | which signifies that two elements are in parallel and changed the scaled microstrip lines in the F4 box to the Manhattan mode. The Manhattan mode allows PUFF to carry out calculations without bothering about the physical size of components. In the F4 box, I have also set Z_d to 5000 Ω and, bearing in mind Figure 5.7, it is readily seen that this is equivalent to having a combined resistance of 2.5 kΩ across the tuned circuit. From Figure 5.8, the –3 dB bandwidth of the circuit is measured to be (506.5 – 494.0) MHz or 12.5 MHz. The Q of the circuit is 500 MHz/12.5 MHz = 40.

5.2.3 Cascading of tuned circuits

Most radio frequency systems use a number of tuned circuits in cascade to achieve the required selectivity (tuning response). One such arrangement commonly used in broadcast receivers is shown in Figure 5.9. You should note that amplifiers are placed in between the tuned circuits so that they do not interact directly with each other.

One way would be to derive the response of each individual tuned circuit, then multiply their individual responses to obtain the overall response. Another way would be to take the individual responses in dB and add them together. I have done this for you. The

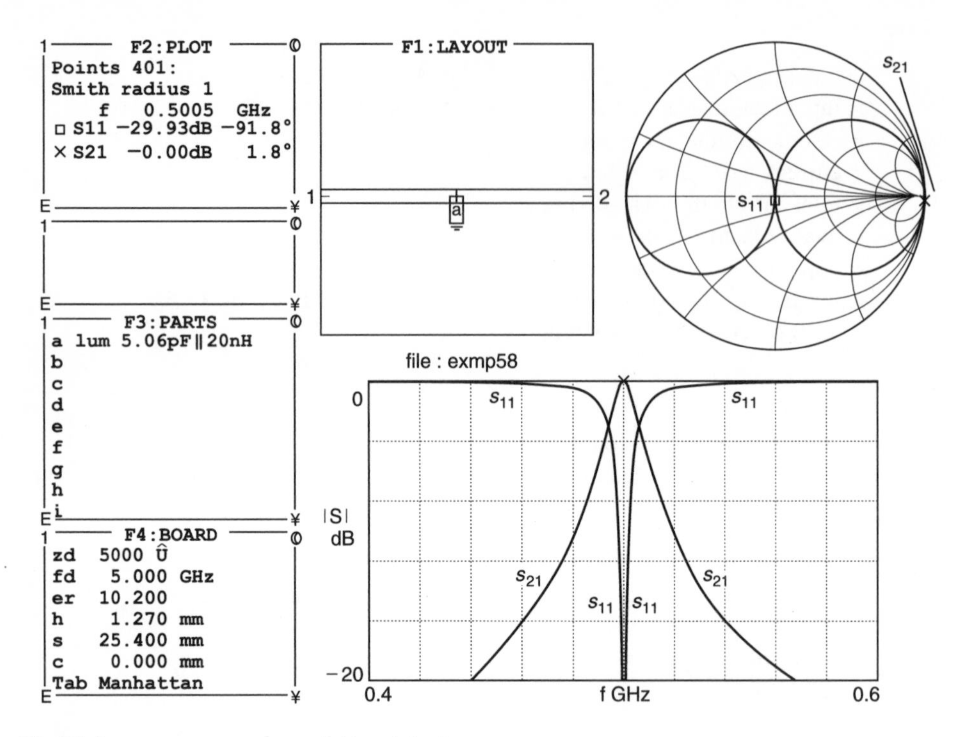

Fig. 5.8 Frequency response of a parallel tuned circuit

calculations are shown in Table 5.1. The results can then be plotted as shown in Figure 5.10. Although there are two tuned circuits each with a Q of 35, I have only plotted one for the sake of clarity. You should also note that the frequency scale (ω/ω_o) has been plotted linearly this time instead of logarithmic. This is to allow a better view of the response near the resonant frequency.

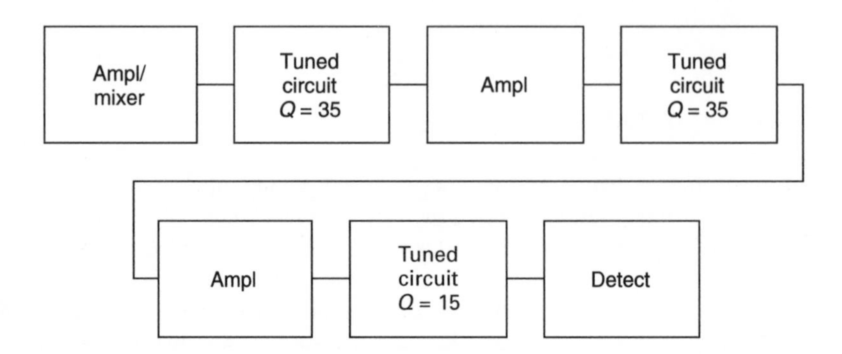

Fig. 5.9 Tuned circuit arrangement of a broadcast radio receiver

Table 5.1 *Q* factor responses

Fractional frequency (rad/s)	Attenuation is given in dB			
ω/ω_o	$Q = 15$	$Q = 35$	$Q = 35$	Total Q
0.85	–13.98	–21.19	–21.19	–56.4
0.9	–10.42	–17.45	–17.45	–45.3
0.93	–7.595	–14.29	–14.29	–36.2
0.95	–5.276	–11.43	–11.43	–28.1
0.97	–2.637	–7.441	–7.441	–17.5
0.98	–1.359	–4.772	–4.772	–10.9
0.99	–0.378	–1.746	–1.746	–3.9
0.995	–0.097	–0.504	–0.504	–1.1
1.0	0.0	0.0	0.0	0.0
1.005	–0.096	–0.5	–0.5	–1.1
1.01	–0.371	–1.718	–1.718	–3.8
1.015	–0.79	–3.194	–3.194	–7.2
1.02	–1.313	–4.656	–4.656	–10.6
1.05	–4.975	–11.03	–11.03	–27.0
1.08	–8.022	–14.78	–14.78	–37.6
1.11	–10.35	–17.37	–17.37	–45.1
1.13	–11.62	–18.72	–18.72	–49.1
1.15	–12.72	–19.88	–19.88	–52.5
1.18	–14.13	–21.35	–21.35	–56.8

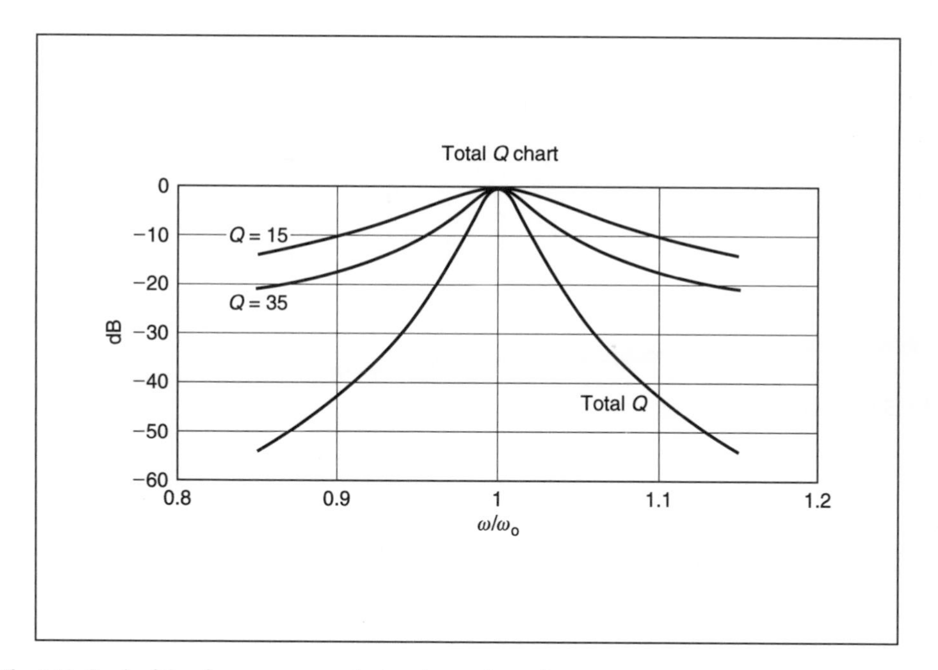

Fig. 5.10 Graph of the r.f. response curve of a broadcast radio receiver

5.3 Filter design

I will now refer you to Figure 5.2 where filters are interspersed between amplifiers to determine the frequency response of an amplifier block. We will commence by discussing the main types of filters and later provide details of how these can be designed.

5.3.1 Introduction

Figure 5.11 shows a multi-element low pass filter which is used at low frequencies. As frequency increases, the circuit elements C_1, L_2, C_3, L_4 and C_5 decrease and at microwave frequencies these element values become very small. In fact, in many cases, these calculated values are simply too small in value to implement as lumped elements and transmission line elements are used to provide the equivalent. Such a microstrip filter is shown in Figure 5.12.

Comparing the two figures, it can be seen that capacitors are represented by low impedance lines while inductors are represented by high impedance lines. You should not be surprised by this innovation because in Section 2.13.3 we have already shown you how transmission lines can be used to construct inductors and capacitors.

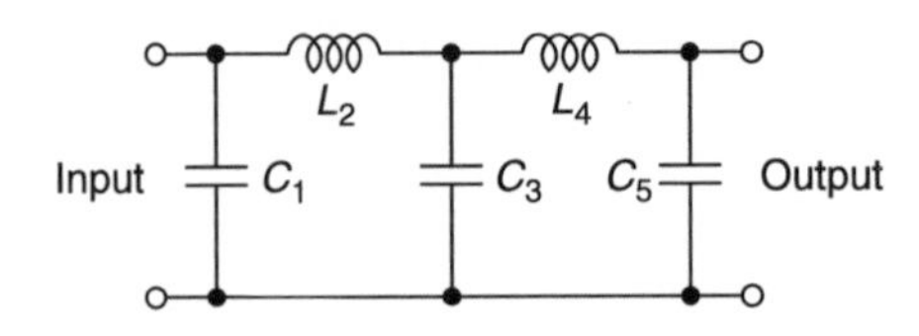

Fig. 5.11 Low pass filter

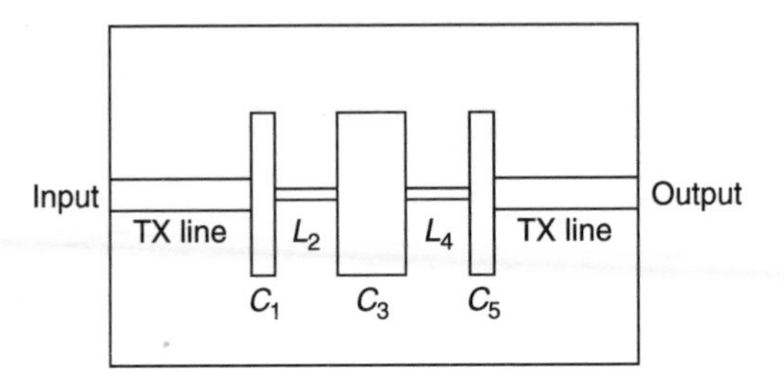

Fig. 5.12 Microstrip low pass filter

An example of how a high pass filter (Figure 5.13) can be implemented in microstrip line is shown in Figure 5.14. In this case similar microstrip type configurations are used to represent inductors and capacitors.

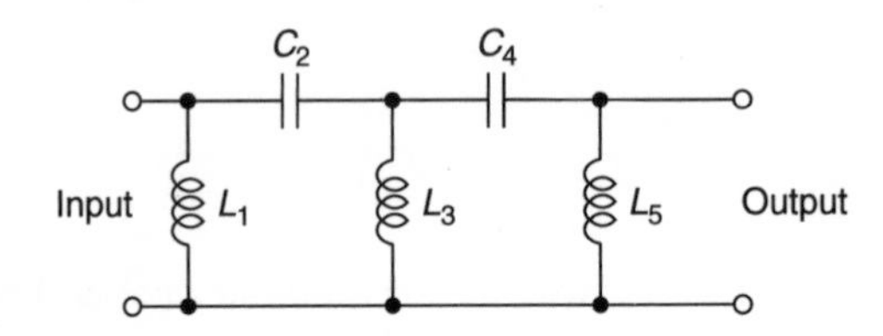

Fig. 5.13 High pass filter

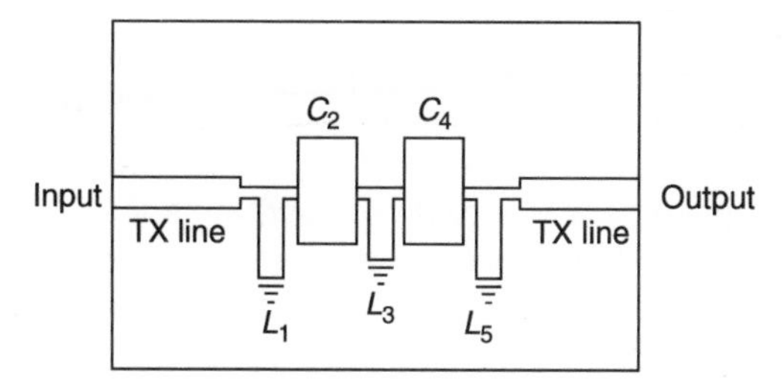

Fig. 5.14 Microstrip high pass filter

Finally in Figures 5.15 and 5.16, we show how coupled filter circuits can be constructed in microstrip configuration.

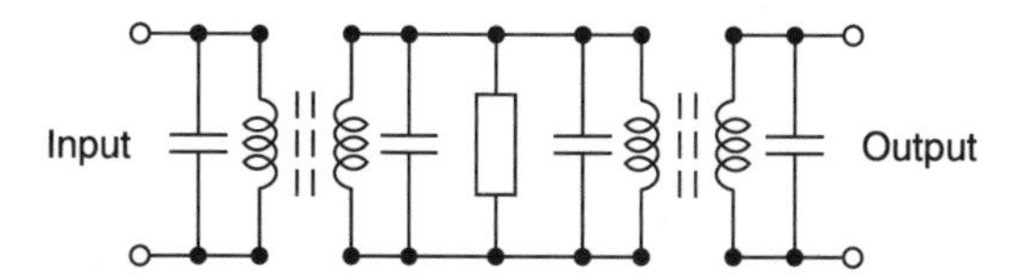

Fig. 5.15 Coupled tuned circuits

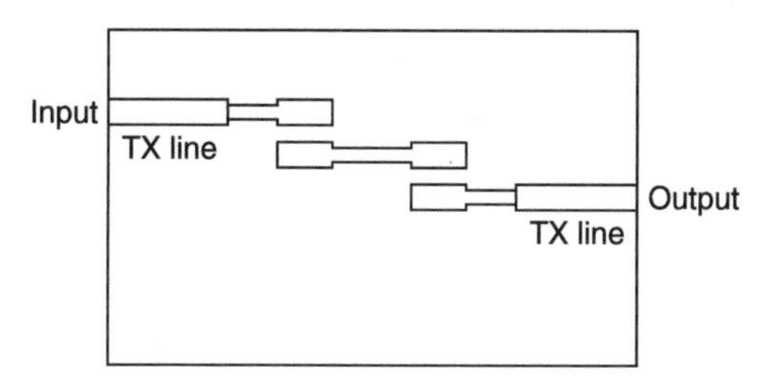

Fig. 5.16 Microstrip coupled circuits

In general, most microwave filters are first designed as conventional filters and then the calculated values are translated into microwave elements. In the above figures, we have only shown microstrip filters but there is no reason why microwave filters cannot be made in other configurations such as transmission lines and waveguides. Microstrip lines are more popular because they can be made easily and are relatively cheap.

5.3.2 Overview of filters

In this section we will show you how to select and design various types of multi-element filters for dedicated purposes. Once these methods have been learnt then it becomes comparatively easy to design microwave filters. Hence the following sections will concentrate on:

- formulating your filter performance requirements;
- deciding which type of filter network you need to meet these requirements;
- calculating or finding out where the normalised element values are published;
- performing simple multiplication and/or division to obtain the component values.

5.3.3 Specifying filters

The important thing to bear in mind is that although the discussion on filters starts off by describing low pass filters, we will show you later by examples how easy it is to change a low pass filter into a high pass, a bandpass or a bandstop filter.

Figure 5.17(a) shows the transmission characteristics of an ideal low pass filter on a normalised frequency scale, i.e. the frequency variable (f) has been divided by the passband line frequency (f_p). Such an ideal filter cannot, of course, be realised in practice. For a practical filter, tolerance limits have to be imposed and it may be represented pictorially as in Figure 5.17(b).

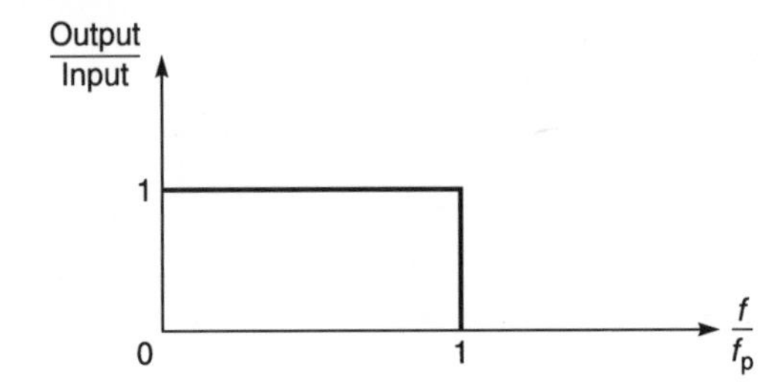

Fig. 5.17(a) Ideal filter

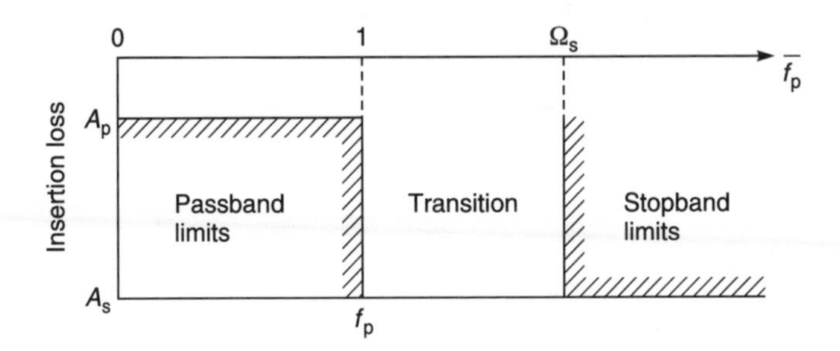

Fig. 5.17(b) Practical filter

The frequency spectrum is divided into three parts, first the passband in which the insertion loss (A_p) is to be less than a prescribed maximum loss up to a prescribed minimum frequency (f_p). The second part is the transition limit of the passband frequency limit f_p and a frequency Ω_s in which the transition band attenuation must be greater than its design attenuation. The third part is the stopband limit in which the insertion loss or attenuation is to be greater than a prescribed minimum number of decibels.

Hence, the performance requirement can be specified by five parameters:

- the filter impedance Z_0
- the passband maximum insertion loss (A_p)
- the passband frequency limit (f_p)
- the stopband minimum attenuation (A_s)
- the lower stopband frequency limit (Ω_s)

Table 5.2 Equivalence between reflection coefficient, RLR, A_p and VSWR

Maximum reflection coefficient ρ%	Minimum return loss ratio RLR(dB)	Maximum passband insertion loss A_p (dB)	$VSWR = \frac{Z_{out}}{Z_{in}}$
1	40	0.00043	1.020
1.7	35	0.001	1.033
2	34	0.0017	1.041
3	30	0.0043	1.062
4	28	0.007	1.083
5	26	0.01	1.105
8	22	0.028	1.174
10	20	0.043	1.222
15	16	0.1	1.353
20	14	0.18	1.50
25	12	0.28	1.667
33	10	0.5	1.984
45	7	1	2.661
50	6	1.25	3
61	4.3	2	4.12
71	3	3	5.8

Sometimes, manufacturers prefer to specify passband loss in terms of return loss ratio (RLR) or reflection coefficient (ρ). We provide Table 5.2 to show you the relationship between these parameters. If the values that you require are not in the table, then use the set of formulae we have provided to calculate your own values.

These parameters are inter-related by the following equations, assuming loss-less reactances:

$$RLR = -20 \log |\rho| \tag{5.12}$$

$$A_p = 10 \log (1 - |\rho|^2) \tag{5.13}$$

and

$$VSWR = \frac{Z_{out}}{Z_{in}} = \frac{1 + |\rho|}{1 - |\rho|} \tag{5.14}$$

5.3.4 Types of filters

There are many types of filter. The more popular ones are:

- Butterworth or maximally flat filter;
- Tchebyscheff (also known as Chebishev) filter;
- Cauer (or elliptical) filter for steeper attenuation slopes;
- Bessel or maximally flat group delay filter.

All of these filters have advantages and disadvantages and the one usually chosen is the filter type that suits the designer's needs best. You should bear in mind that each of these filter types is also available in low pass, high pass, bandpass and stopband configurations. We will discuss in detail the Butterworth and the Tchebyscheff filters.

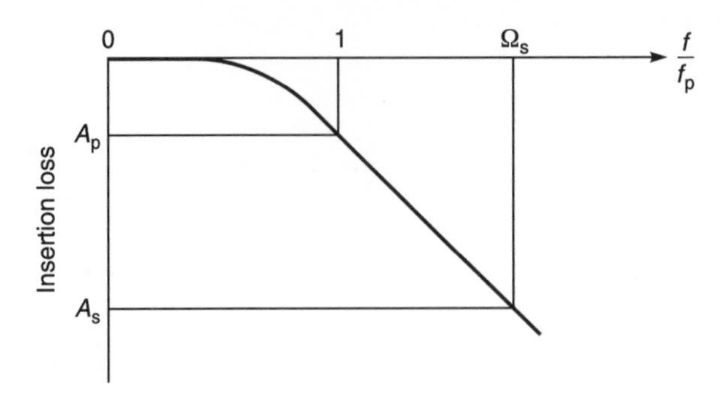

Fig. 5.18 Butterworth filter

5.4 Butterworth filter

Figure 5.18 shows the response of the maximally flat, power law or Butterworth type which is used when a fairly flat attenuation in the passband is required. The Butterworth filter achieves the ideal situation only at the ends of the frequency spectrum. At zero frequency the insertion loss is zero, at low frequencies the attenuation increases very gradually, the curve being virtually flat. With increasing frequency the attenuation rises until it reaches the prescribed limit.

At the 3 dB frequency there is a point of inflexion, and thereafter the cut-off rate increases to an asymptotic value of 6*n* dB/octave, where *n* is the number of arms. As the number of arms is increased the approximation to the ideal improves, the passband response becomes flatter and the transition sharper. For instance, for 3, 5 and 7 arms, the 1 dB loss passband frequencies are 0.8, 0.875 and 0.91 respectively of the 3 dB frequency, and the corresponding 40 dB frequencies are 4.6, 2.6 and 1.9 times the 3 dB frequency.

For values of A_s greater than 20 dB and RLR greater than 10 dB, you can use Equation 5.15 to calculate the number of arms (*n*) required in the filter for a given attenuation at a given frequency:

$$n = \frac{A_s + \text{RLR}}{[20 \log \Omega_s]} \tag{5.15}$$

Alternatively if you prefer, you can use the ABAC of Table 5.3 to get the same result. To use the ABAC, simply lay a ruler across any two parameters and read the third parameter. This is best demonstrated by an example.

Example 5.3

A low pass Butterworth filter is to have a cut-off frequency of 100 MHz. At 260 MHz, the minimum attenuation in the stopband must be greater than 40 dB. Estimate the number of arms required for the filter.

Solution. In terms of normalised units (Ω_s), 260 MHz/100 MHz = 2.6. Using Equation 5.15

$$n = \frac{A_s + \text{RLR}}{20 \log \Omega_s} = \frac{40}{20 \log 2.6} = 4.82 \approx 5 \text{ arms}$$

Alternatively, using the ABAC of Table 5.3 and drawing a straight line between 40 on the left and 2.6 on the right will also give an answer of $n < 5$ arms.

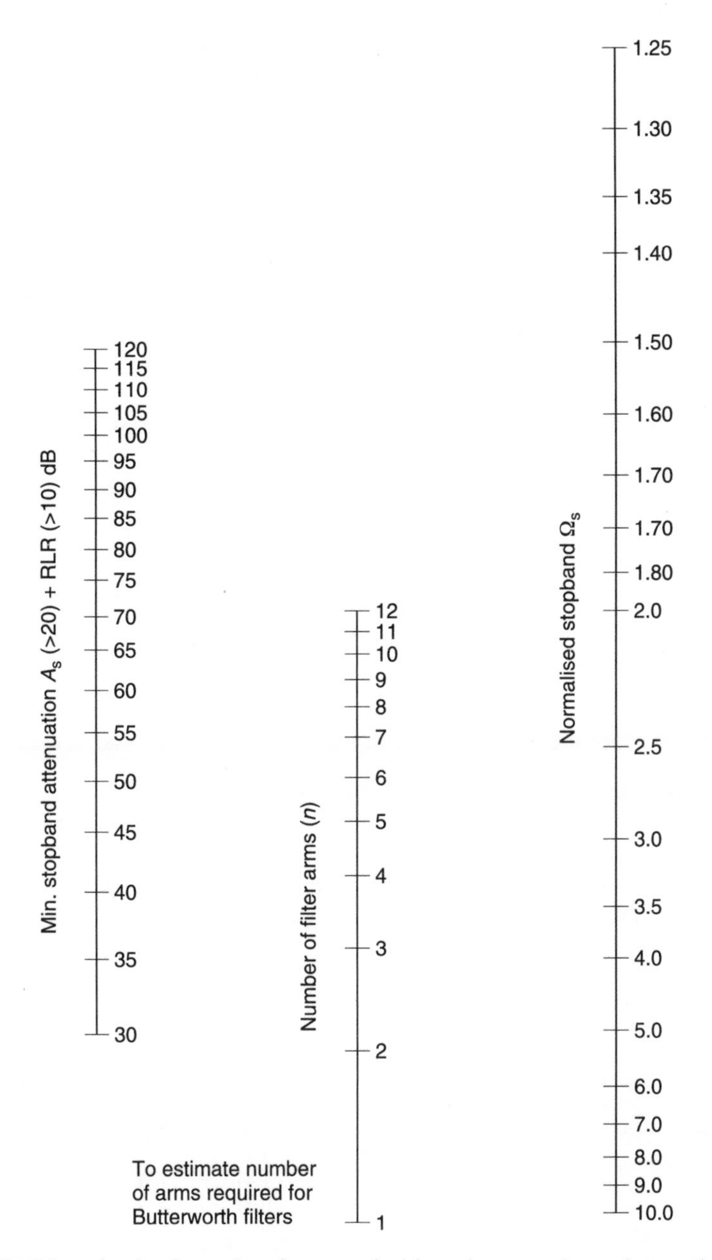

Table 5.3 ABAC for estimating the number of arms required for a given return loss and attenuation

5.4.1 Normalised parameters

Each type of filter also has one or more sets of normalised parameters which are used for calculating its component values. Normalised parameters are values which a low pass filter would assume for its components if it was designed for (i) 1 Ω termination and (ii) an operating angular frequency of 1 rad/s. The reason for choosing 1 Ω and 1 rad/s is that it enables easy scaling for different filter impedances and operating frequencies. Another point you should note about Figure 5.19 is that the same configuration can be used for a high pass filter by simply interchanging *L* with *C*, etc. Bandpass and bandstop filters can also be produced in this manner. We shall carry out all these manipulations later in the design examples.

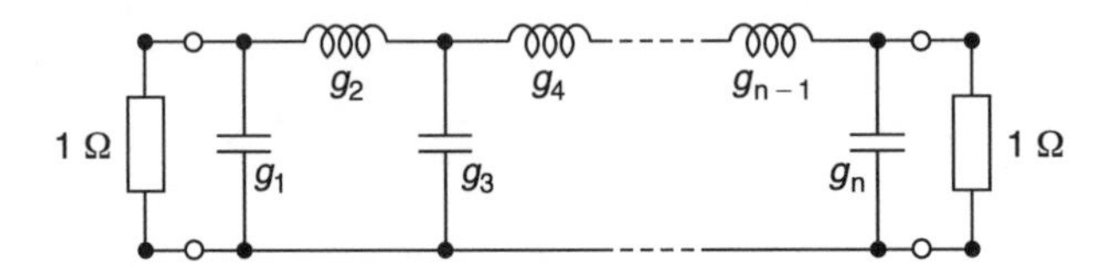

Fig. 5.19 Schematic of a Butterworth normalised low pass filter

The set of normalised parameters for a Butterworth filter can be calculated from Equation 5.16:

$$g_k = 2 \sin \left[\frac{(2k - 1)\pi}{2n} \right] \tag{5.16}$$

where

k = position of element in array [$k = 1, 2, 3, \ldots, n - 1, n$]
n = number of elements for the filter

Note: Before you take the sine value, check that your calculator is set to read radians.

To save you the problem of calculating values, we attach a set of values calculated on a spreadsheet. These are shown in Table 5.4. In this table, *n* signifies the number of components that you are going to use in your filter; *k* signifies the position of the element. You can get a diagrammatic view of the system by referring to Figure 5.19. The values in the tables are constants but they represent Henries or Farads according to

Table 5.4 Butterworth normalised values

k/*n*	2	3	4	5	6	7	8
1	1.4142	1.0000	0.7654	0.6180	0.5176	0.4550	0.3902
2	1.4142	2.0000	1.8478	1.6180	1.4142	1.2470	1.1111
3		1.0000	1.8478	2.0000	1.9319	1.8019	1.6629
4			0.7654	1.6180	1.9319	2.0000	1.9616
5				0.6180	1.4142	1.8019	1.9616
6					0.5176	1.2470	1.6629
7						0.4450	1.1111
8							0.3902

the configuration in which they are used. This is best demonstrated by using simple examples.

5.4.2 Low pass filter design

A typical procedure for low pass filter design is given below, followed by a design example.

Procedure to design a low pass filter

1 Decide the passband and stopband frequencies.
2 Decide on the stopband attenuation.
3 Decide on the type of filter (Butterworth, etc.) you want to use, bearing in mind that some types such as the Butterworth filter give better amplitude characteristics while the Bessel filter gives better group delay.
4 Calculate the number of arms you need to achieve your requirements.
5 Calculate or use normalised tables to find the values of the filter elements. These normalised values (in farads and henries) are the component values required to make a low pass filter with an impedance of 1 Ω and a passband limit frequency (f_p) of one rad/s.
6 To make a filter having a different impedance, e.g. 50 Ω, the *impedance* of each component must be *increased* by the impedance ratio, i.e. all inductances must be multiplied and all capacitances must be divided by the impedance ratio.
7 To make a filter having a higher band limit than the normalised 1 rad/s, divide the value of each component by (2π times the frequency).

Example 5.4

A five element maximally flat (Butterworth) low pass filter is to be designed for use in a 50 Ω circuit. Its 3 dB point is 500 MHz. Calculate its component values.

Given: Five element low pass Butterworth filter, $f_p = 500$ MHz, $Z_0 = 50\ \Omega$.
Required: Calculation of five elements for a low pass filter.

Solution. The required low pass filter circuit is shown in Figure 5.20. The normalised values for this filter will be taken from column 5 of Table 5.4 because we want a five element Butterworth filter.

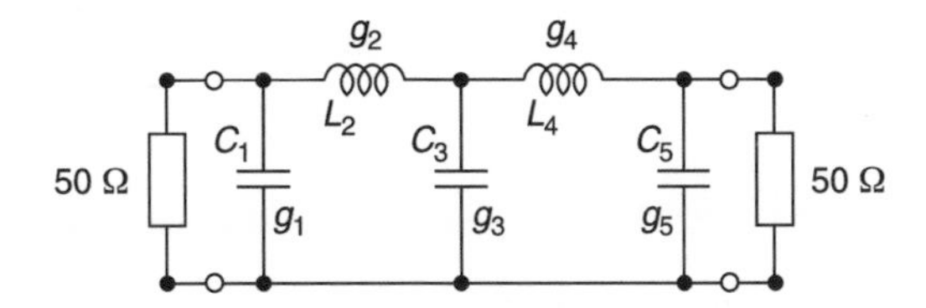

Fig. 5.20 Low pass configuration

Table 5.5 shows how the filter design is carried out.

Table 5.5 Calculated values for a low pass Butterworth filter

Circuit reference	Normalised $Z_0 = 1\ \Omega$ $\omega = 1$ rad/s	$Z_0 = 50\ \Omega$ $f = 1/(2\pi)$ Hz	$Z_0 = 50\ \Omega$ $f_p = 500$ MHz
g_1 or C_1	0.6180 F	$\frac{0.6180}{50}$ F	$\frac{0.6180}{50 \times 2\pi \times 500\ \text{MHz}}$ F or 3.93 pF
g_2 or L_2	1.6180 H	1.6180×50 H	$\frac{1.6180 \times 50}{2\pi \times 500\ \text{MHz}}$ H or 25.75 nH
g_3 or C_3	2.0000 F	$\frac{2.0000}{50}$ F	$\frac{2.0000}{50 \times 2\pi \times 500\ \text{MHz}}$ F or 12.73 pF
g_4 or L_4	1.6180 H	1.6180×50 H	$\frac{1.6180 \times 50}{2\pi \times 500\ \text{MHz}}$ H or 25.75 nH
g_5 or C_5	0.6180 F	$\frac{0.6180}{50}$ F	$\frac{0.6180}{50 \times 2\pi \times 500\ \text{MHz}}$ F or 3.93 pF

Hence, the calculated values for a five element low pass Butterworth filter with a nominal impedance of 50 Ω and a 3 dB cut-off frequency at 500 MHz are:

$$C_1 = 3.93\ \text{pF},\ L_2 = 25.75\ \text{nH},\ C_3 = 12.73\ \text{pF},\ L_4 = 25.75\ \text{nH and } C_5 = 3.93\ \text{pF}$$

The response of the filter designed in the above example is shown in Figure 5.21.

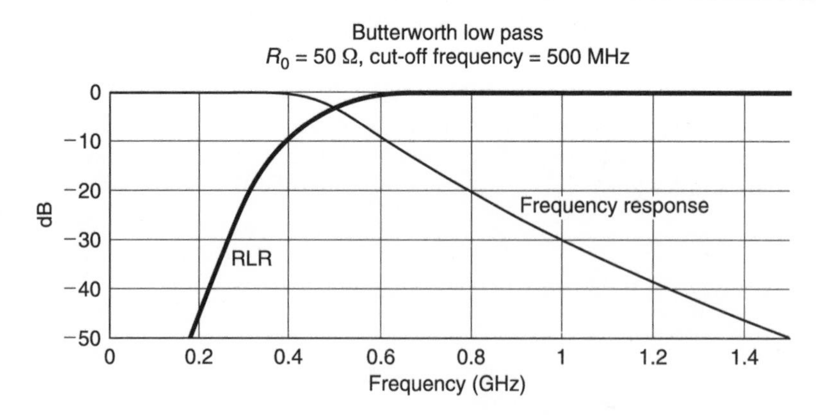

Fig. 5.21 Results of Example 5.4

Alternatively to save yourself work, you can use PUFF for the result. This is shown in Figure 5.22.

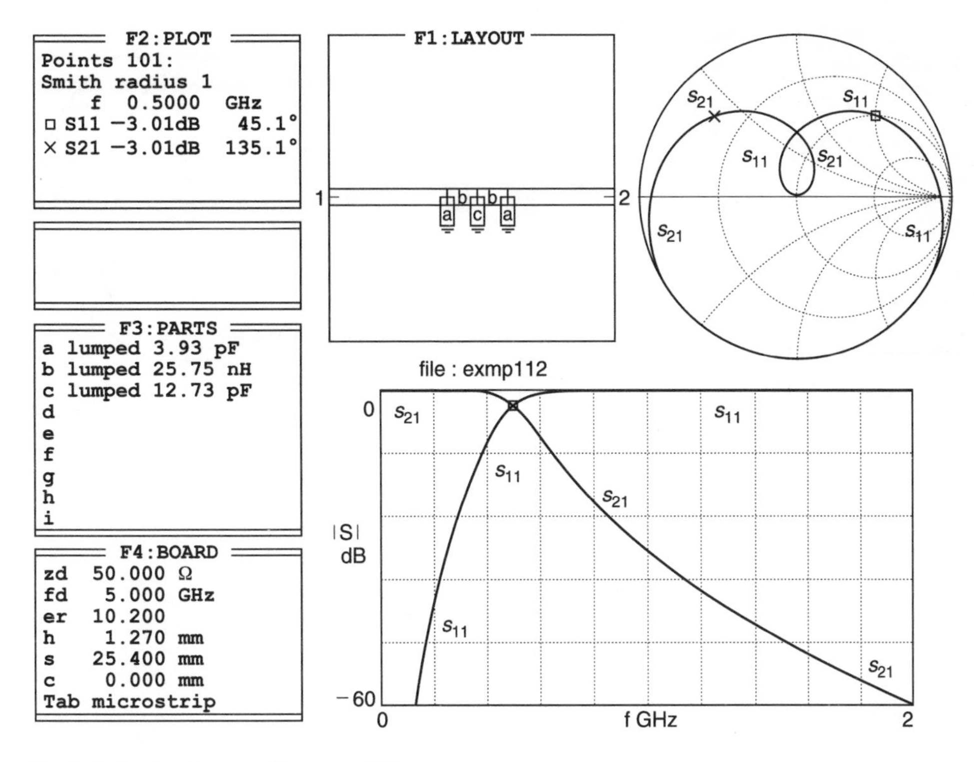

Fig. 5.22 Results of low pass filter using PUFF

5.4.3 Low pass filter example

Low pass filters can also be designed using transmission lines. We will not do it here because (i) the circuit elements must first be translated into electrical line lengths, and their end capacitances and discontinuities must be calculated, (ii) the effect on the other components must be compensated which means altering line lengths again, and (iii) changing line lengths of each element in turn, then again compensating for the effect of each line length on the other elements. The whole process is rather laborious and is best done by computer design. You can find more detail in filter design from two well known books, the first by Matthei, Young and Jones and the other by T. C. Edwards.[2]

5.4.4 Low pass filter using microstrip lines

We use PUFF to show an example of a distributed low pass filter.

Example 5.5

Figure 5.23 shows how a low pass filter can be produced using transmission lines of varying impedances and lengths. You should note that distributed line filters of this type also conduct

[2] G. L. Matthei, L. Young and E. M. T. Jones, *Microwave filters, impedance matching networks and coupling structures*, McGraw-Hill, New York NY, 1964 and T. C. Edwards, *Foundations for microstrip design*, second edition, John Wiley and Sons, 1992.

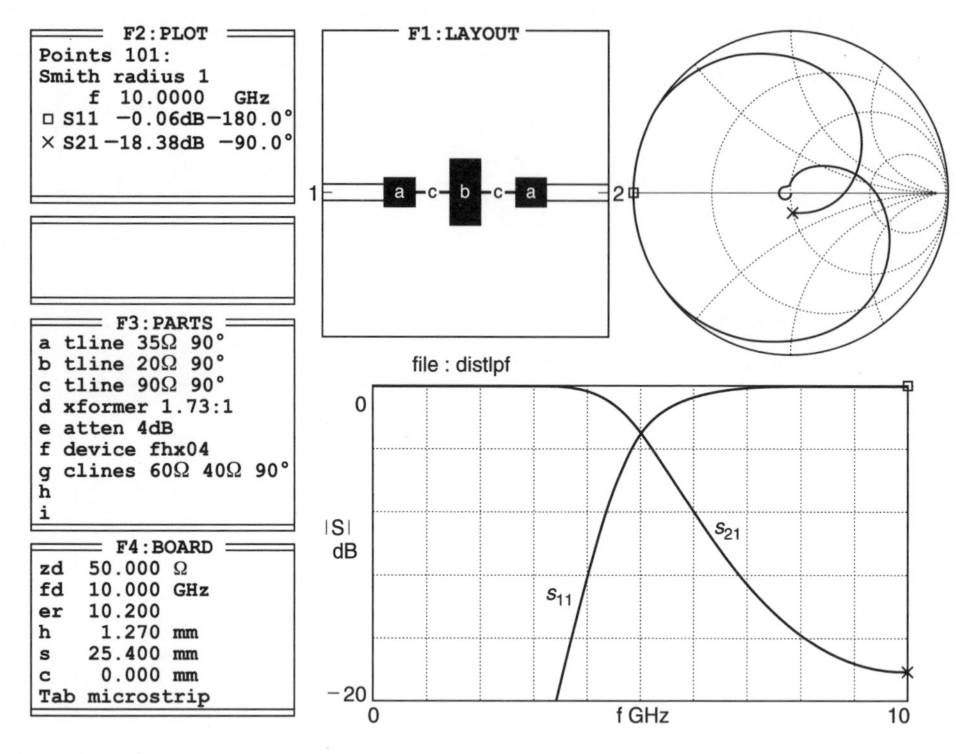

Fig. 5.23 Low pass filter construction using microstrip

d.c. and that if you want to block d.c. then you should use d.c. blocking components such as series capacitors. For details of the filter elements, refer to the F3 box of Figure 5.23.

5.4.5 High pass filter

A typical procedure for high pass filter design is given below. It is immediately followed by a design example.

Procedure to design a high pass filter

1 Decide the passband and stopband frequencies.
2 Decide on the stopband attenuation.
3 Decide on the type of filter (Butterworth, etc.) you want to use, bearing in mind that some types such as the Butterworth filter give better amplitude characteristics while the Bessel filter gives better group delay.
4 Calculate the number of arms you need to achieve your requirements.
5 Calculate or use normalised tables to find the values of the filter elements. These normalised values (in farads and henries) are the component values required to make a low pass filter with an impedance of 1 Ω and a passband limit frequency (f_p) of 1 rad/s.
6 To calculate the corresponding high pass filter, we must (a) replace each capacitor by an inductor and each inductor by a capacitor and (b) give each component a normalised value equal to the *reciprocal* of the normalised component it replaces.

7 To make a filter having a different impedance, e.g. 50 Ω, the *impedance* of each component must be *increased* by the impedance ratio, i.e. all inductances must be multiplied, all capacitances must be divided by the impedance ratio.
8 To make a filter having a higher band limit than the normalised 1 rad/s, divide the value of each component by (2π times the frequency).

Example 5.6 High pass filter design

A five element maximally flat (Butterworth) high pass filter is to be designed for use in a 50 Ω circuit. Its 3 dB point is 500 MHz. Calculate its component values. Hint: note that this is the high pass equivalent of the low pass filter designed previously.

Given: Five element high pass Butterworth filter, $f_p = 500$ MHz, $Z_0 = 50\ \Omega$.
Required: Calculation of five elements for a high pass filter.

Solution. The circuit is shown in Figure 5.24.

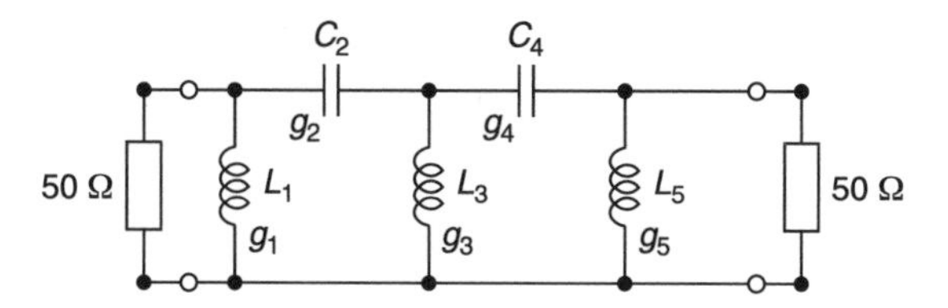

Fig. 5.24 High pass configuration

The normalised values for this filter will be taken from column 5 of Table 5.4 because we want a five element Butterworth filter. Table 5.6a shows how this is carried out.

Table 5.6a Calculated values for a high pass Butterworth filter

Circuit reference	Normalised $Z_0 = 1\ \Omega$ $\omega = 1$ rad/s	$Z_0 = 50\ \Omega$ $f = 1/(2\pi)$ Hz	$Z_0 = 50\ \Omega$ $f_p = 500$ MHz
g_1 or L_1	$\dfrac{1}{0.6180}$ H	$\dfrac{50}{0.6180}$ H	$\dfrac{50 \times 10^9}{0.6180 \times 2\pi \times 500 \times 10^6} = 25.75$ nH
g_2 or C_2	$\dfrac{1}{1.6180}$ F	$\dfrac{1}{1.6180 \times 50}$ F	$\dfrac{1 \times 10^{12}}{1.618 \times 50 \times 2\pi \times 500 \times 10^6} = 3.93$ pF
g_3 or L_3	$\dfrac{1}{2.0000}$ H	$\dfrac{50}{2.0000}$ H	$\dfrac{50 \times 10^9}{2.000 \times 2\pi \times 500 \times 10^6}$ 7.95 nH
g_4 or C_4	$\dfrac{1}{1.6180}$ F	$\dfrac{1}{1.6180 \times 50}$ F	$\dfrac{1 \times 10^{12}}{1.618 \times 50 \times 2\pi \times 500 \times 10^6} = 3.93$ pF
g_5 or L_5	$\dfrac{1}{0.6180}$ H	$\dfrac{50}{0.6180}$ H	$\dfrac{50 \times 10^9}{0.6180 \times 2\pi \times 500 \times 10^6} = 25.75$ nH

Hence, the calculated values for a five element high pass Butterworth filter with a nominal impedance of 50 Ω and a 3 dB cut-off frequency at 500 MHz are:

$$L_1 = 25.75 \text{ nH}, C_2 = 3.93 \text{ pF}, L_3 = 7.95 \text{ nH}, C_4 = 3.93 \text{ pF and } L_5 = 25.75 \text{ nH}$$

The response of the filter designed in the above example is shown in Figure 5.25.

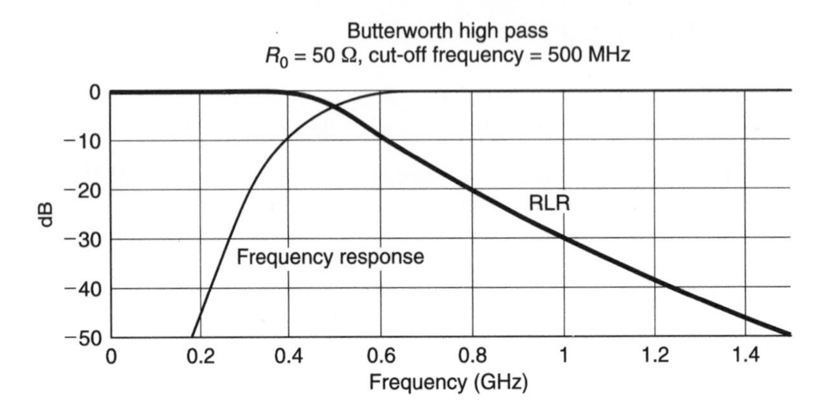

Fig. 5.25 High pass filter

You can verify this design for yourself by using PUFF. Details of this are given in Figure 5.26.

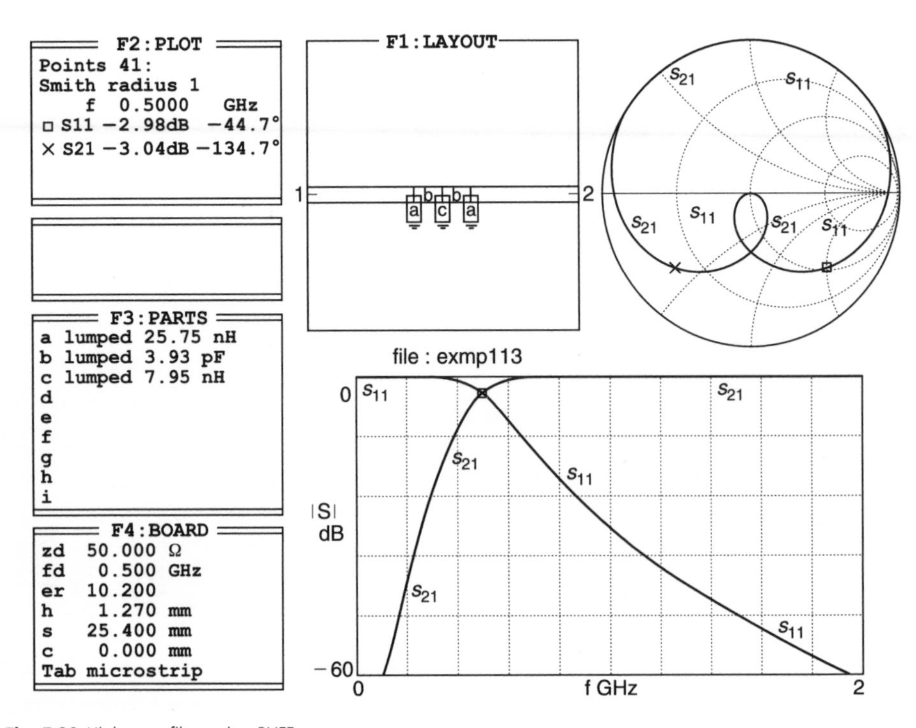

Fig. 5.26 High pass filter using PUFF

5.4.6 High pass filter using microstrip lines

An example of high pass filter design using microstrip lines is given below.

Example 5.7

The construction of a high pass filter using microstrip lines is shown in Figure 5.27. Details of this filter can be found in the same figure.

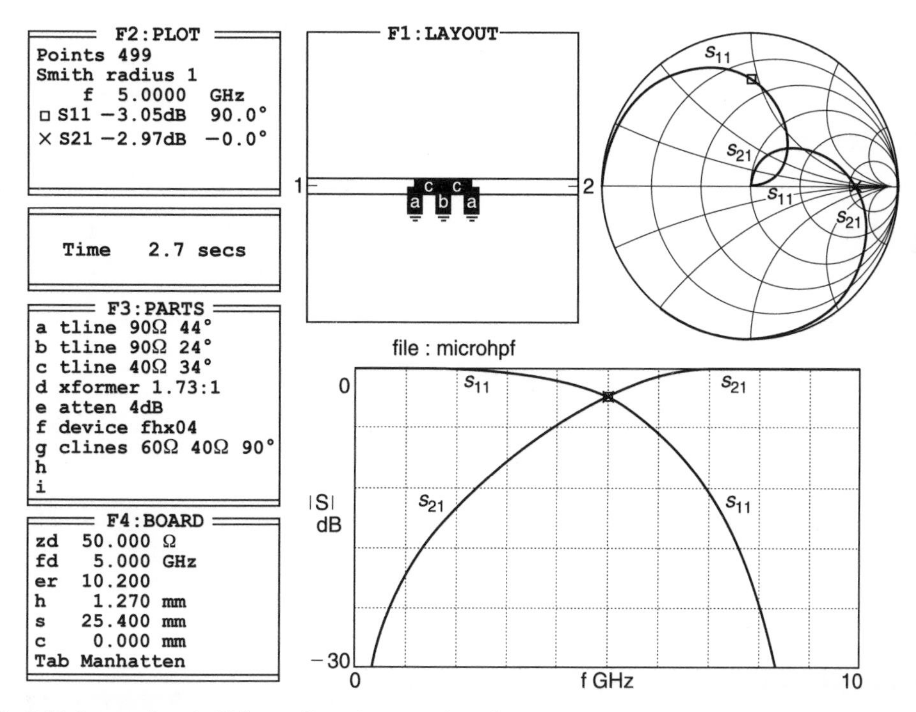

Fig. 5.27 Construction of a high pass filter using transmission lines

5.4.7 Bandpass filter

For a bandpass filter, the performance must be specified in terms of bandwidth (see Figure 5.28). The passband limit (f_p) becomes the difference between the upper frequency limit (f_b) and the lower frequency limit (f_a) of the passband, i.e. $f_p = f_b - f_a$. Similarly the frequency variable (f) becomes the frequency difference between any two points on the response curve at the same level: and the stopband limit (f_s) becomes the frequency difference between the two frequencies (f_x and f_y) outside of which the required minimum stopband attenuation (A_s) is to be achieved.

The response curve will have geometric symmetry about the centre frequency (f_0), i.e. $f_0 = \sqrt{f_a \cdot f_b} = \sqrt{f_x \cdot f_y}$. This means that the cut-off rate in dB/Hz will be greater on the

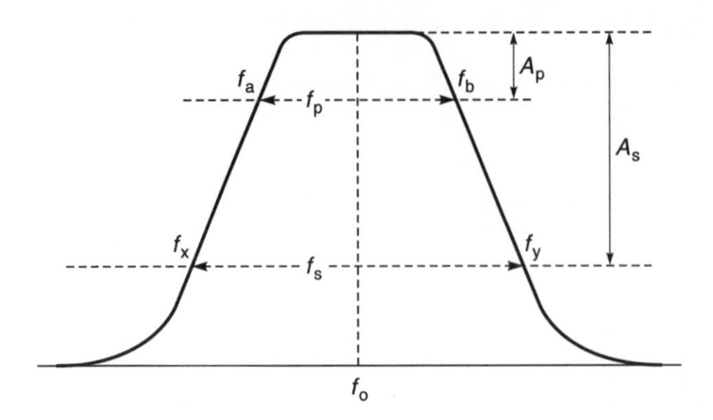

Fig. 5.28 Bandpass characteristics

low frequency side and usually the number of arms required in the filter will be dependent on the cut-off rate of the high frequency side. The normalised stopband limit is given by

$$\Omega_s = (f_s/f_p) = (f_y - f_x)/(f_b - f_a)$$

Procedure to design a bandpass filter

To evaluate a bandpass filter having a passband from f_a to f_b:

1 define the pass bandwidth $f_p = f_b - f_a$;
2 calculate the geometric centre frequency $f_0 = \sqrt{f_a f_b}$;
3 evaluate as previously the lowpass filter having its passband limit frequency equal to f_p;
4 add in series with each inductance (L) a capacitance of value $(1/(\omega_o^2 L)$ and in parallel with each capacitance (C) an inductance of value $(1/(\omega_o^2 C)$, i.e. the added component resonates the original component at the band centre frequency f_0.

Example 5.8

A five element maximally flat (Butterworth) bandpass filter is to be designed for use in a 50 Ω circuit. Its upper passband frequency limit (f_b) is 525 MHz and its lower passband frequency limit is 475 MHz. Calculate its component values. Hint: calculate the low pass filter for the passband design bandwidth, then 'translate' the circuit for operation at $f_0 = \sqrt{f_a \times f_b}$.

Given: Five element bandpass Butterworth filter, $f_p = 50$ MHz, $Z_0 = 50\ \Omega$.
Required: Calculation of five elements for a high pass filter.

Solution. The passband filter components are shown in Figure 5.29.

1 Define the passband frequency;

$$f_p = f_b - f_a = (525 - 475)\text{ MHz} = 50\text{ MHz}$$

2 Calculate the geometric mean frequency;

$$f_0 = \sqrt{f_b \times f_a} = \sqrt{525 \times 475}\text{ MHz} \approx 499.4\text{ MHz}$$

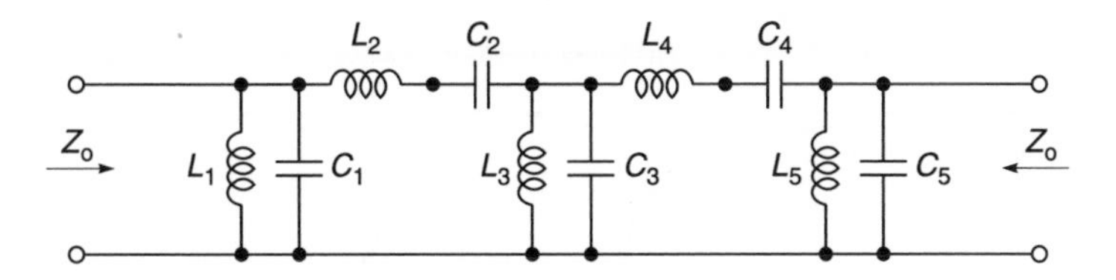

Fig. 5.29 Bandpass configuration

3 Evaluate the low pass filter having its passband limit frequency equal to f_p. Use column 5 of Table 5.4.

Table 5.7 Calculated primary values for a bandpass Butterworth filter

Circuit reference	Normalised $Z_0 = 1\ \Omega$ $\omega = 1$ rad/s	$Z_0 = 50\ \Omega$ $f = 1/(2\pi)$ Hz	$Z_0 = 50\ \Omega$ $f_p = 50$ MHz
g_1 or C_1	0.6180 F	$\frac{0.6180}{50}$ F	$\frac{0.6180}{50 \times 2\pi \times 50\ \text{MHz}}$ F or 39.343 pF
g_2 or L_2	1.6180 H	1.6180×50 H	$\frac{1.6180 \times 50}{2\pi \times 50\ \text{MHz}}$ H or 257.513 nH
g_3 or C_3	2.0000 F	$\frac{2.0000}{50}$ F	$\frac{2.0000}{50 \times 2\pi \times 50\ \text{MHz}}$ F or 127.324 pF
g_4 or L_4	1.6180 H	1.6180×50 H	$\frac{1.6180 \times 50}{2\pi \times 50\ \text{MHz}}$ H or 257.513 nH
g_5 or C_5	0.6180 F	$\frac{0.6180}{50}$ F	$\frac{0.6180}{50 \times 2\pi \times 50\ \text{MHz}}$ F or 39.343 pF

4 Add in series with each inductance (L) a capacitance of value $(1/\omega_0^2L)$ and in parallel with each capacitance (C) an inductance of value $(1/\omega_0^2C)$, i.e. the added component resonates with the original component at the band centre frequency (f_0). In this case, $f_0 = \sqrt{525 \times 475}$ MHz $\approx$ 499.4 MHz.

Table 5.7(a) Resonating values of a bandpass Butterworth filter

Low pass values			Resonating values for f_o.		
C_1	39.343	pF	L_1	2.582	nH
L_2	257.513	nH	C_2	0.394	pF
C_3	127.324	pF	L_3	0.797	nH
L_4	257.513	nH	C_4	0.394	pF
C_5	39.343	pF	L_5	2.582	nH

The response of the filter designed in the above example is shown in Figure 5.30.

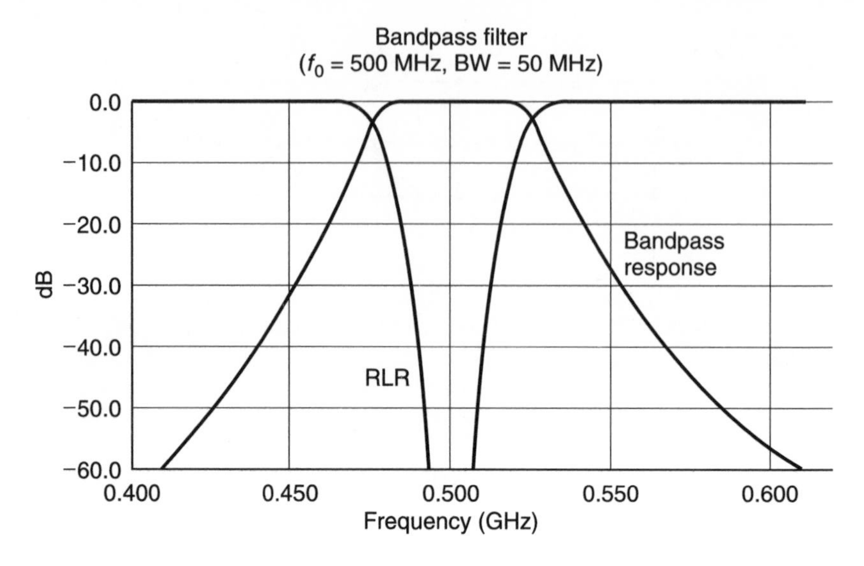

Fig. 5.30 Bandpass filter

Alternatively, you can use PUFF to give you the results shown in Figure 5.31.

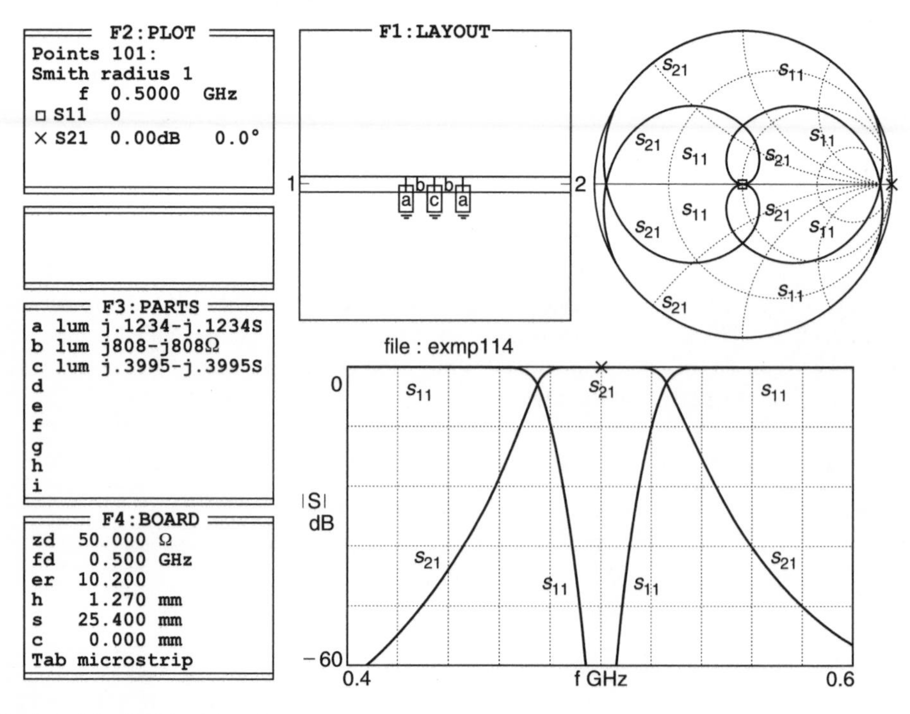

Fig. 5.31 Using PUFF to plot results

5.4.8 Bandpass filter design using microstrip lines

Example 5.9

An example of a coupled line passband filter using microstrip is shown in Figure 5.32. A coupled filter provides d.c. isolation between the input and output ports.

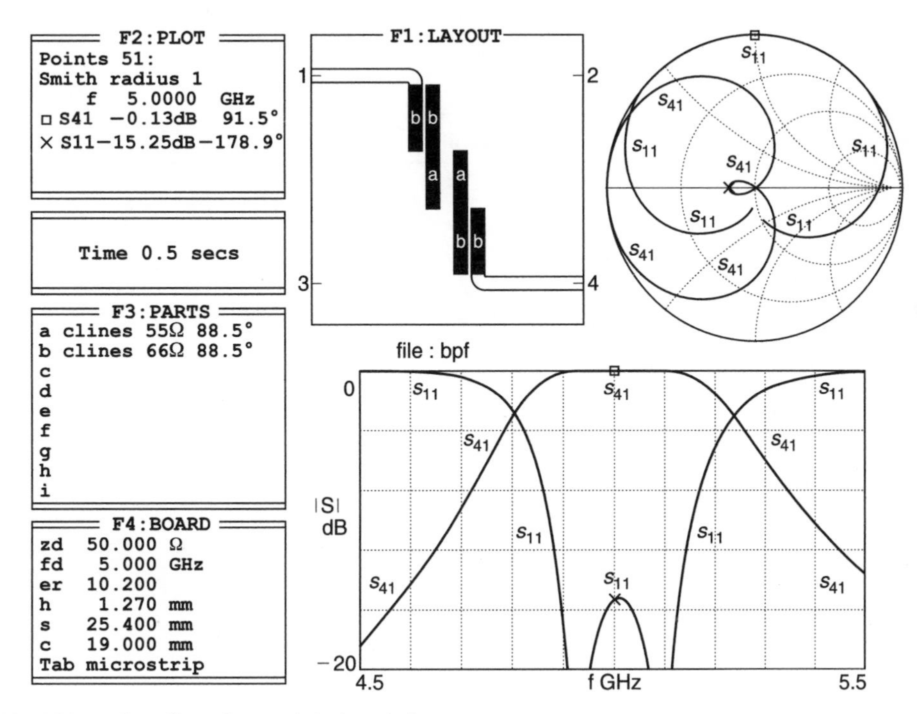

Fig. 5.32 Bandpass filter using coupled microstrip lines

Design modifications

PUFF permits modification and on-screen comparison of different designs. For this demonstration, we will use Figure 5.32 as a template and produce Figure 5.33. This is carried out by using Figure 5.32, going into the F3 box to change values and then re-plotting by pressing the **F2** key followed by pressing **ctrl + p**.

Bandpass filter modification

Figure 5.33 shows clearly how comparisons can be made to a design prior to final choice and fabrication.

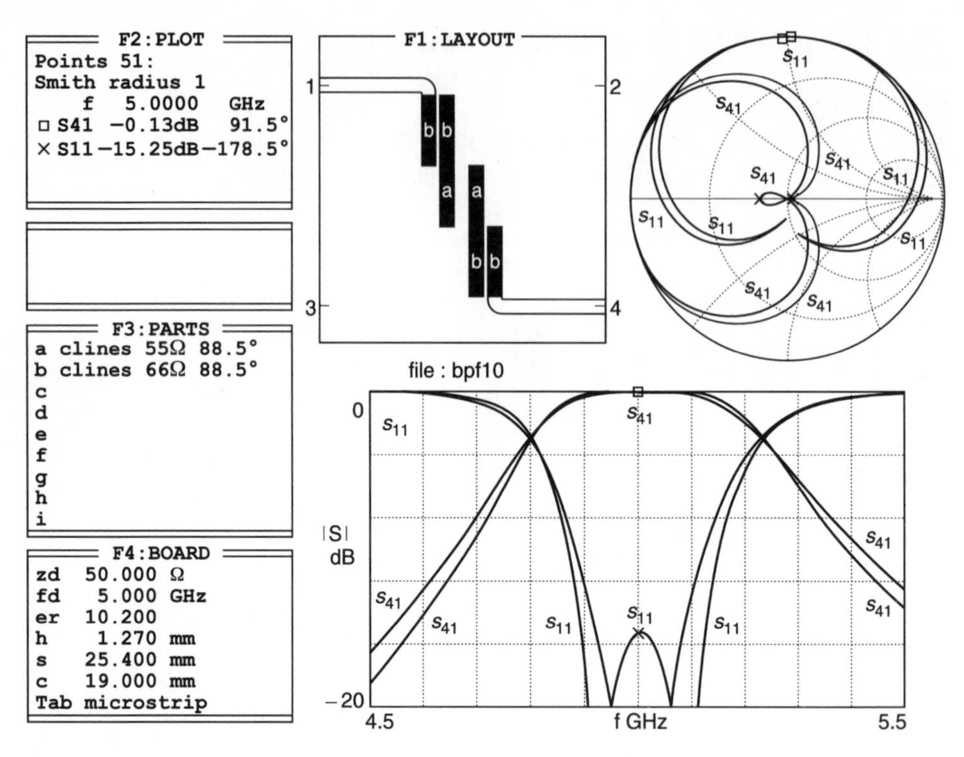

Fig. 5.33 Showing how modifications to a design can be compared to an original design to see if a change is desirable

5.4.9 Bandstop filter

For a bandstop filter, the performance must be specified in terms of bandwidth (see Figure 5.34). The stopband limit (f_p) becomes the difference between the upper frequency limit (f_b) and the lower frequency limit (f_a) of the passband, i.e. $f_p = f_b - f_a$. Similarly the frequency variable (f) becomes the frequency difference between any two points on the response curve at the same level: the stopband limit (f_s) becomes the frequency difference

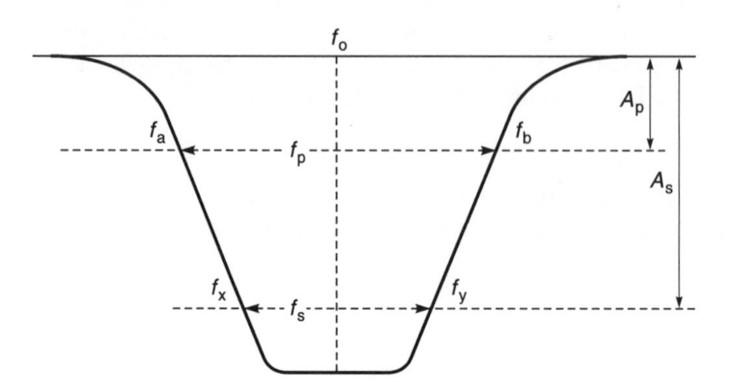

Fig. 5.34 Stoppass characteristics

between the two frequencies (f_x and f_y) inside of which the required minimum stopband attenuation (A_s) is to be achieved.

The response curve will have geometric symmetry about the centre frequency (f_0), i.e. $f_0 = \sqrt{f_a f_b} = \sqrt{f_x f_y}$. This means that the cut-off rate in dB/Hz will be greater on the low frequency side and usually the number of arms required in the filter will be dependent on the cut-off rate of the high frequency side. The normalised stopband limit is given by

$$\Omega_s = (f_s/f_p) = (f_y - f_x)/(f_b - f_a)$$

Procedure to design a bandstop filter

To evaluate a bandstop filter having a stopband from f_a to f_b:

1 define the stop bandwidth $f_p = f_b - f_a$;
2 calculate the geometric centre frequency $f_0 = \sqrt{f_a f_b}$.
3 evaluate as previously the high pass filter having its stopband limit frequency equal to f_p;
4 add in series with each inductance (L) a capacitance of value ($1/\omega_o^2 L$) and in parallel with each capacitance (C) an inductance of value ($1/\omega_o^2 C$), i.e. the added component resonates the original at the band centre f_0.

Example 5.10

A five element maximally flat (Butterworth) bandstop filter is to be designed for use in a 50 Ω circuit. Its upper stopband frequency limit (f_b) is 525 MHz and its lower stopband frequency limit is 475 MHz. Calculate its component values. Hint: calculate the high pass filter for the stopband design bandwidth, then 'translate' the circuit for operation at $f_0 = \sqrt{f_b f_a}$.

Given: Five element band-stop Butterworth filter, f_p = 50 MHz, Z_0 = 50 Ω.
Required: Calculation of ten elements for a bandstop filter.

Solution. The bandstop filter components are shown in Figure 5.35.

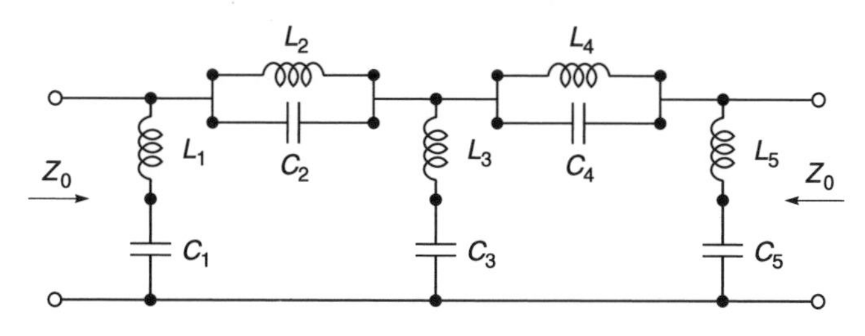

Fig. 5.35 Bandstop configuration

1 Define the stopband frequency:

$$f_p = f_b - f_a = (525 - 475)\text{ MHz} = 50\text{ MHz}$$

2 Calculate the geometric mean frequency

$$f_0 = \sqrt{f_b \times f_a} = \sqrt{525 \times 475}\text{ MHz} \approx 499.4\text{ MHz}$$

3 Evaluate the high pass filter having its passband limit frequency equal to f_p. The normalised values have originally been taken from column 5 of Table 5.4 but because we are evaluating a high pass filter, the reciprocal values have been used.

Table 5.8 Calculated primary values for a bandstop Butterworth filter

Circuit reference	Normalised $Z_0 = 1\ \Omega$ $\omega = 1$ rad/s	$Z_0 = 50\ \Omega$ $f = 1/(2\pi)$ Hz	$Z_0 = 50\ \Omega$ $f_p = 50$ MHz
g_1 or L_1	$\frac{1}{0.6180}$ H	$\frac{50}{0.6180}$ H	$\frac{50 \times 10^9}{0.6180 \times 2\pi \times 50 \times 10^6} = 257.5$ nH
g_2 or C_2	$\frac{1}{1.6180}$ F	$\frac{1}{1.6180 \times 50}$ F	$\frac{1 \times 10^{12}}{1.618 \times 50 \times 2\pi \times 50 \times 10^6} = 39.3$ pF
g_3 or L_3	$\frac{1}{2.0000}$ H	$\frac{50}{2.0000}$ H	$\frac{50 \times 10^9}{2.000 \times 2\pi \times 50 \times 10^6}$ 79.5 nH
g_4 or C_4	$\frac{1}{1.6180}$ F	$\frac{1}{1.6180 \times 50}$ F	$\frac{1 \times 10^{12}}{1.618 \times 50 \times 2\pi \times 50 \times 10^6} = 39.3$ pF
g_5 or L_5	$\frac{1}{0.6180}$ H	$\frac{50}{0.6180}$ H	$\frac{50 \times 10^9}{0.6180 \times 2\pi \times 50 \times 10^6} = 257.5$ nH

4 Add in series with each inductance (L) a capacitance of value $1/(\omega_o^2 L)$ and in parallel with each capacitance (C) an inductance of value $1/(\omega_o^2 C)$, i.e. the added component resonates with the original components at the band centre f_0. In this case, $f_0 = \sqrt{525 \times 475}$ MHz $\approx$ 499.4MHz.

Table 5.9 Resonating values for a bandstop Butterworth filter

High pass values			Resonating values for f_0		
L_1	257.53	nH	C_1	0.39	pF
C_2	39.35	pF	L_2	2.58	nH
L_3	79.58	nH	C_3	1.27	pF
C_4	39.35	pF	L_4	2.58	nH
L_5	257.53	nH	C_5	0.39	pF

The response of the filter designed in the above example is shown in Figure 5.36. Alternatively, you can use PUFF to plot the result as shown in Figure 5.37.

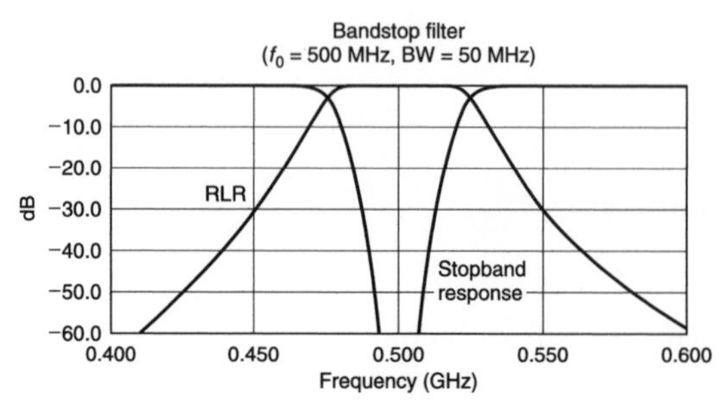

Fig. 5.36 Bandstop filter

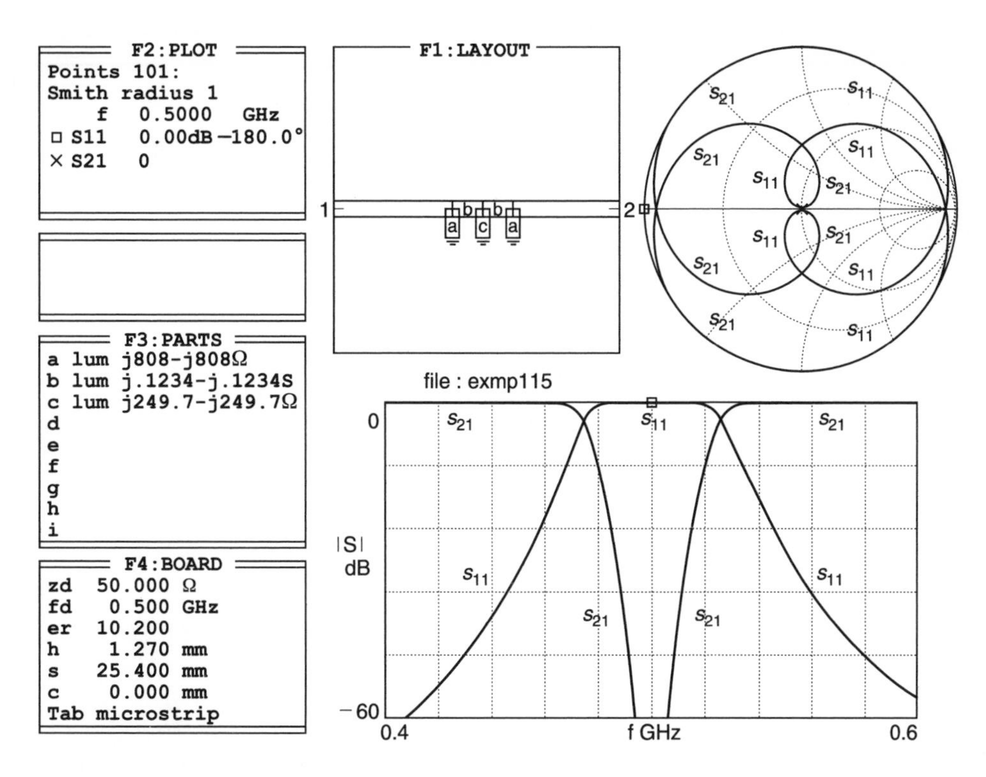

Fig. 5.37 Bandstop filter using PUFF

5.5 Tchebyscheff filter

Figure 5.38 shows the response of a filter with a Tchebyscheff or equal ripple type of characteristic. Its response differs from the Butterworth filter in that (i) there is a ripple in the passband response and (ii) the transition region from passband to stopband is more pronounced. In short, the filter 'trades off' passband ripple (A_m) to achieve a greater skirt loss for the same number of filter components. The penalty for using a Tchebyscheff filter is that there is also greater group-delay distortion.

The passband approaches the ideal filter at a number of frequencies, between which the insertion loss is allowed to reach the design limit (A_p). This results in a better approximation and the transition becomes steeper than that of a Butterworth filter with the same number of arms, e.g. for a filter with 0.1 dB passband ripple and having 3, 5 or 7 arms, the 1 dB frequency is 0.86, 0.945 or 0.97 respectively times the 3 dB frequency and the corresponding 40 dB frequency is 3.8, 2.0 or 1.5 times the 3 dB frequency. For $A_s > 20$ dB and RLR > 10 dB the number of arms required may be estimated from Figure 5.39.

Note that Tchebyscheff type filters, having an even number of arms, may have different terminal impedances, the ratio of which is a function of A_p, as shown in Figure 5.39. These filters are sometimes designated by 'b'. Tchebyscheff filters, having an even number of arms and equal terminating impedances, may be designated by 'c'.

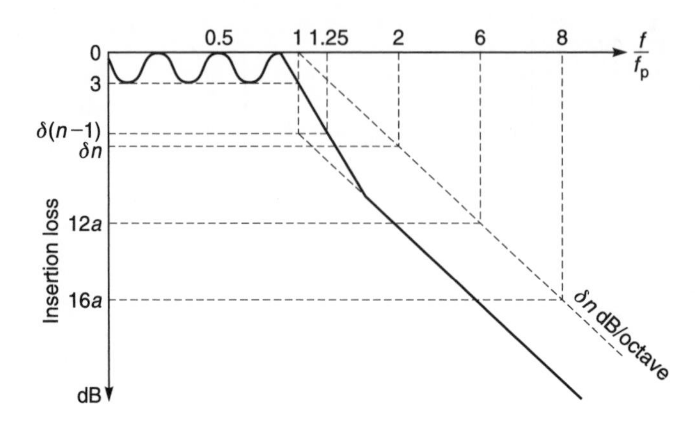

Fig. 5.38 Tchebyscheff filter response

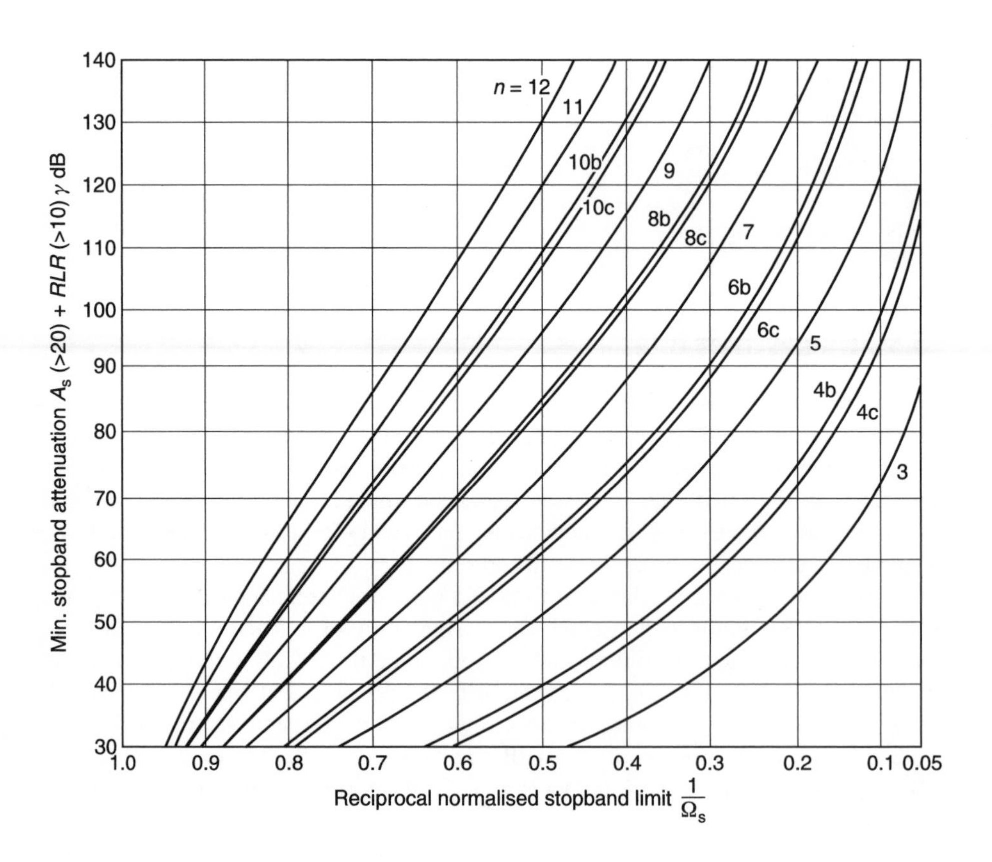

Fig. 5.39 Estimate of number of arms for Tchebyscheff filter

The 'b' sub-type filters have a slightly greater cut-off rate than the corresponding 'c' sub-type.

5.5.1 Normalised Tchebyscheff tables

Normalised tables for the Tchebyscheff filter may be calculated from Equations 5.17 to 5.23. These are:

$$\beta = \ln\left[\coth \frac{A_m}{17.37}\right] \tag{5.17}$$

where A_m = maximum amplitude of passband ripple in dB

$$\gamma = \sinh\left[\frac{\beta}{2n}\right] \tag{5.18}$$

where n = total number of arms in the filter

$$a_k = \sin\left[\frac{(2k-1)\pi}{2n}\right],\ k = 1, 2, \ldots, n \tag{5.19}$$

$$b_k = \gamma^2 + \sin^2\left[\frac{k\pi}{n}\right],\ k = 1, 2, \ldots, n \tag{5.20}$$

$$\gamma = 1 \text{ for } n \text{ odd}, \quad \gamma = \tanh^2(\beta/4) \text{ for } n \text{ even} \tag{5.21}$$

$$g_1 = 2a_1/\gamma \tag{5.22}$$

$$g_k = \frac{4(a_{k-1})(a_k)}{(b_{k-1})(g_{k-1})},\ k = 2, 3, \ldots, n \tag{5.23}$$

As you can see the calculations for normalised Tchebyscheff filters are quite formidable. To save you time, we provide Tables 5.10–5.12.

Table 5.10 Tchebyscheff normalised values (A_m = 0.01 dB)

k\n	2	3	4	5	6	7	8
1	0.4488	0.6291	0.7128	0.7563	0.7813	0.7969	0.8072
2	0.4077	0.9702	1.2003	1.3049	1.3600	1.3924	1.4130
3		0.6291	1.3212	1.5773	1.6896	1.7481	1.7824
4			0.6476	1.3049	1.5350	1.6331	1.6833
5				0.7563	1.4970	1.7481	1.8529
6					0.7098	1.3924	1.6193
7						0.7969	1.5554
8							0.7333

Table 5.11 Tchebyscheff normalised values (A_m = 0.1 dB)

k\n	2	3	4	5	6	7	8
1	0.8430	1.0315	1.1088	1.1468	1.1681	1.1811	1.1897
2	0.6220	1.1474	1.3061	1.3712	1.4039	1.4228	1.4346
3		1.0315	1.7703	1.9750	2.0562	2.0966	2.1199
4			0.8180	1.3712	1.5170	1.5733	1.6010
5				1.1468	1.9029	2.0966	2.1699
6					0.8613	1.4228	1.5640
7						1.1811	1.9444
8							0.8778

Table 5.12 Tchebyscheff normalised values (A_m = 0.25 dB)[3]

k\n	2	3	4	5	6	7	8
1	1.113	1.303	1.378	1.382	1.437	1.447	1.454
2	0.688	1.146	1.269	1.326	1.341	1.356	1.365
3		1.303	2.056	2.209	2.316	2.348	2.367
4			0.851	1.326	1.462	1.469	1.489
5				1.382	2.178	2.348	2.411
6					0.885	1.356	1.462
7						1.447	2.210
8							0.898

5.5.2 Design procedure for Tchebyscheff filters

1 Decide on the number of arms you require to achieve your passband ripple and desired attenuation.
2 Obtain the normalised values from the Tchebyscheff tables.
3 Follow the same procedures as for the Butterworth design examples for low pass, high-pass, bandpass and stopband filters.

Example 5.11

A Tchebyscheff 50 Ω low pass filter is to be designed with its 3 dB cut-off frequency at 50 MHz. The passband ripple is not to exceed 0.1 dB. The filter must offer a minimum of 30 dB attenuation at 100 MHz. Find (a) the number of arms.

Given: 50 Ω low pass Tchebyscheff filter, passband ripple ≤ 0.1 dB, minimum attenuation at 100 MHz ≥ 30 dB.
Required: (a) Number of arms of low pass filter, (b) component values for filter.

Solution

(a) Since 100 MHz/50 MHz = 2 and its reciprocal is 0.5 and assuming a return loss of 20 dB and a passband attenuation of 30 dB from Figure 5.39, it is seen that about five arms will be required.
(b) Therefore the filter follows the configuration shown below in Figure 5.40a. The pertinent values are taken from Table 5.11 and the method of calculation is similar to that of Example 5.4.

[3] Further tables may be obtained from *Reference Data for Radio Engineers*, International Telephone and Telegraph Corporation, 320 Park Avenue, New York 22.

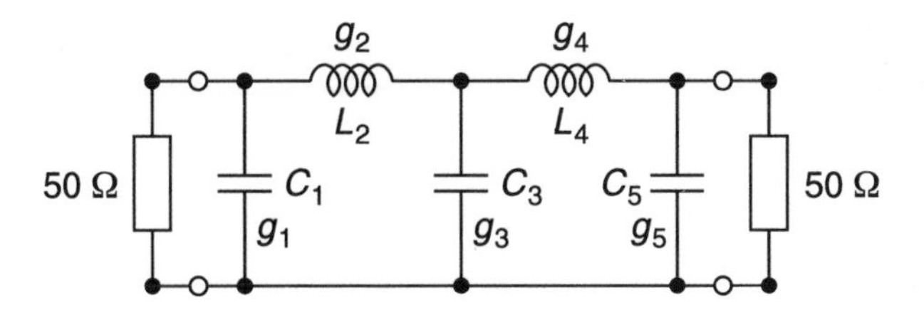

Fig. 5.40a Low pass configuration

Table 5.13 Calculating values for a low pass Tchebyscheff filter

Circuit reference	Normalised $Z_0 = 1\ \Omega$ $\omega = 1$ rad/s	$Z_0 = 50\ \Omega$ $f = 1/(2\pi)$ Hz	$Z_0 = 50\ \Omega$ $f_p = 50$ MHz
g_1 or C_1	0.1468 F	$\frac{1.1468}{50}$ F	$\frac{1.1468}{50 \times 2\pi \times 50\ \text{MHz}}$ F or 73.01 pF
g_2 or L_2	1.3712 H	1.3712×50 H	$\frac{1.3712 \times 50}{2\pi \times 50\ \text{MHz}}$ H or 218.23 nH
g_3 or C_3	1.9760 F	$\frac{1.9760}{50}$ F	$\frac{1.9760}{50 \times 2\pi \times 50\ \text{MHz}}$ F or 125.73 pF
g_4 or L_4	1.3712 H	1.3712×50 H	$\frac{1.3712 \times 50}{2\pi \times 50\ \text{MHz}}$ H or 218.23 nH
g_5 or C_5	1.1468 F	$\frac{1.1468}{50}$ F	$\frac{1.1468}{50 \times 2\pi \times 50\ \text{MHz}}$ F or 73.01 pF

Hence the calculated values for a five element low pass Tchebyscheff filter ($A_m \leq 0.1$ dB) with a nominal impedance of 50 Ω and a 3 dB cut-off frequency at 50 MHz are:

$$C_1 = 73.01\ \text{pF}, L_2 = 218.23\ \text{nH}, C_3 = 125.73\ \text{pF}, L_4 = 218.23\ \text{nH}, C_5 = 73.01\ \text{pF}$$

The response of this filter is shown in Figure 5.40b. It has been obtained by using PUFF. Notice that there is a ripple in the passband but, because of the scale we have used, it is not clear. However, you can try this on PUFF and set a small attenuation scale and see the ripple. Alternatively, you can sweep the frequency and notice the variation in $S21$ in PUFF.

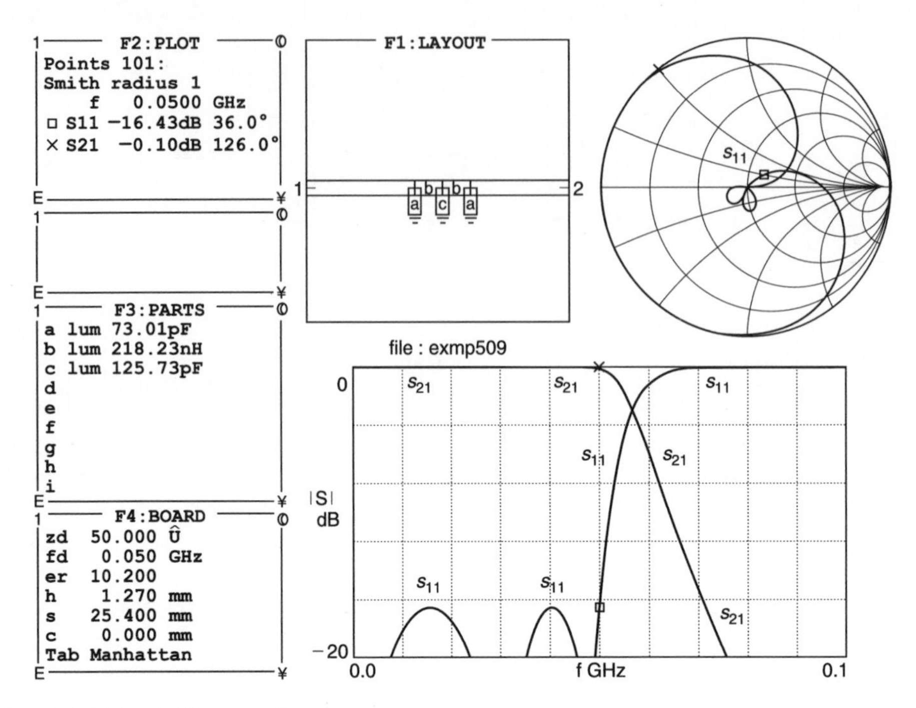

Fig. 5.40b Tchebyscheff low pass filter

Example 5.12

A Tchebyscheff 75 Ω high pass filter is to be designed with its 3 dB cut-off frequency at 500 MHz. The passband ripple is not to exceed 0.25 dB. The filter must offer a minimum of 30 dB passband attenuation at 250 MHz. Find (a) the number of arms required, and (b) the component values.

Given: 75 Ω high pass Tchebyscheff filter, passband ripple ≤ 0.25 dB, minimum attenuation at 250 MHz ≥ 30 dB.
Required: (a) Number of arms of high pass filter, (b) component values for filter.

Solution

(a) Since 500 MHz/250 MHz = 2 and its reciprocal is 0.5 and assuming a return loss of 20 dB and a passband attenuation of 30 dB from Figure 5.39, it is seen that about five arms will be required.
(b) Therefore the filter follows the configuration shown in Figure 5.41.

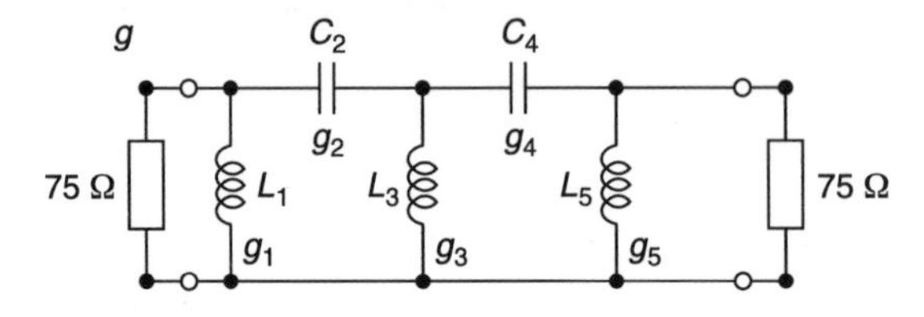

Fig. 5.41 High pass configuration

The pertinent values are taken from Table 5.12 and the method of calculation is similar to that of Example 5.4. Table 5.14 shows how this is carried out.

Table 5.14 Calculated values for a low pass Tchebyscheff filter

Circuit reference	Normalised $Z_0 = 1\ \Omega$ $\omega = 1$ rad/s	$Z_0 = 75\ \Omega$ $Z_0 = 75\ \Omega$ $f = 1/(2\pi)$ Hz	$f_p = 500$ MHz
g_1 or L_1	$\frac{1}{1.382}$ H	$\frac{75}{1.382}$ H	$\frac{75 \times 10^9}{1.382 \times 2\pi \times 500 \times 10^6} = 17.27$ nH
g_2 or C_2	$\frac{1}{1.326}$ F	$\frac{1}{1.326 \times 75}$ F	$\frac{1 \times 10^{12}}{1.326 \times 75 \times 2\pi \times 500 \times 10^6} = 3.20$ pF
g_3 or L_3	$\frac{1}{2.209}$ H	$\frac{75}{2.209}$ H	$\frac{75 \times 10^9}{2.209 \times 2\pi \times 500 \times 10^6} = 10.80$ nH
g_4 or C_4	$\frac{1}{1.326}$ F	$\frac{1}{1.326 \times 75}$ F	$\frac{1 \times 10^{12}}{1.326 \times 75 \times 2\pi \times 500 \times 10^6} = 3.20$ pF
g_5 or L_5	$\frac{1}{1.382}$ H	$\frac{75}{1.382}$ H	$\frac{75 \times 10^9}{1.382 \times 2\pi \times 500 \times 10^6} = 17.27$ nH

Hence the calculated values for a five element high pass Tchebyscheff filter ($A_m \leq 0.25$ dB) with a nominal impedance of 75 Ω and a 3 dB cut-off frequency at 500 MHz are:

$$L_1 = 17.27 \text{ nH},\ C_2 = 3.20 \text{ pF},\ L_3 = 10.80 \text{ nH},\ C_4 = 3.20 \text{ pF, and } L_5 = 17.27 \text{ nH}$$

The response of this filter is shown in Figure 5.42. It has been plotted by using PUFF. Notice the ripple in the passband.

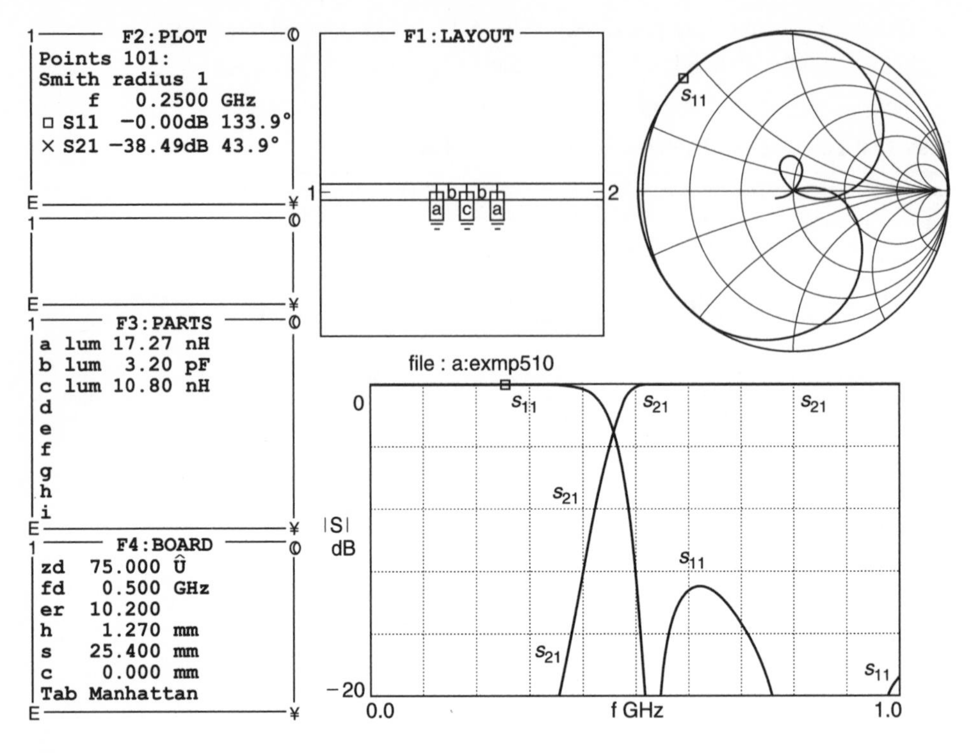

Fig. 5.42 Response of high pass filter using PUFF

5.6 Summary on filters

In the previous sections, we have shown you how to synthesise or design low pass, high pass, bandpass and stopband filters using normalised tables for the Butterworth and Tchebyscheff type filters. We have assumed that the unloaded Q of the elements are relatively high when compared with the loaded Q of the filter.

The calculations for these design examples have been carried out using a spreadsheet. There are computer programs available which will compute filter components and in some cases even produce the microwave circuit layout.

Space limitations prevent us from showing you the synthesis of many other filter types. However, the design procedures are similar. Many filter designers have produced normalised tables for various types of filters. If you wish to pursue this topic further, the classical microwave filter design book is by Matthei, Young and Jones.[4] Last but not least, you should realise that many programs (e.g. SPICE, AppCAD, PUFF) exist to help you with the calculation and response of these filters.

[4] G. Matthei, L. Young and E. Jones, *Design of microwave filters, impedance matching networks and coupling structures*, McGraw-Hill, New York NY, 1964.

5.7 Impedance matching

Impedance matching is a vitally important part of amplifier design. There are four main reasons for impedance matching:

- to match an impedance to the conjugate impedance of a source or load for maximum power transfer;
- to match an amplifier to a certain load value to provide a required transistor gain;
- to match an amplifier to a load that does not cause transistor instability;
- to provide a load for an oscillator that will cause instability and hence oscillations.

5.7.1 Matching methods

There are many means of impedance matching in h.f. and r.f. design. These include:

- quarter-wave transmission line matching
- single stub matching
- double stub matching
- transformer matching
- auto-transformer matching
- capacitive matching
- L network matching
- pi network matching
- T network matching

You have already been shown the first three methods. We will now show you the rest.

5.7.2 Transformer matching

The schematic of a typical i.f. (intermediate frequency) transformer is shown in Figure 5.43. The primary coil (terminals 1 and 2) consists of a number (N_1) turns of wire wound in close magnetic proximity to a secondary winding (terminals 3 and 4) with a number (N_2) turns of wire. Both primary and secondary windings are normally wound on to a coil former consisting of magnetic material (iron, ferrite or special magnetic compounds). Voltage V_1 is the voltage applied to the primary and V_2 is the voltage induced in the secondary by magnetic action. The currents i_1 and i_2 represent the currents flowing in the primary and secondary windings respectively.

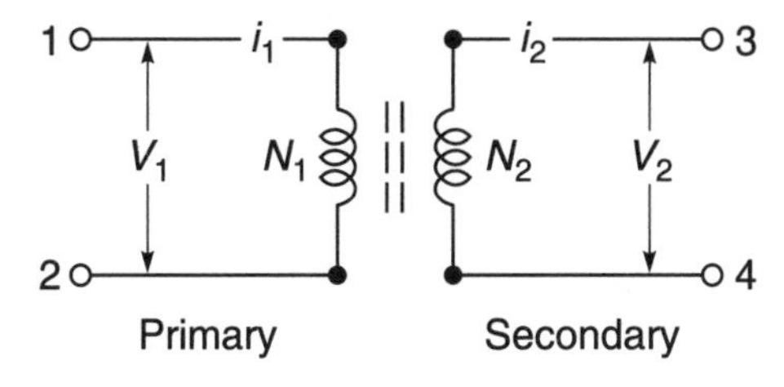

Fig. 5.43 Schematic of a typical i.f. transformer

Magnetic flux linkage between primary and secondary is nearly perfect and for practical purposes the coupling coefficient between primary and secondary coils is unity.[5]

[5] A transformer is said to have a coupling coefficient of 1 if all the flux produced by one winding links with the other windings.

Manufacturers tend to quote coupling coefficients greater than 0.95 for transformers used in 465 kHz i.f. amplifiers. Unless stated otherwise, from now on it will be assumed that the coupling coefficient is unity.

Operating principles

The operating principles of these transformers can be easily understood by using Michael Faraday's law, which states that the voltage (V) induced in a conductor is directly proportional to the rate of change of the effective magnetic flux ($\partial \phi / \partial t$) across it.

If we define N as the number of turns of the conductor, and ϕ as the magnetic field, the induced voltage (V) can be calculated by the following expressions:

$$V_1 = N_1 \frac{\partial \phi}{\partial t} \tag{5.24}$$

and

$$V_2 = N_2 \frac{\partial \phi}{\partial t} \tag{5.25}$$

To calculate the **voltage ratio** of the transformer we simply divide Equation 5.25 by Equation 5.24 giving

$$\frac{V_2}{V_1} = \frac{N_2}{N_1} \tag{5.26}$$

or

$$\frac{V_1}{V_2} = \frac{N_1}{N_2} \tag{5.27}$$

Current ratio

If we define N as the number of turns, i as the current flowing in a circuit and k as a constant of that circuit, then according to Biot Savart's law the magnetic field (ϕ) produced can be written as:

$$\phi_1 = kN_1 i_1 \tag{5.28}$$

and

$$\phi_2 = kN_2 i_2 \tag{5.29}$$

Since the magnetic field is the same for both windings in our transformer, we can combine Equations 5.28 and 5.29 to give

$$N_1 i_1 = N_2 i_2$$

and by transposing we get

$$\frac{i_2}{i_1} = \frac{N_1}{N_2} \tag{5.30}$$

If we define $Z_1 = V_1/i_1$ and $Z_2 = V_2/i_2$, we can obtain the input impedance of a transformer by dividing Equation 5.26 by Equation 5.30 to give

$$\frac{V_2/V_1}{i_2/i_1} = \frac{N_2/N_1}{N_1/N_2}$$

Transposing the above and substituting for $Z_1 = V_1/i_1$ and $Z_2 = V_2/i_2$ yields

$$Z_2 = Z_1\left[\frac{N_2}{N_1}\right]^2 \tag{5.31}$$

Similarly, by transposing

$$Z_1 = Z_2\left[\frac{N_1}{N_2}\right]^2 \tag{5.32}$$

Equations 5.26, 5.30, 5.31 and 5.32 are suitable for use with 'ideal' transformers.[6]

Example 5.12

The primary winding of a two winding transformer is wound with 16 turns while its secondary has 8 turns. The terminating resistance on the secondary is 16 Ω. What is its effective resistance at the primary? Assume that the transformer is 'ideal' and that the coefficient of coupling between the primary and the secondary is 1.

Solution. Since the transformer is 'ideal' with a coupling coefficient of unity, the answer can be found by applying Equation 5.32 and remembering that $Z_p = Z_1$ and $Z_s = Z_2$. This gives

$$Z_p = Z_s\left[\frac{N_1}{N_2}\right]^2 = 16 \times \left[\frac{16}{8}\right]^2 = 64\ \Omega$$

Finally before leaving the subject of transformers for now, I would like to mention the auto-transformer.

5.7.3 Auto-transformers

The auto-transformer is a transformer in which the secondary winding is tapped off the primary winding. It has the advantage that less copper wire is required for the windings but suffers from the fact that the primary and secondary windings are not directly isolated.

Apart from constructional details, Equations 5.24–5.32 apply when the auto-transformer is ideal. The application of transformer action in r.f. design is given in Section 5.7.4.

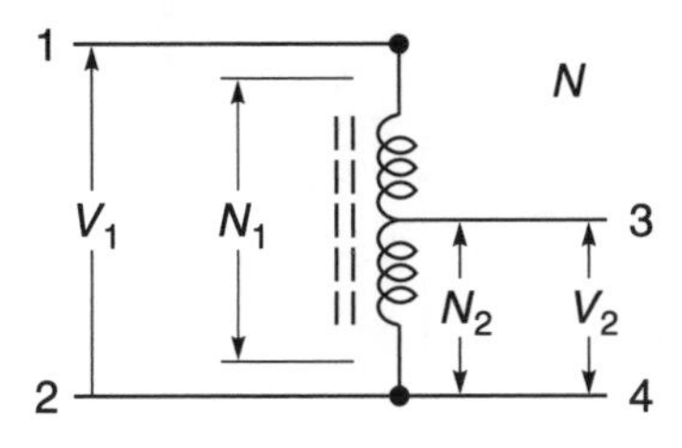

Fig. 5.44 Auto-transformer

[6] An ideal transformer is a transformer which has negligible losses and is one in which all powers coupled between windings is purely due to the magnetic field.

5.7.4 Intermediate frequency (i.f.) amplifier with transformers

A schematic of a typical intermediate frequency (i.f.) amplifier is shown in Figure 5.45. This amplifier is designed to operate efficiently at one frequency. The operational frequency of the amplifier is determined by the tuned circuit components, C_T and L_T. C_T represents the total capacitance of the tuned circuit. It includes circuit tuning capacitor, effective output capacitance of the transistor and all stray capacitances. L_T represents the effective inductance of the circuit. It is mainly due to the primary winding inductance of T_2. The tapping point on the primary winding is at r.f. earth potential because this point is effectively short-circuited to earth through the power supply decoupling capacitors. The position of the tapping point is very important because it determines the working Q and bandwidth of the amplifier circuit. This will be explained shortly.

Intermediate frequency amplifiers are used very extensively in superhet receivers and, to keep costs minimal, standardised construction methods are used in the design of their tuned circuits. Figure 5.46 shows a sectional view of a typical 465 kHz i.f. tuned transformer. Tuned transformers are available in two main base sizes, 7 × 7 mm and 10 × 10 mm. The tuning capacitor (180 pF for the 7 mm size and 150 pF for the 10 mm size) is mounted within the plastic base of the transformer. The primary tuning coil consists of approximately 200 turns of 0.065 mm diameter wire wound on a ferrite bobbin mounted

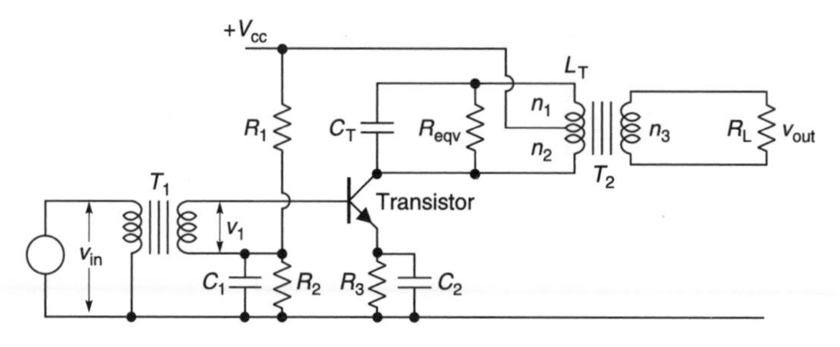

Fig. 5.45 Schematic diagram of a typical 465 kHz i.f. amplifier

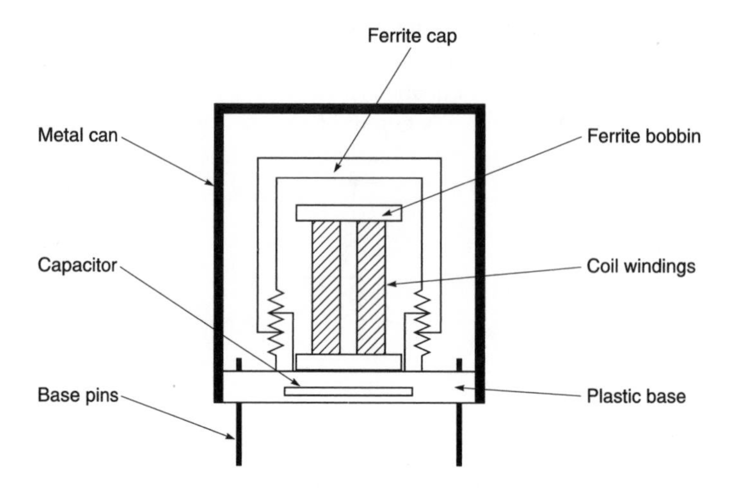

Fig. 5.46 Sectioned view of a typical 465 kHz i.f. transformer

on a plastic base. This bobbin is shaped like a dumbbell. Unloaded primary coil Qs tend to be standardised at 70, 100 or 130. The secondary winding consists of about five to eight turns of wire. Magnetic flux linkage between primary and secondary is nearly perfect and for practical purposes the coupling coefficient between primary and secondary coils is unity.[7] Manufacturers tend to quote coupling coefficients greater than 0.95. Unless stated otherwise, from now on it will be assumed that the coupling coefficient is unity.

All coil leads are welded to pins on the base. Welding is used to ensure that coil connections do not become detached during external soldering operations. Circuit resonance is adjusted by varying tuning inductance. This is done by altering the position of the ferrite cap relative to the winding bobbin.

The tuned circuit load impedance presented by transformer T_2 to its driving transistor is set by using the primary winding of T_2 as an auto-transformer and by careful choice of winding ratios between n_1, n_2 and n_3. See Figures 5.45 and 5.47. Three resistances are reflected into the primary of T_2. These are shown in Figure 5.47.

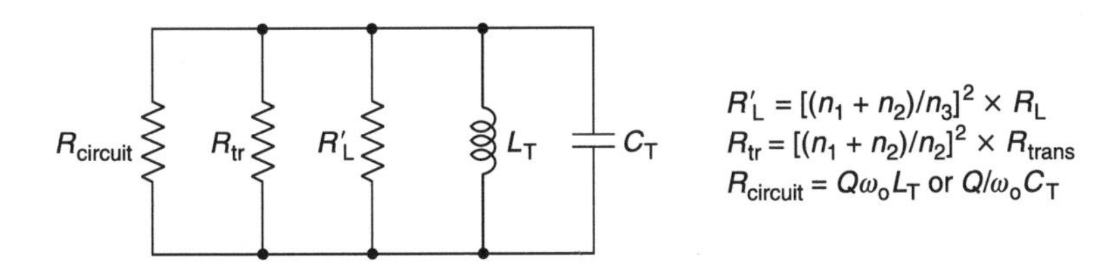

Fig. 5.47 Equivalent tuned circuit of Figure 5.45

The impedance (R'_L) reflected into the primary circuit of T_1 by R_L in the secondary is given by

$$R'_L = \left[\frac{n_1 + n_2}{n_3}\right]^2 \times R_L \tag{5.33}$$

$R_{circuit}$ represents the resistive losses associated with the use of non-perfect capacitors, inductors and transformers. It is

$$R_{circuit} = Q_{unloaded}\, \omega_o L_T \quad \text{or} \quad Q_{unloaded}/\omega_o C_T \tag{5.34}$$

R_{tr} represents the output resistance of the transistor transformed across the tuned circuit, i.e.

$$R_{tr} = [(n_1 + n_2)/n_2]^2 \times R_{transistor} \tag{5.35}$$

These three resistances in parallel form R_{eqv}. Therefore

$$R_{eqv} = R'_L // R_{circuit} // R'_{tr} \tag{5.36}$$

The collector load of the transistor is the ratio $\{n_2/(n_1 + n_2)\}^2 \times R'_L // R_{circuit}$ across the total primary winding. Therefore

[7] A transformer is said to have a coupling coefficient of 1 if all the flux produced by one winding links with the other windings.

$$\text{Transistor load} = \left[\frac{n_2}{n_1 + n_2}\right]^2 \times R_L // R_{circuit} \tag{5.37}$$

Example 5.14

In Figure 5.47, $n_1 = 160$, $n_2 = 40$, $n_3 = 8$ and $R_L = 2$ kΩ. The resistive losses associated with the tuning capacitor and inductors can be assumed to be negligible and the transistor output resistance reflected across the primary of the tuned circuit is so large that it can be neglected. The magnetic coupling coefficients between coils may be assumed to be unity. (a) What is the transistor load impedance at resonance? (b) If the tapping point on L_T is changed so that $n_1 = 150$ and $n_2 = 50$, what is the new transistor load impedance at resonance?

Solution

(a) At resonance, the tuned circuit impedance is very high when compared to the reflected load R'_L from the secondary. Using Equation 5.33

$$R'_L = \left[\frac{n_1 + n_2}{n_3}\right]^2 \times R_L = \left[\frac{160 + 40}{8}\right]^2 \times 2000 = 1\,250\,000\ \Omega$$

Using Equation 5.37 and noting in this case that $R_{eqv} = R'_L$, the transistor load impedance is

$$\left[\frac{n_2}{n_1 + n_2}\right]^2 \times R'_L = \left[\frac{40}{160 + 40}\right]^2 \times 1\,250\,000 = 50\ \text{k}\Omega$$

(b) When $n_1 = 150$ and $n_2 = 50$, using Equation 5.33

$$R'_L = \left[\frac{n_1 + n_2}{n_3}\right]^2 \times R_L = \left[\frac{150 + 50}{8}\right]^2 \times 2000 = 1\,250\,000\ \Omega$$

Using Equation 5.33 and noting in this case that $R_{eqv} = R'_L$, the transistor load impedance is

$$\left[\frac{n_2}{n_1 + n_2}\right]^2 \times R'_L = \left[\frac{50}{150 + 50}\right]^2 \times 1\,250\,000 = 78.5\ \text{k}\Omega$$

Example 5.14 shows clearly that different collector load impedances can be obtained from a standard tuned i.f. transformer simply by altering the tapping point on the primary coil.

For clarity in understanding the previous example, it was assumed that tuned circuits losses ($R_{circuit}$) were negligible and that the reflected output resistance of the transistor (R'_{tr}) was so high that it did not affect the value of R_{eqv}. In practice, the additional resistive losses across the tuned circuit are not negligible and must be taken into account in designing the amplifier. The effect of these additional losses is demonstrated in Example 5.14.

Example 5.15

In Figure 5.45, $n_1 = 160$, $n_2 = 40$, $n_3 = 8$ and $R_L = 2$ kΩ. The unloaded Q of the tuned circuit is 100. The value of the tuning capacitor is 180 pF and the circuit is resonant at 465 kHz. If the coupling coefficient between coils is unity, what is the transistor load impedance at resonance? Assume the output impedance of the transistor to be 100 kΩ.

Solution. At resonance, the effective tuned circuit impedance is the parallel value of the reflected load (R'_L) from the secondary, the equivalent loss resistance of the tuned circuit ($R_{circuit}$) and the reflected output resistance of the transistor (R_{tr}). Using Equation 5.33, the reflected load

$$R'_L = \left[\frac{n_1 + n_2}{n_3}\right]^2 \times R_L = \left[\frac{160 + 40}{8}\right]^2 \times 2000 = 1.25\ \text{M}\Omega$$

The effective loss resistance of the tuned circuit is obtained by using Equation 5.34:

$$R_{circuit} = Q/(\omega_o C_T) = 100\ /(2\pi \times 465 \times 10^3 \times 180 \times 10^{-12}) = 190.143\ \text{k}\Omega$$

Using Equation 5.35

$$R_{tr} = \left[\frac{n_1 + n_2}{n_2}\right]^2 \times R_{transistor} = \left[\frac{160 + 40}{40}\right]^2 \times 100\,000 = 2.5\ \text{M}\Omega$$

Using Equation 5.37, the transistor load impedance is

$$\left[\frac{n_2}{n_1 + n_2}\right]^2 \times R'_L // R_{circuit} = \left[\frac{40}{160 + 40}\right]^2 \times 154.8\ \text{k}\Omega \approx 6.19\ \text{k}\Omega$$

The answer to Example 5.15 clearly indicates that the unloaded Q of the tuned circuit affects the transistor collector load and that the transformation ratio must be changed if the original load impedance of 50 kΩ is desired. Another interesting point is that the loaded Q of the circuit has also fallen drastically because the effective resistance (R_{eqv}) across the tuned circuit has been reduced.

Example 5.16

The unloaded Q of a tuned circuit is 100. The value of its tuning inductance is 650 μH and the circuit is resonant at 465 kHz. When the circuit is loaded by the input resistance of a transistor, the effective resistance across the ends of the tuned circuit is 125 kΩ. What is its loaded Q and 3 dB bandwidth?

Solution. Using Equation 5.10

$$R_{eqv} = Q_{loaded}\, \omega_o L$$

and

$$Q_{loaded} = \frac{R_{eqv}}{\omega_o L} = \frac{125\,000}{2\pi \times 465\,000 \times 650 \times 10^{-6}} = 65.8$$

From Equation 5.11

$$Q_{loaded} = \frac{f_0}{\text{bandwidth}}$$

or

$$\text{bandwidth} = \frac{f_0}{Q_{\text{loaded}}} = \frac{465\,000}{65.8} = 7066 \text{ Hz}$$

Examples 5.15 and 5.16 have brought out a very important point. It is that **transistor collector load impedance** and **circuit bandwidth** can be altered **independently** (within limits) by careful choice of the ratios between n_1, n_2, and n_3. This independence of the two is possible even when standard i.f. transformers are used.

5.7.5 Capacitive matching

For tuned circuits at the higher frequencies, smaller inductance values are required. Smaller inductance values mean fewer turns and trying to make impedance transformers with the correct transformation ratio is difficult. In cases like that, it is often more convenient to use capacitors as the matching element. One such arrangement is shown in Figure 5.48.

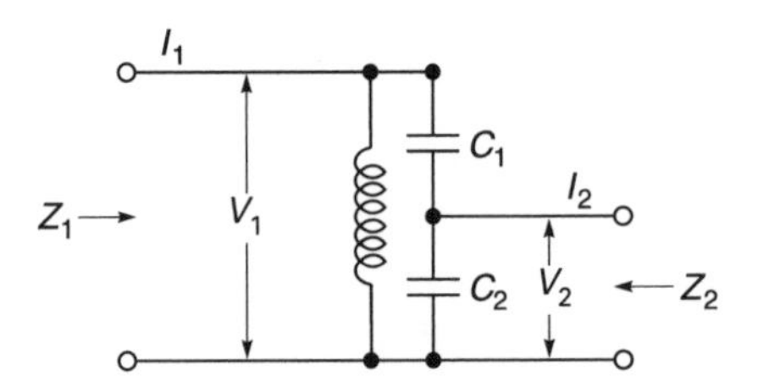

Fig. 5.48 Capacitive divider method

Capacitive divider matching

Capacitive divider matching (Figure 5.48) is particularly useful at very high frequencies (VHF) where the number of turns on an inductor is small and where location of a connection to produce an auto-transformer from the inductor is not practical. The capacitive matching network of Figure 5.48 can be easily obtained by defining

$$Z_1 = V_1/I_1 \quad \text{and} \quad Z_2 = V_2/I_2$$

For ease of analysis, we shall assume that $X_L \gg (X_{c1} + X_{c2})$, and that Z_2 is not loading the circuit. Then by inspection

$$I_1 = \frac{V_1}{X_{c1} + X_{c2}} \quad \text{and} \quad I_2 = \frac{V_2}{X_{c2}}$$

Hence

$$\frac{I_2}{I_1} = \frac{V_2}{X_{c2}} \times \frac{X_{c1} + X_{c2}}{V_1}$$

and transposing

$$\frac{V_1}{I_1} = \frac{V_2}{I_2} \times \frac{X_{c1} + X_{c2}}{X_{c2}}$$

and substituting for Z_1, Z_2 and reactances

$$Z_1 = Z_2 \left[\frac{1/j\omega C_1 + 1/j\omega C_2}{1/j\omega C_2} \right]$$

and multiplying all terms by $j\omega C_1 C_2$

$$Z_1 = Z_2 \left[\frac{C_1 + C_2}{C_1} \right] \tag{5.38}$$

Example 5.17

In the circuit shown in Figure 5.48, $C_1 = 10$ pF, $C_2 = 100$ pF and $Z_1 = 22$ kΩ. If the reactance of the inductor is very much greater than the combined series reactance of C_1 and C_2, calculate the transformed value shown as Z_2.

Solution. Using equation 5.38:

$$Z_2 = Z_1 \left[\frac{C_1}{C_1 + C_2} \right] = 22\,000 \left[\frac{10}{10 + 100} \right] = 2 \text{ k}\Omega$$

5.7.6 Impedance matching using circuit configurations

Many circuit configurations can be used as matching networks. In Figures 5.49, 5.50 and 5.51 we show three circuits which are often used for impedance matching. In Figure 5.49, Z_1 and Z_2 are used to match the load Z_L to the source Z_s. In Figure 5.50, Z_a and part of Z_b are used for matching to Z_1, while the remaining half of Z_b and Z_c are used for matching to Z_2. In Figure 5.51, Z_a and part of Z_b are used for matching to Z_1 while the remaining part of Z_b and Z_c are used to match to Z_2.

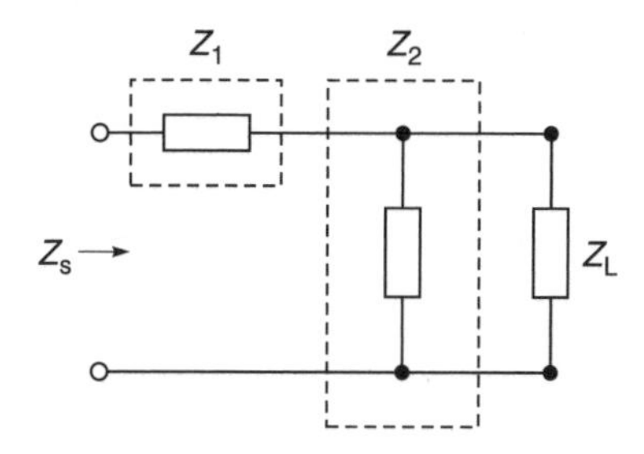

Fig. 5.49 Matching L network

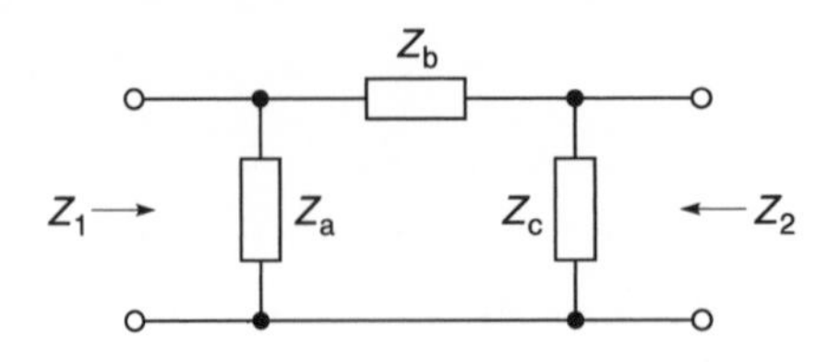

Fig. 5.50 Matching π network

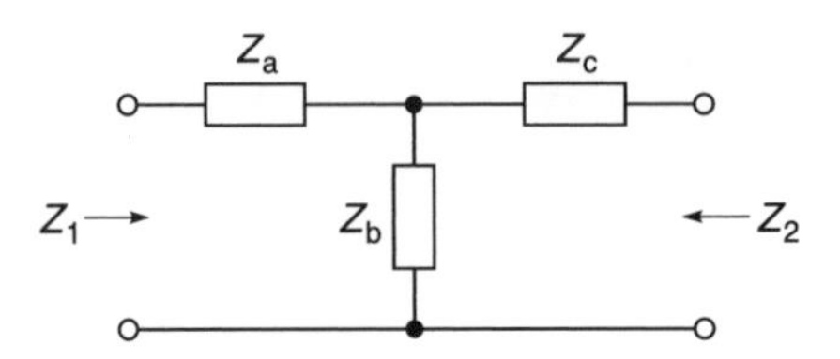

Fig. 5.51 Matching T network

The details of how these circuits can be used to provide matching will be explained shortly but for ease of understanding, we shall first review some fundamental concepts on the series, parallel and Q equivalents of components.

5.7.7 Series and parallel equivalents and quality factor of components

Before we can commence network matching methods, it is best to revise some fundamental concepts on the representation of capacitors and inductors.

Series and parallel forms and Q_s of capacitors

Capacitors are not perfect because their conducting plates contain resistance and their dielectric materials are not perfect insulators. The combined losses can be taken into account by an equivalent series resistance (R_s) in the series case or by an equivalent parallel resistance (R_p) in the parallel case as shown in Figure 5.52. In this diagram, R_s and C_s represent the series equivalent circuit of the capacitor while R_p and C_p represent the parallel equivalent circuit. The parallel representation is preferred when dealing with circuits where elements are connected in parallel.

Quality factor (Q) of a capacitor. We will follow normal convention and define the series quality factor (Q_s) as

$$Q_s = \frac{\text{reactance}}{\text{resistance}} = \frac{1/\omega C_s}{R_s} = \frac{1}{\omega C_s R_s} \tag{5.39}$$

where

ω = angular frequency in radians
C_s = capacitance in Farads
R_s = equivalent series resistance (ESR) of a capacitor in ohms

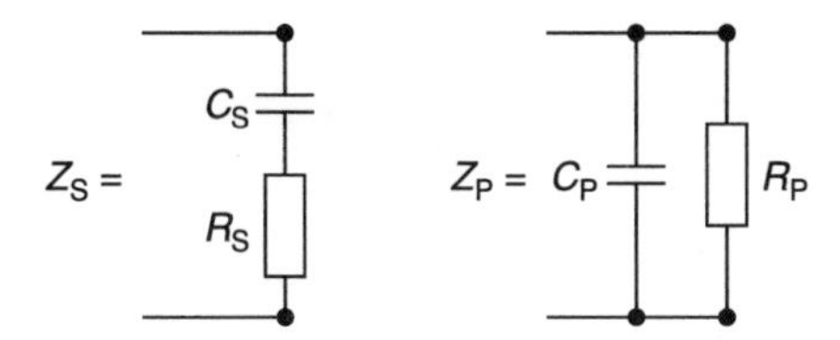

Fig. 5.52 Series and parallel form of a 'practical' capacitor

Similarly, we will define the parallel quality factor (Q_p) as

$$Q_p = \frac{\text{susceptance}}{\text{conductance}} = \frac{\omega C_p}{1/R_p} = \omega C_p R_p \tag{5.40}$$

where

ω = angular frequency in radians
C_p = capacitance in Farads
R_p = equivalent parallel resistance (EPR) of a capacitor in ohms

Example 5.18

A capacitor has an equivalent parallel resistance of 15 000 Ω and a capacity of 100 pF. Calculate its quality factor (Q_p) at 100 MHz.

Solution. Using Equation 5.40:

$$Q_p = \omega C_p R_p = 2\pi \times 100 \text{ MHz} \times 100 \text{ pF} \times 15\,000 = 942.48 \approx 943$$

Equivalence of the series and parallel representations. The purpose of this section is to show the relationships between series and parallel circuit representations. Using the same symbols as before and referring to Figure 5.52

$$Y = \frac{1}{\text{impedance }(Z)} = \frac{1}{R_s + 1/j\omega C_s} \times \left\{ \frac{R_s - 1/j\omega C_s}{R_s - 1/j\omega C_s} \right\}$$

and

$$G + jB = \frac{R_s}{R_s^2 + \dfrac{1}{\omega^2 C_s^2}} - \frac{1/j\omega C_s}{R_s^2 + \dfrac{1}{\omega^2 C_s^2}}$$

Using Equation 5.39:

$$G + jB = \frac{R_s/R_s^2}{1 + Q_s^2} + j\frac{Q_s/R_s}{1 + Q_s^2}$$

Therefore

$$G + jB = \frac{1}{R_s(1 + Q_s^2)} + j\frac{Q_s/R_s}{1 + Q_s^2} \tag{5.41}$$

Equating 'real' parts of Equation 5.41:

$$G = \frac{1}{R_p} = \frac{1}{R_s(1 + Q_s^2)}$$

Transposing

$$R_p = R_s(1 + Q_s^2) \tag{5.42}$$

or

$$Q_s = \sqrt{\frac{R_p}{R_s} - 1} \tag{5.43}$$

Equation 5.43 is used extensively in the matching of L networks which will be discussed shortly.

If $Q > 10$, then from Equation 5.42

$$R_p \approx R_s Q_s^2 \tag{5.44}$$

Equating imaginary parts of Equation 5.41

$$B = \omega C_p = \frac{Q_s/R_s}{1 + Q_s^2}$$

and using Equation 5.42 gives

$$\omega C_p = \frac{Q_s}{\frac{R_p}{(1 + Q_s^2)}(1 + Q_s^2)}$$

Transposing R_p and using Equation 5.40 for Q_p

$$\omega C_p R_p = Q_p = Q_s \tag{5.45}$$

Substituting Equation 5.39 for Q_s in Equation 5.45 and transposing ω and R_p

$$C_p = \frac{1}{\omega^2 C_s R_s R_p} \times \frac{C_s}{C_s}$$

and substituting Equation 5.42 for R_p

$$C_p = \frac{Q_s^2 C_s}{(1 + Q_s^2)} \tag{5.46}$$

If $Q_s > 10$, then from Equation 5.46

$$C_p \approx \frac{Q_s^2}{Q_s^2} C_s \approx C_s \tag{5.47}$$

Series and parallel forms and Q_s of inductors

Quality factor (Q_s) of an inductor. The 'goodness' or quality factor (Q_s) of an inductor is defined as:

$$Q_s = \frac{\omega L_s}{R_s} \tag{5.48}$$

where

ω = angular frequency in radians
L_s = inductance in Henries
R_s = series resistance of inductor in ohms

Example 5.19

An inductor has a series resistance of 8 Ω and an inductance of 365 μH. Calculate its quality factor (Q) at 800 kHz.

Solution. Using Equation 5.48

$$Q_s = \frac{\omega L_s}{R_s} = \frac{2\pi \times 800 \times 10^3 \times 365 \times 10^{-6}}{8} \approx 229$$

Equivalence of the series and parallel representations of inductors. Sometimes, it is more convenient to represent the quality factor (Q_p) of an inductor in its parallel form as shown in Figure 5.53. In this diagram, R_s and L_s represent the series equivalent circuit, while R_p and L_p represent the parallel equivalent circuit.

The equivalent values can be calculated by taking the admittance form of the series circuit:

$$Y = \frac{1}{\text{impedance } (Z)} = \frac{1}{R_s + j\omega L_s} \times \left\{ \frac{R_s - j\omega L_s}{R_s - j\omega L_s} \right\}$$

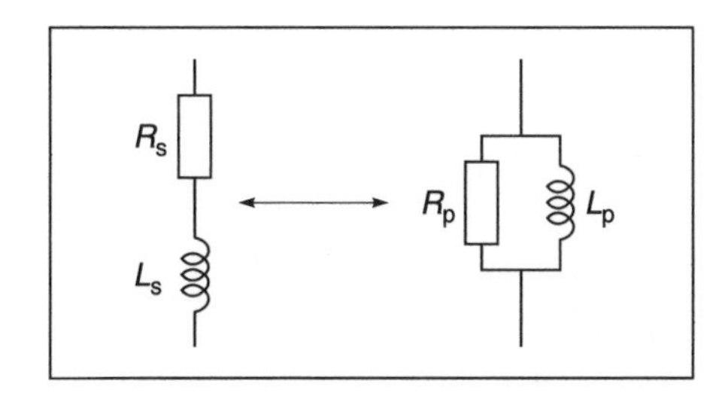

Fig. 5.53 Series and parallel form of an inductor

and

$$G - jB_L = \frac{R_s}{R_s^2 + \omega^2 L_s^2} - \frac{j\omega L_s}{R_s^2 + \omega^2 L_s^2}$$

Using Equation 5.48

$$G - jB_L = \frac{R_s/R_s^2}{1 + Q_s^2} - j\frac{Q_s/R_s}{1 + Q_s^2}$$

Therefore

$$R_p = \frac{1}{G} = \frac{1 + Q_s^2}{1/R_s} = R_s(1 + Q_s^2) \tag{5.49}$$

Alternatively

$$Q_s = \sqrt{\frac{R_p}{R_s} - 1} \tag{5.50}$$

Equation 5.50 is used extensively in the matching of L networks which will be discussed shortly.

If $Q > 10$, then from Equation 5.49

$$R_p \approx R_s Q_s^2 \tag{5.51}$$

$$-jB_L = \frac{-j}{\omega L_p} = \frac{-j\omega L_s}{R_s^2 + \omega^2 L_s^2}$$

Transposing and dividing by ω

$$L_p = \frac{R_s^2 + \omega^2 L_s^2}{\omega^2 L_s}$$

$$= \frac{1 + Q_s^2}{Q_s^2/L_s} = \left\{\frac{1 + Q_s^2}{Q_s^2}\right\} L_s \tag{5.52}$$

If $Q_s > 10$, then from Equation 5.52

$$L_p \approx \frac{Q_s^2}{Q_s^2} L_s \approx L_s \tag{5.53}$$

Finally, defining the parallel equivalent quality factor (Q_p) = susceptance/conductance and using Equations 5.49 and 5.52, it can be shown that $Q_p = Q_s$:

$$Q_p = \frac{R_p}{\omega L_p} = \frac{R_s(1 + Q_s^2)}{\omega\left\{\frac{1 + Q_s^2}{Q_s^2}\right\} L_s} = \frac{Q_s^2}{Q_s} = Q_s \tag{5.54}$$

Example 5.20

The inductor shown in Figure 5.53 has a series resistance (R_s) of 2 Ω and inductance (L_s) of 15 μH. Calculate its equivalent parallel resistance (R_p), parallel inductance (L_p) and its equivalent quality factor (Q_p), at 10 MHz.

Solution. From Equation 5.48

$$Q_s = \frac{\omega L_s}{R_s} = \frac{2\pi \times 10 \times 10^6 \times 15 \times 10^{-6}}{2} = \frac{942.5}{2} \approx 471$$

From Equation 5.49

$$R_p = R_s(1 + Q_s^2) = 2 \times (1 + 471^2) \approx 444 \text{ k}\Omega$$

From Equation 5.52

$$L_p = \left\{\frac{1 + Q_s^2}{Q_s^2}\right\} L_s = 15 \times 10^{-6} \left\{\frac{1 + 471^2}{471^2}\right\} \approx 15 \text{ µH}$$

From Equation 5.54

$$Q_p = Q_s = 471$$

Note: Since Q_s is very high in this case, the approximate formulae in Equations 5.51 and 5.53 could have been used.

5.7.8 L matching network

L matching networks are frequently used to match one impedance to another. They are called L matching networks because the two reactances used (X_a and X_b) are arranged in the form of a letter L.

Figure 5.54 shows a common L type network used when it is desired to match the input impedance of a transistor to a network. In this circuit, C_{in} and R_{in} represent the input impedance of the transistor. C_2 is put in parallel with C_{in} to form a total capacitance C_t; therefore $X_b = 1/j\omega C_t$. $X_a = 1/j\omega C_1$. Z_s is the input impedance looking into the circuit. It follows that

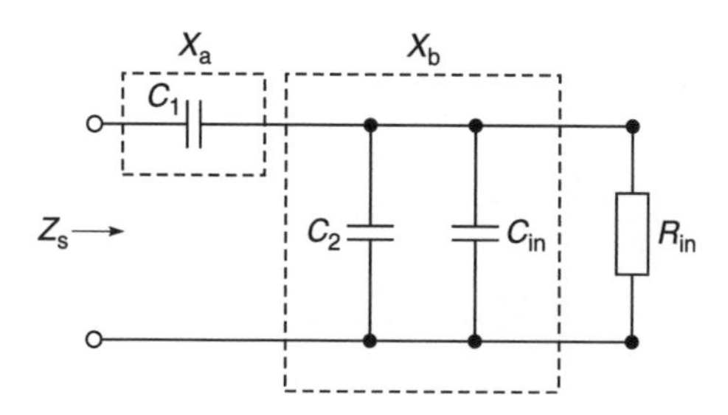

Fig. 5.54 Matching L network

$$Z_s = \frac{1}{j\omega C_1} + \frac{R_{in}(1/j\omega C_t)}{R_{in} + (1/j\omega C_t)}$$

Multiplying the numerator and denominator of the second term by $j\omega C_t$ and normalising gives

$$Z_s = \frac{1}{j\omega C_1} + \frac{R_{in}}{1 + j\omega C_t R_{in}} \times \left[\frac{1 - j\omega C_t R_{in}}{1 - j\omega C_t R_{in}}\right]$$

which when multiplied out results in

$$Z_s = \frac{R_{in}}{1 + (\omega C_t R_{in})^2} - j\left[\frac{1}{\omega C_1} + \frac{\omega C_t R_{in}^2}{1 + (\omega C_t R_{in})^2}\right] \tag{5.55}$$

The first term is real and shows that the resistance R_{in} has been transformed. The reactive term is incorporated into the tuning capacitance of the input tuned circuit. In this particular case, X_a and X_b have turned out to be capacitive, but this is not always the case. In some cases you may find that X_a and/or X_b may be inductive.

Procedure

In Figure 5.55, we wish to use an L network to match a load resistance (R_L) of 500 Ω to a generator resistance (R_g) of 50 Ω.

Method

1 We begin by deciding on the type of reactance (inductive or capacitive) we would like to use for X_p.
2 We then calculate the equivalent series combination of series resistance (R_s) and series reactance (X_s) from the parallel combination of X_p and ($R_p = R_L$) . You already know how to do this. See Figure 5.54 and Equations 5.49 to 5.54. However, here we must make the transformed $R_s = R_{generator} = R_g$ to ensure matching conditions for maximum power transfer from the generator to the transformed load.
3 We evaluate X_a so that it cancels out the transformed reactance (X_s) in the circuit.
4 Then we calculate the value X_b.

An example will help to clarify the method.

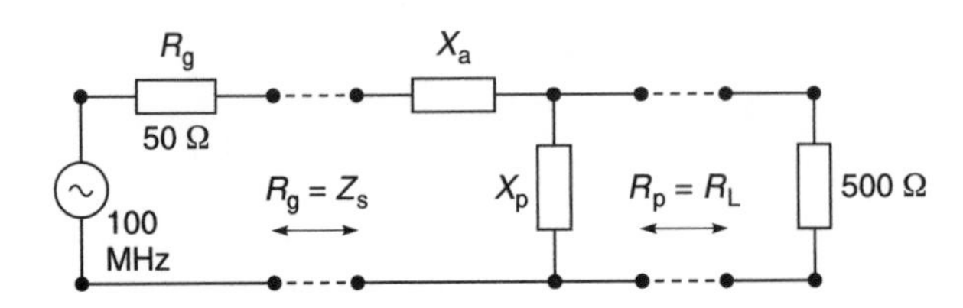

Fig. 5.55 Using a L network for matching

Example 5.21

Calculate a matching network for the circuit shown in Figure 5.55.

Solution. We begin by choosing X_p to be inductive. Next, we use Equation 5.50 which is repeated below for convenience:

$$Q_s = \sqrt{R_p/R_s - 1}$$

Note in this case that R_p is the load resistance (R_L) of 500 Ω and that we want $R_s = 50\ \Omega$ to match the generator resistance (R_g) of 50 Ω. Substituting these values, we obtain,

$$Q_s = \sqrt{500/50 - 1} = 3$$

Using Equation 5.48, and again remembering that $R_g = R_s$, we have

$$Q_s = \omega L_s / R_s$$

and transposing yields

$$L_s = \frac{R_s Q_s}{\omega} = \frac{50 \times 3}{2\pi \times 100\ \text{MHz}} = 239\ \text{nH}$$

$$X_s = \omega L_s = 2\pi \times 100\ \text{MHz} \times 239\ \text{nH} = 150\ \Omega$$

X_a must be chosen to cancel out this inductive reactance of 150 Ω so X_a must be a capacitor whose reactance $X_a = 150\ \Omega = 1/\omega C_a$. Therefore at 100 MHz

$$C_a = 1/(2\pi \times 100\ \text{MHz} \times 150) = 10.6\ \text{pF}$$

All that remains now is to calculate the parallel value of L_p. For this we make use of Equation 5.52, where

$$L_p = \left\{ \frac{1 + Q_s^2}{Q_s^2} \right\} L_s = \left\{ \frac{1 + 3_s^2}{3_s^2} \right\} 239\ \text{nH} = 265\ \text{nH}$$

The L network is now complete and its values are shown in Figure 5.56.

Using PUFF software to check Example 5.21

The results are shown in Figure 5.57. Note that since the matching circuit has been designed to match a 50 Ω generator, it follows that $S11$ will be zero when the transformed network matches the 50 Ω generator at the match frequency.

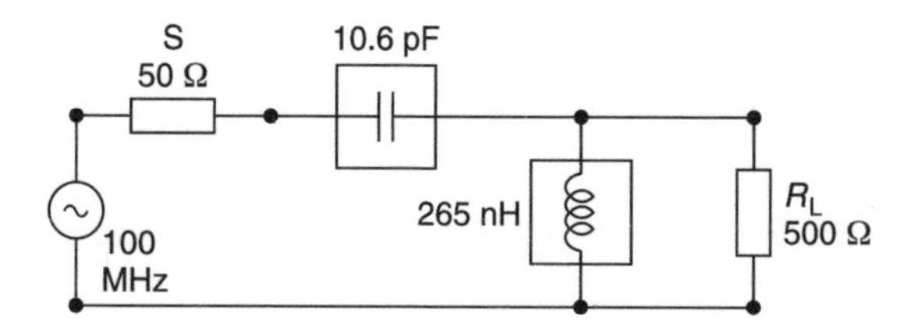

Fig. 5.56 Completed L matching network

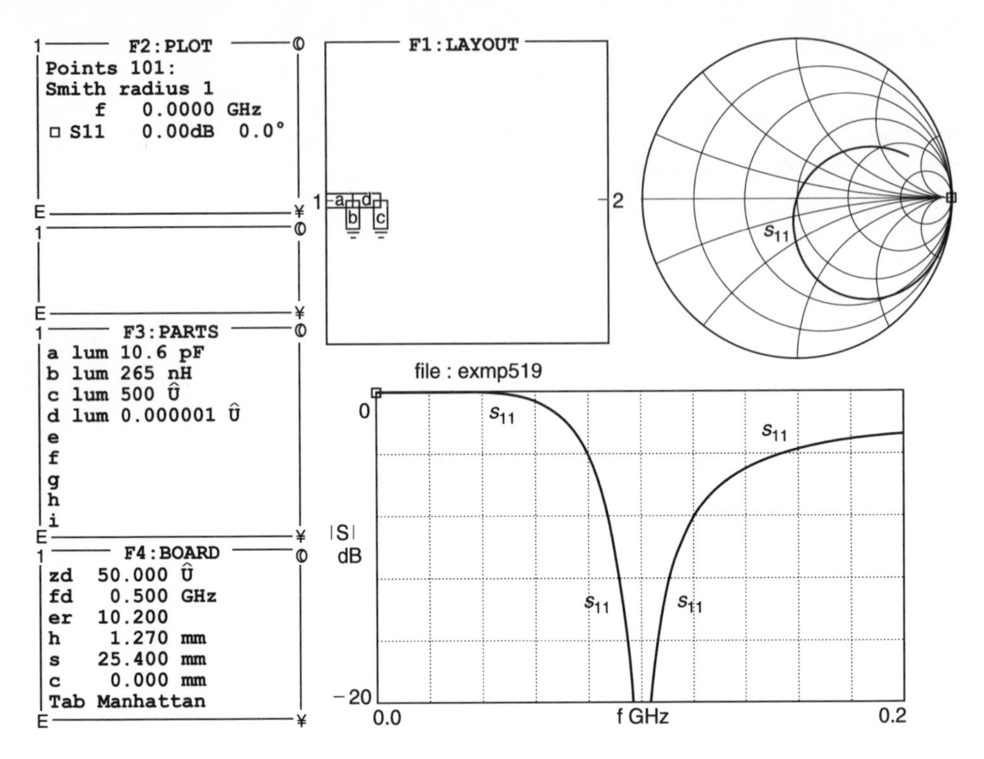

Fig. 5.57 PUFF plot of Example 5.21 matching network

In the real world, R_s and/or R_L has associated reactances. Suppose R_L is 500 Ω with a shunt capacitance of 42 pF. How do we solve this problem? This is best explained by using another example.

Example 5.22

Calculate a matching network for the circuit shown in Figure 5.58.

Solution. This is similar to Example 5.20, except that this time the load resistance (R_L) is shunted by a capacitance of 42 pF (see Figure 5.58). The solution proceeds as follows:

1 Introduce an inductance (L) to negate the effect of C_{shunt} at the frequency of operation, that is, choose L so that $X_L = X_C$ (see Figure 5.59).

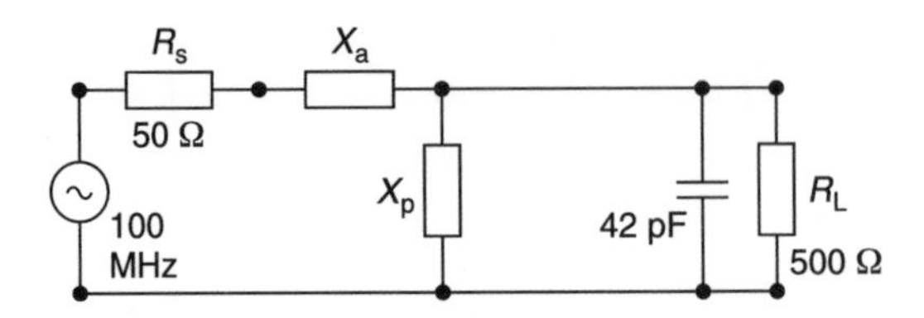

Fig. 5.58 L matching network

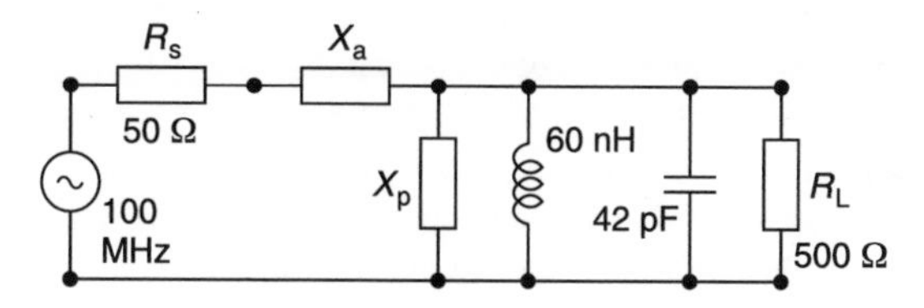

Fig. 5.59 Using an inductor of 60 nH to resonate with the 42 pF

2 Solve for X_a and X_p as in Example 5.21.
3 Solve for the combined value of the two shunt inductors shown in Figure 5.60.

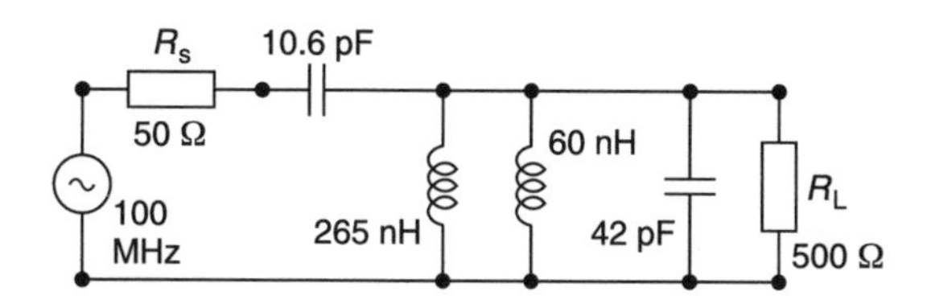

Fig. 5.60 The intermediate L network

The required value of L is

$$L = \frac{1}{\omega^2 C_{\text{shunt}}}$$

$$= \frac{1}{[2\pi(100\text{ MHz})]^2(42\text{ pF})} = 60.3\text{ nH} \approx 60\text{ nH}$$

See Figure 5.59.
4 Calculate the network as in example 5.21. I shall take the values directly from it. See Figure 5.60.
5 Solve for the combined value of the equivalent inductor:

$$L_{\text{combined}} = \frac{(265)(60)}{265 + 60} \approx 49\text{ nH}$$

See Figure 5.61 for the final network.

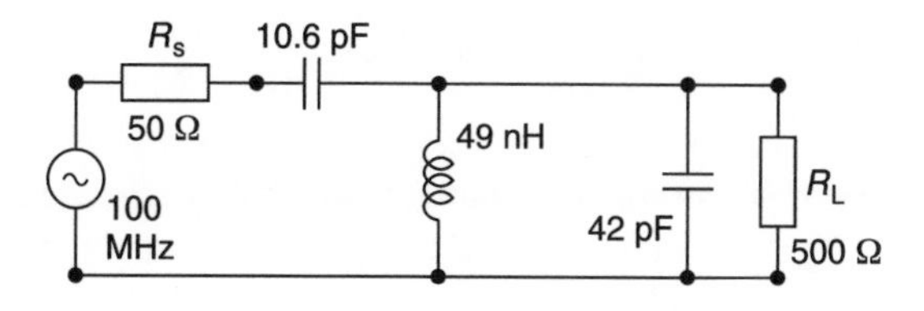

Fig. 5.61 Completed L network

In Example 5.22, X_a has been chosen to be capacitive while X_p has been chosen to be inductive. The network could equally have been designed with X_a inductive and X_p capacitive. In this case, the shunt capacitance of the load should be subtracted from the calculated value of the capacitance forming X_p. For example, if the calculated value of the matching shunt capacitance is 500 pF, then all you need do is subtract the load shunt capacitance (42 pF) from the calculated shunt value (500 pF) and use a value of (500 – 42) pF or 458 pF for the total shunt capacitance and use this value to calculate X_a.

Using PUFF software to check Example 5.22

The results are shown in Figure 5.62. Note that since the matching circuit has been designed to match to a 50 Ω source, it follows that $S11$ will be zero at the match frequency.

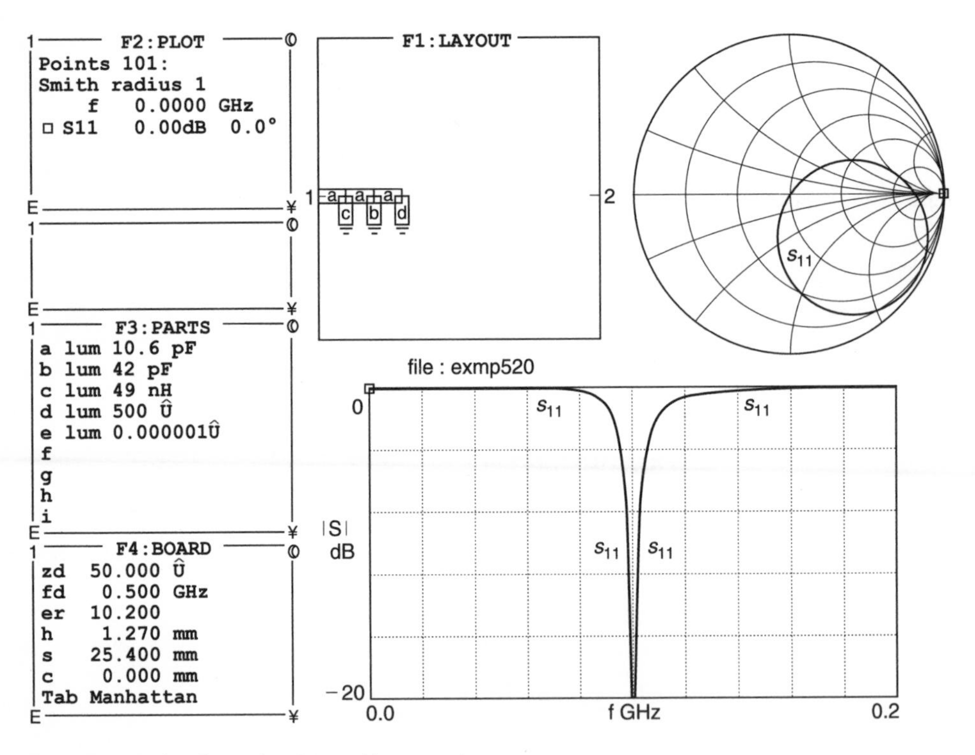

Fig. 5.62 PUFF plot of Example 5.21 matching network

5.8 Three element matching networks

Three or more element matching networks are used when we wish to match and also control the Q of a circuit. If you examine Equation 5.50 which is repeated for convenience, you will see that when R_p and R_s are fixed you are forced to accept the value of Q calculated. However, if you can vary either R_p or R_s then you are in a position to set Q_s. Using Equation 5.50 again

$$Q_s = \sqrt{\frac{R_p}{R_s} - 1} \tag{5.50}$$

Varying either R_s or R_p will no doubt cause a mismatch with your original matching aims. However, you can overcome this by using a second matching L network to match your design back to the source or load.

If you examine the π or T network (Figures 5.63 and 5.71) you will see that these networks are made up from two L type networks. Therefore, it is possible to choose (within limits) the Q of the first L network and match it to a virtual value R_v then use the second L network to match R_v to the load R_L.

5.8.1 The π network

The π network (Figure 5.63) can be described as two 'back to back' L networks (Figure 5.64) that are both configured to match the source and the load to a virtual resistance R_v located at the junction between the two networks.

More details are provided in Figure 5.65. The significance of the negative signs for $-X_{s1}$ and $-X_{s2}$ is symbolic. They are used merely to indicate that the X_s values are the opposite type of reactance from X_{p1} and X_{p2} respectively. Thus, if X_{p1} is a capacitor, X_{s1} must be an inductor and vice-versa. Similarly if X_{p2} is an inductor, X_{s2} must be a capacitor, and vice-versa. They do not indicate negative reactances (capacitors).

The design of each section of the π network proceeds exactly as was done for the L networks in the previous section. The virtual impedance or resistance R_v in Figure 5.65 must be smaller than either Z_1 or Z_2 because it is connected to the series arm of each L section, but otherwise it can be any value of your choice. Most of the time, R_v is determined by the desired loaded Q of the circuit that you specify at the beginning of the design process.

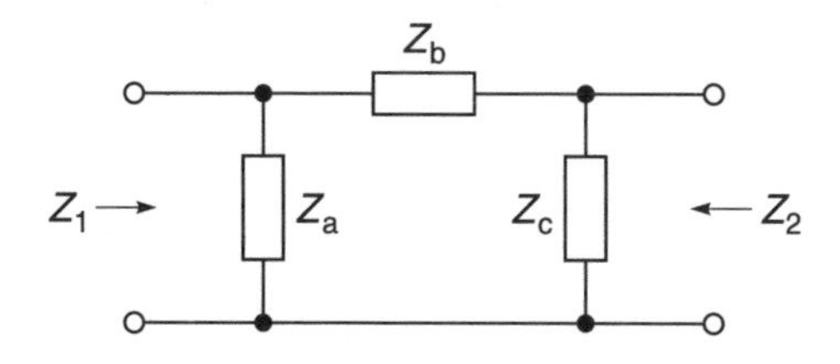

Fig. 5.63 π network

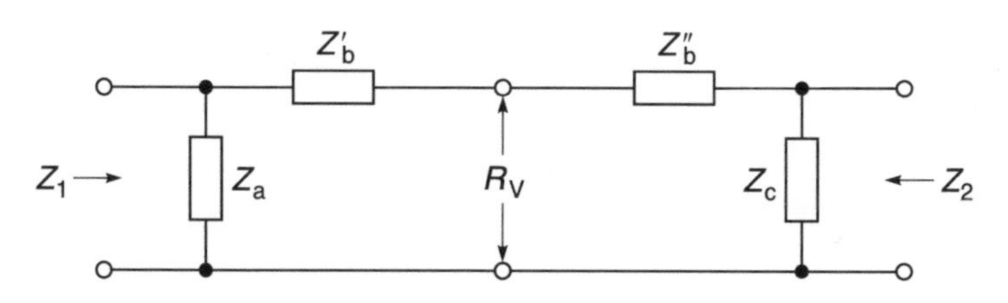

Fig. 5.64 π network made up from two L networks

For our purposes, the loaded Q of the network will be defined as:

$$Q = \sqrt{(R_h / R_v) - 1} \tag{5.56}$$

where

R_h = the largest terminating value of Z_1 or Z_2
R_v = virtual impedance or resistance

Although not entirely accurate, it is a widely accepted Q-determining formula for this circuit, and it is certainly close enough for most practical work. Example 5.23 illustrates the procedure.

Example 5.23

A source impedance of (100 + j0) Ω is to be matched to a load impedance of (1000 + j0) Ω. Design four π networks with a minimum Q of 15 to match the source and load impedances.

Given: R_s = 100 Ω, R_L = 1000 Ω, Q = 15.
Required: To design four types of π matching networks.

Solution. Take the output L network on the load side of the π network. From Equation 5.56, we can find the virtual resistance (R_v) that we will be matching:

$$R_v = \frac{R_h}{Q^2 + 1} = \frac{1000}{15^2 + 1} = 4.425\ \Omega$$

To find X_{p2} we use Equation 5.10:

$$X_{p2} = \frac{R_p}{Q_p} = \frac{R_L}{Q_p} = \frac{1000}{15} = 66.667\ \Omega$$

Similarly to find X_{s2}, we use Equation 5.4:

$$X_{s2} = Q \times R_{series} = 15(R_v) = (15)\,(4.425) = 66.375\ \Omega$$

This completes the design of the L section on the load side of the network. Note that R_{series} in the above equation was substituted for the virtual resistor (R_v) which by definition is in the series arms of the L section.

The Q for the input (source) L section network is defined by the ratio of R_s to R_v, as per Equation 5.56, where:

$$Q_1 = \sqrt{\frac{R_s}{R_v} - 1} = \sqrt{\frac{100}{4.425} - 1} = 4.647$$

Notice here that the source resistor is now considered to be in the shunt leg of the L network. Therefore R_s is defined as R_p, and using Equation 5.10

$$X_{p1} = \frac{R_p}{Q_1} = \frac{100}{4.627} = 21.612\ \Omega$$

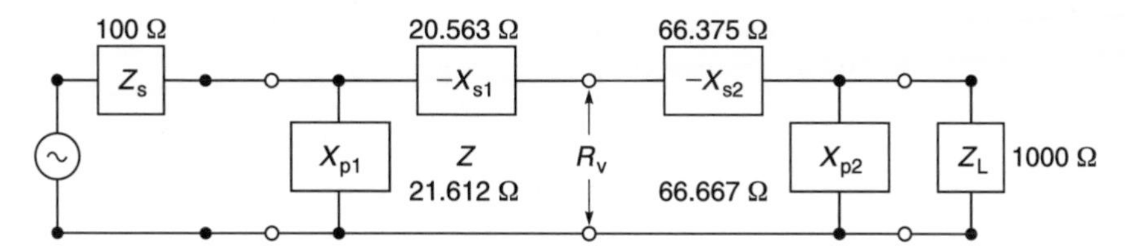

Fig. 5.65 Practical details of a π network

Similarly, using Equation 5.4:

$$X_{s1} = Q_1 R_{series} = (4.647)(4.425) = 20.563\ \Omega$$

The actual network is now complete and is shown in Figure 5.65. Remember that the virtual resistor (R) is not really in the circuit and therefore is not shown. Reactances $-X_{s1}$ and $-X_{s2}$ are now in series and can simply be added together to form a single component.

So far in this design, we have dealt only with reactances and have not yet computed actual component values. This is because of the need to maintain a general design approach so that the four final networks requested in the example can be generated quickly.

Note that X_{p1}, X_{s1}, X_{p2} and X_{s2} can all be either capacitive or inductive reactances. The only constraint is that X_{p1} and X_{s1} are of opposite types, and X_{p2} and X_{s2} are of opposite types. This yields the four networks shown in Figures 5.66 to 5.69. Note that both the source and load have been omitted in these figures. Each component in Figures 5.66 to 5.69 is shown as a reactance in ohms. Therefore to perform the transformation from dual L to π network, the two series components are merely added if they are the same type, and subtracted if the reactances are of opposite type. The final step is to change each reactance into a component value of capacitance and inductance at the frequency of operation.

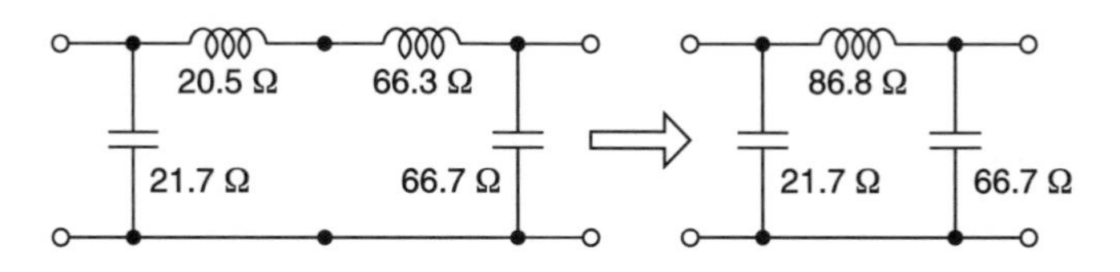

Fig. 5.66 Matching π network used as a low pass filter

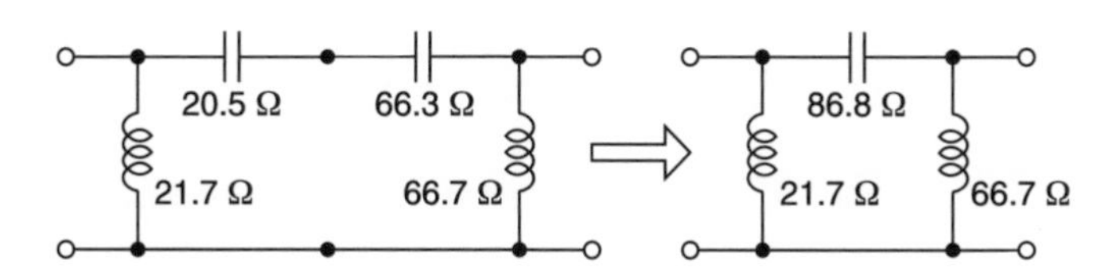

Fig. 5.67 Matching π network used as a high pass filter

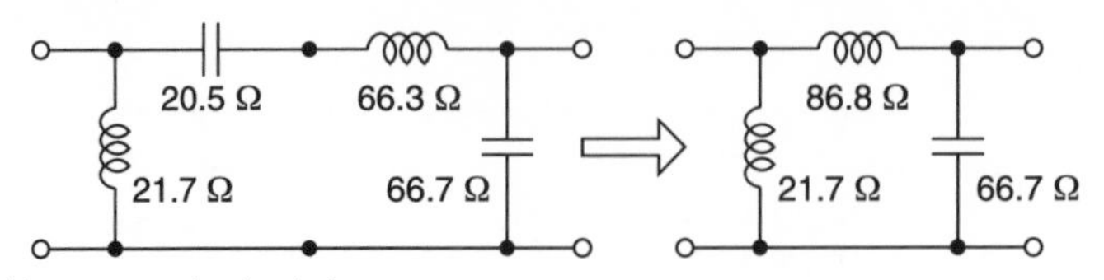

Fig. 5.68 Matching π network using inductors

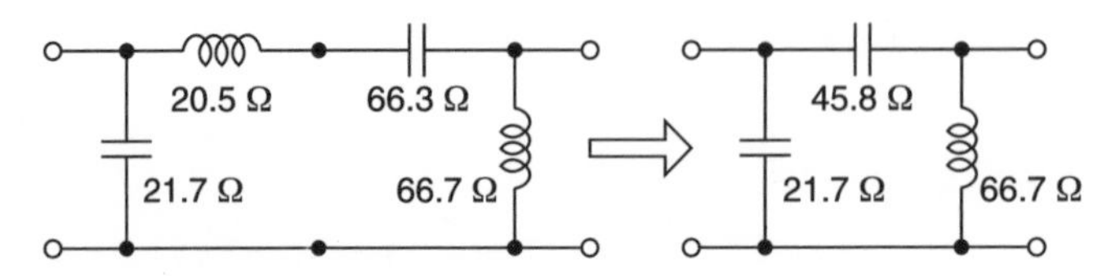

Fig. 5.69 Matching π network using capacitors

Using PUFF software to check Example 5.23

The results are shown in Figure 5.70. For convenience, all four networks have been plotted. *S*11, *S*22, *S*33 and *S*44 are the results of Figures 5.66, 5.67, 5.68 and 5.69 respectively. Note that each of the networks produce the desired input impedance of approximately 100 Ω.[1] You can check this in the Message box of Figure 5.70.

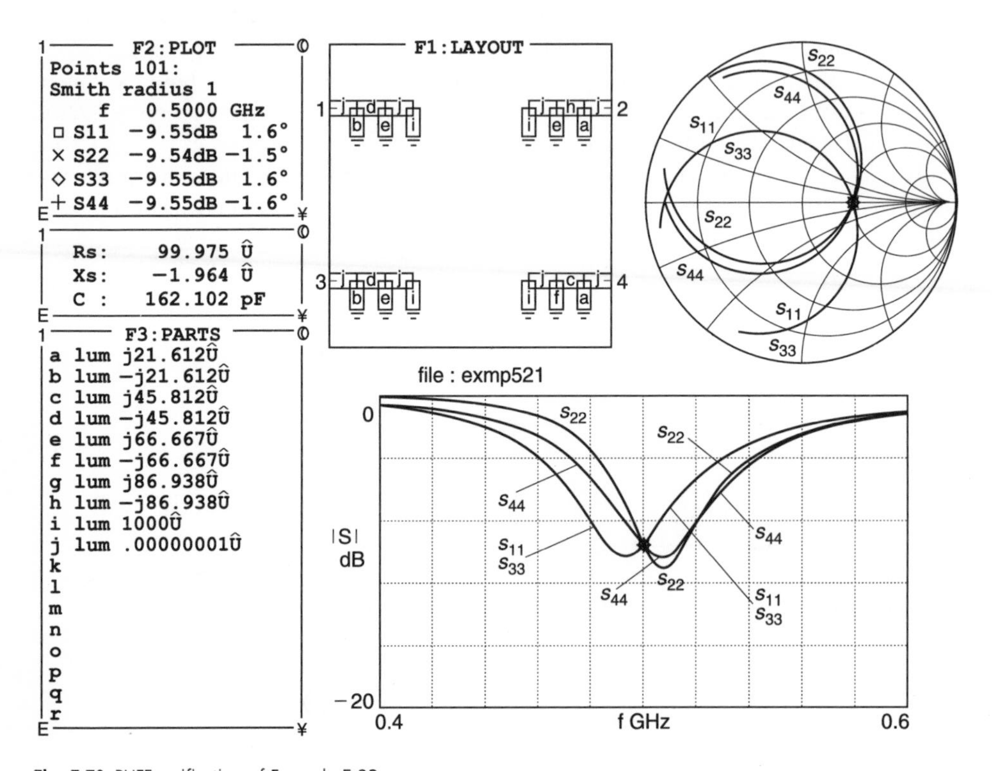

Fig. 5.70 PUFF verification of Example 5.23

[1] In this particular case, I have carried out the solution at 500 MHz but the components can be selected for operation at other frequencies.

5.8.2 The T network

The T network is often used to match two low impedance values when a high Q arrangement is needed. The design of the T network is similar to the design for the π network except that with the T network, you match the source and the load with two L type networks to a virtual resistance (R_v) which is *larger* than either of the load or source resistance. This means that the two L type networks will then have their shunt arms connected together as in Figure 5.71.

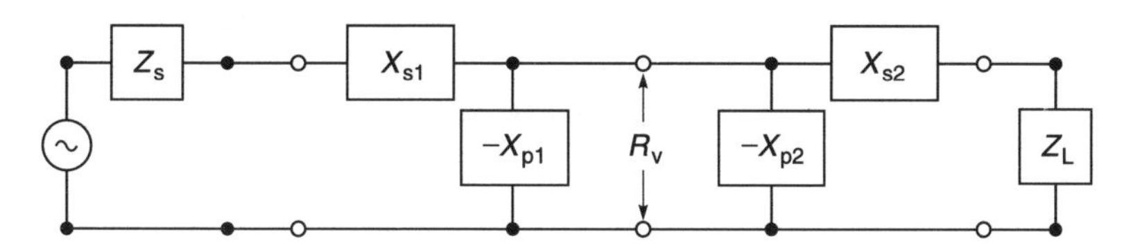

Fig. 5.71 T network shown as two back-to-back L networks

As mentioned earlier, the T network is often used to match two low-valued impedances when a high Q arrangement is desired. The loaded Q of the T network is determined mainly by the L section that has the highest Q. By definition, the L section with the highest Q will occur at the end which has the *smallest* terminating resistor. Each terminating resistor is in the series leg of each network. Therefore the formula for determining the loaded Q of the T network is

$$Q = \sqrt{\frac{R_v}{R_{small}} - 1} \tag{5.57}$$

where

R_v = virtual resistance
R_{small} = the smallest terminating resistance

The above expression is similar to the Q formula that was previously given for the π network. However, since we have reversed the L sections to produce the T network, we must ensure that we redefine the Q expression to account for the new resistor placement in relation to those L networks. In other words, Equations 5.56 and 5.57 are only special applications for the general formula that is given in Equation 5.50 and repeated here for convenience:

$$Q_s = \sqrt{\frac{R_p}{R_s} - 1} \tag{5.58}$$

where
R_p = resistance in the shunt arm of the L network
R_s = resistance in series arm of the L network

Do not be confused with the different definitions of Q because they are all the same in this case. Each L network is calculated in exactly the same manner as was given for the π network previously. We will now show this with Example 5.24.

Example 5.24

Using the configuration shown in Figure 5.71 as a reference, design four different networks to match a 10 Ω source to a 50 Ω load. Design each network for a loaded Q of 10.

Solution. Using Equation 5.57, we can find the required R_v for the match for the required Q:

$$R_v = R_{small}(Q^2 + 1) = 10(10^2 + 1) = 1010\ \Omega$$

Using Equation 5.4

$$X_{s1} = QR_s = 10(10) = 100\ \Omega$$

Using Equation 5.10,

$$X_{p2} = R/Q = 1010/10 = 101\ \Omega$$

Now for the L network on the load end, the Q is defined by the virtual resistor and the load resistor. Thus

$$Q_2 = \sqrt{R/R_L - 1} = \sqrt{1010/50 - 1} = 4.382$$

Therefore

$$X_{p2} = R/Q_2 = 1010/4.382 = 230.488\ \Omega$$

and

$$X_{s2} = Q_sR_L = (4.382)(50) = 219.100\ \Omega$$

The network is now complete and is shown in Figure 5.72 without the virtual resistor. The two shunt reactances of Figure 5.72 can again be combined to form a single element by simply substituting a value that is equal to the combined parallel reactance of the two.

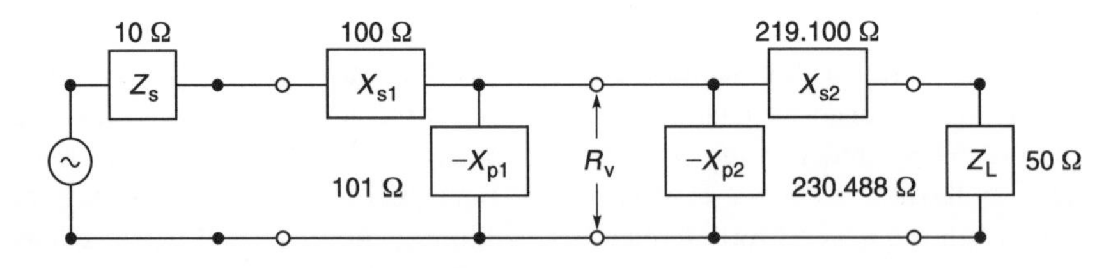

Fig. 5.72 Calculated values for the general T network

The four possible T type networks that can be used are shown in Figures 5.73 to 5.76.

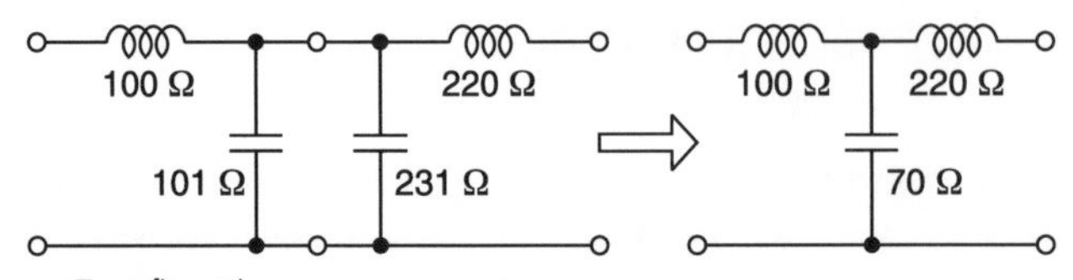

Fig. 5.73 Low pass T configuration

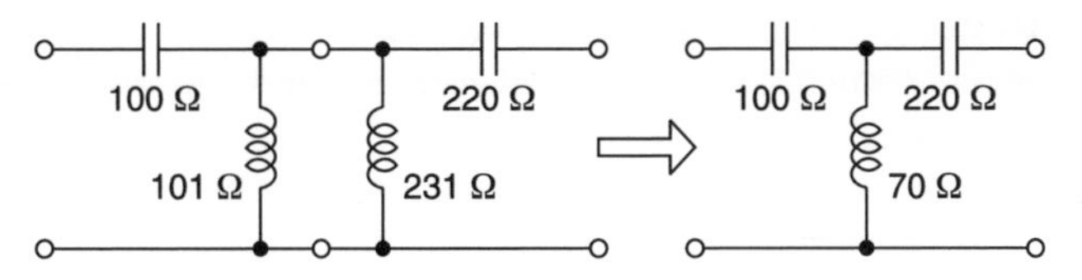

Fig. 5.74 High pass T configuration

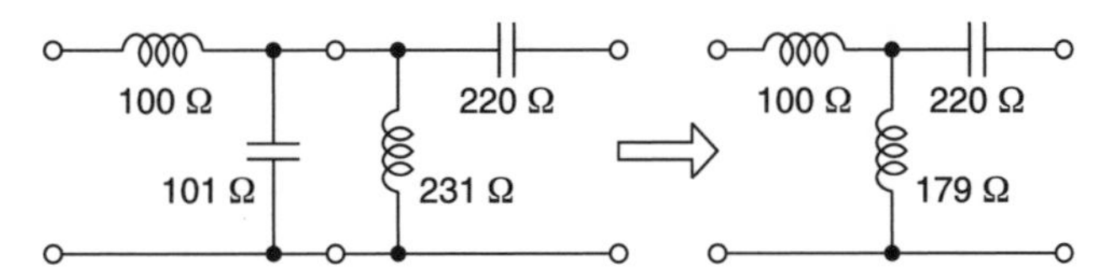

Fig. 5.75 Inductive matched T section

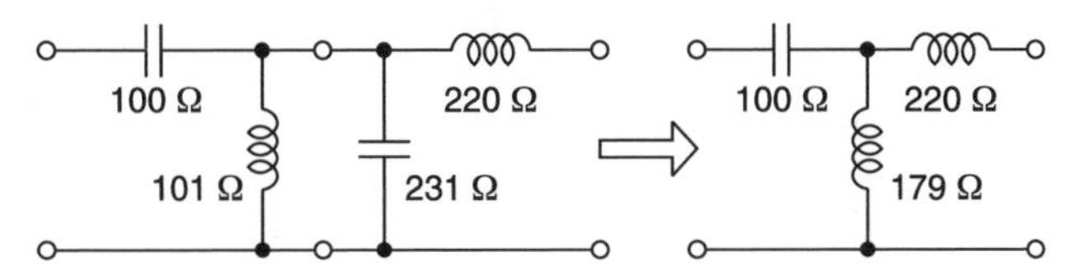

Fig. 5.76 Capacitive matched T section

Using PUFF software to check Example 5.24

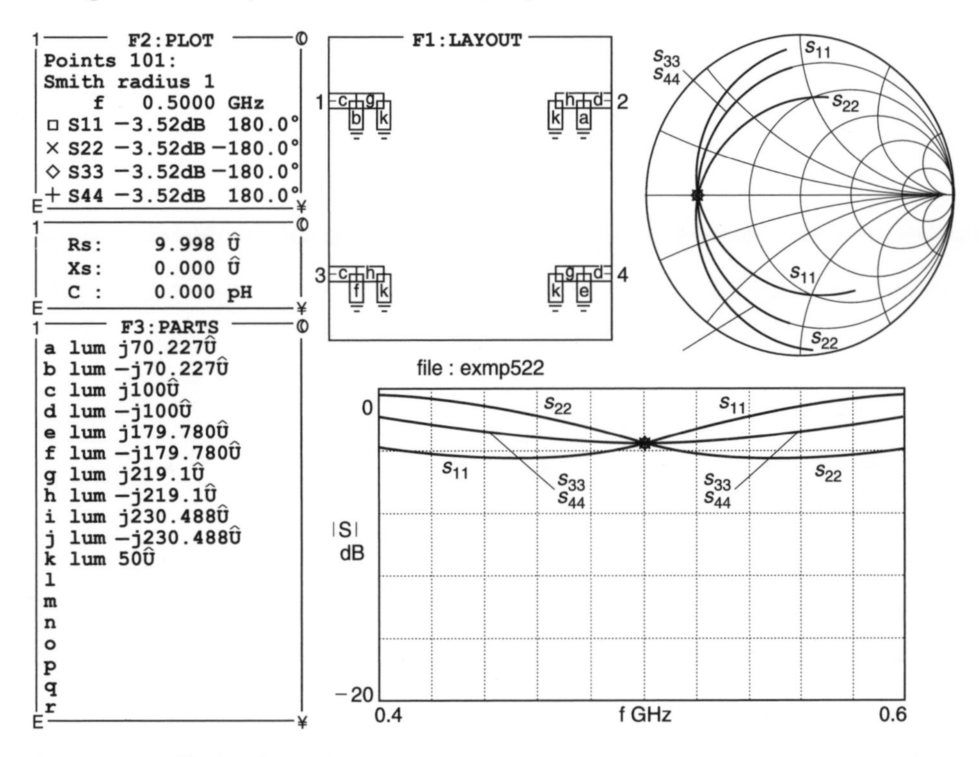

Fig. 5.77 PUFF verification of Example 5.23

Using PUFF software to check Example 5.24

The results are shown in Figure 5.77. For convenience, all four networks have been plotted. $S11$, $S22$, $S33$ and $S44$ are the results of Figures 5.73, 5.74, 5.75 and 5.76 respectively. Note that each of the networks produces the desired input impedance of approximately 10 Ω. You can read this in the Message box of Figure 5.77.

5.9 Broadband matching networks

With regard to the L network, we have noted that the circuit Q is automatically defined when the source and load are selected. With the π and T networks, we can choose a circuit Q provided that the Q chosen is larger than that which is available with the L network. This indicates that the π and T networks are useful for narrow band matching. However, to provide a broadband match, we use two L sections in still another configuration. This is shown in Figures 5.78 and 5.79 where R_v is in the shunt arm of one L section and in the series arm of the other L section. We therefore have two *series-connected* L sections rather than the back-to-back connection of the π and T networks. In this new configuration, the value of R_v must be larger than the smallest termination impedance but also smaller than the largest termination impedance. The net result is a range of loaded Q values that is *less* than the range of Q values obtainable from either a single L section, or the π or T networks previously described.

The maximum bandwidth (minimum Q) available from the networks of Figures 5.78 and 5.79 occurs when R_v is made equal to the geometric mean of the two impedances being matched:

$$R_V = \sqrt{R_S R_L} \tag{5.58}$$

The loaded Q of the above networks is defined as:

$$Q = \sqrt{\frac{R_V}{R_{smaller}} - 1} = \sqrt{\frac{R_{larger}}{R_V} - 1} \tag{5.59}$$

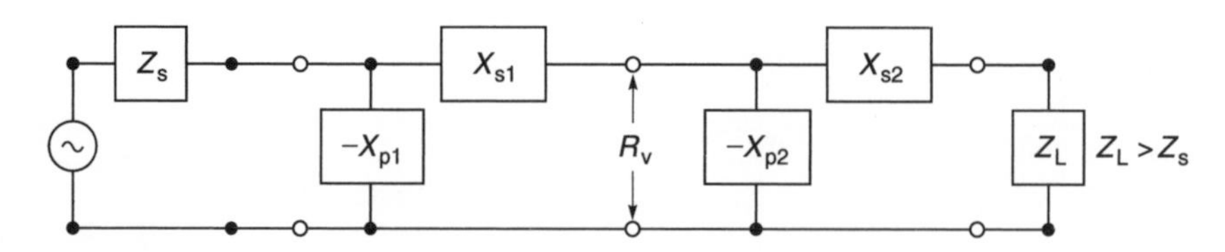

Fig. 5.78 Series connected L networks for lower Q applications. R_v is shunt leg

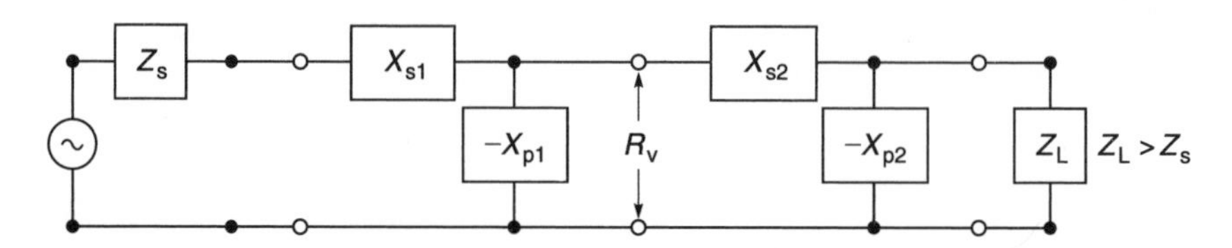

Fig. 5.79 Series connected L networks for lower Q applications. R is series leg

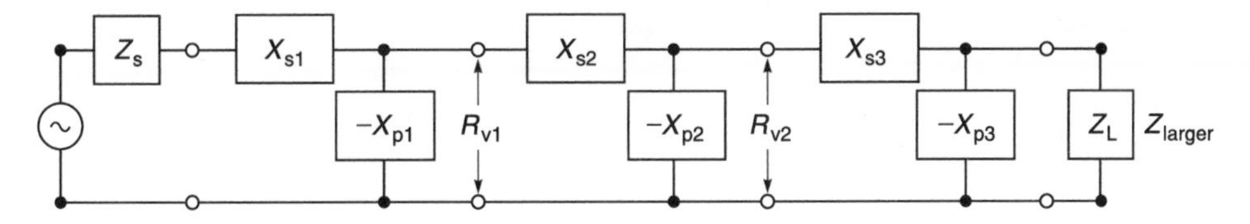

Fig. 5.80 Expanded version of Figure 5.80 for even wider bandwidth

where

R_v = the virtual resistance
$R_{smaller}$ = smallest terminating resistance
R_{larger} = largest terminating resistance

For wider bandwidths, more L networks may be cascaded with virtual resistances between each network as shown in Figure 5.80.

Optimum bandwidths in these cases are obtained if the ratios of each of the two succeeding resistances are equal:

$$\frac{R_{v1}}{R_{smaller}} = \frac{R_{v2}}{R_{v1}} = \frac{R_{v3}}{R_{v2}} = \frac{R_{larger}}{\ldots R_n} \tag{5.60}$$

where

$R_{smaller}$ = smallest terminating resistance
R_{larger} = largest terminating resistance
$R_{v1,}$ $R_{v2}, \ldots, R_{vn}$ = the virtual resistances

The design procedure for these wideband matching networks is much the same as was given for the previous examples. For the configurations of Figures 5.78 and 5.79, use Equation 5.58 to solve for R_v to design for an optimally wideband. For the configurations of Figures 5.78 and 5.79, use Equation 5.58 to solve for R_v to design for a specific low Q. For the configuration of Figure 5.80, use Equation 5.60 to solve for the different values of R_v. In all three cases after you have determined R_v, you can proceed as before.

5.10 Summary of matching networks

You should now be able to use several types of matching networks. These include transformer, capacitor-divider, L, π and T networks. Matching is vitally important in amplifier design because without this ability, it is almost impossible to design good amplifiers and oscillators.

6

High frequency transistor amplifiers

6.1 Introduction

In this part, it is assumed that you are already familiar with transistors and their operation in low frequency circuits. We will introduce basic principles, biasing, and its effects on the a.c. equivalent circuit of transistors, and the understanding of manufacturers' transistor data in the early parts of the chapter. The latter half of the chapter is devoted to the design of amplifier circuits.

6.1.1 Aims

The aims of this part are to review:

- basic principles of transistors
- biasing of transistors
- a.c. equivalent circuit of transistors
- manufacturers' admittance parameters transistor data
- manufacturers' scattering parameters transistor data
- manufacturers' transistor data in graphical form
- manufacturers' transistor data in electronic form

6.1.2 Objectives

After reading this part, you should be able to:

- understand basic operating principles of transistors
- understand manufacturers' transistor data
- apply manufacturers' transistor data in amplifier design
- bias a transistor for proper operation
- check for transistor stability
- design amplifiers using admittance parameters

6.2 Bi-polar transistors

The word **transistor** is an abbreviation of two words **trans**ferring re**sistor**.

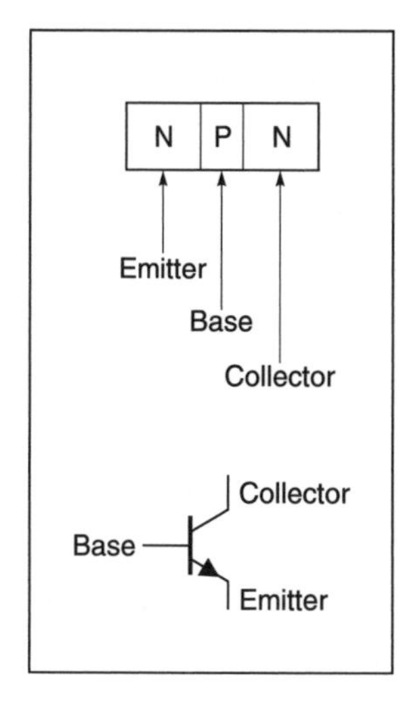

Fig. 6.1(a) Basic construction of an NPN transistor and its symbolic representation

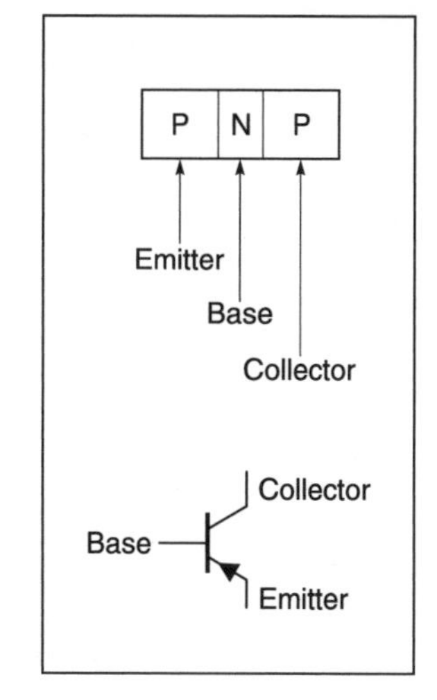

Fig. 6.1(b) Basic construction of a PNP transistor and its symbolic representation

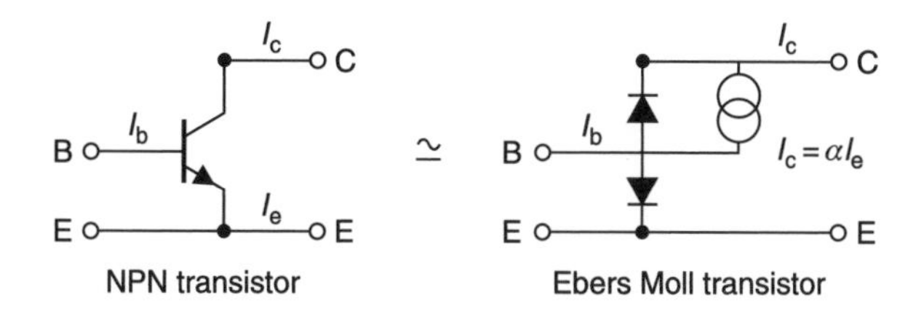

Fig. 6.2 The Ebers Moll model of a transistor

6.2.1 Basic construction

The basic construction of bi-polar transistors and their electrical symbolic representations are shown in Figure 6.1. The arrow indicates the direction of current flow in the transistor. Many transistors are made in complementary pairs. Typical examples are the well known NPN and PNP industrial and military types, 2N2222 and 2N2907, which have been used for over four decades and are still being used in many designs.

6.2.2 Transistor action[1]

For explanation purposes, a transistor may be considered as two diodes connected back to back. This is the well known Ebers Moll model and it is shown in Figure 6.2 for the NPN transistor. In the Ebers Moll model, a current generator is included to show the relationship ($I_c = \alpha I_e$) between the emitter current (I_e) and the collector current (I_c). In a good transistor, α ranges from 0.99 to about 0.999.

[1] Although the description is mainly for NPN transistors, the same principles apply for PNP transistors except that, in the latter case, positive charges (holes) are used instead of electrons.

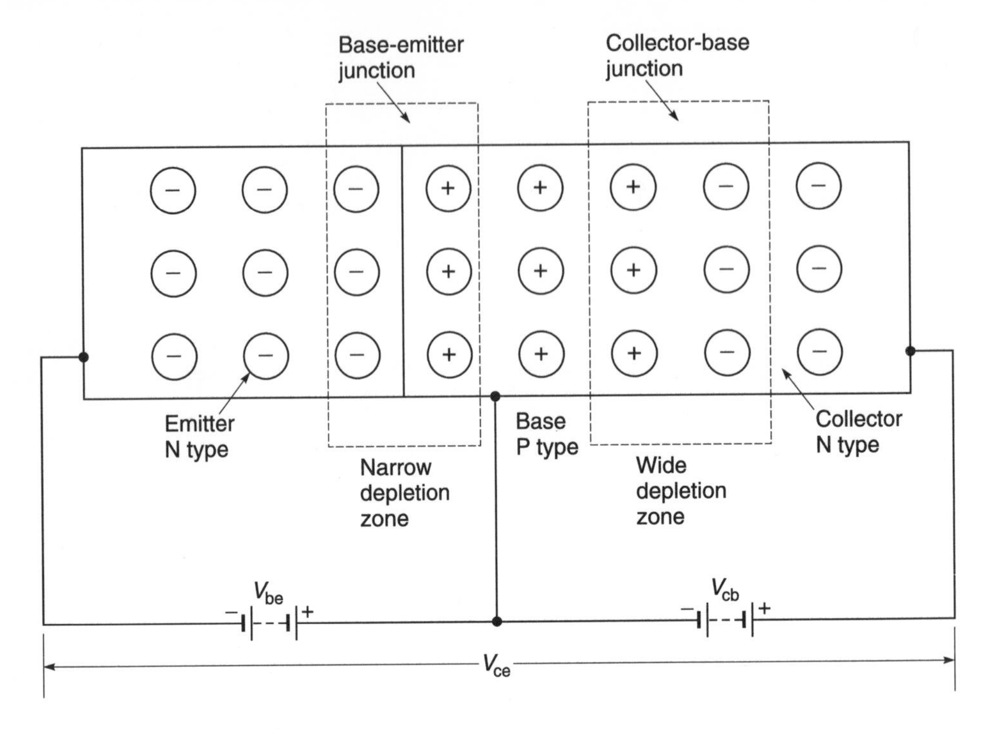

Fig. 6.3 Bi-polar transistor action

The action that takes place for an NPN transistor can be explained by the diagram in Figure 6.3. In this diagram, emitter, base and collector are diffused together and a base–emitter depletion layer is set up between base and emitter, and a collector–base depletion layer is set up between the collector and base. These depletion layers are set up in the same way as p–n junctions.

In normal transistor operation, the base–emitter junction is forward biased and the base–collector junction is reversed biased. This results in a narrow depletion (low resistance) at the base–emitter junction and a wide depletion layer (high resistance) at the collector–base junction.

Electrons from the emitter (I_e) are attracted to the base by the positive potential V_{be}. By the time these electrons arrive in the base region, they will have acquired relatively high mobility and momentum. Some of these electrons will be attracted towards the positive potential of V_{be} but most of them (>99%) will keep moving across the base region which is extremely thin (≈0.2 – 15 microns[2]) and will penetrate the collector–base junction. The electrons (I_c) will be swept into the collector region where they will be attracted by the positive potential of V_{cb}.

Relatively little d.c. energy is required to attract electrons into the base region because it is forward biased (low resistance – R_{be}). Relatively larger amounts of d.c. energy will be required in the collector–base region because the junction is reversed biased (high resistance – R_{cb}).

[2] 1 micron is one millionth of a metre.

Nevertheless we have transferred current flowing in a low resistance region into current flowing in a high resistance region. The ratio of the powers dissipated in these two regions is

$$\frac{\text{power in the collector–base region}}{\text{power in the base–emitter region}} = \frac{I_c^2 R_{cb}}{I_e^2 R_{be}}$$

where

I_c = collector current
I_e = emitter current
R_{cb} = resistance between collector and base
R_{be} = resistance between base and emitter

Since by design, $I_c \approx I_e$ and since $R_{cb} >> R_{be}$

$$\frac{I_c^2 R_{cb}}{I_c^2 R_{be}} \approx \frac{R_{cb}}{R_{be}} \tag{6.1}$$

We have a power gain because $R_{cb} >> R_{be}$. Hence if signal energy is placed between the base–emitter junction, it will appear at a much higher energy level in the collector region and **amplification** has been achieved.

6.2.3 Collector current characteristics

If you were to plot collector current (I_C) of an NPN transistor against collector–emitter (V_{CE}) for various values of the base current (I_B), you will get the graph shown in Figure 6.4. In practice, the graph is either released by transistor manufacturers or you can obtain it by using an automatic transistor curve plotter.

The points to note about these characteristics are:

- the 'knee' of these curves occurs when V_{CE} is at about 0.3–0.7 V;
- collector current (I_C) increases with base current (I_B) above the 'knee' voltage.

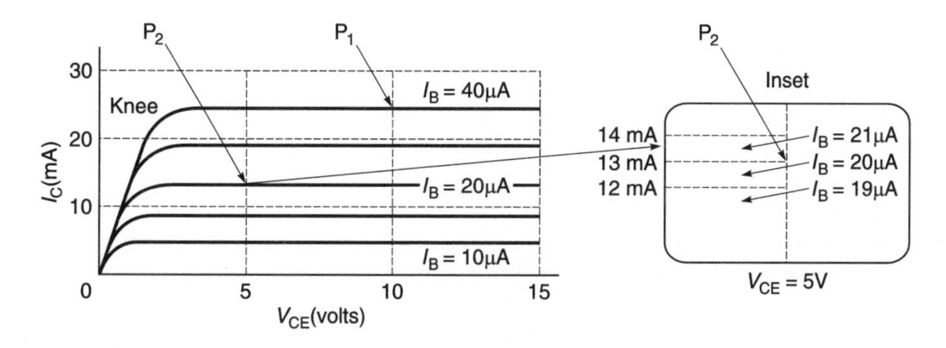

Fig. 6.4 Collector current characteristics

6.2.4 Current gain

Because of the slightly non-linear relation between collector current and base current, there are two ways of specifying the current gain of a transistor in the common emitter circuit.

The d.c. current gain (h_{FE}) is simply obtained by dividing the collector current by the base current. This value is important in switching circuits. At point P_1 of Figure 6.4, when $V_{CE} = 10$ V and $I_B = 40$ μA, $I_c = 25$ mA. Therefore

$$h_{FE} = \frac{\text{collector current}}{\text{base current}} = 25 \text{ mA}/40 \text{ }\mu\text{A} = 625 \tag{6.2}$$

For most amplification purposes, we are only concerned with small variations in collector current, and a more appropriate way of specifying current gain is to divide the *change* in collector current by the *change* in base current and obtain the small signal current gain h_{FE} or β. At point P_2 of Figure 6.4, when the operating point is chosen to be at around V_{CE} = 5 V, and $I = 20$ μA

$$h_{FE} = \frac{\Delta I_c}{\Delta I_b} = \frac{(14 - 12) \text{ mA}}{(21 - 19)\ \mu\text{A}} = 1000 \tag{6.3}$$

6.2.5 Operating point

The point at which a transistor operates is very important. For example at point P_2 of Figure 6.4 (see inset), if we choose the operating point to be at $V_{CE} = 5$ V and $I_C = 13$ mA, it is immediately clear that you will not get V_{CE} excursions ± 5 V because the transistor will not function when $V_{CE} = 0$. The same argument is true with current because you will not get current excursions less than zero. Therefore the operating point must be carefully chosen for your intended purpose. This act of choosing the operating point is called **biasing**. The importance of biasing cannot be over-emphasised because, as you will see later, *d.c. biasing also alters the a.c. parameters* of a transistor. If the a.c. parameters of your transistor cannot be held constant (within limits) then your r.f. design will not be stable.

6.2.6 Transistor biasing

The objectives of transistor biasing are:

- to select a suitable operating point for the transistor;
- to maintain the chosen operating point with changes in temperature;
- to maintain the chosen operating point with changes in transistor current gain with temperature;
- to maintain the chosen operating point to minimise changes in the a.c. parameters of the operating transistor;
- to prevent thermal runaway, where an increase in collector current with temperature causes overheating, burning and self-destruction;
- to try to maintain the chosen operating point with changes in supply voltages – this is particularly true of battery operated equipment where the supply voltage changes considerably as the battery discharges;

- to maintain the chosen operating point with changes in β when a transistor of one type is replaced by another of the same type – it is common to find that β varies from 50% to 300% of its nominal value for the same type of transistor.

There are two basic internal characteristics that have a serious effect upon a transistor's d.c. operating point over temperature. They are changes in the base–emitter voltage (ΔV_{BE}) and changes in current gain ($\Delta\beta$). As temperature *increases*, the required base-emitter voltage (V_{BE}) of a silicon transistor for the same collector current *decreases* at the rate of about 2.3 mV/°C. This means that if V_{BE} was 0.7 V for a given collector current *before* a temperature rise, then the same V_{BE} of 0.7 V after a temperature rise will now produce an increase in base current and more collector current; that in turn causes a further increment in transistor temperature, more base and collector current, and so on, until the transistor eventually overheats and burns itself out in a process known as **thermal runaway**. To prevent this cyclic action we must reduce the effective V_{BE} with temperature.

Sections 6.2.7 to 6.2.9 indicate several ways of biasing bi-polar transistors in order to *increase* bias stability. Complete step-by-step design instructions are included with each circuit configuration. For ease of understanding, a.c. components such as tuned circuits, inductors and capacitors have deliberately been left out of the circuits because they play little part in setting the operating bias point. However, a.c. components will be considered at a later stage when we come to design r.f. amplifiers.

6.2.7 Voltage feedback bias circuit

One circuit that will compensate for V_{BE} is shown in Figure 6.5. Any increase in the quiescent[3] collector current, (I_C) causes a larger voltage drop across R_C which reduces V_C which in turn reduces I_B and I_C. This can be shown by:

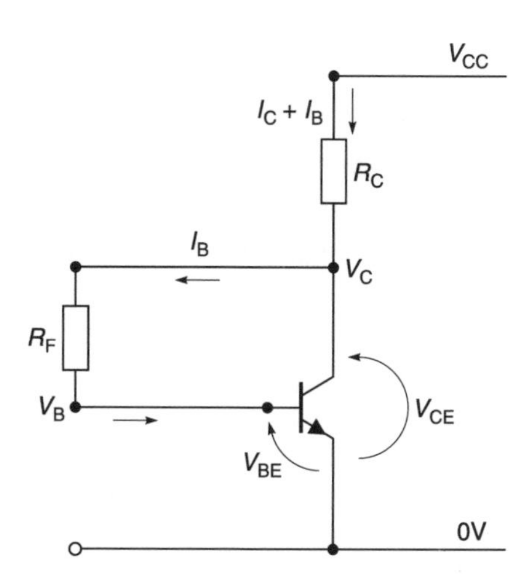

Fig. 6.5 Voltage feedback biasing

[3] Quiescent collector current is defined as the collector current which is desired at a given temperature and with no signal input to the transistor.

$$I_B = (V_C - V_{BE})/R_F \tag{6.4}$$

and

$$V_C = V_{CC} - (I_C + I_B)R_C \tag{6.5}$$

Substituting Equation 6.4 in Equation 6.5 yields

$$V_C = V_{CC} - I_C R_C - R_C(V_C - V_{BE})/R_F$$

Transposing

$$I_C R_C = V_{CC} - V_C - R_C V_C/R_F + R_C V_{BE}/R_F$$

and differentiating I_C with respect to V_{BE} and cancelling R_C from both sides gives

$$\partial I_C = \partial V_{BE}/R_F \tag{6.6}$$

Equation 6.6 shows clearly that the effect of variations in V_{BE} on I_C is reduced by a factor of $1/R_F$ in Figure 6.5.

Example 6.1

Given the transistor circuit of Figure 6.5 with $h_{FE} = 200$ and $V_{CC} = 10$ V find an operating point of $I_C = 1$ mA and $V_C = 5$ V.

Given: Transistor circuit of Figure 6.5 with $h_{FE} = 200$, $V_{CC} = 10$ V.
Required: An operating point of $I_C = 1$ mA and $V_C = 5$ V.

Solution Using Equation 6.2

$$I_B = I_C/h_{FE} = 1000/200 = 5\ \mu\text{A}$$

Assuming V_{BE} and transposing Equation 6.4

$$R_F = (V_C - V_{BE})/I_B = (5 - 0.7)\ \text{V}/\ 5\ \mu\text{A} = 860\ \text{k}\Omega$$

Transposing Equation 6.5

$$R_C = (V_{CC} - V_C)/(I_C + I_B)$$
$$= (10 - 5)\ \text{V}/(1000 + 5)\ \mu\text{A} = 4.97\ \text{k}\Omega$$

Three points should be noted about the solution in Example 6.1.

- The values calculated are theoretical resistor values, so you must use the nearest available commercial values.
- Manufacturers often only quote the minimum and maximum values of h_{FE}. In this case, simply take the geometric mean. For example, if $h_{FE(min)} = 100$ and $h_{FE(max)} = 400$, the geometric mean = $\sqrt{100 \times 400} = 200$.
- Thermal runaway is prevented when the **half-power supply principle** is used; that is when $V_c = 0.5V_{CC}$. This can be shown as follows.

At temperature T_0 collector power (P) is given by

$$V_C I_C = [V_{CC} - I_C R_C]I_C$$
$$= V_{CC} I_C - I_C^2 R_C \tag{6.7}$$

At a higher temperature (T_1) new collector power ($P + \Delta P$) is given by

$$(V_C - \Delta V_C)(I_C + \Delta I_C) \approx V_C(I_C + \Delta I_C)$$

because $\Delta V_C << V_C$

$$P + \Delta P = [V_{CC} - (I_C + \Delta I_C)R_C](I_C + \Delta I_C)$$

$$= V_{CC}(I_C + \Delta I_C) - (I_C + \Delta I_C)^2 R_C \tag{6.7a}$$

Subtracting Equation 6.7 from Equation 6.7a yields

$$\Delta P = V_{CC}I_C + V_{CC}\,\Delta I_C - (I_C^2 + 2I_C\Delta I_C + \Delta I_C^2)R_C - V_{CC}I_C + I_C^2 R_C$$

Simplifying and discarding $\Delta I_C^2 R_C$ because it is very small, gives

$$\Delta P = \Delta I_C(V_{CC} - 2\,I_C R_C)$$

Since $I_C = (V_{CC} - V_C)/R_C$

$$\Delta P = \Delta I_C(-\,V_{CC} + 2V_C)$$

For ΔP to equal zero, we get

$$V_{CC} = 2V_C \text{ or } V_C = 0.5V_{CC} \tag{6.8}$$

Equation 6.8 is the basis for the half-power supply principle and it should be used whenever possible to prevent thermal runaway.

For reasons which will become clearer when we discuss a.c. feedback, you will sometimes find that R_F is split into two resistors, R_{F1} and R_{F2}, with a capacitor C_{F1} connected between its junction and chassis ground as shown in Figure 6.6. The purpose of C_{F1} is to prevent any output a.c. or r.f. signal from travelling back to the input circuit. R_{F1} is used to prevent short-circuiting the collector output signal via C_{F1} and R_{F2} is to prevent short-circuiting the base signal through C_{F1}.

Finally before leaving the bias circuits of Figures 6.5 and 6.6, the great advantage of the

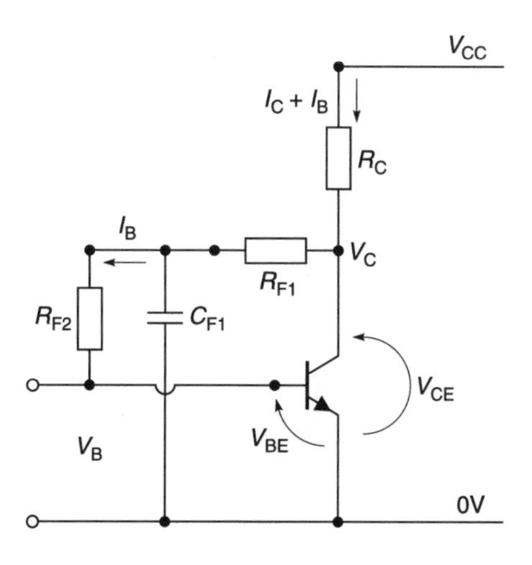

Fig. 6.6 Split feedback bias circuit

voltage feedback circuit is that it enables the emitter to be earthed directly. This is very important at microwave frequencies because it helps to prevent unwanted feedback which may affect amplifier stability.

Example 6.2

A transistor has $h_{FE} = 250$ when $V_C = 12$ V and $I_C = 2$ mA. If your power supply (V_{CC}) is 24 V, design a bias circuit like that shown in Figure 6.5 for operating the transistor with $V_C = 12$ V and $I_C = 2$ mA.

Solution For this solution use the outlines of Example 6.1. Using Equation 6.2, h_{FE} = collector current/base current and

$$I_B = I_C/h_{FE} = 2000\ \mu\text{A}/250 = 8\ \mu\text{A}$$

Transposing Equations 6.4 and 6.5

$$R_F = (V_C - V_{BE})/I_B = (12 - 0.7)\ \text{V}/8\ \mu\text{A} = 1.41\ \text{M}\Omega$$

and

$$R_C = (V_{CC} - V_C)/(I_C + I_B)$$
$$= (24 - 12)\ \text{V}/(2000 + 8)\ \mu\text{A} = 5.97\ \text{k}\Omega$$

6.2.8 Voltage feedback and constant current bias circuit

Another bias circuit frequently used for r.f. amplifiers is shown in Figure 6.7. This circuit is similar to that shown in Figure 6.5 except that the base current is fed from a more stable source. Any increase in collector current (ΔI_C) results in a decrease in V_C, V_{BB} and I_B which in turn counteracts any further increase in I_C. The design of this circuit is shown in Example 6.3.

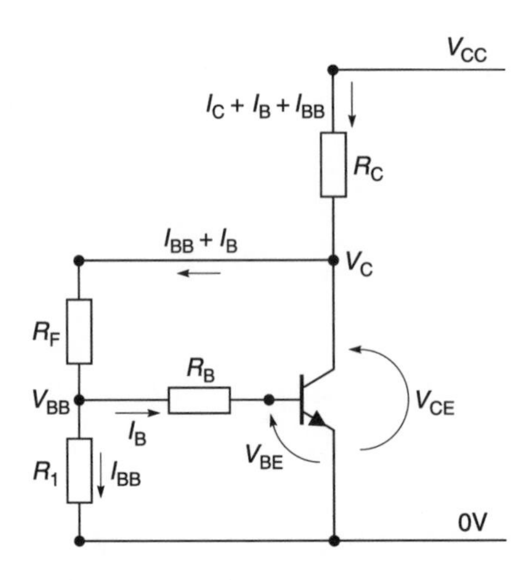

Fig. 6.7 Voltage feedback and constant current bias circuit

Example 6.3

Using the biasing arrangement shown in Figure 6.7, calculate the biasing resistors for a transistor operating with $V_C = 10$ V, $I_C = 5$ mA and a supply voltage $V_{CC} = 20$ V. The transistor has a d.c. gain of $h_{FE} = 150$.

Given: $V_{CC} = 20$ V, $V_C = 10$ V, $I_C = 5$ mA and $h_{FE} = 150$.
Required: R_1, R_B, R_C, R_F.

Solution

1 Assume values for V_{BB} and I_{BB} to supply a constant current I_B:

$$V_{BB} = 2\text{ V} \quad I_{BB} = 1\text{ mA}$$

2 Knowing I_C and h_{FE}, calculate I_B:

$$I_B = \frac{I_C}{h_{FE}} = \frac{5\text{ mA}}{150} = 0.0333\text{ mA}$$

3 Knowing V_{BB} and I_B, and assuming that $V_{BE} = 0.7$ V, calculate R_B:

$$R_B = \frac{V_{BB} - V_{BE}}{I_B} = \frac{(2 - 0.7)\text{ V}}{0.0333\text{ mA}} = 39.39\text{ k}\Omega$$

4 Knowing V_{BB} and I_{BB}, calculate R_1:

$$R_1 = \frac{V_{BB}}{I_{BB}} = \frac{2\text{ V}}{1\text{ mA}} = 2\text{ k}\Omega$$

5 Knowing V_{BB}, I_{BB}, I_B and V_C, calculate R_F:

$$R_F = \frac{V_C - V_{BB}}{I_{BB} + I_B} = \frac{(10 - 2)\text{ V}}{1.033\text{ mA}} = 7.74\text{ k}\Omega$$

6 Knowing V_{CC}, V_C, I_C, I_B and I_{BB}, calculate R_C:

$$R_C = \frac{V_{CC} - V_C}{I_C + I_B + I_{BB}} = \frac{(20 - 10)\text{ V}}{6.033\text{ mA}} = 1.66\text{ k}\Omega$$

Example 6.4

Use the bias circuit shown in Figure 6.7 to set the operating point of a transistor at $I_C =$ 1 mA, $V_C = 6$ V. The current gain of the transistor ranges from 100 to 250. The circuit supply voltage is 12 V.

Solution

Note: In my solution I have chosen $V_{BB} = 1.5$ V and $I_{BB} = 0.5$ mA; within reason you may choose other values but if you do then your solutions will obviously differ from mine.

Using Example 6.3 as a basis for the solution, I have calculated the values of resistors but you must use the closest commercial value in the circuit.

1 The operating point for the transistor is

$$I_C = 1\text{ mA},\ V_C = 6\text{ V},\ V_{CC} = 12\text{ V and}$$

$$h_{FE} = \sqrt{100 \times 250} \approx 158$$

2 Assume values for V_{BB} and I_{BB} to supply a constant current, I_B:

$$V_{BB} = 1.5\text{ V} \qquad I_{BB} = 0.5\text{ mA}$$

3 Knowing I_C and h_{FE}, calculate I_B:

$$I_B = \frac{I_C}{h_{FE}} = \frac{1\text{ mA}}{158} \approx 6.3\ \mu\text{A}$$

4 Knowing V_{BB} and I_B, and assuming that $V_{BE} = 0.7$ V, calculate R_B:

$$R_B = \frac{V_{BB} - V_{BE}}{I_B} = \frac{(1.5 - 0.7)\text{ V}}{6.3\ \mu\text{A}} \approx 126.9\text{ k}\Omega$$

5 Knowing V_{BB} and I_{BB}, calculate R_1:

$$R_1 = \frac{V_{BB}}{I_{BB}} = \frac{1.5\text{ V}}{0.5\text{ mA}} = 3\text{ k}\Omega$$

6 Knowing V_{BB}, I_{BB}, I_B and V_C, calculate R_F:

$$R_F = \frac{V_C - V_{BB}}{I_{BB} + I_B} = \frac{(6 - 1.5)\text{ V}}{(0.5\text{ mA} + 6.3\ \mu\text{A})} = 8.88\text{ k}\Omega$$

7 Knowing V_{CC}, V_C, I_C, I_B and I_{BB}, calculate R_C:

$$R_C = \frac{V_{CC} - V_C}{I_C + I_B + I_{BB}} = \frac{(12 - 6)\text{ V}}{(1 + 0.0063 + 0.5)\text{ mA}} = 3.98\text{ k}\Omega$$

6.2.9 Base-voltage potential divider bias circuit

Another bias circuit that is commonly used is the base-voltage potential divider bias circuit shown in Figure 6.8. In this circuit, V_{BB} is held approximately *constant* by the voltage divider network of R_1 and R_2. V_{BE} is the voltage *difference* between V_{BB} and V_E which is the product of I_E and R_E.

Since $I_E = I_C + I_B$, any collector current rise ΔI_C is followed by a ΔI_E rise which increases V_E. This *increase* in V_E is a form of negative feedback that tends to reduce bias on the base–emitter junction and, therefore, *decrease* the collector current. Example 6.5 shows how to design the bias circuit of Figure 6.8.

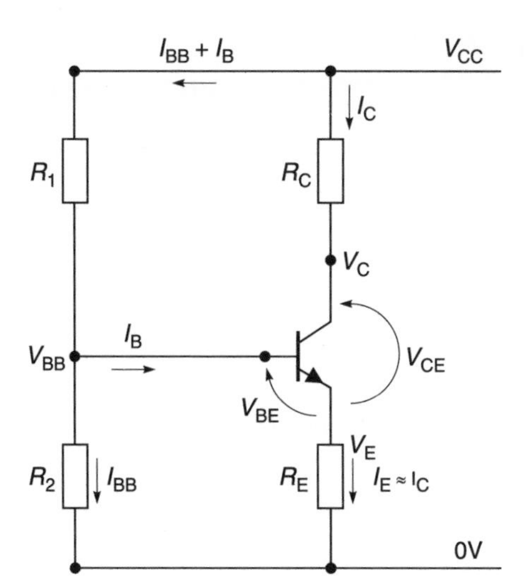

Fig. 6.8 Base-voltage potential divider bias circuit

Example 6.5

Using the biasing arrangement shown in Figure 6.8, calculate the biasing resistors for a transistor operating with $V_C = 10$ V, $I_C = 10$ mA and a supply voltage $V_{CC} = 20$ V. The transistor has a d.c. gain of $h_{FE} = 50$.

Given: $V_{CC} = 20$ V, $V_C = 10$ V, $I_C = 10$ mA and $h_{FE} = 50$.
Required: R_1, R_2, R_C, R_E.

Solution

1 Choose V_E to be approximately 10% of $V_{CC.}$ Make

$$V_E = 10\% \text{ of } 20\text{ V} = 2\text{ V}$$

2 Assume $I_E \approx I_C$ for high gain transistors.
3 Knowing I_E and V_E, calculate R_E:

$$R_E = \frac{2\text{ V}}{10\text{ mA}} = 200\ \Omega$$

4 Knowing V_{CC}, V_C and I_C, calculate R_C:

$$R_C = \frac{V_{CC} - V_C}{I_C} = \frac{(20 - 10)\text{ V}}{10\text{ mA}} = 1000\ \Omega$$

5 Knowing I_C and h_{FE}, calculate I_B:

$$I_B = \frac{I_C}{h_{FE}} = \frac{10\text{ mA}}{50} = 0.2\text{ mA}$$

6 Knowing V_E and V_{BE}, calculate V_{BB}:

$$V_{BB} = V_E + V_{BE} = 2.0 \text{ V} + 0.7 \text{ V} = 2.7 \text{ V}$$

7 Choose a value for I_{BB}; a normal rule of thumb is that $I_{BB} \approx 10I_B$. Hence

$$I_{BB} = 2 \text{ mA}$$

8 Knowing I_{BB} and V_{BB}, calculate R_2:

$$R_2 = \frac{V_{BB}}{I_{BB}} = \frac{2.7 \text{ V}}{2 \text{ mA}} = 1350\ \Omega$$

9 Knowing V_{CC}, V_{BB}, I_{BB} and I_B, calculate R_1:

$$R_1 = \frac{V_{CC} - V_{BB}}{I_{BB} + I_B} = \frac{(20 - 2.7) \text{ V}}{(2 + 0.2)} = 7864\ \Omega$$

Example 6.6

Use the bias circuit shown in Figure 6.8 to set the operating point of a transistor at $I_C = 1$ mA, $V_C = 6$ V. The current gain of the transistor ranges from 100 to 250. The circuit supply voltage is 12 V.

Solution

Note: In my solution I have chosen $V_E = 1.2$ V and $I_{BB} = 0.5$ mA; within reason you may choose other values but if you do then your solutions will obviously differ from mine.

Using Example 6.5 as a basis for the solution, I have calculated the values of resistors but you must use the closest commercial value in the circuit.

1 The operating point for the transistor is:

$$I_C = 1 \text{ mA}, V_C = 6 \text{ V}, V_{CC} = 12 \text{ V and}$$
$$h_{FE} = \sqrt{100 \times 250} \approx 158$$

2 Assume a value for V_E that considers bias stability: choosing V_E to be approximately 10% of V_{CC}

$$V_E = 10\% \text{ of } 12 \text{ V} = 1.2 \text{ V}$$

3 Assume $I_E \approx I_C$ for high h_{FE} transistors.
4 Knowing I_E and V_E, calculate R_E:

$$R_E = \frac{1.2 \text{ V}}{1 \text{ mA}} = 1200\ \Omega$$

5 Knowing V_{CC}, V_C and I_C, calculate R_C:

$$R_C = \frac{V_{CC} - V_C}{I_C} = \frac{(12 - 6) \text{ V}}{1 \text{ mA}} = 6000\ \Omega$$

6 Knowing I_C and h_{FE}, calculate I_B:

$$I_B = \frac{I_C}{h_{FE}} = \frac{1 \text{ mA}}{158} = 6.3 \text{ µA}$$

7 Knowing V_E and V_{BE}, calculate V_{BB}:

$$V_{BB} = V_E + V_{BE} = 1.2 \text{ V} + 0.7 \text{ V} = 1.9 \text{ V}$$

8 I have chosen I_{BB} to be:

$$I_{BB} = 0.5 \text{ mA}$$

9 Knowing I_{BB} and V_{BB}, calculate R_2:

$$R_2 = \frac{V_{BB}}{I_{BB}} = \frac{1.9 \text{ V}}{0.5 \text{ mA}} = 3.8 \text{ k}\Omega$$

10 Knowing V_{CC}, V_{BB}, I_{BB} and I_B, calculate R_1:

$$R_1 = \frac{V_{CC} - V_{BB}}{I_{BB} + I_B} = \frac{(12 - 1.9) \text{ V}}{(0.5 + 0.0063) \text{ mA}} = 19.95 \text{ k}\Omega$$

6.2.10 Summary on biasing of bi-polar transistors

In the voltage feedback circuits of Figures 6.5 and 6.6 and the voltage feedback and constant current circuit of Figure 6.7, increases of collector current with temperature are kept in check by introducing a form of negative feedback to decrease the effective base–emitter voltage (V_{BE}). If we were to use *upward* pointing arrows to indicate *increases* and *downward* pointing arrows to indicate *decreases*, for the circuit of Figures 6.5 to 6.7, we would have

$$I_C \uparrow ; V_{BE} \downarrow ; I_B \downarrow ; \rightarrow I_C \text{ constant}$$

In the base–voltage potential divider circuit of Figure 6.8, the effective base–emitter voltage (V_{BE}) is reduced by increasing the emitter voltage (V_E) against a quasi fixed potential (V_{BB}). It is a better bias circuit than the earlier ones because V_E can be made to almost track and compensate for changes in V_{BE}. Using arrows as before, for the circuit of Figure 6.8, we would have

$$I_C \uparrow ; V_E \uparrow; (V_{BE} = V_{BB} - V_E) \downarrow ; I_B \downarrow ; \rightarrow I_C \text{ constant}$$

The manufacturing tolerance for h_{FE} or β in transistors of the same part number is typically poor. It is not uncommon for a manufacturer to specify a 10:1 range for β on the data sheet (such as 30 to 300). This of course makes it extremely difficult to design a bias network for the device in question when it is used from a production standpoint as well as a temperature standpoint.

However, the base-voltage potential divider circuit works remarkably well on both accounts because a high h_{FE} produces a high V_E which counteracts the quasi fixed

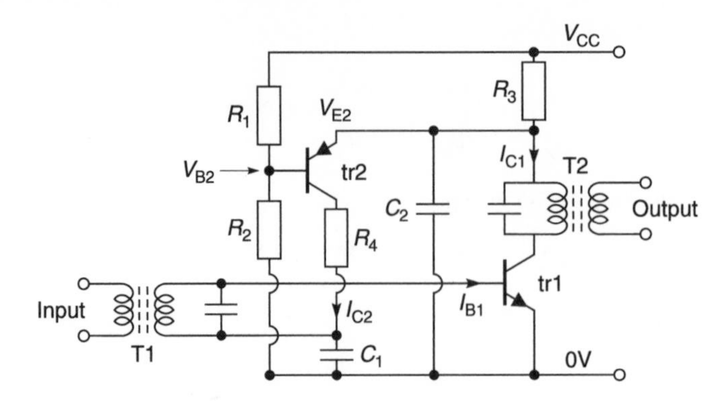

Fig. 6.9 Active bias circuit for an r.f. amplifier

potential V_{BB}. This bias circuit is therefore widely used in production circuits. One drawback of this circuit is that the resistor R_E must be bypassed by a capacitor in order not to affect the a.c. operation of the amplifier. At frequencies less than 100 MHz, this is no problem but at microwave frequencies effective bypassing is difficult and you will see that the earlier circuits are often used.

In the discussions on biasing, I have only concentrated on direct biasing where resistors are used to control I_B and V_{BE}. However, there are **active-bias-circuits** where one or more additional transistors are used to bias the main r.f. amplifier. One well known example is shown in Figure 6.9. In this circuit r.f. signal is applied through the input tuned transformer (T1) to the base of the r.f. amplifier (tr1). The r.f. output signal is taken from the collector tuned transformer (T2). Capacitors C_1 and C_2 are bypass capacitors used to allow r.f. signals to reach the emitter of tr1 easily and to keep r.f. signal out of the biasing circuit.

Biasing is carried out by supplying base current to tr1 via the secondary of T1 from R_4 which in turn is fed from the collector current (I_{C2}) of tr2. The base voltage (V_{B2}) of tr2 is fixed by the potential divider consisting of R_1 and R_2 which sets the base voltage of tr2 approximately 0.7 V below its emitter voltage (V_{E2}). The emitter voltage (V_{E2}) for tr2 is provided by the voltage drop across R_3 which in turn is caused by the collector current (I_{C1}) of tr1. If I_{C1} increases, then V_{E2} decreases and the effective base–emitter voltage (V_{B2}) of tr2 decreases, I_{C2} decreases, which in turn reduces I_{B1}. I_{C1} then decreases to try and maintain its old value. Again, if we were to use *upward* pointing arrows to indicate *increases* and *downward* pointing arrows to indicate *decreases*, for the circuit of Figure 6.9, we would have

$$I_{C1} \uparrow ; V_{E2} \downarrow ; V_{BE2} \downarrow ; I_{C2} \downarrow ; I_{B1} \downarrow ; I_{C1} \downarrow ; \rightarrow I_{C1} \text{ constant}$$

The biasing of this circuit can be made very stable and I have used this arrangement for r.f. amplifers operating from –15°C to +75°C. Another advantage of this biasing circuit is that although tr1 is an expensive r.f. transistor, tr2 can be a cheap low frequency transistor. The only requirement is that tr1 and tr2 should be made from similar material to enable easier tracking for V_{BE} and I_{CO}. For example if tr1 is a silicon transistor then tr2 should also be a silicon transistor.

6.3 Review of field effect transistors

There are two main types of field effect transistors; the junction FET or **JFET**, and the metal oxide silicon FET or **MOSFET**. MOSFETs are sometimes also referred to as 'insulated gate field effect transistors', **IGFET**s. A high electron mobility FET is also known as a **HEMFET**. In this discussion, we will look briefly at FET structures, their operating modes and methods of biasing to enable us to use FETs efficiently at high frequencies.

6.3.1 A brief review of field effect transistor (JFET or FET) construction

An elementary view of field effect transistor construction will aid understanding when we analyse a.c. equivalent circuits and use them in practical amplifier and oscillator circuits.

FET (n-channel)

The basic construction of an n-channel type field effect transistor is shown in Figure 6.10. The device has three terminals, a source terminal (S), a gate terminal (G), and a drain terminal (D). In an n-channel depletion mode FET, n-type material is used as the conducting channel between source and drain. P type material is placed on either side of the channel. The effective electrical width of the channel is dependent on the voltage potential (V_{GS}) between gate and source.

When electrical supplies are connected in the manner shown in Figure 6.11, electrons flow from the source, past the gate, to the drain. If a negative voltage is applied between the gate and source, its negative electric field will try to 'pinch' the electron flow and confine it to a smaller cross-section of the n-channel of the FET. This affects the resistance of the n-channel and restricts the current flowing through it. Hence, by varying the gate–source voltage (V_{GS}), it is possible to control current flow.

Power gain is obtained because very little energy is required to control the input signal (V_{GS}) while relatively large amounts of power can be obtained from the variations in drain–source current (I_{DS}). The symbols for an n-channel depletion mode FET and its output current characteristics are shown in Figure 6.12. Drain–source current (I_D) is maximum when there is zero voltage on V_{GS} and I_D *decreases* as V_{GS} becomes more *negative*. The pinch-off voltage (V_P) is the gate–source voltage required to reduce the effective cross-section of the n-channel to zero. For practical purposes, it is V_{GS} which causes I_{DS} to

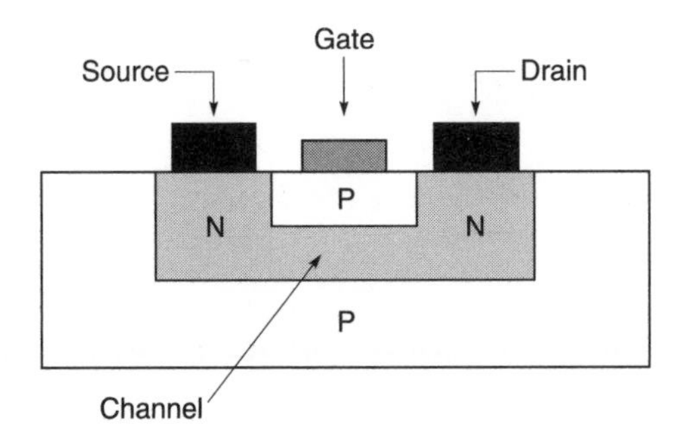

Fig. 6.10 An n-channel depletion mode field effect transistor

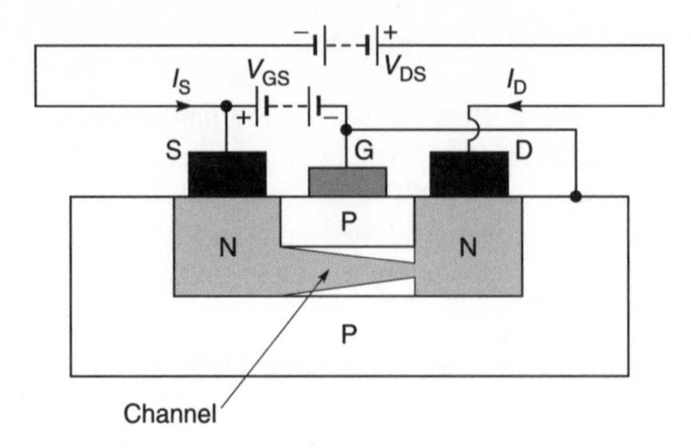

Fig. 6.11 n-channel FET

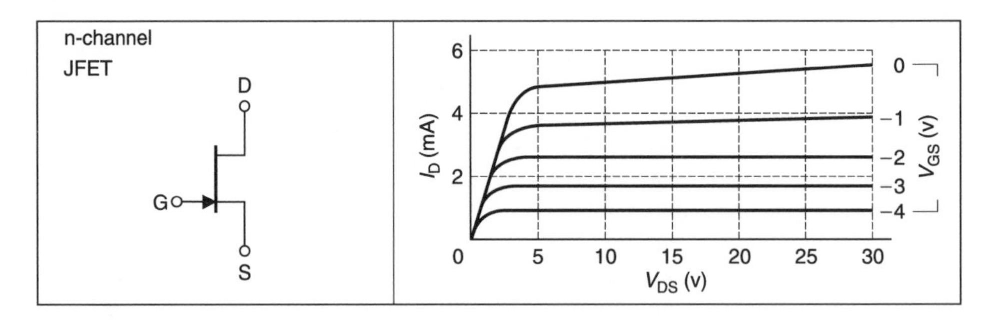

Fig. 6.12 Symbols and output current characteristics for an n-channel FET

become zero. I_{DSS} is the drain–source current when the gate and source are shorted together ($V_{GS} = 0$) for your particular V_{DS}.

Operating point and biasing of an n-channel FET

Selection of the operating point is similar to that explained for the bi-polar transistor, but the biasing method is different. An example of how to bias an n-channel FET is given in Example 6.7.

Example 6.7 Biasing an n-channel FET

Using the biasing arrangement shown in Figure 6.13, calculate the biasing resistors for an FET operating with $V_{DS} = 10$ V, $I_{DS} = 5$ mA and a supply voltage $V_{CC} = 24$ V. From manufacturer's data, for $I_{DS} = 5$ mA and $V_{DS} = 10$ V, $V_{GS} = -2.3$ *V*.

Given: $V_{CC} = 24$ V, $V_{DS} = 10$ V, $I_{DS} = 5$ mA and $V_{GS} = 2.3$ V.
Required: R_s, R_D, R_G, R.

Solution

1 For our particular transistor, the manufacturer's d.c. curves show that for an operating current of 5 mA with $V_{DS} = 10$ V, we require a V_{GS} of −2.3 V which means that the gate must be 2.3 V negative with respect to the source. We do not have a negative supply but we can simulate this supply by making the source *positive* with respect to the gate which

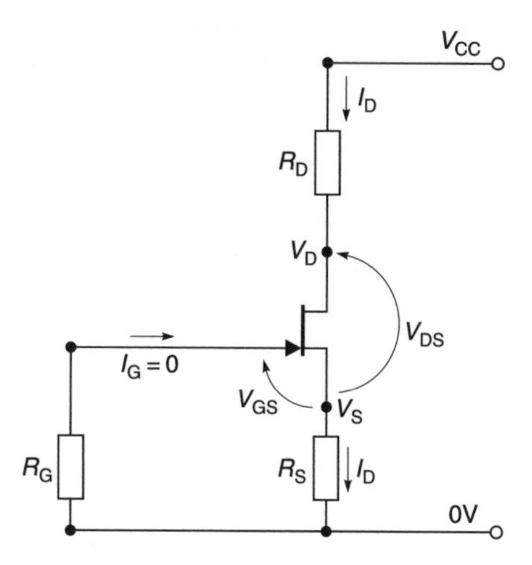

Fig. 6.13 n-channel FET biasing

in turn means that the gate will be *negative* with respect to the source. This is carried out by placing a resistor (R_S) in series with I_D to produce a positive voltage (V_S) which is equal to V_{GS}. We now calculate R_S.

2 Since $I_G = 0$, $|V_{GS}| = |V_S|$, and knowing I_D, calculating R_S gives

$$R_S = \frac{|V_S|}{I_D} = \frac{|V_{GS}|}{I_D}$$

$$= \frac{2.3\text{ V}}{5\text{ mA}} = 460\ \Omega$$

Note particularly that R_S provides a form of negative feedback to stabilise changes in FET parameters with temperature. It also stabilises current when FETs of the same type number are changed, because any *increase* in I_{DS} immediately produces a corresponding *decrease* in V_{GS}. The ratio

$$\frac{\text{change in } I_{DS}(\Delta I_{DS})}{\text{change in } V_{GS}(\Delta V_{GS})}$$

is known as the transconductance (g_m) of the transistor.

3 Since $I_G = 0$, R_G can be chosen to be any convenient large value of resistor – approximately 1 MΩ. This value is useful because it does not appreciably shunt the desirable high input impedance of the transistor.

4 Knowing V_{CC}, V_S, V_{DS} and I_D, we can now calculate V_D and then R_D. Since we require $V_{DS} = 10$ V and V_S has already been chosen as 2.3 V

$$V_D = V_{DS} + V_S = 10\text{ V} + 2.3\text{ V} = 12.3\text{ V}$$

and

$$R_D = \frac{V_{CC} - V_D}{I_D} = \frac{(24 - 12.3)\text{ V}}{5\text{ mA}} = 2340\ \Omega$$

Our bias circuit is now complete.[4]

The above biasing circuit is easy to design but complications often arise when the manufacturer does not supply the d.c. curves for a particular transistor[5] or when you cannot obtain the characteristic curve from a transistor curve plotter. In this case, refer to the manufacturer's FET data sheets for values of V_P and I_{DSS}. With these two values known, you can use the well known FET expression for calculating V_{GS}. It is

$$I_D = I_{DSS}\left[1 - \frac{V_{GS}}{V_P}\right]^2 \tag{6.9}$$

For our particular case, the manufacturer states that $V_P = -8$ V and $I_{DSS} = 10$ mA. Substituting the values in Equation 6.9 and transposing yields

$$V_{GS} = V_P\left\{1 - \sqrt{\frac{I_D}{I_{DSS}}}\right\}$$

$$= -8\text{ V}\left\{1 - \sqrt{\frac{5\text{ mA}}{10\text{ mA}}}\right\}$$

$$= -2.34\text{ V}$$

We can then proceed as in steps 3 to 5 above.

Example 6.8

An n-channel JFET has $V_P = -6$ V and $I_{DSS} = 8$ mA. The desired operating point is $I_D = 2$ mA and $V_{DS} = 12$ V. The supply voltage (V_{CC}) is 24 V. Design a bias circuit for this operating point.

Solution: This circuit will be designed following the method given in Example 6.7.

1 The operating point for the transistor is

$$I_D = 2\text{ mA},\quad V_D = 12\text{ V},\quad V_{CC} = 24\text{ V}$$

2 V_p and I_{DSS} from the data sheet are

$$V_p = -6\text{ V}$$

and

$$I_{DSS} = 8\text{ mA}$$

[4] Older readers might well recognise that this method of biasing is similar to that used for thermionic valves. The only difference is that in thermionic valves, V_S is such a small fraction of the anode–cathode voltage that it may be neglected.

[5] This seems to be the case with many r.f. transistors.

3 Knowing I_D, I_{DSS}, and V_p, V_{GS} can be calculated from:

$$V_{GS} = V_P\left\{1 - \sqrt{\frac{I_D}{I_{DSS}}}\right\}$$

$$= -6\left\{1 - \sqrt{\frac{2\text{ mA}}{8\text{ mA}}}\right\}$$

$$= -3.0\text{ V}$$

4 Since $I_G = 0$, $|V_{GS}| = |V_S|$, and knowing I_D, R_S can be calculated:

$$R_S = \frac{|V_S|}{I_D} = \frac{|V_{GS}|}{I_D} = \frac{3\text{ V}}{2\text{ mA}} = 1500\ \Omega$$

5 Since $I_G = 0$, R_G can be chosen to be any large value of resistor -- approximately 1 M Ω.

6 Knowing V_{CC}, V_S, V_{DS} and I_D, we can now calculate V_D and then R_D. Since we require $V_{DS} = 12$ V and V_S has already been calculated as 3 V

$$V_D = V_{DS} + V_S = 12\text{ V} + 3\text{ V} = 15\text{ V}$$

and

$$R_D = \frac{V_{CC} - V_D}{I_D} = \frac{(24 - 15)\text{ V}}{2\text{ mA}} = 4500\ \Omega$$

The bias circuit is now complete.

FET (p-channel)

It is also possible to make p-channel type depletion mode FETs by substituting p-type material for n-type and vice-versa in the simplified construction illustrated in Figure 6.10. However, the voltage supplies must also be reversed to that shown in Figure 6.11, and this time the current flow is 'hole' current instead of electrons. The net result is the same and gain can be obtained from the FET. Figure 6.14 shows the symbol for a p-channel FET and

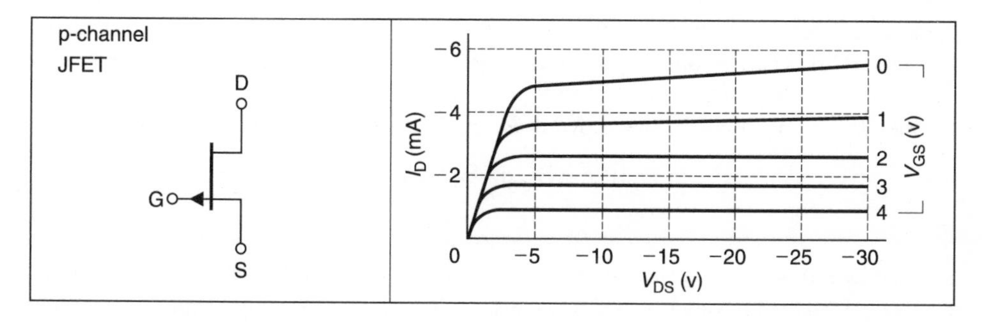

Fig. 6.14 Symbols and output current characteristics for a p-channel FET

its output current characteristics. Note that V_{DS} and I_D are reversed to that for the n-channel characteristics given earlier, and that V_{GS} must be *positive* to *decrease* I_D. Selection of the operating point is similar to that explained for the bi-polar transistor, but the biasing method is different. An example of how to bias a p-channel FET is given in Example 6.9.

Example 6.9

Using the biasing arrangement shown in Figure 6.15, calculate the biasing resistors for a FET operating with $V_{DS} = -10$ V, $I_{DS} = -5$ mA and a supply voltage $V_{CC} = -24$ V. From manufacturer's data, for $I_{DS} = 5$ mA and $V_{DS} = -10$ V, $V_{GS} = +2.3$ V.

Given: $V_{CC} = -24$ V, $V_{DS} = -10$ V, $V_{GS} = +2.3$ V, $I_{DS} = -5$ mA.
Required: R_G, R_D, R_S.

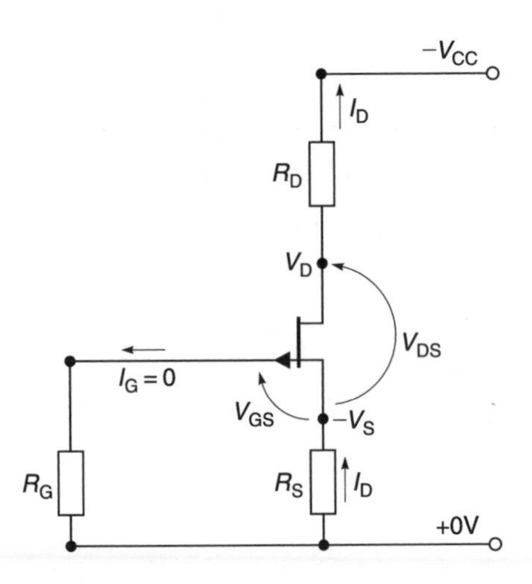

Fig. 6.15 Bias circuit for a p-channel depletion mode FET

Solution

1 For our particular transistor, the manufacturer's d.c. curves show that for an operating current of –5 mA with $V_{DS} = -10$ V, we require a V_{GS} of +2.3 V which means that the gate must be 2.3 V positive with respect to the source. We do not have another power supply but we can simulate this supply by making the source *negative* with respect to the gate which in turn means that the gate will then be *positive* with respect to the source. This is carried out by placing a resistor (R_S) in series with I_D to produce a negative voltage ($-V_S$), which is equal to V_{GS}. We now calculate R_S.
2 Since $I_G = 0$, $|V_{GS}| = |V_S|$, and knowing I_D, calculating R_S gives

$$R_S = \frac{|V_S|}{I_D} = \frac{|V_{GS}|}{I_D} = \frac{-2.3 \text{ V}}{-5 \text{ mA}} = 460\ \Omega$$

3 Since $I_G = 0$, R_G can be chosen to be any convenient large value of resistor – approximately 1 MΩ. This value is useful because it does not appreciably shunt the desirable high input impedance of the transistor.

4 Knowing V_{CC}, V_S, V_{DS} and I_D, we can now calculate V_D and then R_D. Since we require $V_{DS} = -10$ V and V_S has already been chosen as -2.3 V

$$-V_D = -V_{DS} - V_S = -10\text{ V} - 2.3\text{ V} = -12.3\text{ V}$$

and

$$R_D = \frac{V_{CC} - V_D}{I_D} = \frac{(-24\text{ V}) - (-12.3\text{ V})}{-5\text{ mA}} = 2340\ \Omega$$

Our bias circuit is now complete.

The above biasing circuit is easy to design but complications often arise when the manufacturer does not supply the d.c. curves for a particular transistor and when you cannot obtain the characteristic curve from a transistor curve plotter. In this case, refer to the manufacturer's FET data sheets for values of V_P and I_{DSS} and use Equation 6.9 as before. With these two values known, you can calculate V_{GS}.

Example 6.10

A p-channel JFET is to be operated with $I_D = -2$ mA, $V_{DS} = -8$ V. The power supply voltage (V_{CC}) is -14 V. The data sheet gives $V_P = 4$ V and $I_{DSS} = -6$ mA.

Solution: The bias circuit will be designed using Example 6.9 as a guide.

1 The operating point for the transistor is

$$I_D = -2\text{ mA},\quad V_{DS} = -8\text{ V},\quad V_{CC} = -14\text{ V}$$

2 For our particular transistor, $V_P = 5$ V and $I_{DSS} = -6$ mA.
3 Knowing I_D, I_{DSS} and V_P, we can calculate V_{GS}:

$$V_{GS} = V_P\left\{1 - \sqrt{\frac{I_D}{I_{DSS}}}\right\}$$

$$= 5\left\{1 - \sqrt{\frac{-2\text{ mA}}{-6\text{ mA}}}\right\}$$

$$= 2.1\text{ V}$$

4 Since $I_G = 0$, $|V_{GS}| = |V_S|$, and knowing I_D, calculating R_S gives

$$R_S = \frac{|V_S|}{I_D} = \frac{|V_{GS}|}{I_D} = \frac{-2.1\text{ V}}{-2\text{ mA}} = 1050\ \Omega$$

5 Since $I_G = 0$, R_G can be chosen to be any large value of resistor – approximately 1 M Ω.
6 Knowing V_{CC}, V_S, V_{DS} and I_D, we can now calculate V_D and then R_D. Since we require $V_{DS} = -8$ V and V_S has already been chosen as -2.1 V:

$$-V_D = -V_{DS} - V_S = -8\text{ V} - 2.1\text{ V} = -10.1\text{ V}$$

and

$$R_D = \frac{V_{CC} - V_D}{I_D} = \frac{(-14 + 10.1)\ \text{V}}{-2\ \text{mA}} = 1950\ \Omega$$

Our bias circuit is now complete.

6.3.2 A brief review of metal oxide silicon field effect transistors (MOSFET)

MOSFETs (n-channel enhancement-mode type)

In MOSFETs, the drain and source are p–n junctions formed side by side in the surface of a silicon substrate as illustrated in Figure 6.16. This time the gate is a conductor, originally a metal film (hence the name of the device) but nowadays it is usually a layer of well doped silicon. This gate is separated from the silicon substrate by a film of oxide thus forming an input capacitance.

The application of a voltage between gate and source induces carriers in the silicon under the gate – as indicated in Figure 6.17 – the amount of charge induced in the channel being dependent on the gate voltage. When a drain–source voltage is applied (V_{DS}),

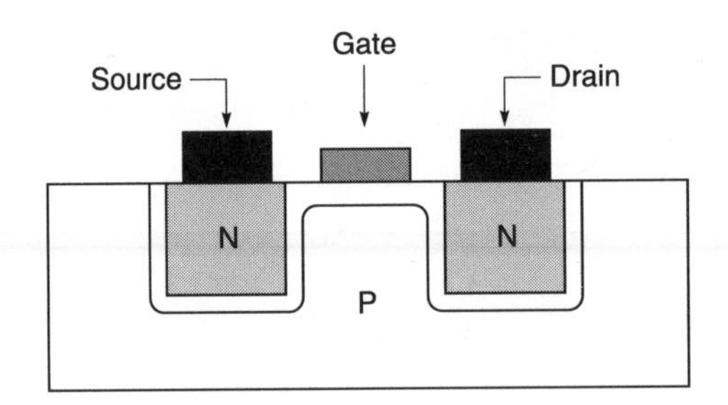

Fig. 6.16 Basic construction of an n-channel MOSFET

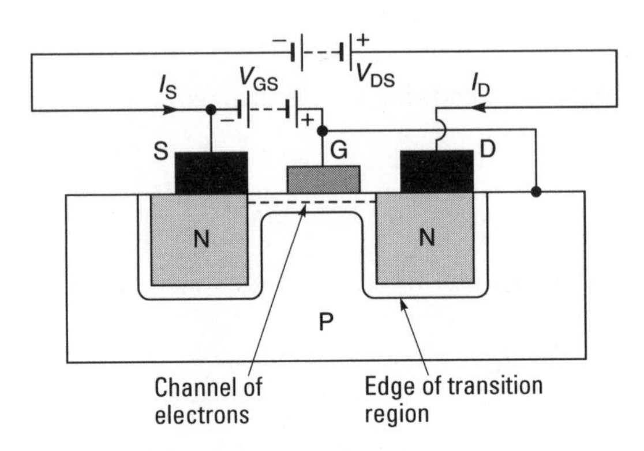

Fig. 6.17 Basic action in an n-channel MOSFET

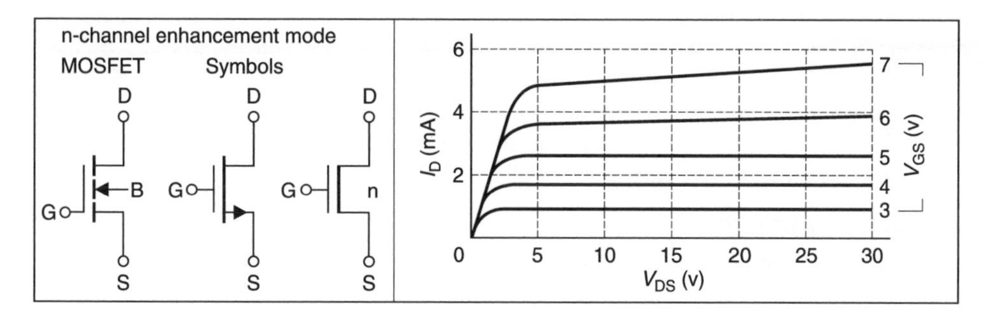

Fig. 6.18 Symbols and characteristic curves for an n-channel enhancement-mode MOSFET

these induced carriers flow between source and drain – the larger the induced charge, the greater the drain current I_D. In short, V_{GS} controls the current flowing through the channel for a fixed value of V_{DS}.

Power gain is obtained because very little energy is required to control the input signal (V_{GS}) while relative large amounts of power can be obtained from the variation in drain–source current (I_{DS}). The symbols for an n-channel enhancement-mode MOSFET and its output current characteristics are shown in Figure 6.18. Note that in the first symbol, the substrate is marked 'B' for bulk to distinguish it from the 'S' for source. Sometimes the substrate is joined to the source internally to give a device with three rather than four terminals. The line connecting source and drain is shown as a dashed line to indicate that in an n-channel enhancement-mode MOSFET, the conduction channel is not established with zero gate–source voltage.

MOSFETs (p-channel enhancement-mode type)

It is also possible to make p-channel enhancement-mode MOSFETs by substituting p-type material for n-type and vice-versa to that shown in Figure 6.16. However, the voltage supplies must also be reversed to that shown in Figure 6.17, and this time the current flow is hole current instead of electrons. The net result is the same and gain can be obtained from the MOSFET. Figure 6.19 shows the symbol for a p-channel MOSFET and its output current characteristics. Note that V_{DS} and I_D are reversed to that for the n-channel characteristics given above.

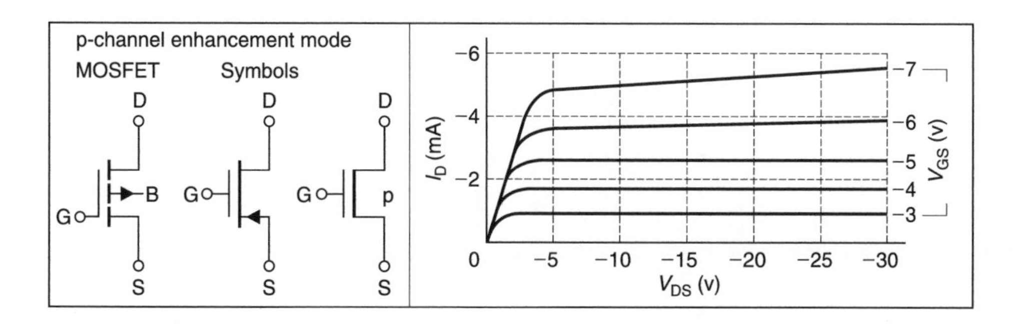

Fig. 6.19 Symbols and characteristic curves for a p-channel enhancement-mode MOSFET

Here again, in the first symbol, the substrate is marked 'B' for bulk to distinguish it from the 'S' for source. However, the arrow is pointing away this time to indicate a p-channel device. Sometimes the substrate is joined to the source internally to give a device with three rather than four terminals. The line connecting source and drain is shown as a dashed line to indicate that in a p-channel enhancement-mode MOSFET, the conduction channel is not established with zero gate–source voltage.

Biasing of MOSFETs

Selection of the operating point is similar to that explained for the bi-polar transistor. The biasing methods are also similar (see section 6.3.1) but the calculations are simpler because $I_G = 0$. One method of biasing is shown in Example 6.11.

Example 6.11

The n-channel enhancement-mode MOSFET shown in Figure 6.20 is to be operated with $I_D = 5$ mA, $V_{DS} = 10$ V with a supply voltage (V_{CC}) of 18 V. Manufacturer's data sheets show that this MOSFET requires a positive bias of 3.2 V for a current of 5 mA when V_{DS} is 10 V. Calculate the values of resistors required for the circuit in Figure 6.20.

Solution

1 The operating point for the transistor is

$$I_D = 5 \text{ mA}, \quad V_{DS} = 10 \text{ V}, \quad V_{CC} = 18 \text{ V}$$

2 From the manufacturer's data sheet $V_{GS} = +3.2$ V for a current of 5 mA.

3 Choose V_S to be approximately 10% of V_{CC}:

$$V_S = 10\% \text{ of } 18 \text{ V} = 1.8 \text{ V}$$

4 Knowing V_S and I_D, calculate R_S:

$$R_S = \frac{V_S}{I_D} = \frac{1.8 \text{ V}}{5 \text{ mA}} = 360\ \Omega$$

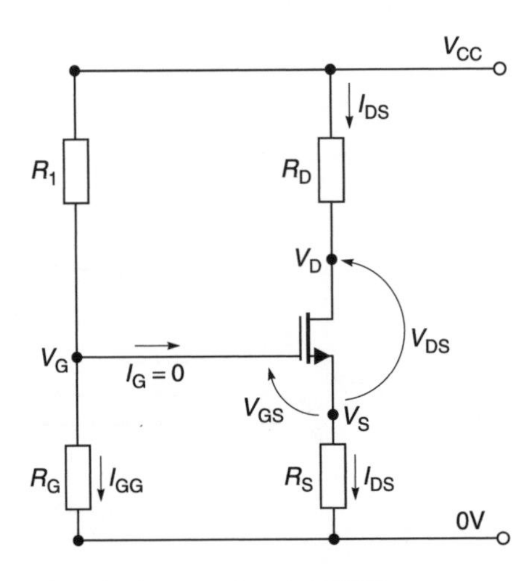

Fig. 6.20 Bias circuit for an n-channel enhancement-mode MOSFET

5 Knowing V_S and V_{GS}, calculate V_G:

$$V_G = V_{GS} + V_S = (3.2 + 1.8)\ \text{V} = 5.0\ \text{V}$$

6 Assume a value for R_2 based upon d.c. input resistance needs:

$$R_2 = 220\ \text{k}\Omega$$

7 Knowing R_2, V_G and V_{CC}, calculate R_1:

$$R_1 = \frac{R_2(V_{CC} - V_G)}{V_G}$$

$$= \frac{220\ \text{k}\Omega\ (18\text{–}5)\ \text{V}}{5\ \text{V}} = 572\ \text{k}\Omega$$

8 Knowing V_S and V_{DS}, calculate V_D and then R_D:

$$V_D = V_S + V_{DS} = 1.8\ \text{V} + 10\ \text{V} = 11.8\ \text{V}$$

and

$$R_D = \frac{V_{CC} - V_D}{I_D} = \frac{(18 - 11.8)\ \text{V}}{5\ \text{mA}} = 1240\ \Omega$$

The bias circuit is now complete.

Example 6.12

An n-channel enhancement-mode MOSFET is to be operated with $I_D = 2$ mA, $V_{DS} = 6$ V with a supply voltage (V_{CC}) of 12 V. Manufacturer's data sheets show that this MOSFET requires a positive bias of 1.8 V for a current of 2 mA when V_{DS} is 6 V. Calculate the values of resistors required for the circuit in Figure 6.20.

Solution

1 The operating point for the transistor is

$$I_D = 2\ \text{mA}, \quad V_{DS} = 6\ \text{V}, \quad V_{CC} = 12\ \text{V}$$

2 From the manufacturer's data sheet $V_{GS} = +1.8$ V for a current of 2 mA.
3 Choose V_S to be approximately 10% of V_{CC}:

$$V_S = 10\%\ \text{of}\ 12\ \text{V} = 1.2\ \text{V}$$

4 Knowing V_S and I_D, calculate R_S:

$$R_S = \frac{V_S}{I_D} = \frac{1.2\ \text{V}}{2\ \text{mA}} = 600\ \Omega$$

5 Knowing V_S and V_{GS}, calculate V_G:

$$V_G = V_{GS} + V_S = (1.8 + 1.2)\ \text{V} = 3.0\ \text{V}$$

6 Assume a value for R_2 based upon d.c. input resistance needs:

$$R_2 = 220\ \text{k}\Omega$$

7 Knowing R_2, V_G and V_{CC}, calculate R_1:

$$R_1 = \frac{R_2(V_{CC} - V_G)}{V_G}$$

$$= \frac{220\ \text{k}\Omega\ (12 - 3)\ \text{V}}{3\ \text{V}} = 660\ \text{k}\Omega$$

8 Knowing V_S and V_{DS}, calculate V_D and then R_D:

$$V_D = V_S + V_{DS} = 1.2\ \text{V} + 6\ \text{V} = 7.2\ \text{V}$$

and

$$R_D = \frac{V_{CC} - V_D}{I_D} = \frac{(12 - 7.2)\ \text{V}}{2\ \text{mA}} = 2400\ \Omega$$

The bias circuit is now complete.

6.3.3 Depletion-mode MOSFETs

Depletion-mode MOSFETs are also known in some books as depletion/enhancement-mode (DE) MOSFETs because, as you will see shortly, they can be biased to operate in both the depletion and enhancement mode. For our purposes, we will refer to them as depletion-mode types to avoid unnecessary confusion.

MOSFETs (n-channel depletion-mode type)

Depletion-mode MOSFETs are made in a similar way to that of enhancement MOSFETs except that a very thin layer of donors is implanted in the surface of the p-type substrate just under the gate as indicated in Figure 6.21. This is done simply by firing donor atoms

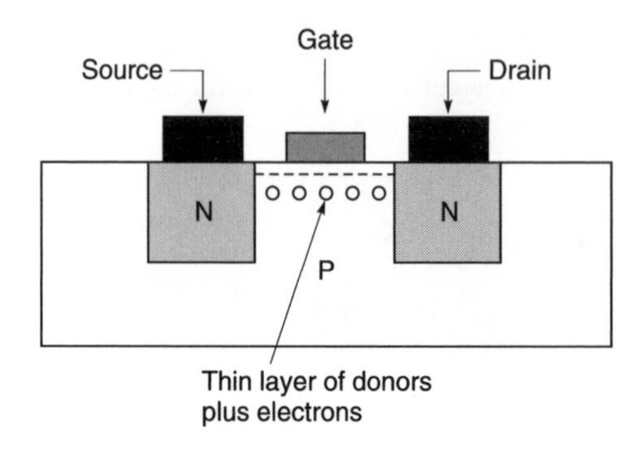

Fig. 6.21 A cross-sectional diagram of an n-channel depletion-mode MOSFET showing the layer of donors implanted in the surface of the p-type substrate to form a channel of electrons even when $V_{GS} = 0$

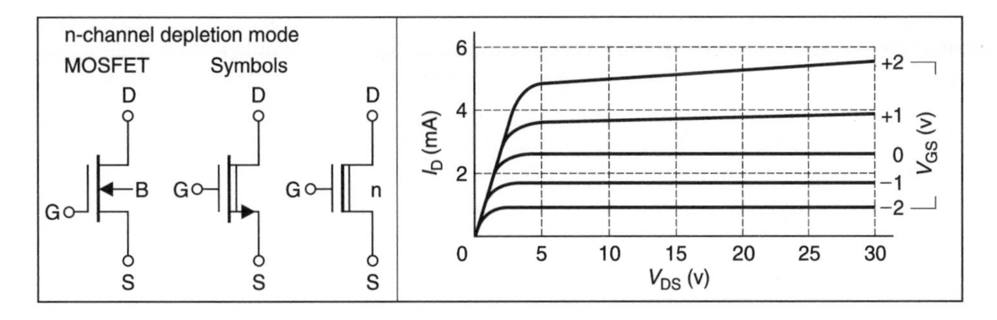

Fig. 6.22 Symbols and output current characteristics for an n-channel depletion-mode MOSFET

in a vacuum at the silicon surface. If the implanted donor density exceeds the density of holes already there, a channel of electrons will be formed even when $V_{GS} = 0$ and a drain current will flow as soon as V_{DS} is applied. The current which flows in a depletion-mode MOSFET when $V_{GS} = 0$ is called I_{DSS} and it is quoted in data sheets.

Figure 6.22 shows the symbols and electrical characteristics of an n-channel depletion-mode MOSFET. Note that V_{GS} can be positive, zero or negative. The V_{GS} required to reduce I_D to zero is called the threshold voltage (V_T). In the first symbol of Figure 6.22, the substrate is marked 'B' for bulk to distinguish it from the 'S' for source. Sometimes the substrate is joined to the source internally to give a device with three rather than four terminals. The line connecting source and drain is shown as a full line to indicate that in an n-channel DE-mode MOSFET, a conduction channel is present even with zero gate–source voltage.

MOSFETs (p-channel depletion-mode type)

MOSFETs (p-channel depletion-mode) are made in a similar manner to that shown for the n-channel depletion-mode MOSFET except that p-type material has been substituted for n-type material and operating voltages must be reversed. The symbols and current characteristics of a p-channel depletion-mode MOSFET are shown in Figure 6.23.

The explanation of the operation of p-channel MOSFETs is identical to that of the n-channel MOSFET just described except that all region types, carrier types, voltages

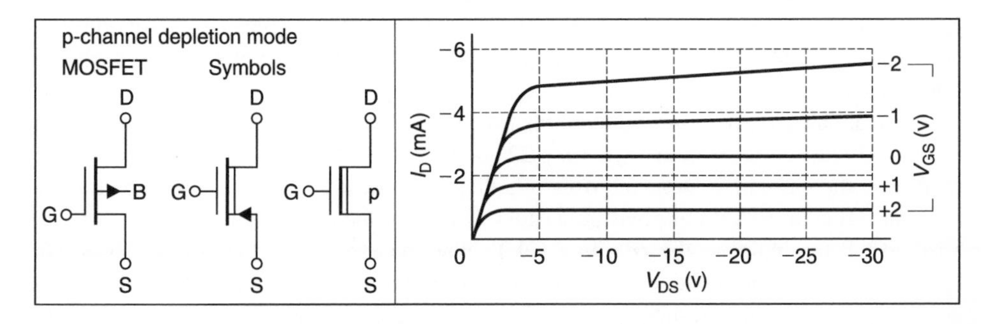

Fig. 6.23 Symbols and output current characteristics for a p-channel depletion-mode MOSFET

and currents are reversed. For example the gate is made increasingly *negative* to cause an increase in drain current, and the threshold voltage of an enhancement-mode p-type channel device is positive. In the first symbol of Figure 6.23, the substrate is marked 'B' for bulk to distinguish it from the 'S' for source. Sometimes the substrate is joined to the source internally to give a device with three rather than four terminals. The line connecting source and drain is shown as a full line to indicate that in a p-channel DE-mode MOSFET, a conduction channel is present even with zero gate–source voltage.

Biasing of depletion-mode MOSFETs

These MOSFETs can be biased according to the methods given in Examples 6.7, 6.9 and 6.11. The circuit you choose will depend on whether you wish to operate the circuit with positive or negative V_{GS}.

6.3.4 Summary of the properties of FETs and MOSFETs

Field effect transistors (FETs) can be regarded as three terminal devices whose terminals are called source, drain and gate. There are two types of field effect transistor; the junction FET or JFET, and the metal oxide silicon FET or MOSFET. In a JFET current flows through a channel of silicon whose cross-sectional area is controlled by a p–n junction whose width is varied by the application of a voltage between gate and source, as illustrated in Figure 6.11. In a MOSFET the drain and source are p–n junctions formed side by side in the surface of a silicon substrate as illustrated in Figure 6.16. The gate is separated from the silicon substrate by a film of oxide. The application of a voltage between gate and source induces carriers in the silicon under the gate which then forms the channel between source and drain. See Figure 6.17.

There are complementary forms of both types of FET: namely n-channel and p-channel devices. In n-channel devices the drain current is carried by electrons, whilst in p-channel devices it is carried by holes. There are two variants of MOSFET called enhancement-mode devices and depletion-mode devices. In enhancement-mode devices $I_D = 0$ when $V_{GS} = 0$. In depletion-mode devices (and in JFETs) $I_D = I_{DSS}$ when $V_{GS} = 0$.

The families of output characteristics of all FETs have the same general form and have been shown in Figures 6.12, 6.14, 6.18, 6.19, 6.22 and 6.23. Note particularly that the polarities of the voltage and directions of currents of p-channel devices are the reverse of those of n-channel ones; and that the differences between the three types of n-channel devices, or between the three p-channel devices, is in their range of values for V_{GS}. The polarity of the threshold voltages (V_T) distinguishes between enhancement-mode and depletion-mode MOSFETs. (The threshold voltages are the gate voltages at which the drain current I_D just begins to flow.)

There are alternative graphical symbols for MOSFETs in common use, which are also shown in Figures 6.12, 6.14, 6.18, 6.19, 6.22 and 6.23. Note that if the (longer) central line represents the piece of silicon, an arrow in a diagram always points at a p-region or away from an n-region, as with bi-polar transistors. However, the arrow indicates the *source* in a MOSFET, but indicates the *gate* in a JFET. Note the extra line added to the MOSFET symbols for depletion-mode operation – it is intended to indicate the existence of a channel when $V_{GS} = 0$. Other symbols may be found in data sheets and in textbooks, so be sure to check the meanings of symbols in other publications. In particular a more complicated standard symbol is used to represent the four terminal nature of MOSFETs, but we shall not be referring to this in this text.

The carriers in the channel of either type of FET flow from source to drain under the influence of an electric field, so the currents are drift currents rather than diffusion currents.

The d.c. output characteristics of a JFET can be calculated by Equation 6.7.

The main electrical advantage of MOSFETs over bi-polar transistors is that their d.c. gate current is virtually zero. This is due to the presence of the oxide layer between the gate electrode and the substrate. The current gain, if it was ever referred to, would be approaching infinity. This means that the input power to the device is very small indeed. The advantage of MOSFETs from a production point of view is that, in general, they are smaller and cheaper to manufacture than bi-polars. Their main disadvantage is that their transconductance[6] (g_m) is normally much less than that of bi-polars at the same operating current; that is, the control of the output current by the input voltage is normally less effective. It is possible, however, using special construction methods to produce MOSFETs which are capable of operating more efficiently than bi-polars at microwave frequencies. This is particularly true of the high electron mobility field effect transistor known as the HEMFET which will be discussed later in the design of microwave amplifiers.

6.4 A.C. equivalent circuits of transistors

Some of the material described in Section 6.4 has already been covered in earlier sections but I need to emphasise some relevant facts to help you understand how a.c. equivalent circuits are derived. In the analyses of a.c. equivalent circuits, apart from a very brief introduction to FETs, I will concentrate mainly on the bi-polar transistor because many of the a.c. equivalent circuits (apart from circuit values) apply to JFETs and MOSFETs as well.

6.4.1 A brief review of bi-polar transistor construction

The NPN type bi-polar transistor shown in Figure 6.24 is constructed by using a crystalline layer of silicon[7] into which carefully controlled amounts of impurities such as arsenic, phosphorus or antimony have been added so that the silicon may be made to provide relatively easy movement for electrons. This layer is known as an n-type material because it contains 'free' *negative* electrical charges (electrons). In the NPN transistor, this layer is called the **emitter** because it has the ability to 'emit' electrons under the influence of a voltage potential. A very thin layer of material about 0.2–10 microns[8] thick is then laid over the emitter. This layer is called the **base** layer. It is usually made of silicon with carefully controlled amounts of impurities such as aluminium, boron, gallium or indium. This

[6] Transconductance (g_m) is defined as

$$\frac{\text{small change in output current } (\Delta I_{DS})}{\text{small change in input voltage } (\Delta V_{GS})}$$

[7] Germanium can also be used but its electrical characteristics with temperature are less stable than that of silicon. Therefore, it is becoming obsolete and is only used for special functions or as replacement transistors for older designs.

[8] 1 micron = 1×10^{-6} metres.

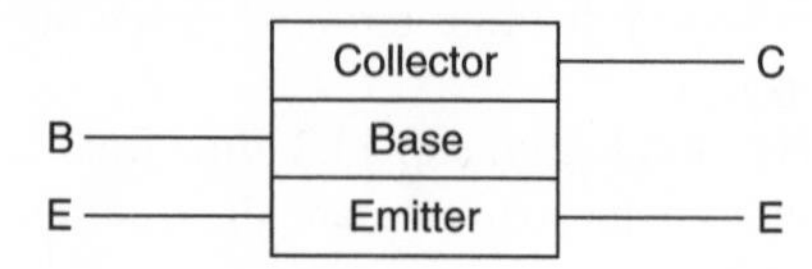

Fig. 6.24 Basic construction of a bi-polar transistor

layer is known as a p-type layer because there are 'free' **positive** electric charges (holes) in the material. Finally another layer of n-type material is placed over the base layer. This layer is known as the **collector** because it collects all the current.

With suitable operating conditions, and when the transistor is connected to a battery, electrons from the emitter are made to pass through the base which controls current flow to the collector. This type of action occurs in an NPN type transistor. The same type of action described above is possible with a transistor made with PNP type construction, that is with the emitter and collector constructed of p-type material and the base of n-type material.

6.4.2 A brief review of field effect transistor construction

The basic construction of an n-channel type field effect transistor is shown in Figure 6.25. In this case, n-type material is used as the conducting channel between source and drain and p-type material is placed on either side of the channel. The effective electrical width of the channel is dependent on the voltage potential between gate and source.

When a power supply is connected in the appropriate manner (+ at the drain and – at the source), electrons flow from the source past the gate to the drain. A negative potential applied to the gate alters the channel width of the transistor. This in turn affects the resistance of the channel and its current flow. Power gain is obtained because the input power applied to the gate is very much less than the output power.

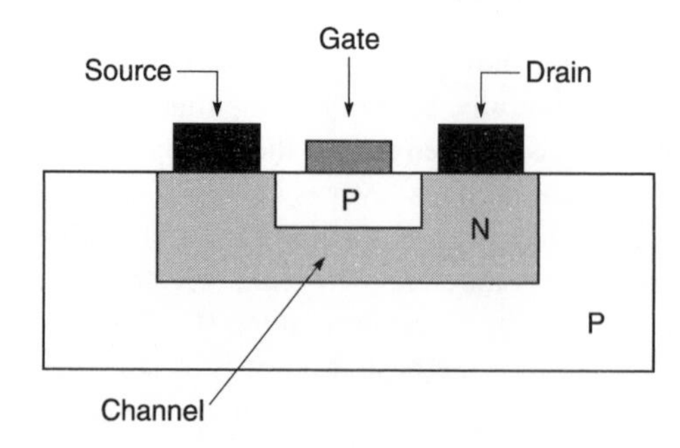

Fig. 6.25 An n-channel depletion-mode field effect transistor

6.4.3 Basic a.c. equivalent circuits

As stated earlier and for the sake of clarity, I will concentrate mainly on the bi-polar transistor. However, you should be aware that much of what is said applies to the FET as well

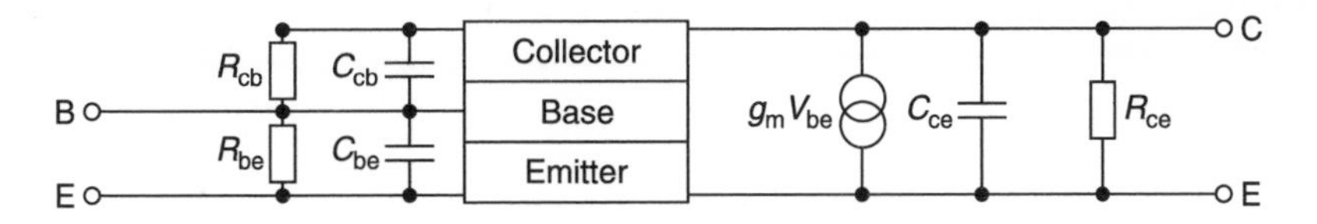

Fig. 6.26 Approximate equivalent electrical circuit of a transistor

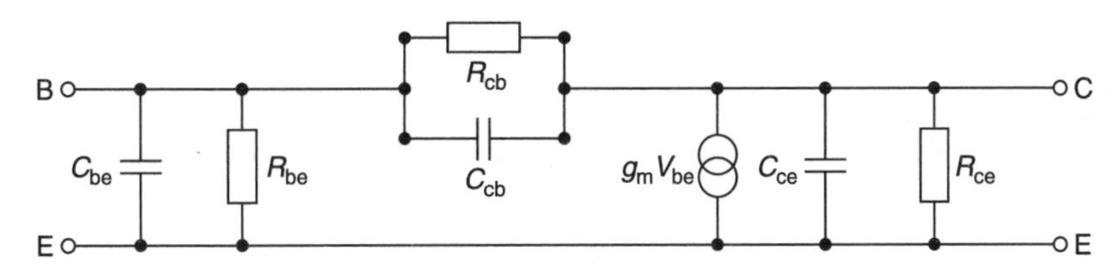

Fig. 6.27 'π' equivalent circuit of a transistor

because its physical construction also produces inter-electrode resistances and capacitances. If you examine Figure 6.24 more closely, you will see that due to the proximity of the emitter, base and collector or the source, gate and drain in Figure 6.25, there is bound to be resistance and capacitance between the layers. The approximate[9] electrical equivalent circuit for Figure 6.24 has been drawn for you in Figure 6.26. The abbreviations used are:

C_{be} = capacitance between the base and emitter
R_{be} = resistance between the base and emitter
C_{cb} = capacitance between the collector and base
R_{cb} = resistance between the collector and base
C_{ce} = capacitance between the collector and emitter
R_{ce} = resistance between the collector and the emitter
$g_m V_{be}$ = a current generator which is controlled by the voltage between base and emitter (V_{be}) – this generator is present because there is current gain in a transistor
g_m = a constant of the transistor and its operating point – it is defined as change in collector current (ΔI_c)/change in base–emitter voltage (ΔV_{be})

In Figure 6.27, I have simplified the circuit by removing the bulk diagram of the transistor and it now becomes more recognisable as the π equivalent circuit of a transistor. The circuit is called a π equivalent circuit because the components appear in the form of the Greek letter π. The resistances and capacitances in Figure 6.27 are not fixed. They are dependent on the d.c. operating conditions of the transistor.[10] For example, if the d.c. current through the transistor increases then R_{be} will decrease and vice-versa. Similarly if the voltage across the transistor increases then C_{ce} will decrease and vice-versa.[11] These variations are inevitable because d.c. operating voltages and currents affect the physical nature of transistor junctions.

[9] The circuit is only approximate because I have not taken into account the resistance and reactances of the lead and its connections.

[10] You will no doubt recall from Section 6.2 that the base–emitter junction of a bi-polar transistor is a forward biased p–n junction diode and that the resistance of this junction (R_{diode}) varies with current.

[11] This is due to the change in the width of the depletion layer described in Section 6.2.

Table 6.1 Typical values for a bi-polar transistor

Resistance value	Capacitance values	Reactances at 100 kHz
$R_{be} \approx$ 1–3 kΩ	$C_{be} \approx$ 10–30 pF	$X_{be} \approx$ 159–53 kΩ
$R_{bc} \approx$ 2–5 MΩ	$C_{bc} \approx$ 2–5 pF	$X_{bc} \approx$ 796–318 kΩ
$R_{ce} \approx$ 20–50 kΩ	$C_{ce} \approx$ 2–10 pF	$X_{ce} \approx$ 796–159 kΩ
g_m 40 mA per volt when the collector current is 1 mA		

In Table 6.1, I have listed some typical values of components in a π equivalent circuit when a bi-polar transistor is operated as a small signal amplifier with a collector–emitter voltage of 6 V and a current of 1 mA. Much of the discussion that follows is dependent on the relative values of these components to each other.

Low frequency equivalent circuit of a transistor

If you examine the first and the third columns of Table 6.1, you will see that at 100 kHz

- $X_{be} >> R_{be}$ so that its effect on R_{be} is negligible.
- $X_{ce} >> R_{ce}$ so that its effect on R_{ce} is negligible.
- If you refer to Figure 6.27, you will see that the fraction of the output voltage between the collector and the emitter (V_{ce}) fed back through the feedback path formed by the parallel combination of R_{bc} and C_{cb} the parallel combination of R_{be} and C_{be} is also very small because the parallel combination $(R_{bc} // X_{cb}) >> (R_{be} // X_{be})$. Therefore it can also be neglected at low frequencies. If you have difficulty understanding this part, refer back to Figure 6.27.

The effects of the above mean that the circuit of Figure 6.27 can be re-drawn at low frequencies to be that shown in Figure 6.28. This equivalent circuit is reasonably accurate for frequencies less than 100 kHz.

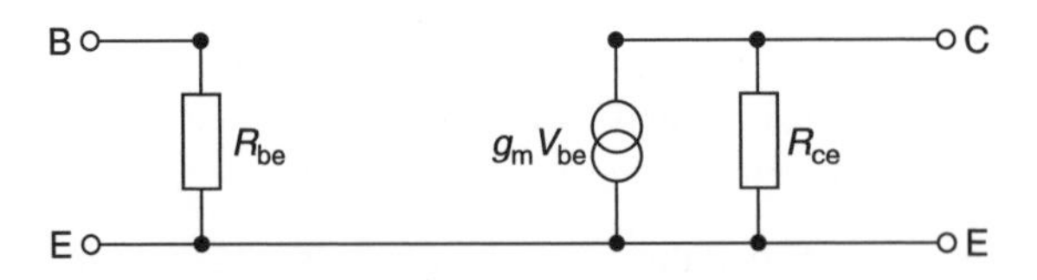

Fig. 6.28 Low frequency equivalent circuit of a transistor

High frequency equivalent circuit of a transistor

Returning to the π equivalent circuit of Figure 6.27, you will recall that I mentioned earlier that the circuit was only approximate because I did not take lead resistances and reactances into account. When these are added the circuit becomes that shown in Figure 6.29. This circuit is called the hybrid configuration because it is a hybrid of the π circuit.

Another resistance (R_{bb}) known as the base spreading resistance emerges. This is the inevitable resistance that occurs at the junction between the base terminal or contact and the semiconductor material that composes the base. Its value is usually in tens of ohms. Smaller transistors tend to exhibit larger values of R_{bb} because of the greater difficulty of

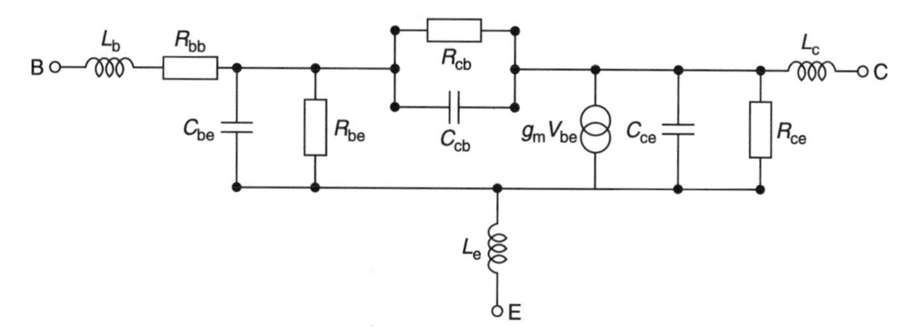

Fig. 6.29 Hybrid π equivalent circuit

connecting leads to smaller surfaces. Inductors L_b, L_c and L_e are the inductances of the base, collector and emitter leads respectively.

Of the three inductors, L_e has the most pronounced effect on circuit performance because of its feedback effect. This is caused by the input current flowing from B via L_b, R_{bb}, C_{be} and R_{be} in parallel, and out via L_e while the output current opposes the input current since it flows outwards via L_c, the external load, and in again via L_e. As frequency increases, the reactance of L_e increases and its effect is to produce a larger ($I_{out} \times X_{Le}$) voltage to oppose input current flow. Manufacturers tend to minimise this effect by providing two leads for the emitter; one for the input current and the other for the output current. This is the reason why some r.f. transistors have two emitter leads.

An increase in operating frequency causes reactances X_{be}, X_{cb} and X_{ce} to decrease and this action will increase their shunting effect on resistances R_{be}, R_{cb} and R_{ce} and eventually the gain of the transistor will begin to fall. The most serious shunting effect is caused by X_{cb} because it affects the negative feedback path from collector to base and a frequency will be reached when the gain of the transistor is reduced to unity. The unity gain frequency (f_T) is also known as the cut-off frequency of the transistor.

6.4.4 Summary

From the foregoing discussions, you should have realised that:

- transistors come in different shapes and sizes;
- they have d.c. parameters as well as a.c. parameters;
- a.c. parameters vary with d.c. operating conditions;
- a.c. parameters vary with frequency;
- a transistor operating under the same d.c. conditions can be represented by different a.c. equivalent circuits at different frequencies;
- transistor data given by manufacturers may appear in several ways, namely π, hybrid π, and other yet to be introduced parameters such as admittance (Y) parameters and scattering (S) parameters.

6.4.5 The transistor as a two-port network

The transistor is obviously a three terminal device consisting of an emitter, base and collector. In most applications, however, one of the terminals is common to both the input

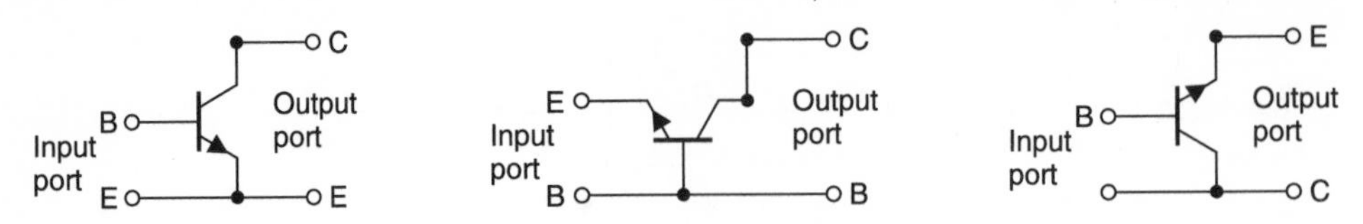

Fig. 6.30 (a) Common emitter configuration; (b) common base configuration; (c) common collector configuration

and the output network as shown in Figure 6.30. In the common emitter configuration of Figure 6.30(a), the emitter is grounded and is common to both the input and output network. So, rather than describe the device as a three terminal network, it is convenient to describe the transistor as a 'black box' by calling it a two port network. One port is described as the input port while the other is described as the output port. These configurations are shown in Figure 6.30(a, b, and c). Once the two port realisation is made, the transistor can be completely characterised by observing its behaviour at the two ports.

6.4.6 Two-port networks

Manufacturers are also aware that in many cases, knowing the actual value of components in a transistor is of little value to the circuit design engineer because there is little that can be done to alter its internal values after the transistor has been manufactured. Therefore, manufacturers resort to giving transistor electrical parameters in another manner, namely the 'two port' approach. This is shown in Figure 6.31. With this approach, manufacturers simply state that for a given transistor operating under certain conditions, what you can expect to find at the input port (port 1) and the output port (port 2), when you apply external voltages (v_1, v_2) and currents (i_1, i_2) to it.

Two port parameters come from manufacturers in different representations. Each type of representation can be described with names such as:

admittance – *Y*-parameters	transfer – *ABCD*-parameters
hybrid – *H*-parameters	impedance – *Z*-parameters
scattering – *S*-parameters	Smith chart information

For radio frequency work, the most favoured parameters are the *y*-, *s*- and *h*-parameters and information on Smith charts which is a convenient graphical display of *y*- and *s*-parameters.

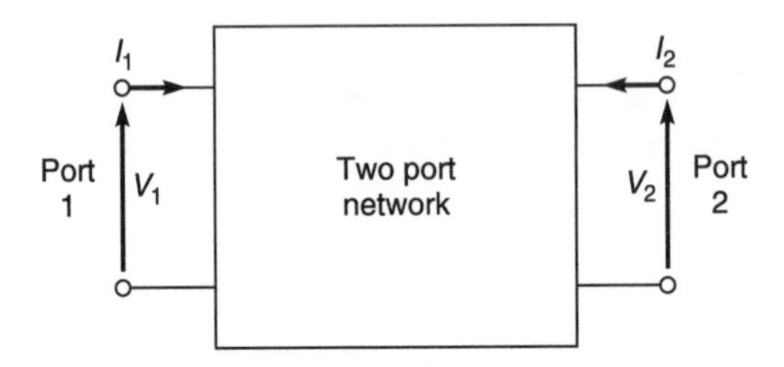

Fig. 6.31 Two port network representation

6.4.7 Radio frequency amplifiers

Radio frequency (r.f.) amplifiers will be investigated by first considering one method of modelling transistors at radio frequencies. This method will then be used to design an aerial distribution amplifier.

R.F. transistor modelling

Transistor modelling serves two main purposes. First, it enables a *transistor* designer to analyse what is happening within a transistor and to design the necessary modifications to improve performance. Second, it enables a *circuit* designer to understand what is happening within a circuit and to carry out the necessary adjustments to achieve optimum circuit performance from the transistor.

The hybrid π model (Figure 6.29) is particularly good for representing the properties of a transistor but as frequencies increase, its shunt reactances cannot be neglected and its equivalent representation becomes increasingly complex. To minimise these complications, electronic circuit designers prefer to treat a transistor as a complete unit or 'black box' and to consider its *performance* characteristics rather than the individual components in its equivalent circuit.

One way to do this is to use **admittance parameters** or ***y*-parameters**. In this approach, the transistor is represented as a two-port network with input port (port 1) and output port (port 2) as shown in Figure 6.32(a). In Figure 6.32(a), the details of the components within the 'box' are not given. The equivalent representation shown in Figure 6.32(b) simply tells us that if the input and output voltages v_1 and v_2 are changing, then the currents i_1 and i_2 must also be changing in accordance with the equations

$$i_1 = y_{11}v_1 + y_{12}v_2 \tag{6.10}$$

and

$$i_2 = y_{21}v_1 + y_{22}v_2 \tag{6.11}$$

where

i_1 is the a.c. current flowing into the input port (port 1)
i_2 is the a.c. current flowing into the output port (port 2)
v_1 is the a.c. voltage at the input port
v_2 is the a.c. voltage at the output port

In practice, the values of the y-parameters, y_{11}, y_{12}, y_{21}, y_{22}, are specified at a particular frequency, in a particular configuration (common base, common emitter, or common

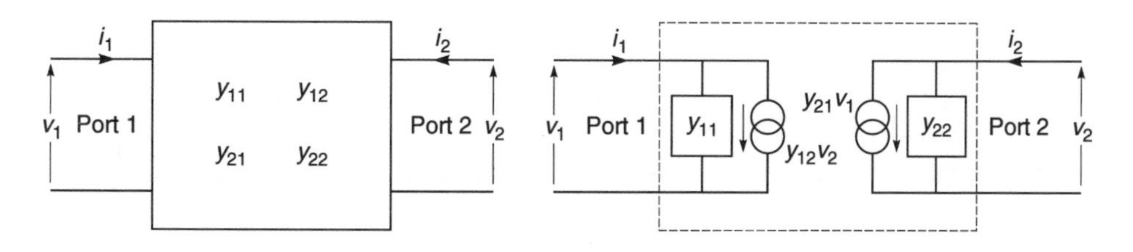

Fig. 6.32 (a) Admittance parameters; (b) equivalent representation

collector) and with stated values of transistor operating voltages and currents. It is important to remember that y-parameters are measured phasor quantities, obtained by measuring *external phasor* voltages and currents for a particular transistor.

From inspection of Equations 6.10 and 6.11, we define

$$y_{11} = \left\{ \frac{i_1}{v_1} \right\}_{v_2=0} \tag{6.12}$$

$$y_{12} = \left\{ \frac{i_1}{v_2} \right\}_{v_1=0} \tag{6.13}$$

$$y_{21} = \left\{ \frac{i_2}{v_1} \right\}_{v_2=0} \tag{6.14}$$

and

$$y_{22} = \left\{ \frac{i_2}{v_2} \right\}_{v_1=0} \tag{6.15}$$

Figure 6.33 shows the same transistor (represented by its 'y'-parameters) being driven by a constant current signal source (i_s) with a source admittance (Y_s). The transistor feeds a load admittance (Y_L). By inspection of the circuit in Figure 6.33, it can be seen that

$$i_s = Y_s v_1 + i_1$$

or

$$i_1 = i_s - Y_s v_1 \tag{6.16}$$

and

$$i_2 = -Y_L v_2 \tag{6.17}$$

Note the minus sign. This is because current is flowing in the *opposite* direction to that indicated in the diagram.

Circuit parameters can be calculated as follows.

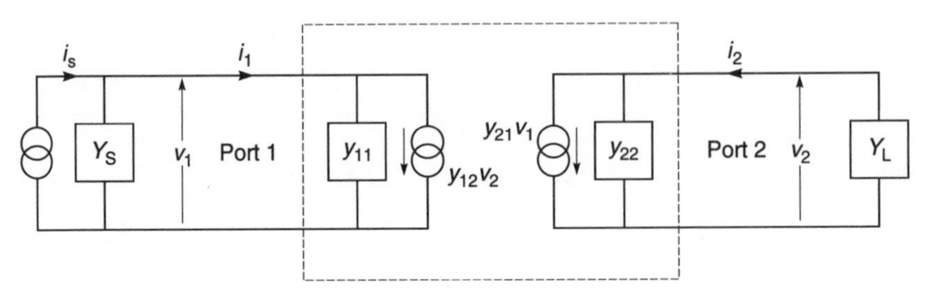

Fig. 6.33 Equivalent circuit of a transistor with source (Y_s) and load (Y_L)

Voltage gain. Voltage gain (A_v) is defined as (v_2/v_1). Substituting Equation 6.17 in Equation 6.11

$$i_2 = y_{21}v_1 + y_{22}v_2 = -Y_L v_2$$

Transposing

$$v_2(Y_L + y_{22}) = -y_{21}v_1$$

and

$$\frac{v_2}{v_1} = \frac{-y_{21}}{Y_L + y_{22}} = A_v \tag{6.18}$$

Input admittance (y_{in}). Input admittance (y_{in}) is defined as i_1/v_1. Substituting Equation 6.18 in Equation 6.10

$$i_1 = y_{11}v_1 + y_{12}v_2 = y_{11}v_1 + y_{12}\left\{\frac{-y_{21}v_1}{y_{22} + Y_L}\right\}$$

and transposing

$$\frac{i_1}{v_1} = y_{11} - \frac{y_{12}y_{21}}{y_{22} + y_L} = Y_{in} \tag{6.19}$$

Current gain (A_i). Current gain (A_i) is defined as i_2/i_1. From Equations 6.19 and 6.17

$$v_1 = \frac{i_1}{Y_{in}} \text{ and } v_2 = \frac{-i_2}{Y_L}$$

and substituting in Equation 6.11

$$i_2 = y_{21}\left[\frac{i_1}{Y_{in}}\right] - y_{22}\left[\frac{i_2}{Y_L}\right]$$

Transposing

$$i_2\left[\frac{Y_L + Y_{22}}{Y_L}\right] = \left[\frac{Y_{21}}{Y_{in}}\right] i_1$$

Transposing

$$\frac{i_2}{i_1} = \frac{y_{21}Y_L}{Y_{in}(y_{22} + Y_L)} = A_i \tag{6.20}$$

Output admittance (Y_{out}). Output admittance (Y_{out}) is defined as $[i_2/v_2]_{i_s=0}$. Substituting Equation 6.16 in Equation 6.10

$$i_1 = y_{11}v_1 + y_{12}v_2 = -Y_s v_1 \text{ (remember } i_s = 0)$$

Transposing

$$v_1 = \left\{ \frac{-y_{12}v_2}{y_{11} + Y_s} \right\}$$

Substituting in Equation 6.11

$$i_2 = y_{21} \left\{ \frac{-y_{12}v_2}{y_{11} + Y_S} \right\} + y_{22}v_2$$

Transposing v_2 results in

$$\frac{i_2}{v_2} = y_{22} - \frac{y_{12}y_{21}}{y_{11} + Y_S} = Y_{out} \tag{6.21}$$

Equations 6.18 to 6.21 enable us to calculate the performance of a transistor circuit. The equations are in a general form and apply to a transistor regardless of whether it is operating in the common emitter, common base or common collector mode. The only stipulations are that you recognise that signal enters and leaves the transistor at port 1 and port 2 respectively and that you use the correct set of '*y*'-parameters in the calculations.

Design case: aerial amplifier design using 'y'-parameters

In this design study (Figure 6.34), the signal is picked up by an aerial whose source impedance is 75 Ω. The signal is then fed into an amplifier whose load is a 300 Ω distribution system which feeds signals to all the domestic VHF/FM receivers in the house. The design was carried out in the following manner.

1 Manufacturers' data sheets were used to find a transistor which will operate satisfactorily at 100 MHz; the approximate centre of the VHF broadcast band. The transistor is assumed to be unconditionally stable.
2 A decision was made on transistor operating conditions. Guidelines are usually given in the data sheets for operating conditions and '*y*'-parameters. Typical operating conditions for a well known transistor with d.c. conditions (V_{ce} = 6 V, I_c = 1 mA) operating in the common emitter mode were found to be:

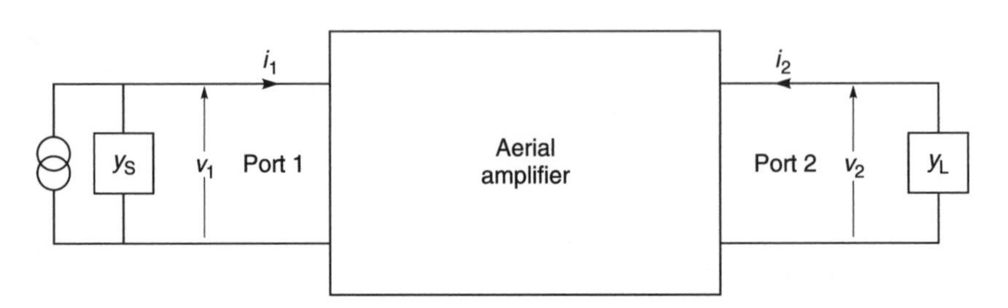

Fig. 6.34 Aerial amplifier design study

$y_{11} = (13.752 + j13.946)$ mS $\quad y_{12} = (-0.146 - j1.148)$ mS
$y_{21} = (1.094 - j17.511)$ mS $\quad y_{22} = (0.3 + j1.571)$ mS

The relevant information is summarised below.

Given

$Z_s = 75\ \Omega$ or $Y_s = 13.33$ mS $\angle 0°$ $\quad Z_L = 300\ \Omega$ or $Y_L = 3.33$ mS $\angle 0°$
$y_{11} = (13.75 + j13.95)$ mS $\quad y_{12} = (-0.15 - j1.15)$ mS
$y_{21} = (1.09 - j17.51)$ mS $\quad y_{22} = (0.3 + j1.57)$ mS

Required: (a) Voltage gain (A_v), (b) input admittance (Y_{in}), (c) output admittance (Y_{out}).

Solution. In this solution, you should concentrate on the method used, rather than the laborious arithmetic which can be easily checked by a calculator. Simpler numerical parameters were not used because they do not reflect realistic design problems.

(a) Use Equation 6.18 to calculate the voltage gain (A_v):

$$A_v = \frac{-y_{21}}{Y_L + y_{22}} = \frac{1\underline{/180°} \times (1.09 - j17.5)\text{ mS}}{3.33\text{ mS}\underline{/0°} + (0.3 + j1.57)\text{ mS}}$$

$$= \frac{17.54\text{ mS}\underline{/93.56°}}{3.95\text{ mS}\underline{/23.39°}} = 4.44\underline{/70.17°}$$

From the above answer, you should note the following.

- The phase relationship between the input and output signals is not always the usual 180° phase reversal expected in a low frequency common emitter amplifier. This is because of transistor feedback caused mainly by the reduced reactance of the internal collector–base capacitance at higher radio frequencies.
- The gain and phase relationship is also dependent on the magnitude and phase of the load.

Figure 6.35 shows the relationships between the input (OA) and output (OB) voltages. It is readily seen that there is a component of the output signal which is in phase with the

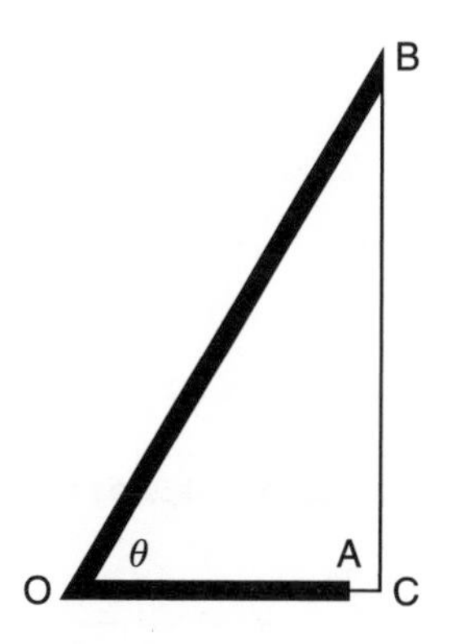

Fig. 6.35 OA and OB represent input and output phasor voltages

input signal. This in-phase component can cause instability problems if it is allowed to stray back into the input port. To keep the two signals apart, good layout, short connecting leads and shielding are essential.

If a tuned circuit is used as an amplifier load, its impedance and phase will vary with tuning. This in turn affects the amplitude and phase relationships between amplifier input and output voltages. These variations must be incorporated into the amplifier design otherwise instability will occur.

Input admittance (Y_{in})

(b) For this calculation, we use Equation 6.19:

$$Y_{in} = Y_{11} - \frac{Y_{12}Y_{21}}{Y_{22} + Y_L}$$

$$= (13.75 + j13.95)\text{ mS} - \frac{(-0.15 - j1.15)\text{ mS }(1.09 - j17.51)\text{ mS}}{(0.3 + j1.57)\text{ mS} + 3.33\text{ mS}\underline{/0°}}$$

$$= (13.75 + j13.95)\text{ mS} - \frac{(1.16\text{ mS}^2 \angle -97.4°)(17.54\text{ mS} \angle -86.44°)}{(3.63 + j1.57)\text{ mS}}$$

$$= (13.75 + j13.95)\text{ mS} - \frac{20.35\ \mu\text{S} \angle -183.84°}{3.95\text{ mS} \angle 23.39°}$$

$$= (13.75 + j13.95)\text{ mS} - 5.15\text{ mS} \angle -207.23°$$

$$= (13.75 + j13.95)\text{ mS} - (-4.58 + j2.36)\text{ mS}$$

$$= (18.33 + j11.59)\text{ mS}$$

From the answer, you should note Y_{in} is also dependent on the phase of the load admittance. This is particularly important in multi-stage amplifiers where the input admittance of the last amplifier provides the load for the amplifier before it. In this case, altering or tuning the load of the last stage amplifier will affect its input admittance and in turn affect the load of the amplifier driving it. This is the reason why the procedure for tuning a multi-stage amplifier usually requires that the last stage be adjusted before the earlier stages.

Output admittance (Y_{out})

(c) For this calculation, we use Equation 6.21:

$$Y_{out} = y_{22} - \frac{y_{12}y_{21}}{y_{11} + Y_s}$$

$$= (0.3 + j1.57)\text{ mS} - \frac{(-0.15 - j1.15)\text{ mS }(1.09 - j17.51)\text{ mS}}{(13.75 + j13.95)\text{ mS} + 13.33\text{ mS} \angle 0°)}$$

$$= (0.3 + j1.57)\text{ mS} - \frac{(1.16 \angle -97.4°)\text{ mS }(17.54 \angle -86.44°)\text{ mS}}{(13.75 + j13.95)\text{ mS} + 13.33\text{ mS} \angle 0°}$$

$$= (0.3 + j1.57)\text{ mS} - \frac{20.35\ \mu\text{S}^2 \angle -183.84°}{(27.08 + j13.95)\text{ mS}}$$

$$= (0.3 + j1.57)\ \text{mS} - \frac{20.35\ \mu S^2 \angle -183.84^\circ}{30.46\ \text{mS} \angle 27.25^\circ}$$

$$= (0.3 + j1.57)\ \text{mS} - 0.67\ \text{mS} \angle -211.09^\circ$$

$$= (0.3 + j1.57)\ \text{mS} - (-0.57 + j0.35)\ \text{mS}$$

$$= (0.87 + j1.22)\ \text{mS}$$

From the answer, you should note Y_{out} is also dependent on the phase of the source admittance. This is particularly important in multi-stage amplifiers where the load admittance of the first amplifier provides the source for the amplifier following it. In this case, altering or tuning the load of the first stage will affect the source admittance of the second stage and so on. This is why, having tuned a multi-stage high frequency amplifier once, you usually have to repeat the tuning again to compensate for the changes in the output admittances.

Theoretically, it would appear that there is no satisfactory way of tuning a multi-stage amplifier because individual amplifiers affect each other. In practice, it is found that after the second tuning, little improvement is obtained if a subsequent re-tune is carried out. Therefore as a compromise between performance and labour costs, most multi-stage amplifiers are considered to be tuned after the second tuning.

Summary of the design case. From this design study, you have learnt how to calculate the voltage gain, input and output admittances of a simple amplifier. You should also have understood why good shielding and layout practices are important in high frequency amplifiers and the reasons for the procedures used in tuning multi-stage high frequency amplifiers.

Example 6.13

Using the parameters given in the case study, calculate the current gain (A_i) of the amplifier.

Solution. Using Equation 6.20, current gain (A_i) is defined as i_2/i_1. Hence

$$A_i = \frac{i_2}{i_1} = \frac{y_{21} Y_L}{Y_{in}(y_{22} + Y_L)}$$

$$= \frac{(1.09 - j17.51)\ \text{mS} \times 3.33\ \text{mS}\underline{/0^\circ}}{(18.33 + j11.59)\ \text{mS}\ [(0.3 + j1.57)\ \text{mS} + 3.33\ \text{mS}\underline{/0^\circ}\ ]}$$

$$= \frac{(17.54 \angle -86.44^\circ)\ \text{mS} \times 3.33 \times \text{mS}\underline{/0^\circ}}{(21.69 \angle 32.31^\circ)\ \text{mS}\ (3.96\ \text{mS} \angle 23.39^\circ)}$$

$$= \frac{(58.41 \angle -86.44^\circ)\ \mu S^2}{(85.89 \angle 55.70^\circ)\ \mu S^2}$$

$$= 0.68 \angle -142.14^\circ$$

Note: Current gain is less than unity. This is because the load admittance is low.

6.5 General r.f. design considerations

The example given in the Design Study has been based on a design methodology, where we have assumed that the transistor is unconditionally stable, that gain is not of paramount importance, and that the inherent electrical noise of the amplifier is not prevalent. In real life, ideal conditions do not exist and we must trade off some properties at the expense of others. The designs that follow show you how these trade-offs can be carried out.

Design of linear r.f. small signal amplifiers is usually based on requirements for specific power gain at specific frequencies. Other design considerations include stability, bandwidth, input–output isolation and production reproducibility. After a basic circuit type is selected, the applicable design equations can be solved.

Many r.f. amplifier designs fail because the incorrect transistor has been chosen for the required purpose. Two of the most important considerations in choosing a transistor for use in any amplifier design are its stability and its **maximum available gain** (MAG). Stability, as it is used here, is a measure of the transistor's tendency to oscillate, that is to provide an output signal with no intended input signal. MAG is a figure-of-merit for a transistor which indicates the maximum theoretical power gain which can be obtained from a transistor when it is **conjugately matched**[12] to its source and load impedances. MAG is never achieved in practice because of resistive losses in a circuit; nevertheless MAG is extremely useful in *evaluating* the initial capabilities of a transistor.

6.5.1 Stability

A major factor in the overall design is the potential stability of the transistor. A transistor is stable if there is no output signal when there is no input signal. There are two main stability factors that concern us in amplifier design, (i) the stability factor of the transistor on its own, and (ii) the stability factor of an amplifier circuit.

Linvill stability factor

The Linvill stability factor is used to determine the stability of a transistor on its own, that is when its input and output ports are open-circuited. Linvill's stability factor (C) can be calculated by using the following expression:

$$C = \frac{|y_f\, y_r|}{2g_i g_o - \mathrm{Re}\,(y_f\, y_r)} \tag{6.22}$$

where

$|y_f\, y_r|$ = magnitude of the product in brackets
y_f = forward-transfer admittance
y_r = reverse-transfer admittance
g_i = input conductance
g_o = output conductance
Re = real part of the product in parentheses

[12] A signal source generator (Z_g) will deliver maximum power to a load (Z_L) when its source impedance ($Z_g = R_g + jX_g$) = ($Z_L = R_L - jX_L$). The circuit is said to be conjugately matched because $R_g = R_L$ and $X_g = -X_L$.

When $C < 1$, the transistor is **unconditionally stable** at the bias point and the frequency which you have chosen. This means that you can choose any possible combination of source and load impedance for your device and the amplifier will remain stable *providing* that no external feedback paths exist between the input and output ports.

When $C > 1$, the transistor is **potentially unstable** and will oscillate for certain values of source and load impedance. However, a C factor greater than 1 does not indicate that the transistor cannot be used as an amplifier. It merely indicates that you must exercise extreme care in selecting your source and load impedances otherwise oscillations may occur. You should also be aware that a potentially unstable transistor at a particular frequency and/or operating point may not necessarily be unstable at another frequency and/or operating point. If for technical or economical reasons, you must use a transistor with $C > 1$, then try using the transistor with a different bias point, and/or mismatch the input and output impedances of the transistor to reduce the gain of the stage.

The Linvill stability factor (C) is useful in predicting a *potential* stability problem. It does not indicate the actual impedance values between which the transistor will go unstable. Obviously, if a transistor is chosen for a particular design problem, and the transistor's C factor is less than 1 (unconditionally stable), that transistor will be much easier to work with than a transistor which is potentially unstable. Bear in mind also that if C is less than *but very close to* 1 for any transistor, then any change in operating point due to temperature variations can cause the transistor to become potentially unstable and most likely oscillate at some frequency. This is because Y-parameters are specified at a particular operating point which varies with temperature. The important rule is: *make C as small as possible.*

Example 6.14

When operated at 500 MHz with a V_{ce} of 5 V and $I_c = 2$ mA, a transistor has the following parameters:

$$y_i = (16 + j11.78)\ \text{mS} \qquad y_r = (1.55 \angle 258°)\ \text{mS}$$
$$y_f = (45 \angle 285°)\ \text{mS} \qquad y_o = (0.19 + j5.97)\ \text{mS}$$

Calculate its Linvill stability factor.

Solution. Using Equation 6.22

$$C = \frac{|y_f\, y_r|}{2g_i g_o - \text{Re}\,(y_f\, y_r)}$$

$$= \frac{|(45\ \text{mS} \angle 285° \times 1.55\ \text{mS} \angle 258°|}{2 \times 16\ \text{mS} \times 0.19\ \text{mS} - \text{Re}\,(45\ \text{mS} \angle 285° \times 1.55\ \text{mS} \angle 258°)}$$

$$= \frac{|(69.75\ \mu\text{S}^2 \angle 183°|}{6.08\ \mu\text{S}^2 - \text{Re}\,(69.75\ \mu\text{S}^2 \angle 183°)} = \frac{|(69.75\ \mu\text{S}^2 \angle 183°|}{6.08\ \mu\text{S}^2 - (-69.65\ \mu\text{S}^2)}$$

$$= \frac{69.75\ \mu\text{S}^2}{75.73\ \mu\text{S}^2} = 0.92$$

Since the Linvill stability factor < 1, the transistor is unconditionally stable. However, it is only just unconditionally stable and, in production, changes in transistor parameters might easily cause instability. If due to costs or the desire to stock a minimum inventory of parts, you cannot change the transistor, try another operating bias point.

The Stern stability factor (K)

The Stern stability factor (K) is used to predict the stability of an amplifier when it is operated with certain values of load and source impedances. The Stern stability factor (K) can be calculated by:

$$K = \frac{2(g_i + G_s)(g_o + G_L)}{|y_f y_r| + \text{Re}(y_f y_r)} \tag{6.23}$$

where

G_s = the source conductance
G_L = the load conductance

If $K > 1$, the circuit is *stable* for that value of source and load impedance. If $K < 1$, the circuit is potentially *unstable* and will most likely oscillate at some frequency or in a production run of the circuit.

Example 6.15

A transistor operating at $V_{CE} = 5$ V, $I_C = 2$ mA at 200 MHz with a source impedance of $(50 + j0)\ \Omega$ and a load impedance of $(1000 + j0)\ \Omega$ has the following y-parameters:

$$y_i = (4.8 + j4.52)\text{ mS} \qquad y_r = (0.90 \angle 265°)\text{ mS}$$
$$y_f = (61 \angle 325°)\text{ mS} \qquad y_o = (0.05 + j2.26)\text{ mS}$$

What is the Stern stability factor of the circuit?

Solution

$$Y_S = 1/(Z_S) = 1/(50 + j0) = 20\text{ mS}$$

and

$$Y_L = 1/(Z_L) = 1/(1000 + j0) = 1\text{ mS}$$

Using Equation 6.23:

$$K = \frac{2(g_i + G_s)(g_o + G_L)}{|y_f y_r| + \text{Re}(y_f y_r)}$$

$$= \frac{2\,(4.8\text{ mS} + 20\text{ mS})\,(0.05\text{ mS} + 1\text{ mS})}{|61\text{ mS} \angle 325° \times 0.9\text{ mS} \angle 265°| + \text{Re}\,(61\text{ mS} \angle 325° \times 0.9\text{ mS} \angle 265°)}$$

$$= \frac{2\,(24.8\text{ mS})\,(1.05\text{ mS})}{54.9\ \mu\text{S}^2 + \text{Re}\,(54.9\ \mu\text{S}^2 \angle 230°)} = \frac{52.08\ \mu\text{S}^2}{54.9\ \mu\text{S}^2 + (-35.29\ \mu\text{S}^2)}$$

$$= \frac{52.08\ \mu S^2}{19.61\ \mu S^2} = 2.656 \approx 2.66$$

Since $K > 1$, the circuit is stable.

Summary of the Linvill and Stern stability factors

The Linvill stability factor (C) is useful in finding stable transistors:

- if $C < 1$, the transistor is unconditionally stable;
- if $C > 1$, the transistor is potentially unstable.

The Stern stability factor (K) is useful for predicting stability problems with circuits:

- if $K > 1$, the circuit is stable for the chosen source and load impedance;
- if $K < 1$, the circuit is potentially unstable for the chosen source and load.

6.5.2 Maximum available gain

The maximum available gain (MAG) of a transistor can be found by using the following equation:

$$\text{MAG} = \frac{|y_f|^2}{4g_i g_o} \tag{6.24}$$

MAG is useful in the initial search for a transistor for a particular application. It gives a good indication as to whether a transistor will provide sufficient gain for a task.

The maximum available gain for a transistor occurs when $y_r = 0$, and when Y_L and Y_S are the complex conjugates of y_o and y_i respectively. The condition that y_r must equal zero for maximum gain to occur is due to the fact that under normal conditions, y_r acts as a *negative* feedback path internal to the transistor. With $y_r = 0$, no feedback is allowed and the gain is at a maximum.

In practical situations, it is physically impossible to reduce y_r to zero and as a result MAG can never be truly obtained. However, it is possible to very nearly achieve the MAG calculated in Equation 6.24 through a **simultaneous conjugate match** of the input and output impedances of the transistor. Therefore, Equation 6.24 remains a valuable tool in the search for a transistor provided you understand its limitation. For example, if your amplifier design calls for a minimum gain of 20 dB at 500 MHz, find a transistor that will give you a small margin of extra gain, preferably at least about 3–6 dB greater than 20 dB. In this case, find a transistor that will give a gain of approximately 23–26 dB. This will compensate for realistic values of y_r, component losses in the matching networks, and variations in bias operating points.

Example 6.16

A transistor has the following Y-parameters:

$$y_i = (16 + j11.78)\ \text{mS} \qquad y_r = (1.55 \angle 258°)\ \text{mS}$$
$$y_f = (45 \angle 285°)\ \text{mS} \qquad y_o = (0.19 + j5.97)\ \text{mS}$$

when it is operated at $V_{CE} = 5$ V and $I_C = 2$ mA at 500 MHz. Calculate its maximum available gain?

Solution. Using Equation 6.24

$$\text{MAG} = \frac{|y_f|^2}{4g_i g_o} = \frac{|45 \text{ mS}|^2}{4 \times 16 \text{ mS} \times 0.19 \text{ mS}}$$

$$= \frac{2025 \ \mu\text{S}^2}{12.16 \ \mu\text{S}^2} = 166.53 \text{ or } 22.21 \text{ dB}$$

6.5.3 Simultaneous conjugate matching

Optimum power gain is obtained from a transistor when y_i and y_o are conjugately matched to Y_s and Y_L respectively. However the reverse-transfer admittance (y_r) associated with each transistor tends to reflect[13] any **immittance** (impedance or admittance) changes made at one port back to the other port, causing a change in that port's immittance characteristics. This makes it difficult to design good matching networks for a transistor while using only its input and output admittances, and totally ignoring the contribution that y_r makes to the transistor's immittance characteristics. Although Y_L affects the input admittance of the transistor and Y_S affects its output admittance, it is still possible to provide the transistor with a simultaneous conjugate match for maximum power transfer (from source to load) by using the following design equations:

$$G_s = \frac{\sqrt{[2g_i g_o - \text{Re}(y_f y_r)]^2 - |y_f y_r|^2}}{2g_o} \tag{6.25}$$

when $y_r = 0$

$$G_s = g_i \tag{6.25a}$$

$$B_s = -jb_i + \frac{\text{Im}\,(y_f\, y_r)}{2g_o} \tag{6.26}$$

and

$$G_L = \frac{\sqrt{[2g_i g_o - \text{Re}(y_f y_r)]^2 - |y_f y_r|^2}}{2g_i} \tag{6.27}$$

or by using Equation 6.25 for the numerator

$$G_L = \frac{G_s g_o}{g_i} \tag{6.27a}$$

and

[13] If you have forgotten this effect, refer to Equations 6.19 and 6.21.

$$B_L = -jb_o + \frac{\mathrm{Im}(y_f y_r)}{2g_i} \qquad (6.28)$$

where

G_s = source conductance
B_s = source susceptance
G_L = load conductance
B_L = load susceptance
Im = imaginary part of the product in parenthesis

The above equations may look formidable but actually they are not because the numerators in these sets of equations are similar and need not be calculated twice. A case study of how to apply these equations is shown in Example 6.17.

Example 6.17

Design an amplifier which will provide maximum gain for conjugate matching of source and load at 300 MHz. The transistor used has the following parameters at 300 MHz with V_{CE} = 5 V and I_C = 2 mA:

y_i = 17.37 + j11.28 mS y_r = 1.17 mS ∠ –91°
y_o = 0.95 + j3.11 mS y_f = 130.50 mS ∠ –69°

What are the admittance values which must be provided for the transistor at (a) its input and (b) its output?

Given: y_i = 17.37 + j11.28 mS y_r = 1.17 mS ∠ – 91°
y_o = 0.95 + j3.11 mS y_f = 130.50 mS ∠ – 69°
f = 300 MHz, V_{CE} = 5 V and I_C = 2 mA
Required: (a) Its input and (b) its output admittances for conjugate match.

Solution

1 Calculate the Linvill stability factor (C) using Equation 6.22:

$$C = \frac{|y_f y_r|}{2g_i g_o - \mathrm{Re}(y_f y_r)}$$

$$= \frac{|(130.5 \text{ mS} \angle -69°)(1.17 \text{ mS} \angle -91°)|}{2(17.37 \text{ mS})(0.95 \text{ mS}) - \mathrm{Re}[(130.5 \text{ mS} \angle -69°)(1.17 \text{ mS} \angle -91°)]}$$

$$= \frac{152.69\ \mu\text{S}^2}{33.00\ \mu\text{S}^2 - (-143.48\ \mu\text{S}^2)} = \frac{152.69\ \mu\text{S}^2}{176.48\ \mu\text{S}^2} = 0.87$$

Since $C < 1$, the device is unconditionally stable and we may proceed with the design. If $C > 1$, we would have to be extremely careful in matching the transistor to the source and load as instability could occur.

2 Calculate the maximum available gain (MAG) using Equation 6.24:

$$\text{MAG} = \frac{|y_f|^2}{4g_i g_o} = \frac{|130.5 \text{ mS} \angle -69°|^2}{4(17.37 \text{ mS})(0.95 \text{ mS})}$$

$$= \frac{17\,030.15 \ \mu\text{S}^2}{66.01 \ \mu\text{S}^2} = 258 \text{ or } 24.12 \text{ dB}$$

The actual gain achieved will be less due to y_r and component losses.

3 Determine the conjugate values to match transistor input admittance using Equation 6.25:

$$G_s = \frac{\sqrt{[2g_i g_o - \text{Re}(y_f y_r)]^2 - |y_f y_r|^2}}{2g_o}$$

$$= \frac{\sqrt{[33 \ \mu\text{S}^2 - \text{Re}(152.69 \ \mu\text{S}^2 \angle -160°)]^2 - |152.69 \ \mu\text{S}^2|^2}}{1.9 \text{ mS}}$$

$$= \frac{\sqrt{[33 \ \mu\text{S}^2 - (-143.48 \ \mu\text{S}^2)]^2 - |152.69 \ \mu\text{S}^2|^2}}{1.9 \text{ mS}}$$

$$= \frac{\sqrt{[176.48 \ \mu\text{S}^2)]^2 - |152.69 \ \mu\text{S}^2|^2}}{1.9 \text{ mS}} = \frac{\sqrt{[13\,145.19 - 23\,314.24] \text{ pS}^4}}{1.9 \text{ mS}}$$

$$= \frac{\sqrt{[7830.95] \text{ pS}^4}}{1.9 \text{ mS}} = \frac{88.49 \ \mu\text{S}^2}{1.9 \ \mu\text{S}} = 46.57 \text{ mS}$$

Using Equation 6.26

$$B_s = -jb_i + \frac{\text{Im}(y_f y_r)}{2g_o}$$

$$= -j11.28 \text{ mS} + j\frac{-52.22 \ \mu\text{S}^2}{2(0.95 \text{ mS})} = -j38.76 \text{ mS}$$

Therefore the source admittance for the transistor is (46.57 – j38.76) mS. The transistor input admittance is (46.57 + j38.76) mS.

4 Determine the conjugate values to match transistor output admittance using Equation 6.27(a):

$$G_L = \frac{G_s g_o}{g_i} = \frac{(46.57)(0.95) \ \mu\text{S}^2}{17.37 \text{ mS}} = 2.55 \text{ mS}$$

Using Equation 6.28

$$B_L = -jb_o + \frac{\text{Im}(y_f\, y_r)}{2g_i}$$

$$= -j3.11 \text{ mS} + j\frac{-52.22\ \mu S^2}{2\ (17.37\text{ mS})} = -j4.61 \text{ mS}$$

Therefore, the load admittance required for the transistor is (2.55 – j4.61) mS. The transistor output admittance is (2.55 + j4.61) mS.

5 Calculate the Stern stability factor (K) using Equation 6.23:

$$K = \frac{2(g_i + G_s)(g_o + G_L)}{|y_f\, y_r| + \text{Re}(y_f\, y_r)}$$

$$= \frac{2(17.37 + 46.57)(0.95 + 2.55)\ \mu S^2}{|152.69|\ \mu S^2 + \text{Re}(152.69\ \angle -160°)\ \mu S^2}$$

$$= \frac{(2)(63.94)(3.50)\ \mu S^2}{152.69\ \mu S^2 + (-143.48)\ \mu S^2} = 48.60$$

Since $K > 1$, the circuit is stable.

After you have satisfied yourself with the design, you need to design networks which will give the transistor its required source and load impedances and, within reason, also its operating bandwidth. This can be done by using the filter and matching techniques described in Chapter 5 and/or by using the Smith chart and transmission line techniques explained in Chapter 3. Smith chart and transmission line techniques will be expanded in the microwave amplifiers which will be designed in Chapter 7.

Summary. The calculated parameters of Example 6.17 are:

- $C = 0.87$
- MAG = 24.12 dB
- conjugate input admittance = (46.57 – j38.76) mS
- conjugate output admittance = (2.55 – j4.61) mS
- $K = 4.45$

6.5.4 Transducer gain (G_T)

Transducer gain (G_T) of an amplifier stage is the gain achieved after taking into account the gain of the device and the actual input and output impedances used. This is the term most often used in r.f. amplifier design work. Transducer gain includes the effects of input and output impedance matching as well as the contribution that the transistor makes to the overall gain of the amplifier stage. Component resistive losses are neglected. Transducer gain (G_T) can be calculated from:

$$G_T = \frac{4G_s G_L |y_f|^2}{|(y_i + Y_s)(y_o + Y_L) - y_f\, y_r|^2} \tag{6.29}$$

where Y_s and Y_L are respectively the source and load admittances used to terminate the transistor.

Example 6.18

Find the gain of the circuit that was designed in Example 6.17. Disregard any component losses.

Solution. The transducer gain for the amplifier is determined by substituting the values given in Example 6.17 into Equation 6.29:

$$G_T = \frac{4G_sG_L|y_f|^2}{|(y_i + Y_s)(y_o + Y_L) - y_f y_r|^2}$$

$$= \frac{4(46.57)(2.55)|130.50|^2 \times 10^{-12}}{|(63.94 - j27.48)(3.50 - j1.50) \times 10^{-6} - (152.69 \angle -160°) \times 10^{-6}|^2}$$

$$= \frac{8\,089\,607.17 \times 10^{-12}}{|69.60 \times 10^{-6} \angle -23.26° \times 3.81 \times 10^{-6} \angle -23.20° - 152.69 \times 10^{-6} \angle -160°|^2}$$

$$= \frac{8\,089\,607.17 \times 10^{-12}}{|265 \times 10^{-6} \angle -46.46° - 152.69 \times 10^{-6} \angle -160°|^2}$$

$$= \frac{8\,089\,607.17 \times 10^{-12}}{|182.55 \times 10^{-6} - j192.10 \times 10^{-6} - (-143.48 \times 10^{-6} - j52.22 \times 10^{-6})|^2}$$

$$= \frac{8\,089\,607.17 \times 10^{-12}}{|326.03 - j139.88|^2 \times 10^{-12}}$$

$$= \frac{8\,089\,607.17}{|354.74 \angle -23.22|^2} = 64.28 \text{ or } 18.08 \text{ dB}$$

The transistor gain calculated in Example 6.18 is approximately 6 dB less than the MAG that was calculated in Example 6.17. In this particular case, the reverse-transfer admittance (y_r) of the transistor has taken an appreciable toll on gain. It is best to calculate G_T immediately after the transistor's load and source admittances are determined to see if the gain is sufficient for your purpose.

If you cannot tolerate the lower gain, the alternatives are:

- increase the operating current to increase g_m and hopefully achieve more gain;
- unilaterise or neutralise the transistor to increase gain (this is explained shortly);
- find a transistor with a higher f_T, to reduce the effect of y_r.

If you carry out one or more of the items above, you will have to go through all the calculations in Examples 6.17 and 6.18 again.

6.5.5 Designing amplifiers with conditionally stable transistors

If the Linvill stability factor (C) calculated with Equation 6.22 is greater than 1, the transistor chosen is potentially unstable and may oscillate under certain conditions of source and load impedance. If this is the case, there are several options available that will enable use of the transistor in a stable amplifier configuration:

- select a new bias point for the transistor; this will alter g_i and g_o;
- unilaterise or neutralise the transistor; this is explained shortly;
- mismatch the input and output impedance of the transistor to reduce stage gain.

Alternative bias point

The simplest solution is probably a new bias point, as any change in a transistor's biasing point will affect its r.f. parameters. If this approach is taken, it is absolutely critical that the bias point be temperature-stable over the operational temperature range especially if C is close to unity.

Unilaterisation and neutralisation

Unilaterisation consists of providing an external feedback circuit (C_n and R_n in Figure 6.36) from the output to the input. The external current is designed to be equal but opposite to the internal y_r current so that the net current feedback is zero. Stated mathematically, $I_{(R_nC_n)} = -I_{(y_r)}$ and the effective composite reverse-transfer admittance is zero. With this condition, the device is unconditionally stable. This can be verified by substituting $y_r = 0$ in Equation 6.22. The Linvill stability factor in this case becomes zero, indicating unconditional stability.

One method of applying unilaterisation is shown in Figure 6.36(a). The principle of operation is explained in Figure 6.36(b). Referring to the latter figure, V_t is the total voltage across the tuned circuit, n_1 and n_2 form the arms of one side of the bridge while C_n and R_n (external components) and the y_r components (R_r and C_r) form the other arm of the bridge. C_n and R_n are adjusted until the bridge is balanced for zero feedback, i.e. $V_{BE} = 0$. It follows that at balance

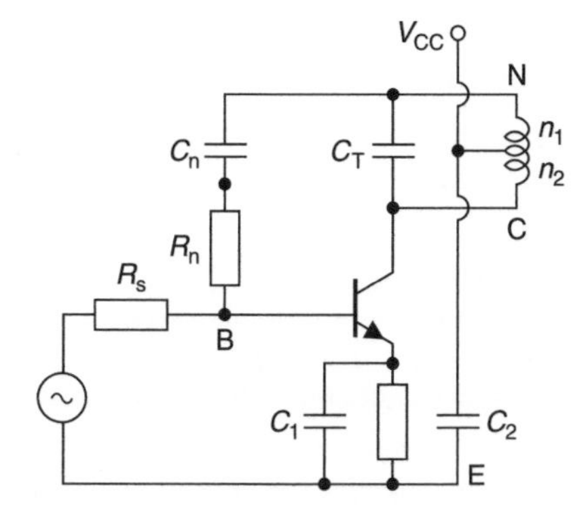

Fig. 6.36 (a) Transformer unilaterisation circuit

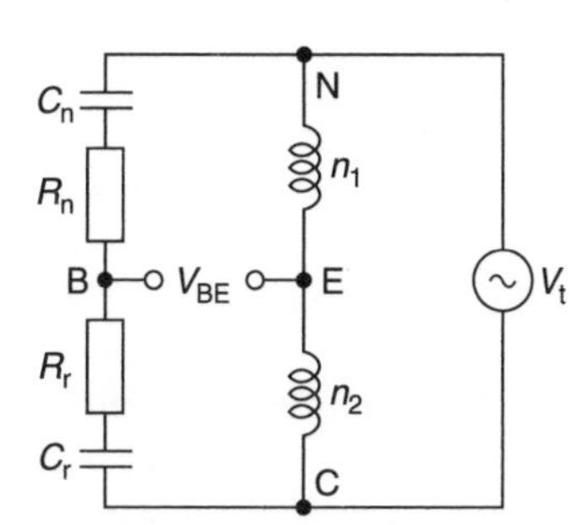

Fig. 6.36 (b) Equivalent unilaterisation circuit

$$V_{EC} = \frac{n_2}{n_1 + n_2} \qquad V_t = \frac{Z_r}{Z_r + Z_n} V_t$$

and after cross-multiplication

$$Z_n = Z_r \frac{n_1}{n_2} \tag{6.30}$$

Often when y_r is a complex admittance consisting of $g_r \pm jb_r$, it becomes very difficult to provide the correct *external* reverse admittance needed to totally eliminate the effect of y_r. In such cases **neutralisation** is often used. Neutralisation is similar to unilaterisation except that only the *imaginary* component of y_r is counteracted. An external feedback path is constructed as before, from output to input such that $B_f = b_r (n_1/n_2)$. Thus, the composite reverse-transfer susceptance is effectively zero. Neutralisation is very helpful in stabilising amplifiers because in most transistors, g_r is negligible when compared to b_r. The effective cancellation of b_r very nearly cancels out y_r. In practical cases, you will find that neutralization is used instead of unilaterisation. However, be warned: the addition of external components increases the costs and the complexity of a circuit. Also, most neutralisation circuits tend to neutralise the amplifier at the operating frequency only and may cause problems (instability) at other frequencies.

Summing up, unilaterisation/neutralisation is an effective way of minimising the effects of y_r and increasing amplifier gain, but it costs more and is inherently a narrow-band compensation method.

Mismatching techniques

A more economical method stabilising an amplifier is to use **selective mismatching**. Another look at the Stern stability factor (K) in Equation 6.23 will reveal how this can be done. If G_s and G_L are made large enough to increase the numerator sufficiently, it is possible to make K greater than 1, and the amplifier will then become stable for those terminations. This suggests selectively mismatching the transistor to achieve stability. The price you pay is that the gain of the amplifier will be less than that which would have been possible with a simultaneous conjugate match.

Procedure for amplifier design using conditionally stable transistors

The procedure for a design using conditionally-stable devices is as follows.

1 Choose G_s based on some other criteria such as convenience of input-network, Q factor. Alternatively, from the transistor's data sheet, choose G_s to be that value which gives you transistor operation with minimal noise figure.
2 Select a value of K that will assure a stable amplifier ($K > 1$).
3 Substitute the above values for K and G_s into Equation 6.2. and solve for G_L.
4 Now that G_s and G_L are known, all that remains is to find B_s and B_L. Choose a value of B_L equal to $-b_o$ of the transistor. The corresponding Y_L which results will then be *very close* to the true Y_L that is theoretically needed to complete the design.
5 Next calculate the transistor input admittance (Y_{in}) using the load chosen in step 4 and Equation 6.19:

$$y_{in} = y_{11} - \frac{y_{12}\, y_{21}}{y_{22} + Y_L}$$

where $Y_L = G_L \pm jB_L$ (found in steps 3 and 4).

6 Once Y_{in} is known, set b_s equal to the *negative* of the imaginary part of Y_{in} or $B_s = -B_{in}$.
7 Calculate the gain of the stage using Equation 6.29.

From this point forward, it is only necessary to produce input and output admittance networks that will present the calculated Y_s and Y_L to the transistor. Example 6.19 shows how the procedure outlined above can be carried out.

Example 6.19

A transistor has the following *y*-parameters at 200 MHz:

$$y_i = 2.25 + j7.2 \qquad y_r = 0.70 \angle -85.9°$$
$$y_f = 44.72 \angle -26.6° \qquad y_o = 0.4 + j1.9$$

All of the above parameters are in mS. Find the source and load admittances that will assure you of a stable design. Find the gain of the amplifier.

Given: $y_i = 2.25 + j7.2$ $\quad y_r = 0.70 \angle -85.9°$
$y_f = 44.72 \angle -26.6°$ $\quad y_o = 0.4 + j1.9$

Required: (a) Load admittance, (b) source admittance and (c) gain when the circuit is designed for a Stern stability factor (*K*) of 3.

Solution. If you were to use Equation 6.22 to calculate the Linvill stability factor (*C*) for the transistor, you will find $C = 2.27$. Therefore the device is potentially unstable and you must exercise extreme caution in choosing source and load admittances for the transistor.

1 The data sheet for the transistor states that the source resistance for optimum noise figure is 250 Ω. Choosing this value results in $G_s = 1/R_s = 4$ mS.
2 For an adequate safety margin choose a Stern stability factor of $K = 3$.
3 Substituting G_S and K into Equation 6.23 and solving for G_L yields

$$K = \frac{2(g_i + G_S)(g_o + G_L)}{|y_f\, y_r| + \mathrm{Re}(y_f\, y_r)}$$

Thus

$$3 = \frac{(2)(2.25 + 4)(0.4 + G_L)}{|31.35| + \mathrm{Re}(-12)}$$

and

$$G_L = 4.24 \text{ mS}$$

4 Setting $B_L = -b_o$ of the transistor

$$B_L = -j1.9 \text{ mS}$$

The load admittance is now defined as

$$Y_L = (4.24 - j1.9) \text{ mS}$$

5 The input admittance of the transistor is calculated using Equation 6.19:

$$y_{in} = y_{11} - \frac{y_{12}\, y_{21}}{y_{22} + Y_L}$$

$$= 2.25 + j7.2 - \frac{(0.701 \angle -85.9°)(44.72 \angle -26.6°)}{0.4 + j1.9 + 4.24 - j1.9}$$

$$= (4.84 + j13.44) \text{ mS}$$

6 Setting B_s equal to the negative of the imaginary part of Y_{in}, yields

$$B_s = -j13.44 \text{ mS}$$

The source admittance needed for the design is now defined as:

$$Y_s = (4 - j13.44) \text{ mS}$$

7 Now that Y_s and Y_L are known, we can use Equation 6.29 to calculate the gain of the amplifier:

$$G_T = \frac{4G_s G_L |y_f|^2}{|(y_i + Y_s)(y_o + Y_L) - y_f y_r|^2}$$

$$= \frac{4(4)(4.24)|44.72|^2}{|(6.25 - j6.24)(4.64) - (-12 - j28.96)|^2}$$

$$= \frac{135\,671.7}{1681} = 80.71 \text{ or } 19.1 \text{ dB}$$

Therefore even though the transistor is not conjugately matched, you can still realise a respectable amount of gain while maintaining a stable amplifier.

After you have satisfied yourself with the design, you need to design networks which will give the transistor its required source and load impedances and, within reason, also its operating bandwidth. This can be done by using the filter and matching techniques described in Part 5 and/or by using Smith chart and transmission line techniques explained in Part 3. Smith chart and transmission line techniques will be expanded in the microwave amplifiers which will be designed in Part 7.

6.6 Transistor operating configurations

6.6.1 Introduction

Sometimes you will find that you want one set of *y*-parameters (e.g. common base parameters) while the manufacturer has only supplied *y*-parameters for the common emitter configuration. What do you do? Well, you simply use the **indefinite admittance matrix** to convert from one set of parameters to another.

6.6.2 The indefinite admittance matrix

Admittance parameters provide an easy way of changing the operating configuration of a transistor. For example, if y-parameters for the common emitter configuration are known, it is easy to derive the parameters for common base and common collector configurations. These derivations are carried out using the indefinite admittance matrix method. The use of this matrix is best shown by example but to avoid confusion in the discussions which follow, it is best to first clarify the meaning of the suffixes attached to y-parameters.

Each y-parameter is associated with two sets of suffixes. The first set, y_i, y_r, y_f and y_o, refer to y_{11}, y_{12}, y_{21}, y_{22} respectively. (The symbols i, r, f, and o stand for input, reverse transconductance, forward transconductance and output respectively.) The second set, e, b, c, refer to the emitter, base or collector configuration respectively. For example, y_{ie} refers to y_{11} in the common emitter mode, y_{rb} refers to y_{12} in the common base mode, y_{fc} refers to y_{21} in the common collector mode and so on.

An admittance matrix for a transistor is made as follows.

1 Construct Table 6.2.
2. Insert the appropriate set of y-parameters into the correct places in the table. If the common emitter parameters for a transistor are:

$$y_{ie} = (13.75 + j13.95) \times 10^{-3}\ \text{S} \qquad y_{re} = (-0.15 - j1.15) \times 10^{-3}\ \text{S}$$
$$y_{fe} = (1.09 - j17.51) \times 10^{-3}\ \text{S} \qquad y_{oe} = (0.30 + j1.57) \times 10^{-3}\ \text{S}$$

   This set should be inserted as shown in Table 6.3.

   Note: No entries are made in the emitter row and column. Similarly, if a set of common base parameters is used instead, no entries will be made in the base row and column. The same applies for a set of common collector parameters; no entries are made in the collector row and column.

3. Sum real and imaginary parts of all rows and columns to zero. See Table 6.4.
4. Extract the required set of parameters.

Table 6.2 Blank indefinite admittance matrix

	Base	Emitter	Collector
Base			
Emitter			
Collector			

Table 6.3 Indefinite admittance matrix with common emitter entries

	Base	Emitter	Collector
Base	$(13.75 + j13.95)10^{-3}$ S		$(-0.15 - j1.15)10^{-3}$ S
Emitter			
Collector	$(1.09 - j1.75)10^{-3}$ S		$(0.30 + j1.57)10^{-3}$ S

Table 6.4 Indefinite admittance matrix with rows and columns summed to zero

	Base	Emitter	Collector
Base	$(13.75 + j13.95)10^{-3}$ S	$(-13.60 - j12.80)10^{-3}$ S	$(-0.15 - j1.15)10^{-3}$ S
Emitter	$(-14.84 - j12.20)10^{-3}$ S	$(14.99 + j12.62)10^{-3}$ S	$(-0.15 - j0.42)10^{-3}$ S
Collector	$(1.09 - j1.75)10^{-3}$ S	$(-1.39 + j0.18)10^{-3}$ S	$(0.30 + j1.57)10^{-3}$ S

To obtain the y-parameters for the common base configuration, ignore all data in the base row and base column but extract the remaining four parameters. These are:

$$y_{ib} = (14.99 + j12.62) \times 10^{-3}\ S \qquad y_{rb} = (-0.15 - j0.42) \times 10^{-3}\ S$$
$$y_{fb} = (-1.39 + j0.18) \times 10^{-3}\ S \qquad y_{ob} = (0.30 + j1.57) \times 10^{-3}\ S$$

To obtain the y-parameters for the common collector configuration, ignore all data in the collector row and collector column but extract the remaining four parameters. These are:

$$y_{ic} = (13.75 + j13.95) \times 10^{-3}\ S \qquad y_{rc} = (-13.60 - j12.80) \times 10^{-3}\ S$$
$$y_{fc} = (-14.84 - j12.20) \times 10^{-3}\ S \qquad y_{oc} = (14.99 + j12.62) \times 10^{-3}\ S$$

This information can now be applied to the general Equations 6.18 to 6.21 and subsequent equations.

The information given above has been shown for a bi-polar transistor but the method is general and applies to other transistor types as well. For FETS, replace the words base, emitter and collector in Tables 6.2 to 6.4 by gate, source and drain respectively.

Example 6.20

A transistor operating with the d.c. conditions of $V_{CE} = 5$ V, $I_C = 2$ mA and at a frequency of 500 MHz is stated by the manufacturer to have the following y-parameters in the common emitter mode.

$$y_{11} = (16 + j12)\ mS \qquad y_{12} = 1.55\ mS \angle 258°$$
$$y_{21} = 45\ mS \angle 285° \qquad y_{22} = (0.19 + j6)\ mS$$

Calculate its equivalent y-parameters for the base configuration when the transistor is operating with the same d.c. operating conditions and at the same frequency.

Solution. We will use the indefinite admittance matrix but, before doing so, the parameters y_{12} and y_{21} must be converted from its polar form:

$$y_{12} = 1.55\ mS \angle 258° = (-0.32 - j1.52)\ mS$$
$$y_{21} = 45\ mS \angle 285° = (11.65 - j43.47)\ mS$$

Fill in the indefinite admittance matrix and sum all rows and columns to zero results in Table 6.5.

Table 6.5

	Base (mS)	Emitter (mS)	Collector (mS)
Base	(16 + j12)	(–15.68 – j10.48)	(–0.32 – j1.52)
Emitter	(–27.65 + 31.47)	(27.52 – j26.99)	(0.13 – j4.48)
Collector	(11.65 – j43.47)	(–11.84 + j37.47)	(0.19 + j6)

Extracting the common base parameters yields:

$$y_{ib} = (27.52 - j26.99)\ mS \qquad y_{rb} = (0.13 - j4.48)\ mS$$
$$y_{fb} = (-11.84 + j37.47)\ mS \qquad y_{ob} = (0.19 + j6)\ mS$$

6.7 Summary

In Chapter 6, we have reviewed the operating principle of bi-polar transistors FETs and MOSFETs. We have also reviewed some transistor biasing methods. A brief resumé of a.c. equivalent circuits was introduced. Admittance parameters (y) were re-introduced, derived and applied to the design of amplifiers. You were shown methods on how to design amplifiers, conjugate matched amplifiers and amplifiers using conditionally-stable transistors.

Our software program, PUFF, has no facilities for employing y-parameters directly. However, if you want, you can use Table 3.1 of Chapter 3 to convert all the y-parameters (or at least the results) into s-parameters. You can then use the PUFF techniques of Chapter 7 for amplifier design. Details of this technique are explained in Examples 7.1 and 7.2 followed by its PUFF design results in Figure 7.4 in the next chapter.

7

Microwave amplifiers

7.1 Introduction

In Part 2, we introduced transmission lines. Part 3 was devoted to Smith charts and scattering parameters while Part 4 covered the use of PUFF as a computing aid. In Part 5, we discussed the behaviour of **passive devices** such as capacitors and inductors at radio frequencies and investigated the use of these elements in the design of resonant tuned circuits, filters, transformers and impedance matching networks. We showed how the indefinite admittance matrix can be used to convert transistor parameters given in one configuration to another configuration. In Chapter 6, we investigated biasing techniques, the a.c. equivalent circuit of transistors, admittance parameters, and their use in high frequency amplifier design. We will now combine all of this information and use it in the design of microwave amplifiers.

7.1.1 Aim

The main aim of this chapter is to show how microwave amplifiers can be designed using scattering parameters.

7.1.2 Objectives

After you have read this chapter, you should be able to:

- calculate transistor stability;
- calculate maximum available gain of an amplifying device;
- design amplifiers with conjugate matching impedances;
- design amplifiers using conditionally stable transistors;
- design amplifiers for a specific gain;
- understand and calculate stability circles;
- design amplifiers for optimum noise figure;
- understand broadband matching amplifier techniques;
- design broadband amplifiers;
- understand feedback amplifier techniques;
- design feedback amplifiers.

7.2 Transistors and s-parameters

7.2.1 Introduction

The purpose of this section is to show you how s-parameters can be used in the design of transistor amplifiers. It has already been shown that transistors can be characterised by their s-parameters. Smith charts have been introduced, therefore it is now time to apply these parameters to produce practical design amplifiers.

7.2.2 Transistor stability

Before designing a circuit, it is important to check whether the active device which we will use is (i) unconditionally stable or (ii) conditionally stable. This is necessary because different conditions require the appropriate design method. It is possible to calculate potential instabilities in transistors even before an amplifier is built. This calculation serves as a useful aid in finding a suitable transistor for a particular application.

To calculate a transistor's stability with s-parameters, we first calculate an intermediate quantity D_s where

$$D_s = s_{11}s_{22} - s_{12}s_{21} \tag{7.1}$$

We do this because in the expressions that follow, you will find that the quantity D_s is used many times and we can save ourselves considerable work by doing this.

The Rollett stability factor (K) is calculated as:

$$K = \frac{1 + |D_s|^2 - |s_{11}|^2 - |s_{22}|^2}{(2)(|s_{21}|)\,(|s_{12}|)} \tag{7.2}$$

If K is *greater* than 1, the transistor is **unconditionally stable** for any combination of source and load impedance. If K is *less* than 1, the transistor is **potentially unstable** and will most likely oscillate with certain combinations of source and load impedances. With K less than 1, we must be extremely careful in choosing source and load impedances for the transistor. It does not mean that the transistor cannot be used for a particular application; it merely indicates that the transistor will have to be used with more care.

If K is less than 1, there are several approaches that we can take to complete the design:

- select another bias point for the transistor;
- choose a different transistor;
- design the amplifier heeding carefully detailed procedures that we will introduce shortly.

7.2.3 Maximum available gain

The maximum gain we can ever get from a transistor under conjugately matched conditions is called the **maximum available gain** (MAG). Maximum available gain is calculated in two steps.

(1) Calculate an intermediate quantity called B_1, where

$$B_1 = 1 + |s_{11}|^2 - |s_{22}|^2 - |D_s|^2 \tag{7.3}$$

D_s is calculated from Equation 7.1.

Note: The reason B_1 has to be calculated first is because its polarity determines which sign (+ or –) to use before the radical in Equation 7.4 which follows shortly.

(2) Calculate MAG using the result from Equation 7.2:

$$\text{MAG} = 10 \log \frac{|s_{21}|}{|s_{12}|} + 10 \log \left| K \pm \sqrt{K^2 - 1} \right| \tag{7.4}$$

where

MAG = maximum available gain in dB
K = stability factor from Equation 7.2

Note: K must be greater than 1 (unconditionally stable) otherwise Equation 7.4 will be undefined because the radical sign will become an imaginary number rendering the equation invalid. Thus, MAG is undefined for unstable transistors.

7.3 Design of amplifiers with conjugately matched impedances

This method of design is only applicable to transistors which are *stable* and give sufficient gain for our design aims. For other conditions we will have to use other design methods. This design procedure results in load and source reflection coefficients which provide a conjugate match for the *actual* output and input impedances of the transistor. However, remember that the actual *output* impedance of a transistor is dependent on its source impedance and vice-versa. This dependency is caused by the reverse gain (s_{12}) of the transistor. If s_{12} was zero, then of course the load and source impedances would have no effect on the transistor's input and output impedances.

7.3.1 Output reflection coefficient

To find the desired **load reflection coefficient** for a conjugate match

$$C_2 = s_{22} - (D_s s_{11}{}^*) \tag{7.5}$$

where the asterisk indicates the complex conjugate of s_{11} (same magnitude but opposite angle). The quantity D_s is the quantity calculated in Equation 7.1.

Next we calculate B_2:

$$B_2 = 1 + |s_{22}|^2 - |s_{11}|^2 - |D_s|^2 \tag{7.6}$$

The **magnitude** of the reflection coefficient is found from the equation

$$|\Gamma_L| = \frac{B_2 \pm \sqrt{B_2^2 - 4|C_2|^2}}{2|C_2|} \tag{7.7}$$

The sign preceding the radical is opposite to the sign of B_2 previously calculated in Equation 7.6. The angle of the load reflection coefficient is simply the negative of the angle of C_2 calculated in Equation 7.5.

After the desired load reflection coefficient is found, we can either (a) plot Γ_L on a Smith chart to find the load impedance (Z) directly, or (b) substitute Γ_L from the equation

$$\Gamma_L = \frac{Z_L - Z_o}{Z_L + Z_o} \tag{7.8}$$

You have encountered the above equation in Chapter 2 when we were looking at transmission lines.

7.3.2 Input reflection coefficient

With the desired load reflection coefficient specified, the source reflection coefficient needed to terminate the transistor's input can now be calculated:

$$\Gamma_s = \left[s_{11} + \frac{s_{12}s_{21}\Gamma_L}{1 - s_{22}\Gamma_L} \right]^* \tag{7.9}$$

The asterisk sign again indicates the conjugate of the quantity in brackets (same magnitude but opposite sign for the angle). In other words, once we complete the calculation within brackets of Equation 7.9, we will have the correct magnitude but the incorrect angle sign and will have to change the sign of the angle.

As before when Γ_s is found, it can be plotted on a Smith chart or we can again use the equation

$$\Gamma_s = -\frac{Z_o - Z_s}{Z_o + Z_s} \tag{7.10}$$

All the foregoing is best clarified by an example.

Example 7.1

A transistor has the following *s*-parameters at 150 MHz with a $V_{ce} = 12$ V and $I_c = 8$ mA:

$$s_{11} = 0.3 \angle 160^\circ \qquad s_{12} = 0.03 \angle 62^\circ$$
$$s_{21} = 6.1 \angle 65^\circ \qquad s_{22} = 0.40 \angle -38^\circ$$

The amplifier must operate between 50 Ω terminations. Design (a) input and (b) output matching networks for simultaneously conjugate matching of the transistor for maximum gain.

Given: $f = 150$ MHz, $V_{ce} = 12$ V, $I_c = 8$ mA, $s_{11} = 0.3 \angle 160^\circ$, $s_{12} = 0.03 \angle 62^\circ$, $s_{21} = 6.1 \angle 65^\circ$, $s_{22} = 0.40 \angle -38^\circ$.

Required: Conjugate input and output matching networks for maximum gain to 50 Ω source and load impedances.

Solution. Using Equations 7.1 and 7.2, check for stability:

$$\begin{aligned} D_s &= s_{11}s_{22} - s_{12}s_{21} \\ &= (0.3 \angle 160^\circ)(0.4 \angle -38^\circ) - (0.03 \angle 62^\circ)(6.1 \angle 65^\circ) \\ &= (0.120 \angle 122^\circ) - (0.183 \angle 127^\circ) \end{aligned}$$

$= (-0.064 + j0.102) - (-0.110 + j0.146)$
$= (0.046 - j0.044)$
$= (0.064 \angle -43.73°)$

Using the magnitude of D_s, calculate K:

$$K = \frac{1 + |D_s|^2 - |s_{11}|^2 - |s_{22}|^2}{(2)(|s_{21}|)(|s_{12}|)}$$

$$= \frac{1 + (0.064)^2 - (0.3)^2 - (0.4)^2}{2(6.1)(0.03)}$$

$$= 2.06$$

Since $K > 1$, the transistor is unconditionally stable and we may proceed with the design.

Using Equation 7.3, calculate B_1:

$$B_1 = 1 + |s_{11}|^2 - |s_{22}|^2 - |D_s|^2$$
$$= 1 + (0.3)^2 - (0.4)^2 - (0.064)^2$$
$$= 0.926$$

Using Equation 7.4, calculate maximum available gain (MAG):

$$\text{MAG} = 10 \log \frac{|s_{21}|}{|s_{12}|} + 10 \log \left| K \pm \sqrt{K^2 - 1} \right|$$

Since B_1 is positive, the negative sign will be used in front of the square root sign and

$$\text{MAG} = 10 \log \frac{6.1}{0.03} + 10 \log \left| 2.06 - \sqrt{(2.06)^2 - 1} \right|$$
$$= 23.08 + (-5.87)$$
$$= 17.21 \text{ dB}$$

We will consider 17.21 dB to be adequate for our design gain of 16 dB minimum. If the design had called for a minimum gain greater than 16 dB, a different transistor would be needed.

To find the load reflection coefficient for a conjugate match, the two intermediate quantities (C_2 and B_2) must first be found. Using Equation 7.5:

$$C_2 = s_{22} - (D_s s_{11}{}^*)$$
$$= (0.4 \angle -38°) - [(0.064 \angle -43.73°)(0.3 \angle -160°)]$$
$$= 0.315 - j0.246 - [-0.018 + j0.008]$$
$$= (0.419 \angle -37.33°)$$

Using Equation 7.6

$$B_2 = 1 + |s_{22}|^2 - |s_{11}|^2 - |D_s|^2$$
$$= 1 + (0.4)^2 - (0.3)^2 - (0.064)^2$$
$$= 1.066$$

Therefore the magnitude of the load reflection coefficient can now be found using Equation 7.7:

$$|\Gamma_L| = \frac{B_2 \pm \sqrt{B_2^2 - 4|C_2|^2}}{2|C_2|}$$

$$= \frac{1.066 - \sqrt{(1.066)^2 - 4(0.419)^2}}{2(0.419)}$$

$$= 0.486$$

The angle of the load reflection coefficient is simply equal to the negative of the angle of C_2 or +37.33°. Thus

$$\Gamma_L = 0.486 \angle 37.33°$$

Γ_L can now be substituted in Equation 7.9 to calculate Γ_s:

$$\Gamma_s = \left[s_{11} + \frac{s_{12}s_{21}\Gamma_L}{1 - s_{22}\Gamma_L} \right]^*$$

$$= \left[0.3 \angle 160° + \frac{(0.03 \angle 62°)(6.1 \angle 65°)(0.486 \angle 37.33°)}{1 - (0.4 \angle -38°)(0.486 \angle 37.33°)} \right]^*$$

$$= \left[(-0.282 + j0.103) + \frac{(0.089 \angle 164.33°)}{(1 - 0.194 \angle -0.670°)} \right]^*$$

$$= \left[(-0.282 + j0.103) + \frac{(0.089 \angle 164.33°)}{(0.806 \angle 0.142°)} \right]^*$$

$$= [(-0.282 + j0.103) + (-0.160 + j0.030)]^*$$

$$= [0.463 \angle 163.29°]^*$$

$$= 0.463 \angle -163.29°$$

Once the desired Γ_s and Γ_L are known, all that remains is to provide the transistor with components which 'look like' Γ_s and Γ_L.

(a) Input matching network. The input matching network design is shown on the Smith chart of Figure 7.1. The object of the design is to force the 50 Ω source to present a reflection coefficient[1] of 0.463 ∠ –163.29° to the transistor input. With Γ_s plotted as shown, the corresponding desired and normalised impedance is read directly from the chart as Z_s = (0.36 – j0.12) Ω. Remember this is a normalised impedance because the chart has been normalised to 50 Ω. The actual impedance represented by Γ_s is equal to 50(0.36 – j0.12) Ω = (18.6 – j6) Ω.

Now we must transform the 50 Ω source to (18.6 – j6) Ω impedance. The most common circuit used is a low pass filter configuration consisting of a shunt C and a series

[1] In all examples containing reflection coefficients and Smith charts, the reflection coefficients are plotted on the Smith chart and the resultant values are read from it. Theoretically, the Smith chart should give you an exact answer. In practice, reading difficulties and interpolations must be made, so expect slightly different answers if you are using mathematics to derive these values. Strictly speaking, this is not a problem because transistor characteristics change, even when the same type number is used. Hence perfect match with one transistor does not mean perfect match with another transistor. The most difficult part of amplifier design is choosing loads that will produce the same circuit characteristics in spite of transistor changes.

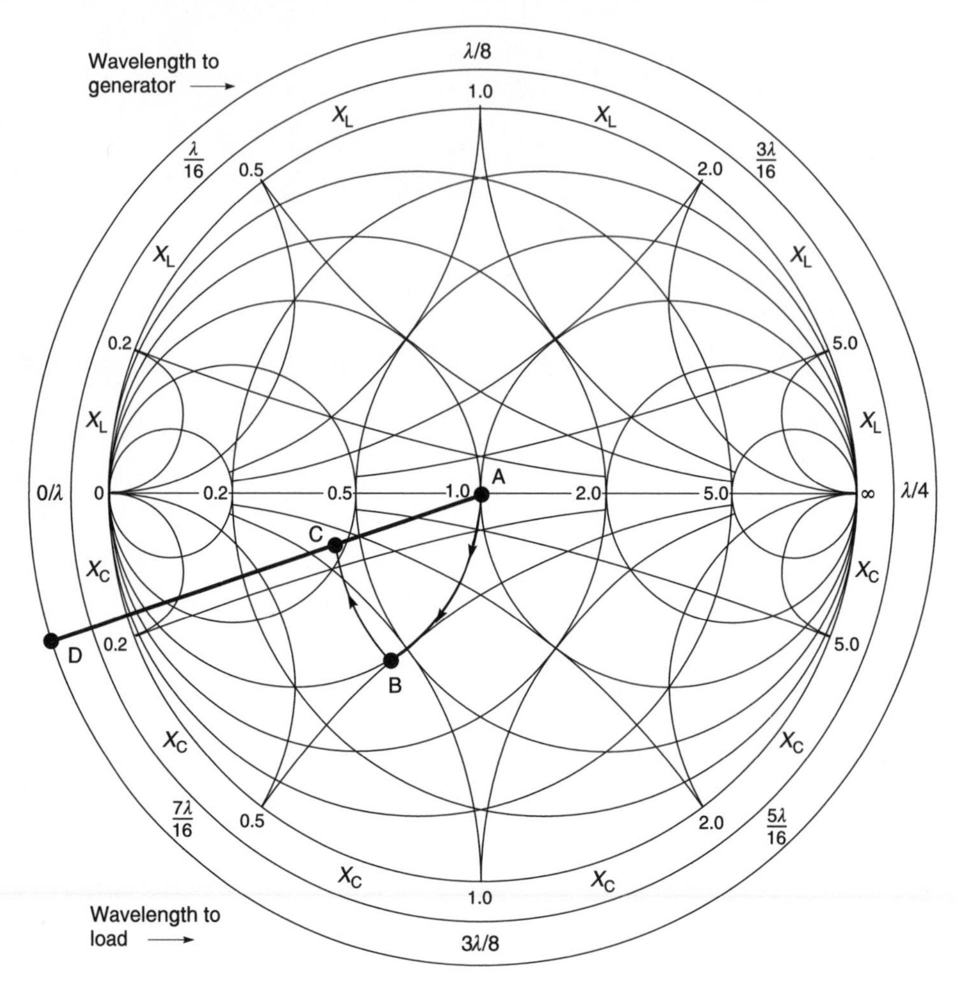

Fig. 7.1 Input matching network A = 1 + j0, B = 0.43 – j0.5, C = 0.43 – j0.14

L. Remember that when using a Smith chart with shunt elements, you use it as an admittance chart, and when using the chart for series elements, you use it as an impedance chart. For ease of transformation, we use the Smith chart type ZY-10-N which was introduced in Chapter 3. Proceeding from the source, we have:

$$\text{Arc AB} = \text{shunt } C = \text{j}1.33 \text{ S}$$
$$\text{Arc BC} = \text{series } L = \text{j}0.34\ \Omega$$

If you have difficulty following the above construction, look first at Figure 7.3 where you will see the schematic of the amplifier. It starts off with a 50 Ω source (point A on Figure 7.1). Across this point, we have a shunt capacitor; therefore we must use the admittance part of the chart. This shunt capacitance moves us to point B in Figure 7.1, i.e. along arc AB. The next element in Figure 7.1 is a series inductor. We must therefore use the impedance part of the Smith chart in Figure 7.1. Our final destination is the required source impedance for transistor (point C in Figure 7.1) – hence the arc BC. The Smith chart values are read according to the part of the chart being used; admittance values for the shunt components and impedance values for the series components.

The actual component values are found using Equations 7.11 to 7.14 which are:

$$C_{\text{series}} = \frac{1}{\omega XN} \tag{7.11}$$

$$C_{\text{shunt}} = \frac{B}{\omega N} \tag{7.12}$$

$$L_{\text{series}} = \frac{XN}{\omega} \tag{7.13}$$

and

$$L_{\text{shunt}} = \frac{N}{\omega B} \tag{7.14}$$

where

N = normalisation value
B = susceptance in Siemens
X = reactance in ohms
ω = frequency in radians/second

For this example

$$C_1 = \frac{1.33}{2\pi(150 \text{ MHz})(50)} = 28.22 \text{ pF} \approx 28 \text{ pF}$$

and

$$L_1 = \frac{(0.34)(50)}{2\pi(150 \text{ MHz})} = 18.04 \text{ nH} \approx 18 \text{ nH}$$

This completes the input matching network.

(b) Output matching network. The load reflection coefficient is plotted in Figure 7.2 and after plotting in $\Gamma_L = 0.486 \angle 37.33°$ the Smith chart shows a normalised load impedance of (1.649 + j1.272) Ω. After re-normalisation, it represents a load impedance Z_L = 50 (1.649 + j1.272) Ω or (82.430 + j63.611) Ω. The matching network is designed as follows. Proceeding from the load:

Arc AB = series C = –j1.1 Ω
Arc BC = shunt L = –j0.8 S

If you have difficulty with the Smith chart, look at the schematic on Figure 7.3. The final load is 50 Ω (point A in Figure 7.2). The transistor load is point C in Figure 7.2.

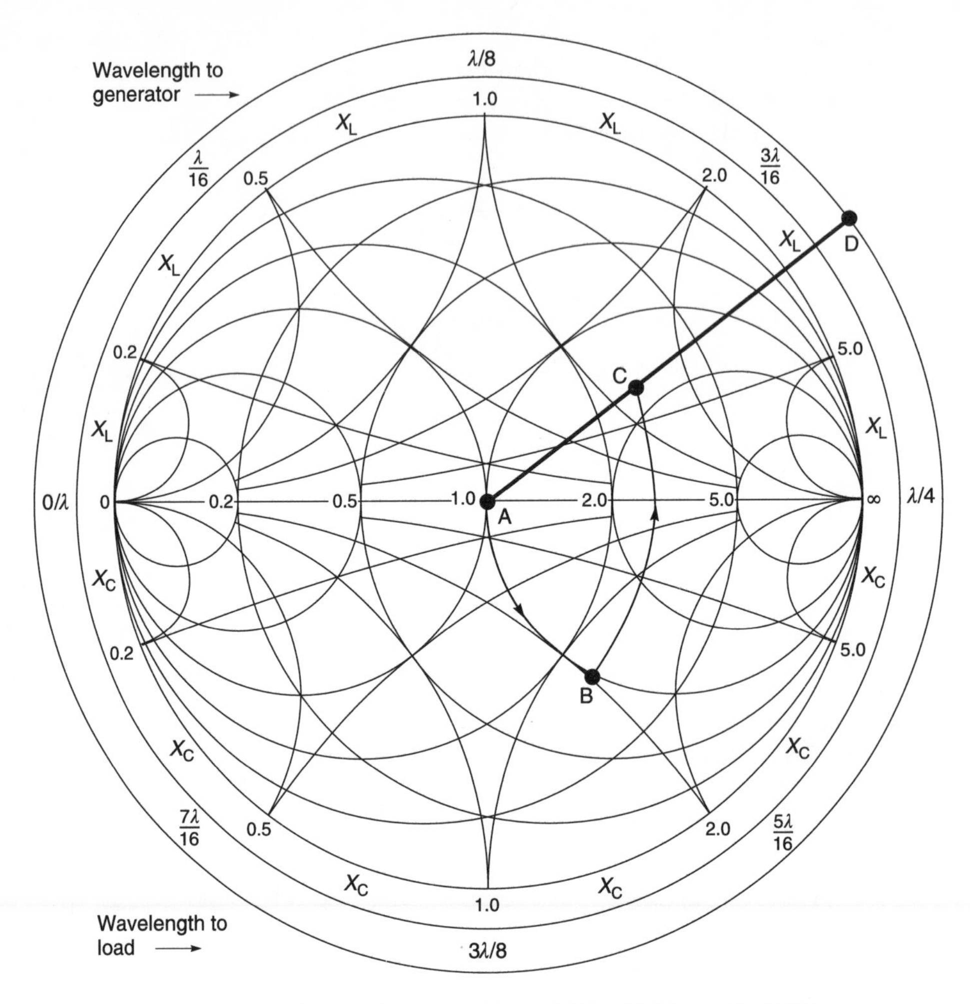

Fig. 7.2 Output matching network A = 1 + j0, B = 1 –j1.28, C = 0.486 ∠ 37.33° or 1.65 + j1.272

Starting from the 50 Ω point (point A in Figure 7.2), you encounter a series capacitor. Therefore the series part of the chart must be used. This takes you to point B on the chart, along arc AB. Next you have a shunt inductor (see Figure 7.3); therefore you must use the admittance part of the chart to get to your destination (point C), that is along arc BC in Figure 7.2. The Smith chart values are read according to the part of the chart being used; admittance values for the shunt components and impedance values for the series components.

The actual component values are found using Equations 7.11 and 7.14 which are for this example:

$$C_2 = \frac{1}{2\,\pi(150\text{ MHz})(1.1)(50)} = 19.29\text{ pF} \approx 19\text{ pF}$$

and

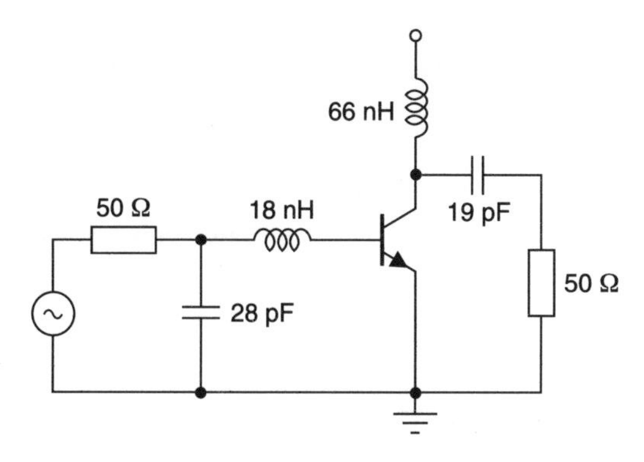

Fig. 7.3 R.F. circuit for Example 7.1

$$L_2 = \frac{50}{2\,\pi(150\text{ MHz})(0.8)} = 66.31\text{ nH} \approx 66\text{ nH}$$

The completed design (minus biasing network) is shown in Figure 7.3.

Transducer gain (G_T)

The transducer gain is the actual gain of an amplifier stage including the effects of input and output matching and device gain. It does not include losses attributed to power dissipation in imperfect components. Transducer gain is calculated as follows:

$$G_T = \frac{|s_{21}|^2(1 - |\Gamma_s|^2)(1 - |\Gamma_L|^2)}{|(1 - s_{11}\Gamma_s)(1 - s_{22}\Gamma_L) - s_{12}s_{21}\Gamma_L\Gamma_s|^2} \tag{7.15}$$

where

Γ_s and Γ_L are the source and load reflection coefficients respectively

Calculation of G_T is a useful method of checking the power gain of an amplifier *before* it is built. This is shown by Example 7.2.

Example 7.2

Calculate the transducer gain of the amplifier that was designed in Example 7.1.

Given: As in Example 7.1.
Required: Transducer gain.

Solution. Using Equation 7.15

$$G_T = \frac{|s_{21}|^2(1 - |\Gamma_s|^2)(1 - |\Gamma_L|^2)}{|(1 - s_{11}\Gamma_s)(1 - s_{22}\Gamma_L) - s_{12}s_{21}\Gamma_L\Gamma_s|^2}$$

$$= \frac{(6.1)^2(1 - (0.463)^2)(1 - (0.486)^2)}{|(1 - 0.139\angle 3.3°)(1 - 0.194\angle 01.7°) - (0.03\angle 62°)(6.1\angle 65°)(0.486\angle 37.33°)(0.463\angle -163.29°)|^2}$$

$$\approx \frac{22.329}{|0.694 - (0.041\angle 1.04°)|^2} = \frac{22.329}{|0.694 - (0.041 + j0.001)|^2}$$

$$= \frac{22.329}{0.653^2}$$

$$= 52.365 \text{ or } 17.19 \text{ dB}$$

Note: The transducer gain calculates to be very close to MAG. This is due to the fact that s_{12} is not equal to zero and is therefore providing some internal transistor feedback.

PUFF results

The results of Examples 7.1 and 7.2 using PUFF are shown in Figure 7.4. However, you should be aware of how PUFF was modified for the results.

- The amplitude range and the frequency range in the rectangular plot had to be modified as explained in Chapter 4 to obtain the required amplitude and frequency ranges.
- There is no transistor device in PUFF that meets the requirements of the transistor in the examples. A new transistor template called 'ed701' was generated by copying the FHX04.device template and modifying it to suit the requirements of the example.
- A resistor of 1 pΩ was generated within PUFF to provide a spacer for ease in laying out the circuit.
- The input matching circuit of Figure 7.1 was plotted at port 1. At the match frequency, the input match reflection coefficient is shown as –35.39 dB in Figure 7.4.
- The output matching circuit of Figure 7.2 was plotted at port 2. At the match frequency,

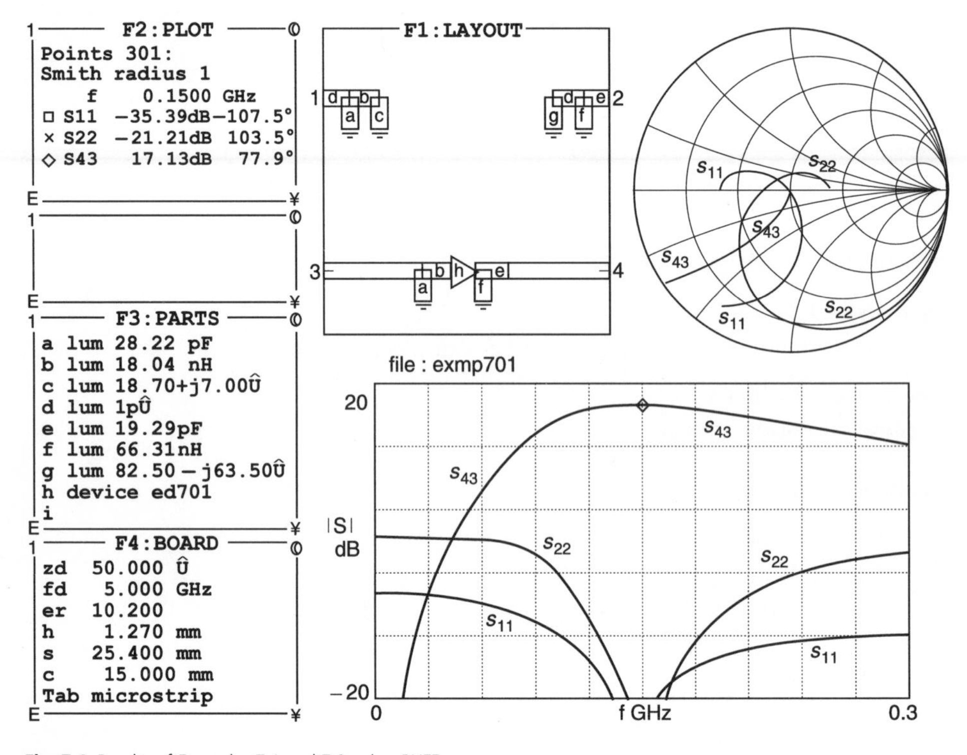

Fig. 7.4 Results of Examples 7.1 and 7.2 using PUFF

the output match reflection coefficient is shown as –21.21 dB in Figure 7.4.
- The gain of the circuit is given by PUFF as 17.13 dB. We calculated the gain as 17.19 dB.

Note: There is a very slight discrepancy between the results but this is to be expected because the examples were carried out graphically. Nonetheless, this should convince you that our design methods are reliable!

Example 7.3

A MESFET has the following *S*-parameters at 5 GHz with $V_{ce} = 15$ V and $I_c = 10$ mA:

$$s_{11} = 0.3 \angle 140^\circ \qquad s_{12} = 0.03 \angle 65^\circ$$
$$s_{21} = 2.1 \angle 62^\circ \qquad s_{22} = 0.40 \angle -38^\circ$$

Calculate the maximum available gain (MAG) for the transistor under these operating conditions.

Given: $f = 5$ GHz, $V_{ce} = 15$ V, $I_c = 10$ mA, $s_{11} = 0.3 \angle 140^\circ$, $s_{12} = 0.03 \angle 65^\circ$, $s_{21} = 2.1 \angle 62^\circ$, $s_{22} = 0.40 \angle -38^\circ$.
Required: MAG for the MESFET at 5 GHz.

Solution. Using Equations 7.1 and 7.2, check for stability:

$$\begin{aligned} D_s &= s_{11}s_{22} - s_{12}s_{22} \\ &= (0.3 \angle 140^\circ)(0.4 \angle -38^\circ) - (0.03 \angle 65^\circ)(2.1 \angle 62^\circ) \\ &= (0.120 \angle 102^\circ) - (0.063 \angle 127^\circ) \\ &= (-0.025 + j0.117) - (-0.038 + j0.050) \\ &= (0.013 + j0.067) \\ &= (0.068 \angle 79.06^\circ) \end{aligned}$$

Using the magnitude of D_s, calculate K:

$$\begin{aligned} K &= \frac{1 + |D_s|^2 - |s_{11}|^2 - |s_{22}|^2}{(2)(|s_{21}|)(|s_{12}|)} \\ &= \frac{1 + (0.068)^2 - (0.3)^2 - (0.4)^2}{2(2.1)(0.03)} \\ &= 5.99 \end{aligned}$$

Since $K > 1$, the transistor is unconditionally stable and we may proceed with the design.
Using Equation 7.3, calculate B_1:

$$\begin{aligned} B_1 &= 1 + |s_{11}|^2 - |s_{22}|^2 - |D_s|^2 \\ &= 1 + (0.3)^2 - (0.4)^2 - (0.068)^2 \\ &= 0.925 \end{aligned}$$

Using Equation 7.4, calculate maximum available gain (MAG):

Since B_1 is positive, the negative sign will be used in front of the square root sign:

$$\text{MAG} = 10 \log \frac{|s_{21}|}{|s_{12}|} + 10 \log \left| K \pm \sqrt{K^2 - 1} \right|$$

If you wish, check the answer by using PUFF.

$$\begin{aligned}\text{MAG} &= 10\log\frac{2.1}{0.03} + 10\log\left|5.99 - \sqrt{(5.99)^2 - 1}\right| \\ &= 18.45 + (-10.75) \\ &\approx 7.7\text{ dB}\end{aligned}$$

Example 7.4

An integrated circuit has the following *S*-parameters:

$$s_{11} = 0.3 \angle 140° \qquad s_{12} = 0.03 \angle 65°$$
$$s_{21} = 2.1 \angle 62° \qquad s_{22} = 0.40 \angle -38°$$

If its source reflection coefficient $\Gamma_s = 0.463 \angle -164°$ and its load reflection coefficient $\Gamma_L = 0.486 \angle 38°$, calculate the transducer gain of the amplifier.

Given: $\Gamma_s = 0.463 \angle -140°$ $\quad \Gamma_L = 0.486 \angle 38°$
$s_{11} = 0.3 \angle 140°$ $\quad s_{12} = 0.03 \angle 65°$
$s_{21} = 2.1 \angle 62°$ $\quad s_{22} = 0.40 \angle -38°$
Required: Amplifier transducer gain.

Solution. Using Equation 7.15

$$G_T = \frac{|s_{21}|^2(1 - |\Gamma_s|^2)(1 - |\Gamma_L|^2)}{|(1 - s_{11}\Gamma_s)(1 - s_{22}\Gamma_L) - s_{12}s_{21}\Gamma_L\Gamma_s|^2}$$

$$= \frac{(2.1)^2(1 - (0.463)^2)(1 - (0.486)^2)}{|(1 - 0.139)(1 - 0.194) - (0.03 \angle 65°)(2.1 \angle 62°)(0.486 \angle 38°)(0.463 \angle -140°)|^2}$$

$$= \frac{2.646}{|0.694 - (0.014 \angle 25°)|^2} = \frac{2.646}{|0.694 - (0.014 - j0.002)|^2}$$

$$= \frac{2.646}{0.68^2}$$

$$= 5.708 \text{ or } 7.6\text{ dB}$$

If you wish, check the answer by using PUFF.

7.4 Design of amplifiers for a specific gain

In cases where a specific gain is required, it is normal practice to provide **selective mismatching** so that transistor gain can be reduced to the desired gain. Selective mismatching is a relatively inexpensive method used to decrease gain by not matching a transistor to its conjugate load.

One of the easiest ways of selective mismatching is through the use of a **constant gain circle** plotted on the Smith chart. A constant gain circle is merely a circle, the circumference of which represents a locus of points (load impedances) that will force the amplifier gain to a specific value. For instance, any of the infinite number of impedances located on the circumference of a 12 dB constant gain circle would force the amplifier stage gain to 12 dB. Once the circle is drawn on a Smith chart, you can see the load impedances that will provide a desired gain.

7.4.1 Constant gain circles

To plot a constant gain circle on a Smith chart, it is necessary to know (i) where the centre of the circle is located and (ii) its radius. The procedure for calculating a constant gain circle is as follows.

1 Calculate D_s as in Equation 7.1.
2 Calculate D_2:

$$D_2 = |s_{22}|^2 - |D_s|^2 \tag{7.16}$$

3 Calculate C_2:

$$C_2 = s_{22} - D_s s_{11}{}^* \tag{7.17}$$

4 Calculate G:

$$G = \frac{|\text{desired gain}|}{|s_{21}|^2} \tag{7.18}$$

5 Calculate centre location of constant gain circle:
6 Calculate radius of the circle:

$$r_o = \frac{GC_2^*}{1 + D_2 G} \qquad (7.19$$

Equation 7.19 produces a complex number in magnitude–angle format similar to that of a

$$p_o = \frac{\sqrt{1 - 2K|s_{12}s_{21}|G + |s_{12}s_{21}|^2 G^2}}{1 + D_2 G} \qquad (7.20$$

reflection coefficient. This number is plotted on the Smith chart exactly as you would plot a value of reflection coefficient.

The radius of the circle that is calculated with Equation 7.20 is simply a fractional number between 0 and 1 which represents the size of that circle in relation to a Smith chart. A circle with a radius of 1 has the same radius as a Smith chart; a radius of 0.5 represents half the radius of a Smith chart and so on.

After the load reflection coefficient (in effect the load impedance) is chosen by the designer, the next step will be to determine the source reflection coefficient that is required to prevent any further decrease in gain. This value is of course the **conjugate** of the actual input reflection coefficient of the transistor with the specified load calculated by Equation 7.9. To clarify the procedure, we now present Example 7.5.

Example 7.5

A transistor has the following *S*-parameters at 1 GHz, with V_{ce} = 15 V and I_c = 5 mA:

$$s_{11} = 0.28 \angle -58^\circ \qquad s_{12} = 0.08 \angle 92^\circ$$
$$s_{21} = 2.1 \angle 65^\circ \qquad s_{22} = 0.8 \angle -30^\circ$$

Design an amplifier to present 9 dB of gain at 1 GHz. The source impedance $Z_s = (35 - j60)\ \Omega$ and the load impedance $Z_L = (50 - j50)\ \Omega$. The transistor is unconditionally stable with $K = 1.168$. Design the output and input networks.

Given: $f = 1$ GHz $V_{ce} = 15$ V $I_c = 5$ mA $s_{11} = 0.28 \angle -58^\circ$
$s_{12} = 0.08 \angle 92^\circ$ $s_{21} = 2.1 \angle 65^\circ$ $s_{22} = 0.8 \angle -30^\circ$
$Z_s = (35 - j60)\ \Omega$ $Z_L = (50 - j50)$ $K = 1.168$
Required: 9 dB amplifier, output network, input network.

Solution. Using Equation 7.1, find D_s for substitution in Equations 7.16 and 7.17:

$$\begin{aligned} D_s &= s_{11}s_{22} - s_{12}s_{21} \\ &= (0.28 \angle -58^\circ)(0.8 \angle -30^\circ) - (0.08 \angle 92^\circ)(2.1 \angle 65^\circ) \\ &= (0.224 \angle -88^\circ) - (0.168 \angle 157^\circ) \\ &= 0.008 - j0.224 + 0.155 - j0.066 \\ &= 0.333 \angle -60.66^\circ \end{aligned}$$

Using Equation 7.16, find D_2 for subsequent insertion in Equation 7.19:

$$\begin{aligned} D_2 &= |s_{22}|^2 - |D_s|^2 \\ &= (0.8)^2 - (0.333)^2 \\ &= 0.529 \end{aligned}$$

Using Equation 7.17, find C_2 for subsequent insertion in Equation 7.19:

$$\begin{aligned} C_2 &= s_{22} - D_s s_{11}{}^* \\ &= (0.8 \angle -30^\circ) - (0.333 \angle -60.66^\circ)(0.28 \angle 58^\circ) \\ &= (0.693 - j0.400) - (0.093 - j0.004) \\ &= (0.719 \angle -33.42^\circ) \end{aligned}$$

Bearing in mind that a power ratio of 9 dB is a power ratio of 7.94, use Equation 7.18 to find G for subsequent insertion in Equation 7.19:

$$G = \frac{|\text{desired gain}|}{|s_{21}|^2} = \frac{7.94}{(2.1)^2} = 1.80$$

Using Equation 7.19, find the centre of the constant gain circle:

$$r_o = \frac{GC_2{}^*}{1 + D_2 G} = \frac{1.80\ (0.719 \angle 33.42^\circ)}{1 + (0.529)(1.80)} = 0.663 \angle 33.42^\circ$$

Using Equation 7.20, find the radius for the 9 dB constant gain circle:

$$p_o = \frac{\sqrt{1 - 2K|s_{12}s_{21}|G + |s_{12}s_{21}|^2 G^2}}{1 + D_2 G}$$

$$= \frac{\sqrt{1 - 2(1.168)(0.08)(2.1)(1.80) + (0.08 \times 2.1)^2 (1.80)^2}}{1 + (0.529)(1.80)}$$

$$= \frac{\sqrt{1 - 0.706 + 0.091}}{1 + 0.952}$$

$$= 0.318$$

The Smith chart construction is shown in Figure 7.5. Note that any load impedance located on the circumference of the circle will produce an amplifier gain of 9 dB if the input impedance of the transistor is *conjugately* matched. The actual load impedance we have to

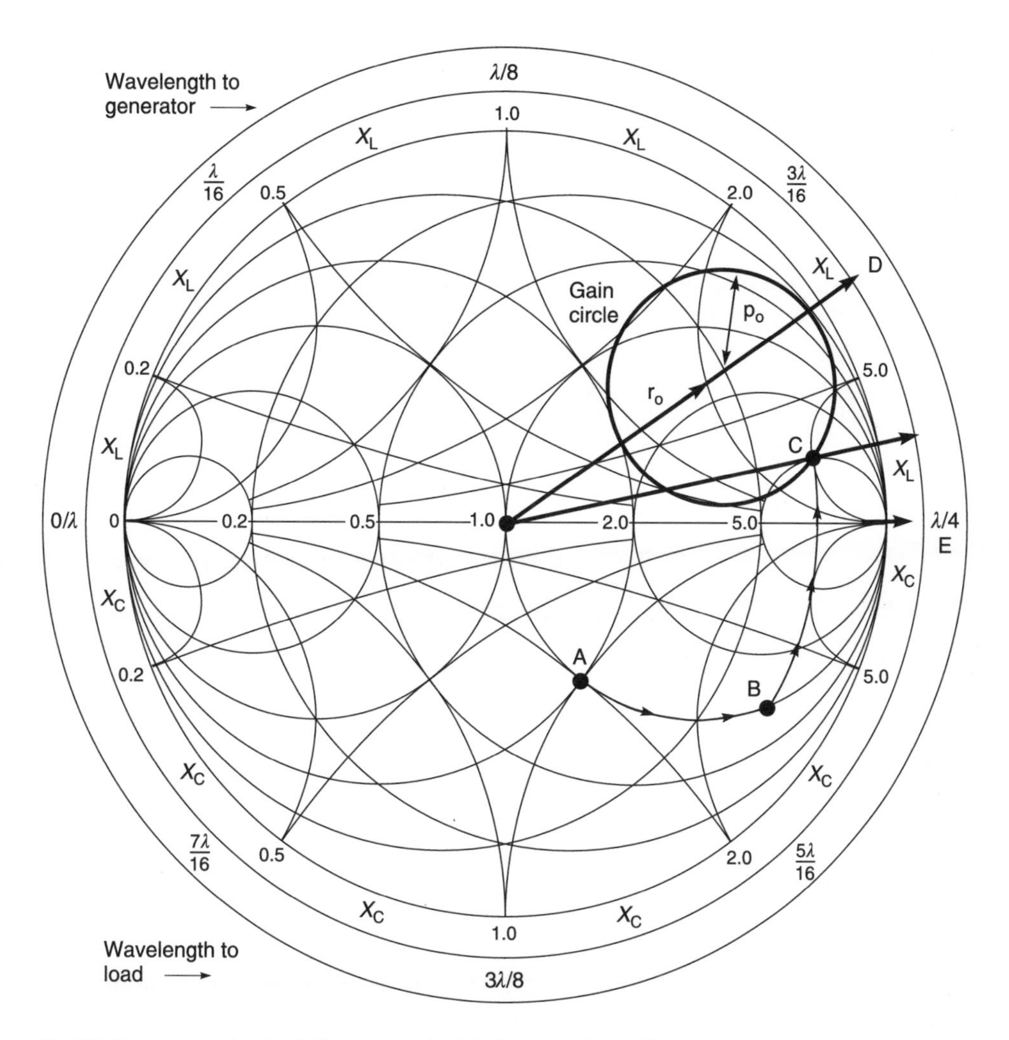

Fig. 7.5 Output network using 9 dB constant gain circle A = 1 – j1, B = 1 – j3, C = 0.1 – j0.11, $r_o = 0.663 \angle 33.43°$, $p_o = 0.318$. Angle between point D and E = 33.43°

work with is (50 – j50) which in its normalised form is (1 – j1) Ω on the Smith chart and denoted by point A.

Output network. The transistor's output network must transform the actual load impedance into a value that falls on the constant gain 9 dB circle. Obviously there are many circuit configurations which will satisfy these conditions. The configuration shown in Figure 7.5 has been chosen for convenience. Proceeding from the load:

Arc AB = series C = –j2.0 Ω
Arc BC = shunt L = –j0.41 S

Using Equation 7.11 for a series C:

$$C_1 = \frac{1}{2\pi(1\ \text{GHz})(2)(50)} \approx 1.6\ \text{pF}$$

Using Equation 7.14 for a shunt L:

$$L_1 = \frac{50}{2\pi(1\ \text{GHz})(0.41)} \approx 19.4\ \text{nH}$$

Input network. For a conjugate match at the input to the transistor with $\Gamma_L = 0.82 \angle 13°$ (point C in Figure 7.5), the desired source reflection coefficient must be (using Equation 7.9):

$$\Gamma_s = \left[(s_{11} + \frac{s_{12}s_{21}\Gamma_L}{1 - s_{22}\Gamma_L}\right]^*$$

$$= \left[0.28 \angle -58° + \frac{(0.08 \angle 92°)(2.1 \angle 65°)(0.82 \angle 13°)}{1 - (0.8 \angle -30°)(0.82 \angle 13°)}\right]^*$$

$$= \left[0.28 \angle -58° + \frac{(0.138 \angle 170.00°)}{1 - (0.656 \angle -17°)}\right]^*$$

$$= \left[0.148 - j0.237 + \frac{(0.138 \angle 170.00°)}{(0.420 \angle 27.24°)}\right]^*$$

$$= [0.148 - j0.237 + (-0.262 + j0.199]^*$$

$$= [0.120 \angle -161.56°]^*$$

$$= 0.120 \angle 161.56°$$

The point is plotted as point D in Figure 7.6. The actual normalised source impedance is plotted at point A (0.7 – j1.2) Ω. The input network must transform the actual impedance at point A to the desired impedance at point D. We have used a three element matching network this time:

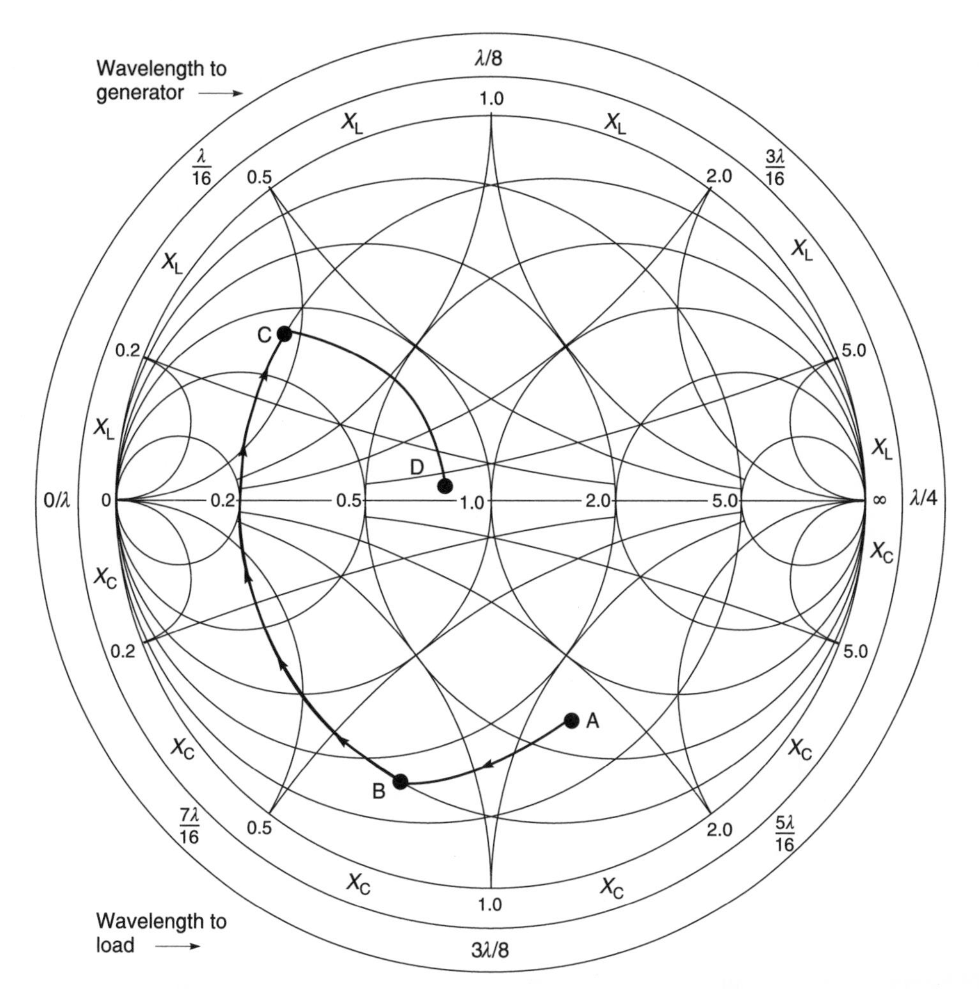

Fig. 7.6 Input network of Example 7.5 A = 0.7 – j1.2, B = 0.37 + j1.25, C = 0.2 + j0.33, D = 1.25 – j0.1

$$\text{Arc AB} = \text{shunt } C_2 = \text{j}0.63 \text{ S}$$
$$\text{Arc BC} = \text{series } L_2 = \text{j}1.08\ \Omega$$
$$\text{Arc CD} = \text{shunt } C_3 = \text{j}2.15 \text{ S}$$

Using Equation 7.12 for a shunt capacitance:

$$C_2 = \frac{(0.63)}{2\pi(1 \text{ GHz})(50)} \approx 2 \text{ pF}$$

Using Equation 7.13 for a series inductance:

$$L_2 = \frac{(1.08)(50)}{2\pi(1 \text{ GHz})} \approx 8.5 \text{ nH}$$

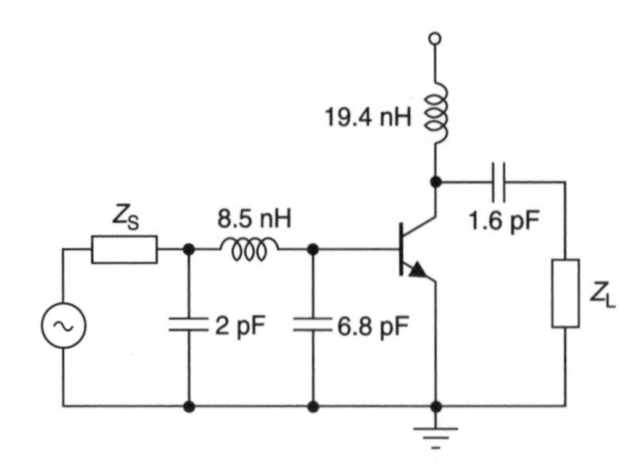

Fig. 7.7 R.F. circuit for Example 7.5

Using Equation 7.12 for a shunt capacitance:

$$C_3 = \frac{(2.15)}{2\pi(1 \text{ GHz})(50)} \approx 6.8 \text{ pF}$$

The completed design (minus biasing network) is shown in Figure 7.7.

Example 7.6

Use the information of Example 7.5 to calculate a constant gain circle of 8 dB.

Given: $f = 1$ GHz $V_{ce} = 15$ V $I_c = 5$ mA $s_{11} = 0.28 \angle -58°$
$s_{12} = 0.08 \angle 92°$ $s_{21} = 2.1 \angle 65°$ $s_{22} = 0.8 \angle -30°$
$D_s = 0.333 \angle -60.66°$ $D_2 = 0.529°$ $C_2 = 0.719 \angle -33.42°$
$K = 1.168$

Required: A constant gain circle of 8 dB.

Solution. Bearing in mind that a power ratio of 8 dB is a power ratio of 6.31, use equation 7.18 to find G for subsequent insertion in Equation 7.19:

$$G = \frac{|\text{desired gain}|}{|s_{21}|^2} = \frac{6.31}{(2.1)^2} = 1.43$$

Using Equation 7.19, find the centre of the constant gain circle:

$$r_o = \frac{GC_2^*}{1 + D_2 G} = \frac{1.43(0.719 \angle 33.42°)}{1 + (0.529)(1.43)} = 0.585 \angle 33.42°$$

Using Equation 7.20, find the radius for the 9 dB constant gain circle:

$$p_o = \frac{\sqrt{1 - 2K|s_{12}s_{21}|G + |s_{12}s_{21}|^2 G^2}}{1 + D_2 G}$$

$$= \frac{\sqrt{1 - 2(1.168)(0.08)(2.1)(1.43) + (0.08 \times 2.1)^2 (1.43)^2}}{1 + (0.529)(1.43)}$$

$$= \frac{\sqrt{1 - 0.561 + 0.058}}{1 + 0.756}$$

$$= 0.401$$

7.4.2 Design of amplifiers with conditionally stable devices

When the Rollett stability factor (K) calculates to be less than unity, it is a certainty that with some combinations of source and load impedances, the transistor will oscillate. To prevent oscillation the source and load impedances must be chosen very carefully. One of the best methods of determining those source and load impedances that will cause the transistor to go unstable is to plot **stability circles** on a Smith chart.

7.4.3 Stability circles

A stability circle is simply a circle on a Smith chart which represents the boundary between those values of source or load impedance that cause instability and those that do not. The circumference of the circle represents the locus of points which forces $K = 1$. Either the inside or the outside of the circle may represent the unstable region and that determination must be made after the circles are drawn.

The location and radii of the input and output stability circles are found as follows.

1 Calculate D_s using Equation 7.1.
2 Calculate C_1:

$$C_1 = s_{11} - D_s s_{22}^* \tag{7.21}$$

3 Calculate C_2 using Equation 7.5.
4 Calculate the *centre location* of the *input* stability circle:

$$\Gamma_{s1} = \frac{C_1^*}{|s_{11}|^2 - |D_s|^2} \tag{7.22}$$

5 Calculate the *radius* of the input stability circle:

$$p_{s1} = \left| \frac{s_{12} s_{21}}{|s_{11}|^2 - |D_s|^2} \right| \tag{7.23}$$

6 Calculate the *centre location* of the *output* stability circle:

$$\Gamma_{s2} = \frac{C_2^*}{|s_{22}|^2 - |D_s|^2} \tag{7.24}$$

7 Calculate the *radius* of the output stability circle:

$$p_{s2} = \left| \frac{s_{11} s_{21}}{|s_{22}|^2 - |D_s|^2} \right| \tag{7.25}$$

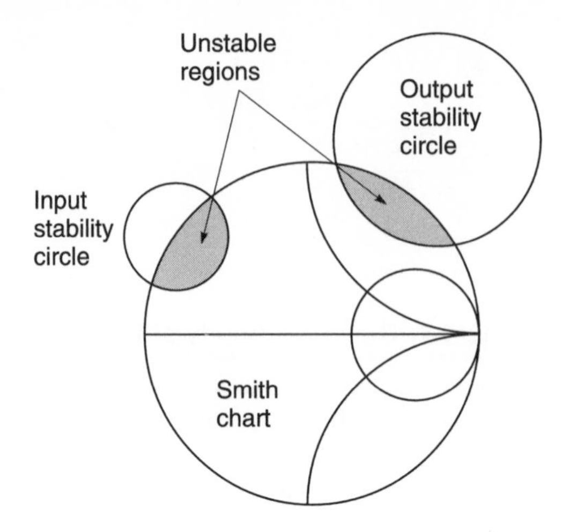

Fig. 7.8 Unstable regions for a potentially unstable transistor

Once the calculations are made, circles can be plotted directly on the Smith chart. For a potentially unstable transistor, the stability circles might resemble those shown in Figure 7.8.

After the stability circles are plotted on the chart, the next step is to determine which side of the stability circles (inside or outside) represents the stable region. This is easily done if s_{11} and s_{22} for the transistor are less than 1. If the stability circles do not enclose the centre of the Smith chart, then regions inside the stability circles are unstable and all regions *outside* the stability circles on the Smith chart are *stable* regions. See Figure 7.8. If one of the stability circles encloses the centre of the Smith chart, then the region inside that stability circle is stable. This is because the *S*-parameters were measured with a 50 Ω source and load, and since the transistor remained stable (s_{11} and $s_{22} < |1|$) under these conditions, then *the centre of the Smith chart must be part of the stable regions*. Example 7.7 will help to clarify this.

Example 7.7

The *S*-parameters for a transistor at 200 MHz with $V_{ce} = 6$ V and $I_c = 5$ mA are:

$$s_{11} = 0.4 \angle 280° \qquad s_{12} = 0.048 \angle 65°$$
$$s_{21} = 5.4 \angle 103° \qquad s_{22} = 0.78 \angle 345°$$

Design and choose stable load and source reflection coefficients that will provide a power gain of 12 dB at 200 MHz.

Given: $f = 200$ MHz $V_{ce} = 6$ V $I_c = 5$ mA
$s_{11} = 0.4 \angle 280°$ $s_{12} = 0.048 \angle 65°$
$s_{21} = 5.4 \angle 103°$ $s_{22} = 0.78 \angle 345°$
Required: Stable load reflection coefficient, stable source reflection coefficient, and power gain of 12 dB at 200 MHz.

Solution

1 Using Equation 7.1, find D_s:

$$\begin{aligned} D_s &= s_{11}s_{22} - s_{12}s_{21} \\ &= (0.4 \angle 280°)(0.78 \angle 345°) - (0.048 \angle 65°)(5.4 \angle 103°) \\ &= (-0.027 - j0.311) - (-0.254 + j0.054) \\ &= (0.429 \angle -58.2°) \end{aligned}$$

2 Using Equation 7.2, calculate Rollett's stability factor (K):

$$K = \frac{1 + |D_s|^2 - |s_{11}|^2 - |s_{22}|^2}{(2)(|s_{21}|)\ (|s_{12}|)}$$

$$= \frac{1 + (0.429)^2 - (0.4)^2 - (0.78)^2}{(2)(5.4)(0.048)}$$

$$= 0.802$$

Since $K < 1$, we must exercise extreme care in choosing source and load impedances otherwise the transistor will oscillate so stability circles must be plotted.

3 Using Equation 7.21, calculate C_1:

$$\begin{aligned} C_1 &= s_{11} - D_s s_{22}{}^* \\ &= (0.4 \angle 280°) - (0.429 \angle -58.2°)(0.78 \angle -345°) \\ &= (0.069 - j0.394) - (0.244 - j0.229) \\ &= (0.241 \angle -136.7°) \end{aligned}$$

4 Using Equation 7.5, calculate C_2:

$$\begin{aligned} C_2 &= s_{22} - (D_s s_{11}{}^*) \\ &= (0.78 \angle 345°) - (0.429 \angle -58.18°)(0.4 \angle -280°) \\ &= (0.753 - j0.202) - (0.159 + j0.064) \\ &= 0.651 \angle -24.1° \end{aligned}$$

5 Using Equation 7.22, calculate the *centre location* of the *input* stability circle:

$$\Gamma_{s1} = \frac{C_1{}^*}{|s_{11}|^2 - |D_s|^2}$$

$$= \frac{(0.241 \angle 136.7°)}{(0.4)^2 - (0.429)^2}$$

$$= -10 \angle 136.7° \text{ or } 10 \angle -43.4°$$

6 Using Equation 7.23, calculate the *radius* of the input stability circle:

$$p_{s1} = \left| \frac{s_{12}s_{21}}{|s_{11}|^2 - |D_s|^2} \right|$$

$$= \left| \frac{(0.048 \angle 65°)(5.4 \angle 103°)}{(0.4)^2 - (0.429)^2} \right|$$

$$= 10.78$$

7 Using Equation 7.24, calculate the *centre location* of the *output* stability circle:

$$\Gamma_{s2} = \frac{C_2^*}{|s_{22}|^2 - |D_s|^2}$$

$$= \frac{(0.651 \angle 24.1°)}{(0.78)^2 - (0.429)^2}$$

$$= 1.534 \angle 24.1°$$

8 Using Equation 7.25, calculate the *radius* of the output stability circle:

$$p_{s2} = \left| \frac{s_{12}s_{21}}{|s_{22}|^2 - |D_s|^2} \right|$$

$$= \left| \frac{(0.048 \angle 65°)(5.4 \angle 103°)}{(0.78)^2 - (0.429)^2} \right|$$

$$= 0.611$$

These circles are shown in Figure 7.9. Note that the input stability circle is so large that it is actually drawn as a straight line on the Smith chart. Since s_{11} and s_{22} are both < 1, we can deduce that the *inside* of the *input* stability circle represents the region of stable source impedances, while the *outside* of the *output* stability circle represents the region of stable load impedances for the device.

The 12 dB gain circle is also shown plotted in Figure 7.9. It is found using Equations 7.1 and 7.16 to 7.20 in a manner similar to that given in Example 7.7. The centre location for the 12 dB gain circle is $\Gamma_O = 0.287 \angle 24°$. The radius for the 12 dB gain circle is p_o = 0.724.

The only load impedances we may not select for the transistor are located *inside* of the *output* stability circle. Any other load impedance located on the 12 dB gain circle will provide the needed gain provided the input of the transistor is conjugately matched and as long as the input impedance required for a conjugate match falls inside of the input stability circle.

A convenient point on the 12 dB gain circle will be selected and for this example, we choose

$$\Gamma_L = 0.89 \angle 70°$$

Using Equation 7.9 to calculate the source reflection coefficient for a conjugate match and plotting this point on the Smith chart

$$\Gamma_s = 0.678 \angle 79.4°$$

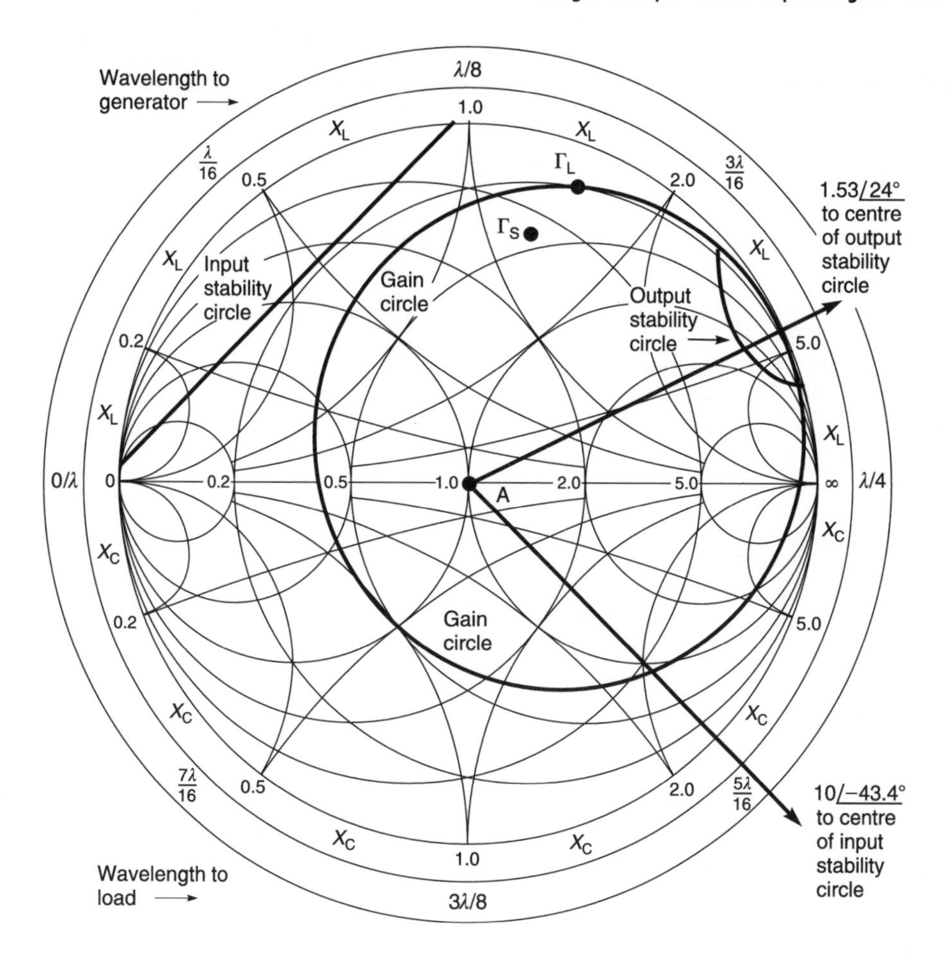

Fig. 7.9 Stability parameters A = 1 + j0, $\Gamma_L = 0.89 \angle 70°$, $\Gamma_S = 0.678 \angle 79.4°$, $\Gamma_o = 0.287 \angle 24°$

Notice that Γ_s falls within the stable region of the input stability circle and therefore represents a stable termination for the transistor. The input and output matching networks are then designed in the manner detailed in Example 7.5.

Example 7.8

In Example 7.7, the centre location for the 12 dB gain circle is given as $r_o = 0.287 \angle 24°$ and the radius for the 12 dB circle is given as $p_o = 0.724$. Show that these values are correct.

Given: Values of Example 7.7.
Required: r_o and p_o.

Solution. Bearing in mind that a power gain of 12 dB represents a power ratio of 15.85 and using Equation 7.18

$$G = \frac{\text{gain desired}}{|s_{21}|^2} = \frac{15.85}{5.4^2} = 0.544$$

Using Equation 7.16, find D_2 for subsequent insertion in Equation 7.19:

$$\begin{aligned} D_2 &= |s_{22}|^2 - |D_s|^2 \\ &= |0.78|^2 - |0.429|^2 \\ &= 0.424 \end{aligned}$$

Using Equation 7.19, find the centre of the constant gain circle:

$$r_o = \frac{GC_2^*}{1 + D_2G} = \frac{0.544(0.651 \angle 24.1°)}{1 + (0.424)(0.544)} = 0.288 \angle 24.1°$$

$$\approx 0.287 \angle 24°$$

Using Equation 7.20, find the radius for the 12 dB constant gain circle:

$$p_o = \frac{\sqrt{1 - 2K|s_{12}s_{21}|G + |s_{12}s_{21}|^2 G^2}}{1 + D_2G}$$

$$= \frac{\sqrt{1 - 2(0.802)(0.048)(5.4)(0.544) + (0.048 \times 5.4)^2 (0.544)^2}}{1 + (0.424)(0.544)}$$

$$= \frac{\sqrt{1 - 0.266 + 0.020}}{1 + 0.231}$$

$$= 0.724$$

Example 7.9

In Example 7.8, it is stated that Γ_s is 0.678 ∠ 79.4°. Show that this value is correct for a Γ_L of 0.89 ∠ 70°.

Given: Values of Example 7.8.
Required: Γ_s.

Solution. Using Equation 7.9

$$\Gamma_s = \left[s_{11} + \frac{s_{12}s_{21}\Gamma_L}{1 - s_{22}\Gamma_L} \right]^*$$

$$= \left[0.4 \angle 280° + \frac{(0.048 \angle 65°)(5.4 \angle 103°)(0.89 \angle 70°)}{1 - (0.78 \angle 345°)(0.89 \angle 70°)} \right]^*$$

$$= \left[(0.069 - j0.394) + \frac{(0.231 \angle 238°)}{(1 - 0.694 \angle 55°)} \right]^*$$

$$= \left[(0.069 - j0.394) + \frac{(0.231 \angle 238°)}{(0.828 \angle -43.36°)} \right]^*$$

$$= [(0.069 - j0.394) + (0.055 - j0.274)]^*$$

$$= [0.679 \angle -79.48°]^*$$

$$\approx 0.678 \angle 79.5°$$

7.5 Design of amplifiers for optimum noise figure

Many manufacturers specify optimum driving resistances and operating currents on their data sheets for their transistors to operate with minimum noise figures.[2] Designing amplifiers for a minimum noise figure then becomes simply a matter of setting the optimum conditions for a particular transistor. In practice, it means that the input network must be made to transform the input source generator impedance (generally 50 Ω) to that of the optimum driving resistance for the transistor to achieve its minimum noise operating conditions.

After providing the transistor with its optimum source impedance, the next step is to determine the optimum load reflection coefficient needed to properly terminate the transistor's output. This is given by:

$$\Gamma_L = \left[s_{22} + \frac{s_{12}s_{21}\Gamma_s}{1 - s_{11}\Gamma_s} \right]^* \tag{7.26}$$

where

Γ_s is the source reflection coefficient for minimum noise figure

The rest of the design then follows the conventional design methods as you will see in Example 7.10.

Example 7.10

The optimum source reflection coefficient (Γ_s) for a transistor under minimum noise figure operating conditions is $\Gamma_s = 0.68 \angle 142°$. Its *s*-parameters under the same conditions are:

$$s_{11} = 0.35 \angle 165° \qquad s_{12} = 0.035 \angle 58°$$
$$s_{21} = 5.9 \angle 66° \qquad s_{22} = 0.46 \angle -31°$$

The d.c parameters for the transistor are $V_{ce} = 15$ V, $I_c = 4$ mA, and the operating frequency is 300 MHz. Design a low noise amplifier to operate between a 75 Ω source and a 100 Ω load at 300 MHz.

[2] All devices produce electrical noise. In a transistor, noise figure is defined as the ratio of (signal-to-noise ratio at the input) to (signal-to-noise ratio at the output). It follows that if the transistor has a noise figure of 0 dB (power ratio = 1) then the signal-to-noise ratio at both the output and input remains the same and we are said to have a 'noise free' transistor. This does not happen in practice although at the time of writing noise figures of 0.5 dB are now being achieved.

Given: $s_{11} = 0.35 \angle 165°$ $s_{12} = 0.035 \angle 58°$ $s_{21} = 5.9 \angle 66°$ $s_{22} = 0.46 \angle -31°$
$f = 300$ MHz $V_{ce} = 12$ V $I_c = 4$ mA h_{FE} gain = 100
$\Gamma_s = 0.68 \angle 142°$ $R_L = 100\ \Omega$
Required: Low noise amplifier with $Z_s = 75\ \Omega$, $R_L = 100\ \Omega$.

Solution

1 Using Equation 7.1

$$\begin{aligned} D_s &= s_{11}s_{22} - s_{12}s_{21} \\ &= (0.35 \angle 165°)(0.46 \angle -31°) - (0.035 \angle 58°)(5.9 \angle 66°) \\ &= (-0.122 + j0.116) - (-0.115 + j0.171) \\ &= (0.056 \angle -86.25°) \end{aligned}$$

2 Using Equation 7.2, calculate Rollett's stability factor (K):

$$K = \frac{1 + |D_s|^2 - |s_{11}|^2 - |s_{22}|^2}{(2)(|s_{21}|)(|s_{12}|)}$$

$$= \frac{1 + (0.056)^2 - (0.35)^2 - (0.46)^2}{(2)(5.9)(0.035)}$$

$$= 1.620$$

The Rollett stability factor (K) calculates to be 1.62 which indicates unconditional stability. Therefore we may proceed with the design.

3 Input Matching Network
The design values of the matching network are shown in Figures 7.10 and 7.12. Here the normalised 75 Ω source resistance is transformed to Γ_s using two components:

Arc AB = shunt C = j1.65 S
Arc BC = series L = j0.85 Ω

4 Using Equation 7.12

$$C_1 = \frac{1.65}{2\pi(300 \text{ MHz})(50)} \ 17.5 \text{ pF}$$

5 Using Equation 7.13

$$L_1 = \frac{(0.85)(50)}{2\pi(300 \text{ MHz})} \approx 22.5 \text{ nH}$$

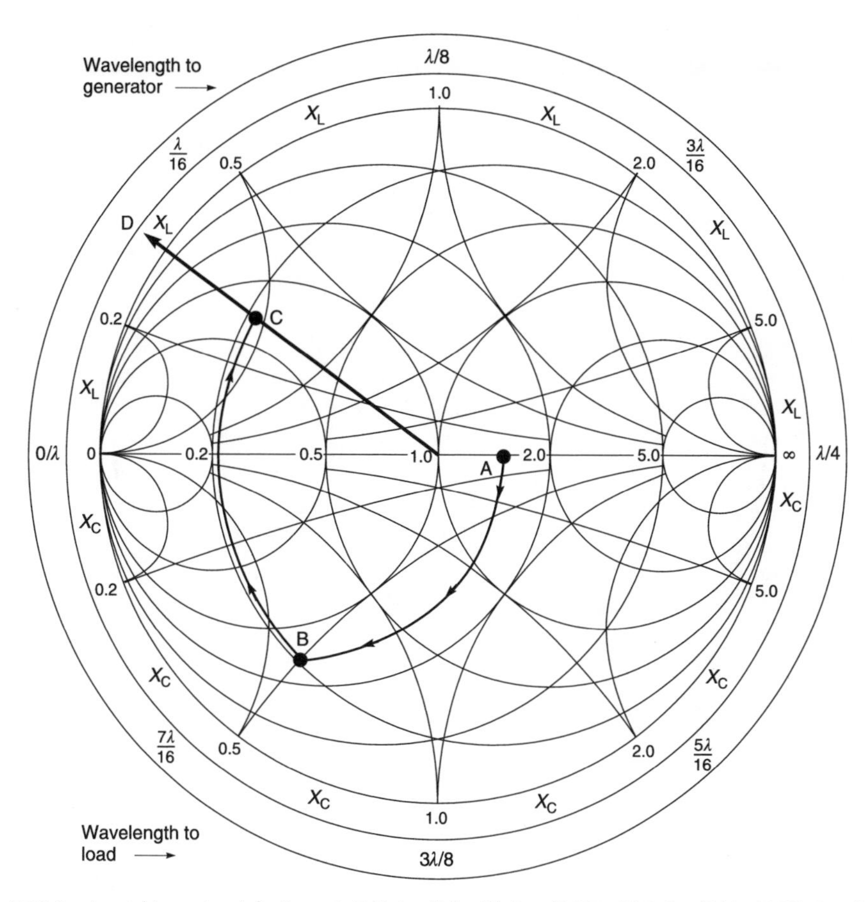

Fig. 7.10 Input matching network for Example 7.10 A = (1.5 + j0) Ω or (0.667 – j0) S, B = (0.21 – j0.52) Ω or (0.667 + j1.653) S, C = (0.212 + j0.33) Ω or (1.433 – j2.184) S

6 Output Matching Network

The load reflection coefficient needed to properly terminate the transistor is found from Equation 7.26:

$$\Gamma_L = \left[s_{22} + \frac{s_{12}s_{21}\Gamma_s}{1 - s_{11}\Gamma_s} \right]^*$$

$$= \left[0.46 \angle -31° + \frac{(0.035 \angle 58°)(5.9 \angle 66°)\,(0.68 \angle 142°)}{1 - (0.35 \angle 165°)(0.68 \angle 142°)} \right]^*$$

$$= \left[0.46 \angle -31° + \frac{(0.140 \angle 266°)}{1 - (0.143 - j0.190)} \right]^*$$

$$= [(0.394 - j0.237) + (-0.045 - j0.152)]^*$$

$$= 0.523 \angle 48.2°$$

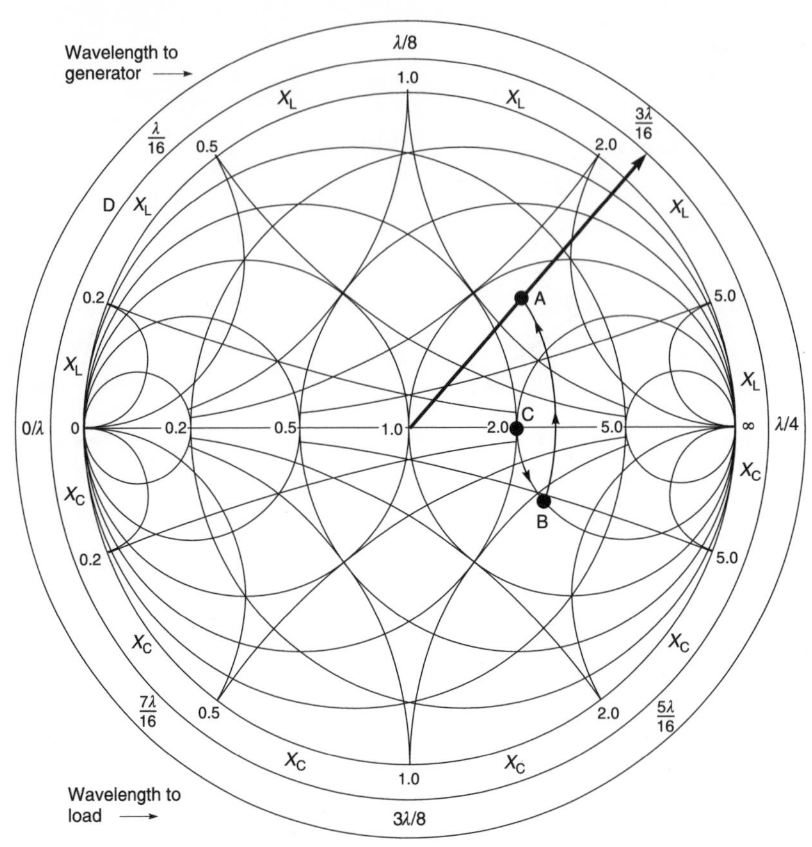

Fig. 7.11 Output matching network for Example 7.10 A = 1.252 + j1.234, B = 2.0 + j1.2, C = 2.0 + j0, D = 0.524 ∠–48.4°

This value along with the normalised load resistance value is plotted in Figure 7.11. The 100 Ω load must be transformed into Γ_L. One possible method is shown in Figure 7.11:

Arc AB = shunt L = –j0.72 S
Arc BC = series C = –j1.07 Ω

Using Equation 7.14, the inductor's value is

$$L_2 = \frac{50}{2\pi\ (300\ \text{MHz})\ (0.62)} \approx 43\ \text{nH}$$

Using Equation 7.11, the series capacitance is

$$C_2 = \frac{1}{2\pi\ (300\ \text{MHz})\ (1.2)\ (50)} \approx 8.8\ \text{pF}$$

The final design including a typical bias network is shown in Figure 7.12. The 0.1 μF capacitors are used only as bypass and coupling elements. The gain of the amplifier can be calculated with Equation 7.15.

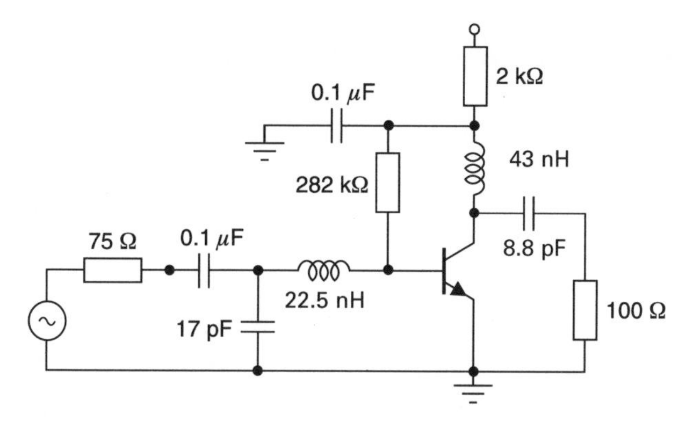

Fig. 7.12 Final circuit for Example 7.10

Using PUFF. Here again, you can use PUFF software to design or verify the amplifier circuit.

7.6 Design of broadband amplifiers

7.6.1 Design methods

There are many approaches to broadband amplifier design. We can use amplifier mismatching, feedback amplifiers and distributed amplifiers. We show how amplifier mismatching can be used in Example 7.11.

7.6.2 Broadband design using mismatch techniques

This method is explained and illustrated by Example 7.11.

Example 7.11

A broadband amplifier is to be designed to operate in the 1.5–2.5 GHz frequency range, with a 12 dB transducer power gain, using the HP-Avantek 41410 BJT. The *S*-parameters of the transistor at the operating range are shown in Table 7.1.

Solution. For the purposes of the example we will assume that $s_{12} \approx 0$ and therefore the unilateral case is considered. The expression for the transducer power gain in the unilateral case is given as Equation 7.27:

Table 7.1 Scattering parameters of HP-AVANTEK BJT

f (GHz)	s_{11}		s_{12}		s_{21}		s_{22}	
1.5	0.6	169°	0.04	58°	5.21	58°	0.41	–40°
2.0	0.6	157°	0.05	55°	3.94	55°	0.41	–45°
2.5	0.61	151°	0.06	55°	3.20	50°	0.4	–49°

$$G_{TU} = G_S G_O G_L \tag{7.27}$$

where

G_{TU} = gain of amplifier circuit
G_S = 'gain' of source network
G_O = gain of transistor
G_L = 'gain' of load network

$$G_S = \frac{1 - |\Gamma_S|^2}{|1 - S_{11}\Gamma_S|^2} \tag{7.28}$$

$$G_O = |S_{21}|^2 \tag{7.29}$$

and

$$G_L = \frac{1 - |\Gamma_L|^2}{|1 - S_{22}\Gamma_L|^2} \tag{7.30}$$

Table 7.1 shows that there is a considerable variation of *s*-parameters with frequency and the degree of variation can be calculated by Equations 7.27 to 7.30.

The circuit gain is given by Equation 7.27 and it is dependent on G_S, G_O and G_L. If the individual gains are calculated for a conjugate match at the input and output ports, we will get the results shown in Table 7.2. The maximum gain we can ever hope to achieve over the bandwidth 1.5–2.5 GHz is limited by the minimum gain of the three frequencies, i.e. 12.88 dB at 2.5 GHz. The other two frequencies (1.5 GHz and 2.0 GHz) show gains of 17.07 dB and 14.64 dB respectively but these gains can be reduced to 12.88 dB by mismatching of the ports. Hence, by designing for circuit losses, it is realistic to expect gains of approximately 12 dB over the three frequencies. Inspection of Table 7.2 reveals that little variation of G_L occurs with frequency. It is therefore easier to manipulate G_s to achieve the required controlled loss.

Table 7.3 shows the gain characteristic required for G_s to achieve an overall average gain of 12 dB over the frequency range. The input circuit should now be designed to produce the required response for G_s shown in Table 7.3. This is carried out by using

Table 7.2 Circuit gain vs frequency (when the input and output circuits are conjugately matched)

f (GHz)	$G_{S,max}$ (dB)	G_O (dB)	$G_{L,max}$ (dB)	G_{TU} (dB)
1.5	1.94	14.34	0.79	17.07
2.0	1.94	11.91	0.79	14.64
2.5	2.02	10.1	0.76	12.88

Table 7.3 Expected gains for an average overall gain of 12 dB

f (GHz)	$G_{S,max}$ (dB)	G_O (dB)	$G_{L,max}$ (dB)	G_{TU} (dB)
1.5	– 3.13	14.34	0.79	12
2.0	– 0.7	11.91	0.79	12
2.5	+1.14	10.1	0.76	12

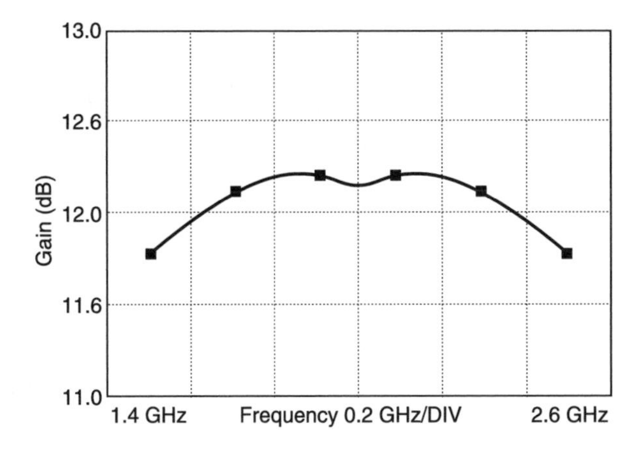

Fig. 7.13 Gain of the broadband amplifier

constant gain circles for the three frequencies and by choosing a network that will satisfy the response for G_s in Table 7.3. An exact response is not always possible and a compromise is often the case.

The process of design is one of trial and error and as such is greatly assisted by optimising software. As the necessary software is not available with this book, no attempt will be made to do the necessary input broadband matching. Instead we will simply look at the results to see what can be achieved after CAD matching.

In Figure 7.13, we show how the gain-frequency response of the amplifier has been improved after optimisation. The amplifier now has a nominal gain of 12 dB ± 0.25 dB over the band 1.5–2.5 GHz, instead of the original 4 dB gain fall-off in gain as calculated in Table 7.2.

This levelling of gain has been achieved by using a T network as the input matching circuit. This circuit is shown in Figure 7.14. However, the penalty paid for this levelling of gain is poor matching at the input circuit.

The return loss of the matching networks of this amplifier is shown in Figure 7.15. Note that the return loss of the input circuit is poor at 1.5 GHz (about 2 dB) but gradually improves towards about 11 dB at 2.5 GHz. The return loss of the output circuit is more even, and ranges from 7.5 dB at 1.5 GHz to about 5 dB at 2.5 GHz.

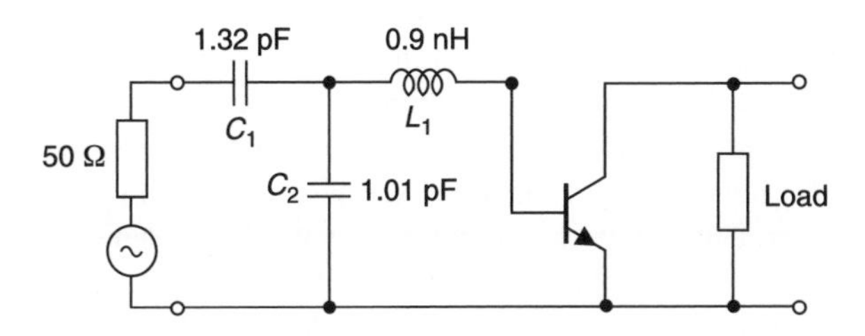

Fig. 7.14 Circuit diagram of the broadband amplifier

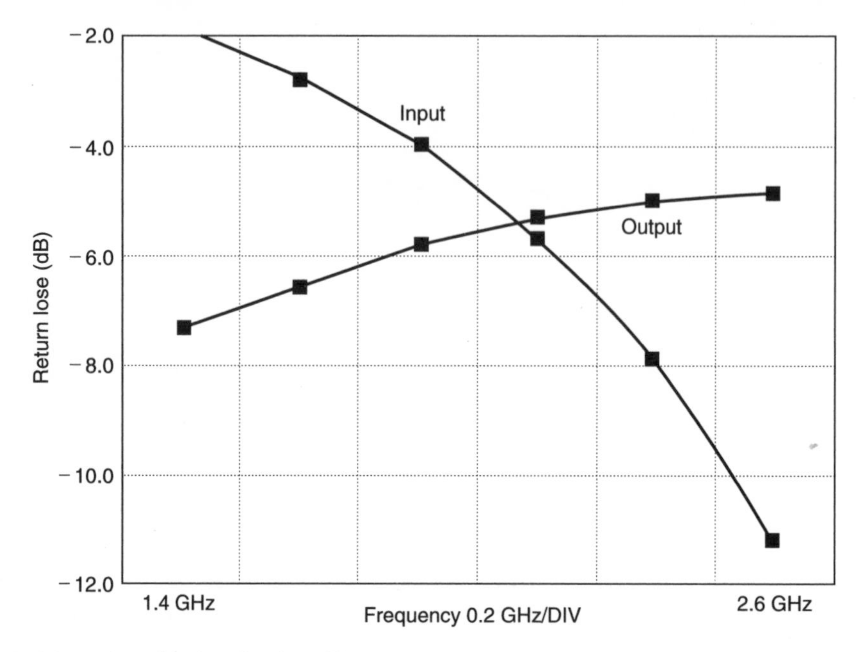

Fig. 7.15 Return loss of the broadband amplifier

7.7 Feedback amplifiers

7.7.1 Introduction

R.F. feedback amplifiers are used in much the same way as feedback elements are introduced in operational amplifier circuits to produce constant gain over a desired bandwidth. In this section we shall show you how these amplifiers can be designed. Feedback amplifiers are usually designed by first decomposing the combined circuit into individual subsystems. They are then re-combined into a composite amplifier and its parameters are then calculated to yield the desired results.

7.7.2 Design of feedback amplifiers

Consider the basic feedback circuit shown in Figure 7.16. If you look at it closely, you will find that it consists of two basic parts; the feedback circuit which comprises R_{FB}, L_{FB} and its d.c. blocking capacitor C_{FB} situated between points A and B, and the transistor circuit and inductor L_D which is also situated between the same two points. Since both circuits are in parallel, we can draw them as shown in Figure 7.17. For this example, we will make considerable use of Y-parameters which were originally introduced in Chapter 6.

In Figure 7.17(a), Y_{FB} now represents the feedback network, R_{FB}, L_{FB} and its d.c. blocking capacitor C_{FB} situated between points A and B. Block Y_A represents the amplifier and the inductor L_D. Each network is subject to the same voltage across its terminals; therefore it follows that the currents of each network can be added together to form a composite network Y_C. This is also shown diagramatically in Figure 7.17(b). From the composite Y_C

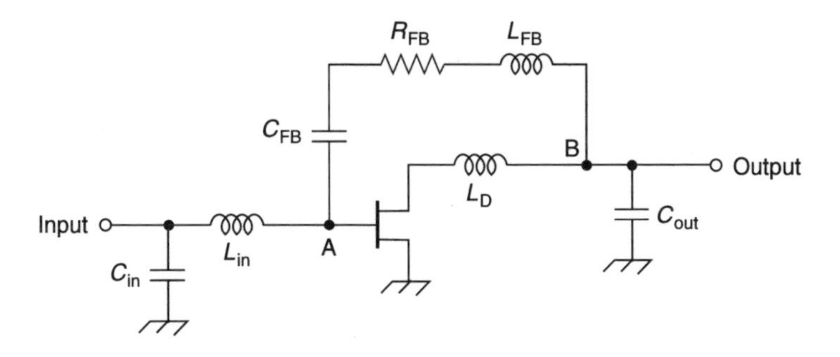

Fig. 7.16 A radio frequency feedback amplifier

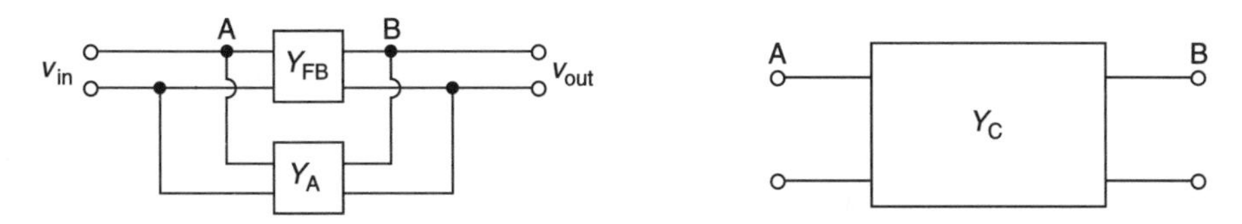

Fig. 7.17 Block diagram of the feedback amplifier of Figure 7.16: (a) composite sections, Y_A and Y_{FB}; (b) combined network, Y_C

network, it is now possible to calculate the circuit gain, input and output admittance as a single circuit in Y-parameters or if you wish you may change them into s-parameters and carry out the calculations using s-parameters. The conversion tables for changing from one type of parameter to another are given in Table 3.1. Hence, the parameters of the circuit can be evaluated using the system above.

To clarify the design method, we will show you a very simple example where this technique is used. We will assume that the values for the circuit of Figure 7.18 have already been chosen. Furthermore in order to simplify matters, we will assume that the values are already in Y-parameters and that only resistances are used in the network. The last assumption simplifies the mathematics considerably yet it does not obscure the principles which we are trying to use.

Example 7.12

In the circuit of Figure 7.18, the open-circuit generator voltage is 200 mV. Calculate (a) the input impedance (Z_{in}), (b) the gain (A_v) of the circuit and (c) V_{out}. You can assume that the d.c. blocking capacitor in the feedback chain has negligible reactance. The transistor Y-parameters for the given frequency of operation are:

$$\begin{bmatrix} 1/1200 & 0 \\ 70/1200 & 1/40\,000 \end{bmatrix}$$

Solution. The Y-parameters of the transistor Y_A are:

$$Y_A = \begin{bmatrix} 1/1200 & 0 \\ 70/1200 & 1/40\,000 \end{bmatrix} \text{S} = \begin{bmatrix} 833.3 & 0 \\ 58\,333.3 & 25 \end{bmatrix} \mu\text{S}$$

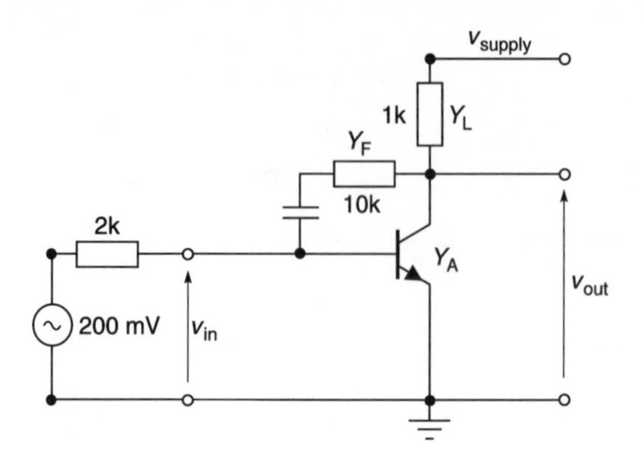

Fig. 7.18 Negative feedback amplifier

From inspection of the circuit, the Y-parameters of the feedback element Y_F are:

$$Y_F = \begin{bmatrix} 1/10\text{k} & -1/10\text{k} \\ -1/10\text{k} & 1/10\text{k} \end{bmatrix} = \begin{bmatrix} 100 & -100 \\ -100 & 100 \end{bmatrix} \mu\text{S}$$

The composite admittance matrix $[Y_C] = [Y_A] + [Y_F]$. Hence

$$[Y_C] = \begin{bmatrix} 833.3 & 0 \\ 58\,333.3 & 25 \end{bmatrix} \mu\text{S} + \begin{bmatrix} 100 & -100 \\ -100 & 100 \end{bmatrix} \mu\text{S}$$

$$= \begin{bmatrix} 933.3 & -100 \\ 58\,233.3 & 125 \end{bmatrix} \mu\text{S}$$

We will now obtain the answers.

(a) Defining $[\Delta_y]$ as $[y_{11}y_{22} - y_{12}y_{21}]$

$$[\Delta_y] = [y_{11}y_{22} - y_{12}y_{21}] = [9.33 \times 1.25 + 1 \times 582.3] \times 10^{-8} = 593.96 \times 10^{-8}$$

Using Equation 6.19:

$$y_{in} = y_{11} - \frac{y_{12}y_{21}}{y_{22} + y_L} = \frac{\Delta y + y_{11}Y_L}{y_{22} + Y_L}$$

and

$$Z_{in} = \frac{y_{22} + Y_L}{\Delta y + y_{11}Y_L} = \frac{[1.25 + 10] \times 10^{-4}}{[593.96 + 9.33 \times 10] \times 10^{-8}}$$

$$= \frac{11.25 \times 10^{-4}}{687.26 \times 10^{-8}} = 163.9\ \Omega$$

(b) First find v_{in}:

$$v_{in} = \frac{Z_{in} V_g}{Z_{in} + Z_g} = \frac{163.69 \times 200}{2163.7} = 15.13 \text{ mV}$$

Using Equation 6.18:

$$A_v = \frac{v_{out}}{v_{in}} = \frac{-y_{21}}{y_{22} + y_L} = \frac{-582.3}{1.25 + 10}$$

$$= \frac{-582.3}{11.25} = -51.76$$

(c) Output voltage is given by

$$\begin{aligned} v_{out} &= v_{in} \times A_v \\ &= 15.13 \times [-51.76] \\ &= -783 \text{ mV} \end{aligned}$$

PUFF results. If you wish, you can carry out this example on PUFF by converting the transistor admittance parameters into scattering parameters and generating a transistor device as described in Section 4.9. You can then insert your feedback components and vary them accordingly.

7.7.3 Summary of feedback amplifiers

Example 7.12 should now convince you that the procedure used above is useful for evaluating feedback amplifiers. However, this method is laborious especially without the use of a computer program.

The disadvantage of this method is that each block, Y_A, Y_F and Y_C, is only applicable for one frequency at one time. Thus, if you were designing a broadband circuit, you would have to calculate the parameters for each frequency and then sum up the results. This involves considerable work if hand calculators are used.

Another great disadvantage is that the component values that you may have chosen in the first instance may not produce the desired result. Therefore you must carry out the complete procedure again and again until the desired result is achieved. However, it is fortunate that good computer programs, such as 'SUPERCOMPACT, SPICE, etc.', provide optimisation facilities and allow you to design the circuit quickly and efficiently.

7.8 R.F. power transistors

The design of r.f. power transistors is treated differently from that of the low power linear transistors described in the early part of this chapter. The reason for this is because r.f. power transistors are normally operated in a non-linear mode. This means that manufacturers tend to only specify output power and output capacitance for a given input power and input capacitance. A typical example is shown in Table 7.4 where values of input

Table 7.4 Typical optimum input and conjugate of load impedances for MRF658. P_{out}= 65 W, V_{dc} = 12.5 V

Frequency (MHz)	Z_{in} Ω	Z_{out} Ω
400	0.620 + j0.28	1.2 + j2.5
440	0.720 + 0j3.1	1.1 + j2.8
470	0.790 + 0j3.3	0.98 + j3.0
490	0.84 + j3.4	0.91 + j3.2
512	0.88 + j3.5	0.84 + j3.3
520	0.90 + j3.6	0.80 + j3.4

impedance and conjugate of load impedances are specified for a given output power and operating d.c. voltage. However, once these values are known, the matching networks are designed in a similar way. You can find a good introduction to the design of power amplifiers by consulting Baeten.[3]

7.9 Summary

Many of the commonly used techniques in amplifier design have been covered in this chapter. The circuit topics discussed included transistor stability, maximum available gain and matching techniques. In addition, we produced design examples of conjugate matched amplifiers, conditionally stable transistor amplifiers, optimum noise figure amplifiers, and amplifiers designed for a specific gain.

The design techniques of broadband amplifiers, feedback amplifiers and power amplifiers were investigated. We also showed how some designs can be carried out using the PUFF software supplied with this book.

You should now have a good knowledge of microwave engineering principles that will allow you to do simple amplifier designs and to understand more complicated devices and circuits.

I would like to remind you of the article 'Practical Circuit Design' which has been reproduced on the disk supplied with this book. This article provides many more examples of how PUFF can be used in practical circuit design. There are particularly interesting sections on components, earthing techniques, biasing, passive, active and circuit layout techniques. The article is crowned by the complete design of a 5 GHz microwave amplifier from its conception as transistor data to final layout. R.F. design calculations are carried out in Appendix A, input and output line matching and layout using PUFF are shown. Frequency response and gain are checked with PUFF. Bias design for this amplifier is also given in Appendix B. Calculations for the design of input and output matching filters are shown in Appendix C.

[3] R. Baeten, CAD of a broadband class C 65 watt UHF amplifier, *RF Design*, March 1993, 132–9.

8

Oscillators and frequency synthesizers

8.1 Introduction

Prior to the invention of an amplifying device (vacuum tube, transistor, special negative-resistance device, etc.) great difficulty was experienced in producing an undamped radio signal. The early radio transmitters used a high frequency a.c. generator to produce a high voltage which was increased by a step-up transformer. The output voltage was applied to a series resonant circuit and the Q of the circuit produced a high enough voltage to jump across a 'capacitor gap' to produce a spark[1] which in turn produced a radio signal. This principle is still used in a petrol engine today where the spark is used to ignite the petrol mixture. You can frequently hear it on your car radio when the engine cover is removed. Such a radio signal is damped, i.e. it decays exponentially and it produces many harmonics which interfere with other communication systems. It is now illegal to transmit a damped oscillation.

There are many criteria in choosing an oscillator, but the main ones are:

- frequency stability
- amplitude stability
- low noise
- low power consumption
- size.

Frequency stability is important because it enables narrow-band communication systems to be accurately fixed within a frequency band. An unstable frequency oscillator also behaves like an unstable f.m. modulator and produces unwanted f.m. noise. An amplitude unstable oscillator behaves like an amplitude modulated modulator because it produces unwanted a.m. modulation noise. Even if the oscillation frequency and amplitude can be held precisely, it is inevitable that noise will be produced in an oscillator because of transistor noise which includes 'flicker noise', 'shot noise' and '1/frequency' noise. In other words, oscillator noise is inevitable but it should be kept as low as possible. Low power consumption and small size are specially important in portable equipment.

[1] This is the reason why radio operators are often called sparkies.

8.1.1 Aim

The aims of this chapter are to explain radio frequency oscillators and frequency synthesizers. Radio frequency oscillators produce radio frequency signals without an input signal. Frequency synthesizers are used to control and vary the frequency of an oscillator very precisely.

8.1.2 Objectives

After reading this chapter, you should be able to:

- understand the criteria for oscillation
- calculate the criteria (gain and frequency) of
- Hartley oscillators
- Colpitts oscillators
- Clapp oscillators
- crystal oscillators
- voltage controlled oscillators
- phase locked loops
- frequency synthesizers

8.2 Sine wave type oscillators

An oscillator is a device which produces an output signal without requiring an external input signal. An amplifier can be made into an oscillator if its output signal is fed back into its own input terminals to provide an input signal of the correct amplitude and phase.

One easy way of producing an r.f. oscillator is to use an r.f. amplifier and to feed its output signal (with the correct amplitude and phase) back to its input. Figure 8.1 shows a typical common emitter r.f. amplifier. The important things to note about this circuit are its waveforms. V_{in} is the input sinusoidal applied to the amplifier. V_{tc} is the inverted voltage appearing across the collector, and V_{out} is the voltage appearing across the output. The phasor relationship between V_{tc} and V_{out} is dependent on the manner in which the secondary winding of T_1 is connected. In Figure 8.1, the output winding has been earthed in a manner that will cause V_{out} to appear with a similar phase to V_{in}.

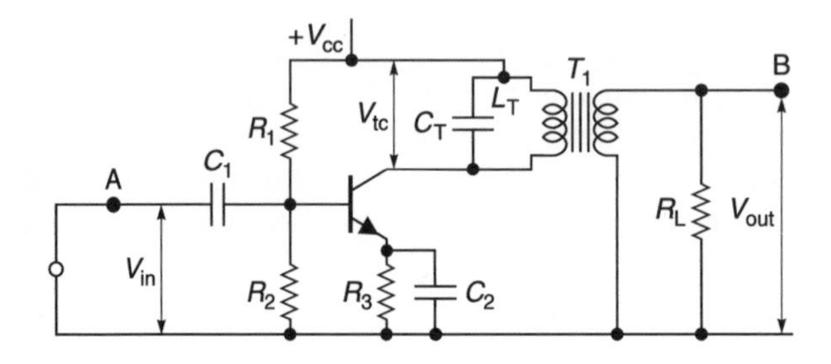

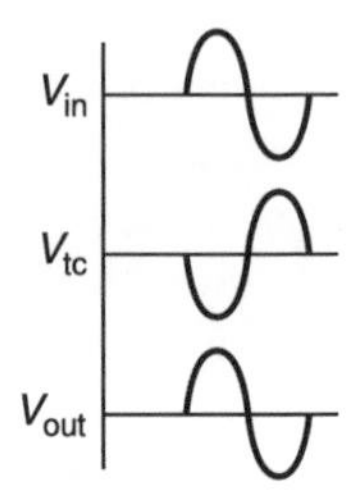

Fig. 8.1 An r.f. amplifier with associated waveforms

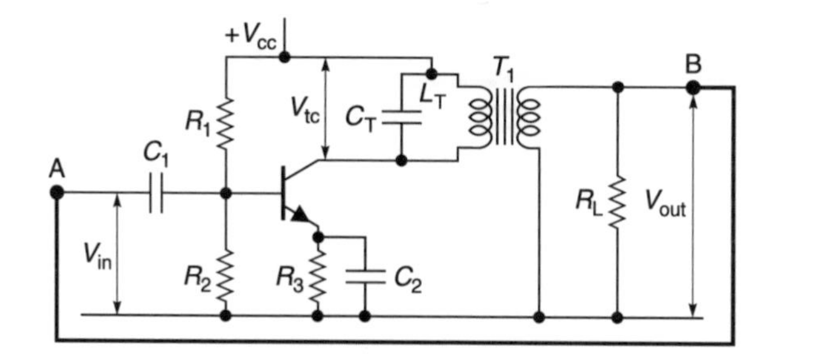

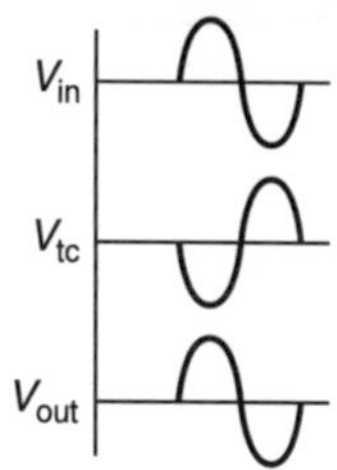

Fig. 8.2 An r.f. oscillator constructed by feeding output to input

As V_{in} and V_{out} both have similar phases, there is no reason why V_{out} cannot be connected back to the amplifier input to supply its own input voltage. This is shown schematically in Figure 8.2 where a connection (thick line) has been made between points A and B. Examination of these waveforms shows clearly that if, in addition to the phase requirements, $V_{out} > V_{in}$, then the amplifier will supply its own input and no external signal source will be needed. Therefore the amplifier will produce an output on its own and will become an oscillator!

One question still remains unanswered. How do we produce V_{out} in the first instance without an external V_{in}? Any operating amplifier produces inherent wideband noise which contains an almost infinite number of frequencies. The collector tuned circuit selects only its resonant frequency for amplification and rejects all other frequencies; therefore only the resonant frequency of the tuned circuit will appear as V_{out}. Initially, V_{out} will probably have insufficient amplitude to cause oscillation but as it is fed back around again and again to the amplifier input terminals, V_{in} will increase in amplitude and, if the circuit has been designed properly, V_{in} will soon be large enough to cause oscillation.

Example 8.1

The tuned circuit of the oscillator circuit shown in Figure 8.2 has an effective inductance of 630 nH and a total capacitance (C_T) of 400 pF. If conditions are set so that oscillations can take place, what is its frequency of oscillation (f_{osc})?

Solution. In Figure 8.2, the frequency of oscillation is determined by the resonant frequency of the tuned circuit. For the values given

$$f_{osc} = \frac{1}{2\pi\sqrt{LC}} = \frac{1}{2\pi\sqrt{630\ \text{nH} \times 400\ \text{pF}}} = 10.026\ \text{MHz}$$

8.2.1 Barkhausen criteria

The introduction to oscillators above was to provide you with an elementary idea of oscillator requirements. To design oscillators, we need a more systematic method. Consider an amplifier with a positive feedback loop (Figure 8.3).

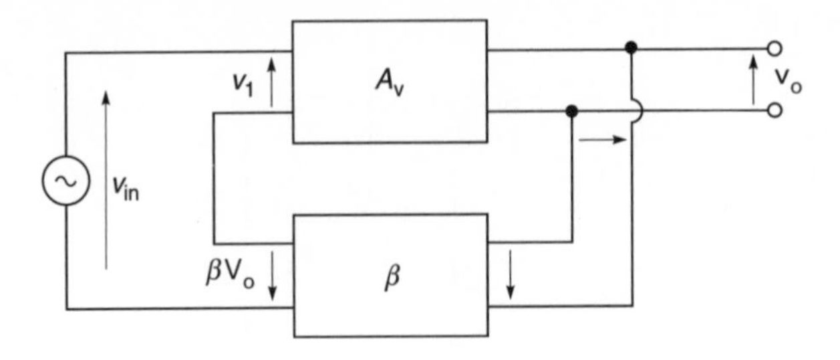

Fig. 8.3 An amplifier with positive feedback signal

The following terms are defined.
Voltage gain (A_v) of the amplifier on its own is defined as

$$A_v = \frac{v_o}{v_1} \tag{8.1}$$

Voltage gain (A_{vf}) of the amplifier with feedback applied is

$$A_{vf} = \frac{v_o}{v_{in}} \tag{8.2}$$

By inspection of Figure 8.3

$$\beta = \text{output voltage fraction fed back} \tag{8.3}$$

and

$$v_1 = v_{in} + \beta v_o \tag{8.4}$$

or

$$v_{in} = v_1 - \beta v_o \tag{8.4a}$$

Using Equations 8.2 and 8.4a and dividing each term by v_1

$$A_{vf} = \frac{v_o}{v_1 - \beta v_o} = \frac{v_o/v_1}{1 - \beta v_o/v_1}$$

Therefore

$$A_{vf} = \frac{A_v}{1 - A_v\beta} \tag{8.5}$$

If β is positive, and if $A_v\beta$ (defined as loop gain) = 1, then the denominator $(1–A_v\beta) = 0$, or

$$A_v\beta = 1 \tag{8.6}$$

Substituting Equation 8.6 into Equation 8.5 yields

$$A_{vf} = \frac{v_o}{0} = \infty \tag{8.7}$$

which means that there is an output (v_o) in spite of there being no input signal. This system is known as an oscillator.

The two main requirements for oscillation are:

$$\text{loop gain amplitude } (A_v\beta) = 1 \quad (8.8)$$

and

$$\text{loop gain phase} = 0° \text{ or } n360° \qquad n \text{ is any integer} \quad (8.9)$$

Equations 8.8 and 8.9 are known as the **Barkhausen criteria** for oscillation.

Note: $A_v\beta = 1$ is the minimum condition for oscillation. If $A_v\beta > 1$, it merely means that the oscillation will start more easily but then, due to non-linearity in the amplifier, $A_v\beta$ will revert back to 1.

For ease of understanding the above explanation, I have assumed that there is no phase change in A_v and β. In practice, A_v is a phasor quantity and if it produces a phase shift of, say, 170°, then β must produce a complementary phase shift of 190° to make the total phase shift of the signal feedback equal to 360° (or any multiple of it). This enables the returned feedback (input) signal to be in the correct phase to aid oscillation.

8.2.2 Summary

The Barkhausen criteria state that for an oscillator, the loop gain ($A_v\beta$) must equal unity and the loop gain phase must be 0° of any integer multiple of 360°.

It follows that if the Barkhausen criteria can be met then any amplifier may be made into an oscillator. It is relatively easy to calculate the conditions required for oscillation, but it is important to realise that when oscillation occurs, linear theory no longer applies because the transistor is no longer working in its linear mode. In the discussion that follows, we will show how (i) the conditions for oscillation and (ii) the desired frequency of oscillation may be achieved with various circuits.

8.3 Low frequency sine wave oscillators

At frequencies less than about 2 MHz, oscillators are often made using resistances and capacitances as the frequency determining elements instead of LC circuits. This is because at these frequencies, LC elements are physically larger, more expensive, and more difficult to control in production. Two main types of RC oscillators will be previewed. One is the well known **Wien bridge** oscillator which is used extensively in instruments and the other is the **Phase-Shift** oscillator.

8.4 Wien bridge oscillator

A block diagram of the Wien bridge oscillator is shown in Figure 8.4. The basic parts of this oscillator consist of a **non-inverting** amplifier[2] and an RC network which determines its frequency of operation.

[2] This non-inverting amplifier can consist of either two common emitter amplifiers in cascade (one following the other) or a common base or operational amplifier.

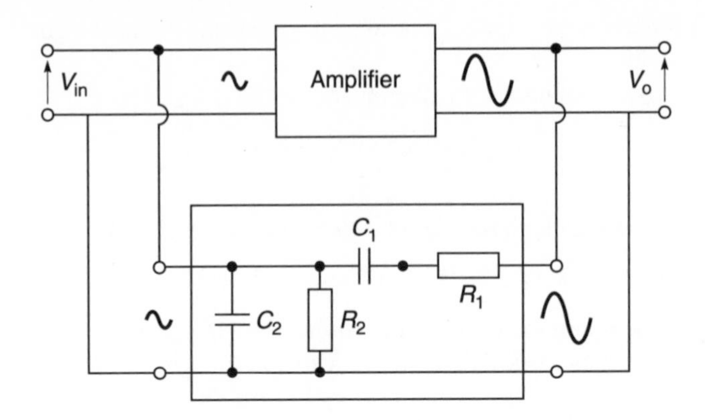

Fig. 8.4 Wien bridge oscillator

8.4.1 Operation

When power is applied to the circuit, currents (including inherent noise currents and voltages) appear in the amplifier. This noise voltage is fed back through the Wien (RC network) back to the input of the amplifier. The circuit is designed to allow sufficient feedback voltage to satisfy the Barkhausen criterion on loop gain ($A_v\beta = 1$), but only noise frequencies which satisfy the second Barkhausen criterion ($\angle A_v\beta = n360°$) will cause the oscillation. The circuit will oscillate at a frequency $\omega^2 = 1/(R_1R_2C_1C_2)$ radians per second.

8.4.2 Wien bridge oscillator analysis

In this analysis, it is assumed that the amplifier does not load the Wien bridge network shown in Figure 8.5.

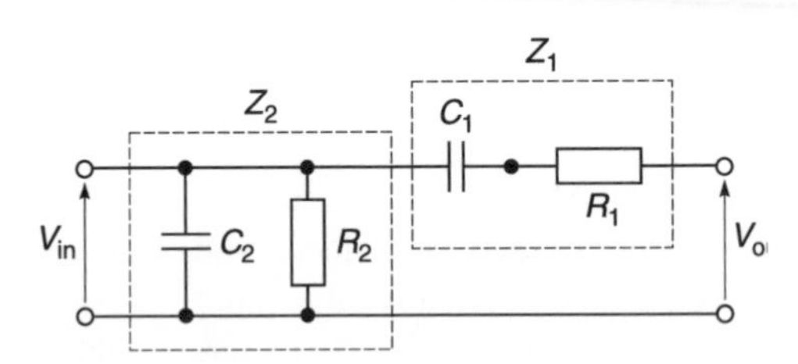

Fig. 8.5 Wien bridge network

By inspection

$$Z_1 = R_1 + 1/(j\omega C_1)$$

$$Z_2 = \frac{1}{Y_2} = \frac{1}{1/R_2 + j\omega C_2} = \frac{R_2}{1 + j\omega C_2R_2}$$

and

$$v_{in} = \frac{Z_2}{Z_1 + Z_2}(v_o)$$

Transposing

$$\frac{v_o}{v_{in}} = \frac{Z_1 + Z_2}{Z_2} = 1 + Z_1Y_2$$

Substituting for Z_1 and Y_2

$$\frac{v_o}{v_{in}} = 1 + \frac{(R_1 + 1/j\omega C_1)(1 + j\omega C_2R_2)}{R_2}$$

Multiplying out and sorting the real and imaginary terms

$$\frac{v_o}{v_{in}} = 1 + \frac{R_1}{R_2} + \frac{C_2}{C_1} + j(\omega C_2R_1 - 1/(\omega C_1R_2)) \tag{8.10}$$

For the phase to equal zero, the quadrature or j terms = 0 which gives

$$\omega C_2R_1 = 1/(\omega C_1R_2)$$

Transposing

$$\omega^2 = \frac{1}{C_1C_2R_1R_2} \tag{8.11}$$

From Equation 8.10, the real part of the equation indicates that the gain of the amplifier (A_v) at the oscillation frequency must be

$$A_v = \frac{|v_o|}{|v_{in}|} = 1 + \frac{R_1}{R_2} + \frac{C_2}{C_1} \tag{8.12}$$

From Figure 8.4, the fraction of the voltage fed-back (β) = $|v_{in}|/|v_{out}|$. From Equation 8.10

$$\beta = \frac{|v_{in}|}{|v_{out}|} = \frac{1}{1 + R_1/R_2 + C_2/C_1} \tag{8.13}$$

For oscillation, the Barkhausen criteria is $A_v\beta = 1$. Using Equations 8.12 and 8.13

$$A_v\beta = \frac{|v_o|}{|v_{in}|} \times \frac{|v_{in}|}{|v_o|}$$

$$= (1 + R_1/R_2 + C_2/C_1) \times \frac{1}{(1 + R_1/R_2 + C_2/C_1)} = 1$$

Therefore the Barkhausen gain criteria are satisfied and the circuit will oscillate.

8.4.3 Practical Wien bridge oscillator circuits

In practical designs and for reasons of economy, variable frequency Wien bridge oscillators use twin-gang[3] variable capacitors ($C_1 = C_2 = C$) or twin-gang resistors ($R_1 = R_2 = R$) to vary the oscillation frequency.

For the case where $C_1 = C_2 = C$ and $R_1 = R_2 = R$, Equations 8.11 and 8.12 become

$$\omega = \frac{1}{CR} \text{ rads s}^{-1} \tag{8.11a}$$

and

$$A_v = \frac{|v_o|}{|v_{in}|} = 1 + 1 + 1 = 3 \tag{8.12a}$$

Example 8.2

In the Wien bridge oscillator of Figure 8.4, $R_1 = 100$ kΩ, $R_2 = 10$ kΩ, $C_1 = 10$ nF and $C_2 = 100$ nF. Calculate (a) the frequency of oscillation and (b) the minimum gain of the amplifier for oscillation.

Solution

(a) Using Equation 8.11

$$\omega^2 = 1/(100 \text{ k}\Omega \times 10 \text{ k}\Omega \times 10 \text{ nF} \times 100 \text{ nF}) = 1\,000\,000$$
$$\omega = 1000 \text{ rads}^{-1}$$

and

$$f_{osc} = 159.15 \text{ Hz}$$

(b) Using Equation 8.12, the minimum gain of the amplifier is

$$\frac{|v_o|}{|v_{in}|} = 1 + \frac{R_1}{R_2} + \frac{C_2}{C_1}$$

$$= 1 + \frac{100 \text{ k}\Omega}{10 \text{ k}\Omega} + \frac{100 \text{ nF}}{10 \text{ nF}} = 1 + 10 + 10 = 21$$

8.5 Phase shift oscillators

8.5.1 Introduction

The circuit of a phase shift oscillator is shown in Figure 8.6. In this circuit, an inverting amplifier (180° phase shift) is used. To feed the signal back in the correct phase, RC

[3] Twin-gang capacitors or resistors are variable elements whose values are changed by the same rotating shaft.

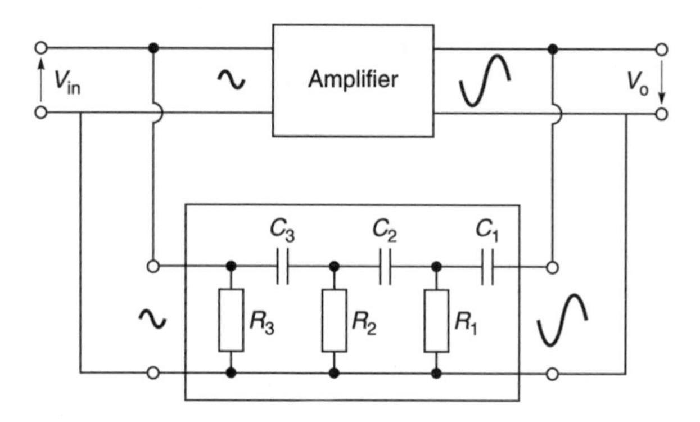

Fig. 8.6 Phase shift oscillator

networks are used to produce an additional nominal 180° phase shift. The theoretical maximum phase shift for one RC section is 90° but this is not easily obtained in practice so three RC stages are used to produce the required phase shift. The transmission (gain or loss) analysis of a three section RC circuit can be difficult unless some simplifying methods are employed. To do this, I shall use matrix methods and assume that you are familiar with matrix addition, subtraction, multiplication and division.

8.5.2 Analysis of the phase shift network

In the analysis that follows, it is assumed that the input and output impedances of the transistor are sufficiently large so that they do not load the phase shifting network. It can be shown[4] that the two port transmission matrix for a series impedance (Z) is:

$$\mathbf{Z} = \begin{bmatrix} 1 & Z \\ 0 & 1 \end{bmatrix} \tag{8.14}$$

It can also be shown[5] that the two port transmission matrix for a shunt admittance (Y) is:

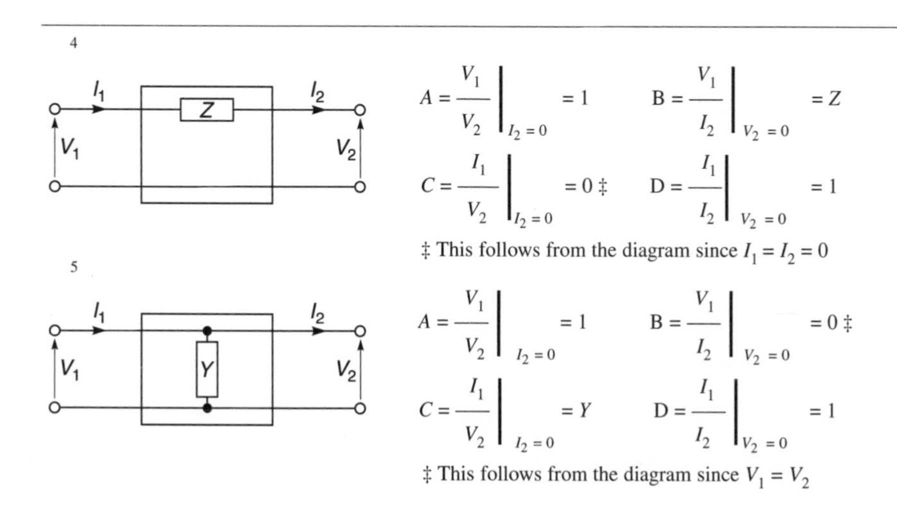

$$\boldsymbol{Y} = \begin{bmatrix} 1 & 0 \\ Y & 1 \end{bmatrix} \tag{8.15}$$

The transmission parameters for the six element network shown in Figure 8.7 can be easily obtained by multiplying out the matrices of the individual components. I will simplify the arithmetic by making $Z_1 = Z_2 = Z_3 = Z$ and $Y_1 = Y_2 = Y_3 = Y$. I have also drawn v_o and v_{in} in the conventional manner but the analysis will show that the amplifier gain must be inverted. By inspection of Figure 8.7

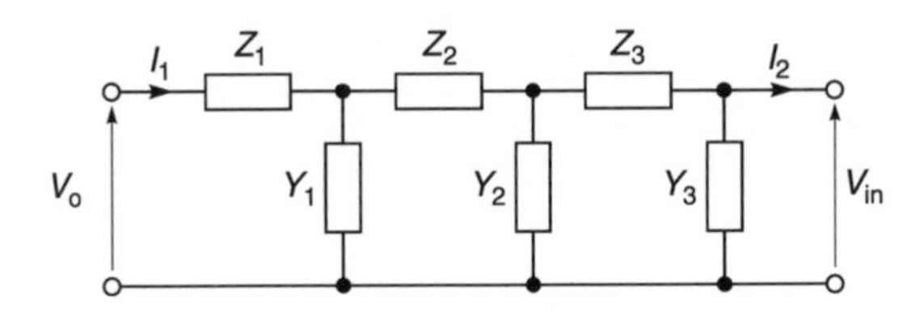

Fig. 8.7 Six element network

$$\begin{bmatrix} v_o \\ I_1 \end{bmatrix} = \begin{bmatrix} 1 & Z \\ 0 & 1 \end{bmatrix} \begin{bmatrix} 1 & 0 \\ Y & 1 \end{bmatrix} \begin{bmatrix} 1 & Z \\ 0 & 1 \end{bmatrix} \begin{bmatrix} 1 & 0 \\ Y & 1 \end{bmatrix} \begin{bmatrix} 1 & Z \\ 0 & 1 \end{bmatrix} \begin{bmatrix} 1 & 0 \\ Y & 1 \end{bmatrix} \begin{bmatrix} v_{in} \\ I_2 \end{bmatrix}$$

$$\begin{bmatrix} v_o \\ I_1 \end{bmatrix} = \begin{bmatrix} 1+YZ & Z \\ Y & 1 \end{bmatrix} \begin{bmatrix} 1+YZ & Z \\ Y & 1 \end{bmatrix} \begin{bmatrix} 1+YZ & Z \\ Y & 1 \end{bmatrix} \begin{bmatrix} v_{in} \\ I_2 \end{bmatrix}$$

$$\begin{bmatrix} v_o \\ I_1 \end{bmatrix} = \begin{bmatrix} 1+YZ & Z \\ Y & 1 \end{bmatrix} \begin{bmatrix} 1+3YZ+Y^2Z^2 & YZ^2+2Z \\ Y^2Z+2Y & YZ+1 \end{bmatrix} \begin{bmatrix} v_{in} \\ I_2 \end{bmatrix}$$

$$\begin{bmatrix} v_o \\ I_1 \end{bmatrix} = \begin{bmatrix} Y^3Z^3+5Y^2Z^2+6YZ+1 & Y^2Z^3+4YZ^2+3Z \\ Y^3Z^2+4Y^2Z+3Y & Y^2Z^2+3YZ+1 \end{bmatrix} \begin{bmatrix} v_{in} \\ I_2 \end{bmatrix}$$

Since we have assumed that the transistor impedance does not load the circuit, $I_2 = 0$. Therefore

$$\left. \frac{v_o}{v_{in}} \right|_{I_2=0} = A = Y^3Z^3 + 5Y^2Z^2 + 6YZ + 1$$

Substituting for Y and Z

$$\left. \frac{v_o}{v_{in}} \right|_{I_2=0} = \frac{1}{(j\omega CR)^3} + \frac{5}{(j\omega CR)^2} + \frac{6}{(j\omega CR)} + 1 \tag{8.16}$$

Sorting out real and imaginary terms

$$\left. \frac{v_o}{v_{in}} \right|_{I_2=0} = \frac{1}{(j\omega CR)^3} + \frac{5}{(j\omega CR)^2} + \frac{6}{(j\omega CR)} + 1$$

$$= \left[1 - \frac{5}{(\omega CR)^2}\right] + \mathrm{j}\left[\frac{1}{(\omega CR)^3} - \frac{6}{\omega CR}\right] \tag{18.6a}$$

[real part] [imaginary part]

The Barkhausen criterion for oscillation is that the voltage through the network must undergo a phase change of 180°, i.e. imaginary or quadrature terms are zero. Therefore,

$$\frac{-6}{\omega CR} + \frac{1}{(\omega CR)^3} = 0$$

Hence

$$6 = \frac{1}{(\omega CR)^2} \quad \text{and} \quad \omega^2 = \frac{1}{6(CR)^2} \tag{8.17}$$

The real part of Equation 8.16a at resonance is:

$$\frac{v_o}{v_{in}} = \frac{-5}{(\omega CR)^2} + 1 \tag{8.18}$$

and using Equation 8.17 to substitute for $1/(\omega CR)^2$

$$\frac{v_o}{v_{in}} = -5 \times 6 + 1 = -29 = 29 \angle 180°$$

Since $v_o/v_{in} = A_v$

$$A_v = 29 \angle 180° \tag{8.19}$$

and

$$\frac{|v_o|}{|v_{in}|} = 29 \tag{8.19a}$$

From Figure 8.6, the fraction of the voltage fed back $\beta = |v_{in}|/|v_o|$. Using Equation 8.19

$$\beta = \frac{|v_{in}|}{|v_o|} = \frac{1}{29} \tag{8.20}$$

To check for oscillation at resonance, the Barkhausen criterion is $A_v\beta = 1$. Using Equations 8.19a and 8.20

$$A_v\beta = \frac{|v_o|}{|v_{in}|} \times \frac{|v_{in}|}{|v_o|} = 29 \times \frac{1}{29} = 1$$

Therefore the Barkhausen criteria are met and the circuit will oscillate.

8.6 Radio frequency (LC) oscillators

8.6.1 Introduction

Oscillators operating at frequencies greater than 500 kHz tend to use inductors and capacitors as their frequency controlling elements because:

- RC values are beginning to get inconveniently small;
- LC values are beginning to assume practical and economical values.

8.6.2 General analysis of (LC) oscillators

In the analysis of the oscillators within this section, it must be realised that all calculations to establish the conditions and frequency of oscillation are based on linear theory. When oscillation occurs, the transistor no longer operates in a linear mode and some modification (particularly bias) is inevitable.

The simplified solutions derived for each type of oscillator are based on the following assumptions.

- The input impedance of the transistor does not load the feedback circuit.
- The output impedance of the transistor does not load the feedback circuit.
- The collector–emitter voltage (V_{CE}) is the output voltage.
- The emitter–base voltage (V_{EB}) is the input voltage.
- The feedback circuit is purely reactive (no resistive losses).
- If the transistor is operated in the common emitter configuration, a positive base input voltage will result in an inverted collector voltage. If the transistor is operated in the common base configuration, there is zero phase shift through the transistor.
- There is 0 or 2π radians (360°) shift through the loop gain circuit. In the case of a common emitter amplifier, if π radians (180°) shift is caused by the transistor when its collector load is resistive and/or resonant, then a further π radians shift will be required in the feedback circuit to return the feedback signal in the correct phase. With a common-base amplifier, if 0 radians phase shift is produced by the transistor, then zero phase shift through the feedback network is required to return the feedback signal in the correct phase.
- For clarity, oscillator outputs are not shown in Figures 8.8, 8.9 and 8.10. Outputs are taken from either the collector or emitter via capacitance coupling or magnetic coupling from the inductor.

The above assumptions are *approximately* true in practice and form a reasonably accurate starting point for oscillator design.

8.7 Colpitts oscillator

A schematic diagram of the Colpitts oscillator circuit is shown in Figure 8.8. Two capacitors, C_1 and C_2, are connected in series to provide a divider network for the voltage developed across points C and B. The tuned circuit is formed by the series equivalent capacitance of C_1 and C_2 and the inductor L.

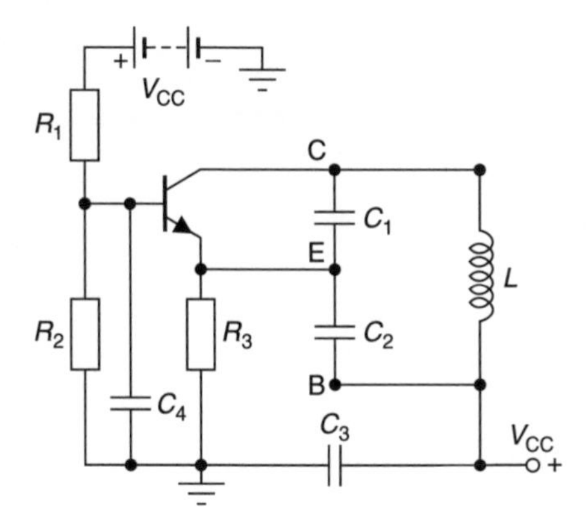

Fig. 8.8 Colpitts oscillator

The transistor is operated in the common base configuration. The base is a.c. earthed via C_4. The point B is a.c. earthed through capacitor C_3. If the voltage at point C is positive with respect to earth, then point E is also positive with respect to earth. Hence the transistor supplies its own input voltage in the correct phase.

8.7.1 Frequency of oscillation

Using the assumptions of Section 8.6.2 and since we assume that there is no loading of the tuned circuit at resonance, $X_L = X_C$, yielding

$$\frac{1}{\omega C_1} + \frac{1}{\omega C_2} = \omega L$$

$$\omega^2(L) = \frac{1}{C_1} + \frac{1}{C_2}$$

and

$$\omega^2 = \frac{1}{L}\left[\frac{1}{C_1} + \frac{1}{C_2}\right] \tag{8.21}$$

8.7.2 Conditions for oscillation

From Figure 8.8 and using the assumptions of Section 8.6.2

$$V_{EB} = v_{in} \quad \text{and} \quad V_{CB} = v_o$$

By inspection of Figure 8.8

$$v_{in} = \frac{X_{c2}}{X_{c1} + X_{c2}} v_o \tag{8.22}$$

By definition, $\beta = v_{in}/v_o$, therefore

$$\beta = \frac{X_{c2}}{X_{c1} + X_{c2}} \tag{8.23}$$

By definition, $A_v = v_o/v_{in}$ and using Equation 8.22

$$A_v = \frac{X_{c1} + X_{c2}}{X_{c2}} = \frac{C_2}{C_1} + 1 \quad \text{or} \quad 1 + \frac{C_2}{C_1} \tag{8.24}$$

For oscillation, we must satisfy the Barkhausen criteria and ensure that the loop gain $A_v\beta$ = 1. Using Equations 8.23 and 8.24

$$A_v\beta = \frac{X_{c1} + X_{c2}}{X_{c2}} \times \frac{X_{c2}}{X_{c1} + X_{c2}} = 1$$

Therefore the circuit will oscillate provided

$$A_v \geq \left[1 + \frac{C_2}{C_1}\right] \tag{8.25}$$

Summing up Equation 8.21 determines the frequency of oscillation and Equation 8.25 determines the minimum gain of the amplifier for oscillation.

Example 8.3

If in Figure 8.8 C_1 = 10 pF and C_2 = 100 pF and the desired oscillation frequency is 100 MHz, calculate (a) the value of the inductor and (b) the minimum voltage gain of the amplifier. Assume that the transistor does not load the tuned circuit.

Solution

(a) From Equation 8.21

$$\omega^2 = \frac{1}{L}\left[\frac{1}{C_1} + \frac{1}{C_2}\right]$$

Transposing and substituting for C_1 and C_2

$$L = \frac{1}{(2\pi \times 100\ \text{MHz})^2}\left[\frac{1}{10\ \text{pF}} + \frac{1}{100\ \text{pF}}\right]\ \text{H}$$

$$= \frac{1}{3.948 \times 10^{17}}\left[\frac{11 \times 10^{12}\ \text{F}}{100}\right]\ \text{H} = 278.6\ \text{nH}$$

(b) From Equation 8.24 the minimum voltage gain of the amplifier is

$$\frac{C_2}{C_1} = 1 + \frac{100\ \text{pF}}{10\ \text{pF}} = 11$$

8.8 Hartley oscillator

A schematic diagram of the Hartley oscillator circuit is shown in Figure 8.9. Note that the Hartley circuit is the dual of the Colpitts circuit where inductors and capacitors have been interchanged. Inductor L in conjunction with capacitor C forms the tuned circuit. Inductor L also serves as an auto-transformer. In an auto-transformer, the voltage developed[6] across points E and B is proportional to the number of turns (n_2) between E and B. Similarly, the voltage developed across points C and B is proportional to the number of turns ($n_1 + n_2$) between points C and B.

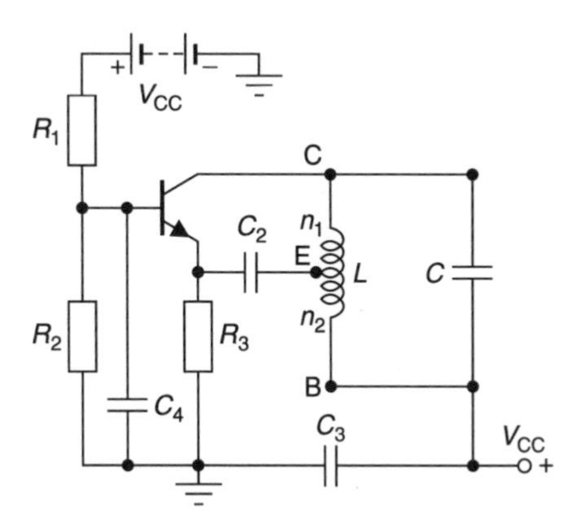

Fig. 8.9 Hartley oscillator

The transistor is operated in the common base configuration. The base is a.c. earthed via capacitor C_4. The point B is a.c. earthed through capacitor C_3. C_2 is a d.c. blocking capacitor. If the voltage at point C is positive with respect to earth, then point E is also positive with respect to earth. Hence the transistor supplies its own input voltage in the correct phase.

8.8.1 Frequency of oscillation

Using the assumptions of Section 8.6.2 and since we assume that there is no loading of the tuned circuit at resonance, $X_L = X_C$, yielding

$$\omega L = \frac{1}{\omega C}$$

and

$$\omega^2 = \frac{1}{LC} \tag{8.26}$$

8.8.2 Conditions for oscillation

From Figure 8.9 and using the assumptions of Section 8.6.2

$$V_{EB} = v_{in} \quad \text{and} \quad V_{CB} = v_o$$

[6] This assumes that the same flux embraces both parts of the auto-transformer.

From Figure 8.9, since inductor L serves as an auto-transformer

$$v_{in} = \frac{n_2}{n_1 + n_2} v_o \tag{8.27}$$

By definition, $\beta = v_{in}/v_o$. Therefore

$$\beta = \frac{n_2}{n_1 + n_2} \tag{8.28}$$

By definition, $A_v = v_o/v_{in}$ and using Equation 8.27

$$A_v = \frac{n_1 + n_2}{n_2} = \frac{n_1}{n_2} + 1 \quad \text{or} \quad 1 + \frac{n_1}{n_2} \tag{8.29}$$

For oscillation, we must satisfy the Barkhausen criteria and ensure that the loop gain $A_v \beta = 1$. Using Equations 8.28 and 8.29

$$A_v\beta = \frac{n_1 + n_2}{n_2} \times \frac{n_2}{n_1 + n_2} = 1$$

Therefore the circuit will oscillate provided

$$A_v \geq \left[1 + \frac{n_1}{n_2}\right] \tag{8.30}$$

Summing up Equation 8.26 determines the frequency of oscillation and Equation 8.30 determines the minimum gain of the amplifier.

8.9 Clapp oscillator

A schematic diagram of the Clapp oscillator circuit is shown in Figure 8.10. Two capacitors, C_1 and C_2, are connected in series to provide a divider network for the voltage developed across points C and B. The tuned circuit is formed by the equivalent series capacitance of C_1, C_2 and C_T and the inductor L. The Clapp oscillator is a later development of the Colpitts oscillator except that an additional capacitance C_T has been added to improve frequency stability and facilitate design.

The transistor is operated in the common base configuration. An r.f. choke is used to feed d.c. power to the collector. The reactance of the r.f. choke is made deliberately high so that it does not shunt the tuned circuit. The base is a.c. earthed via capacitor C_3. The point B is earthed directly. If the voltage at point C is positive with respect to earth, then point E is also positive with respect to earth. Hence the transistor supplies its own input voltage in the correct phase.

8.9.1 Frequency of oscillation

Using the assumptions of Section 8.6.2 and since we assume that there is no loading of the tuned circuit at resonance, $X_L = X_C$, yielding

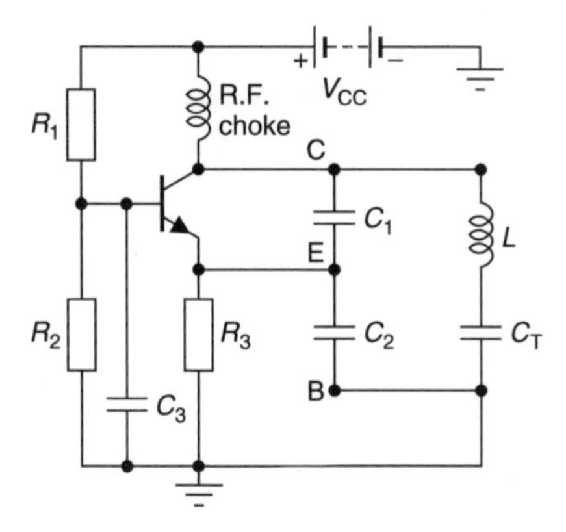

Fig. 8.10 Clapp oscillator

$$\omega L = \frac{1}{\omega C_1} + \frac{1}{\omega C_2} + \frac{1}{\omega C_T}$$

$$\omega^2 L = \frac{1}{C_1} + \frac{1}{C_2} + \frac{1}{C_T}$$

and

$$\omega^2 = \frac{1}{LC_T}\left[1 + \frac{C_T}{C_1} + \frac{C_T}{C_2}\right] \tag{8.31}$$

If $[C_T/C_1 + C_T/C_2] << 1$, then Equation 8.31 becomes

$$\omega^2 = \frac{1}{LC_T} \tag{8.31a}$$

Equation 8.31a is the preferred mode of operation for the Clapp oscillator for the following reasons.

- It allows C_T and L to be the main contributors for determining the oscillation frequency. This is particularly useful when the oscillation is to be set to another frequency because only one control is needed. In many cases, C_T is a varactor (capacitance diode) whose capacitance can be changed electronically by applying a d.c. control voltage. This is particularly usefully in crystal oscillators which we shall be describing shortly.
- It provides freedom for setting C_1 and C_2 to get the required values for easy oscillation. C_1 and C_2 can be made reasonably large provided their ratio remains the same.
- Larger values of C_1 and C_2 help to swamp transistor inter-electrode capacitances which change with operating bias and temperature.

8.9.2 Conditions for oscillation

From Figure 8.10 and using the assumptions of Section 8.6.2

$$V_{EB} = v_{in} \quad \text{and} \quad V_{CB} = v_o$$

By inspection of Figure 8.10

$$v_{in} = \frac{X_{c2}}{X_{c1} + X_{c2}} v_o \tag{8.32}$$

By definition, $\beta = v_{in}/v_o$. Therefore

$$\beta = \frac{X_{c2}}{X_{c1} + X_{c2}} \tag{8.33}$$

By definition, $A_v = v_o/v_{in}$ and using Equation 8.32

$$A_v = \frac{X_{c1} + X_{c2}}{X_{c2}} = \frac{C_2}{C_1} + 1 \quad \text{or} \quad 1 + \frac{C_2}{C_1} \tag{8.34}$$

For oscillation, we must satisfy the Barkhausen criteria and ensure that the loop gain $A_v\beta$ = 1. Using Equations 8.33 and 8.34

$$A_v\beta = \frac{X_{c1} + X_{c2}}{X_{c2}} \times \frac{X_{c2}}{X_{c1} + X_{c2}} = 1$$

Therefore the circuit will oscillate provided

$$A_v \geq \left[1 + \frac{C_2}{C_1}\right] \tag{8.35}$$

Summing up Equation 8.31 determines the frequency of oscillation and Equation 8.35 determines the minimum gain of the amplifier for oscillation.

Example 8.4

Calculate the approximate frequency of oscillation for the Clapp oscillator circuit of Figure 8.10 when C_T = 15 pF, C_1 = 47 pF, C_2 = 100 pF and L = 300 nH.

Solution. Using Equation 8.31

$$\omega^2 = \frac{1}{LC_T}\left[1 + \frac{C_T}{C_1} + \frac{C_T}{C_2}\right] = \frac{1}{300\text{ nH} \times 15\text{ pF}}\left[1 + \frac{15}{47} + \frac{15}{100}\right]$$

$$= 2.222 \times 10^{17} \times 1.469 = 3.264 \times 10^{17}$$

Hence

$$\omega = 571\,314\,274.3 \text{ radians/s}$$

and

$$f_{osc} = 90.93 \text{ MHz}$$

An alternative approach is to calculate the combined series capacitance of the circuit:

$$C_{total} = [1/15 \text{ pF} + 1/47 \text{ pF} + 1/100 \text{ pF}]^{-1} = 10.21 \text{ pF}$$

and

$$f_{osc} = \frac{1}{\sqrt{300 \text{ nH} \times 10.21 \text{ pF}}} = 90.93 \text{ MHz}$$

8.10 Voltage-controlled oscillator

A voltage-controlled oscillator is shown in Figure 8.11(a). If you examine this circuit, you will find that it is almost identical to that of the Clapp oscillator (Figure 8.10). The exception is that C_T has been replaced by C_V which is a variable capacitance diode or varactor. The voltage across the varactor and hence its capacitance is controlled by varying the varactor voltage. C_3 is a d.c. blocking capacitor used to isolate the varactor voltage from the collector voltage.

All the conditions relating to the Clapp oscillator apply here except that C_T in all the equations must be replaced by C_v.

8.10.1 Frequency of oscillation

The frequency of oscillation is that calculated by Equation 8.31 except that C_T must be replaced by C_v.

8.10.2 Oscillator gain

If you plotted the frequency of oscillation against the varactor voltage, you would get a curve similar to that of Figure 8.11(b). The frequency sensitivity or frequency gain of the oscillator is defined as

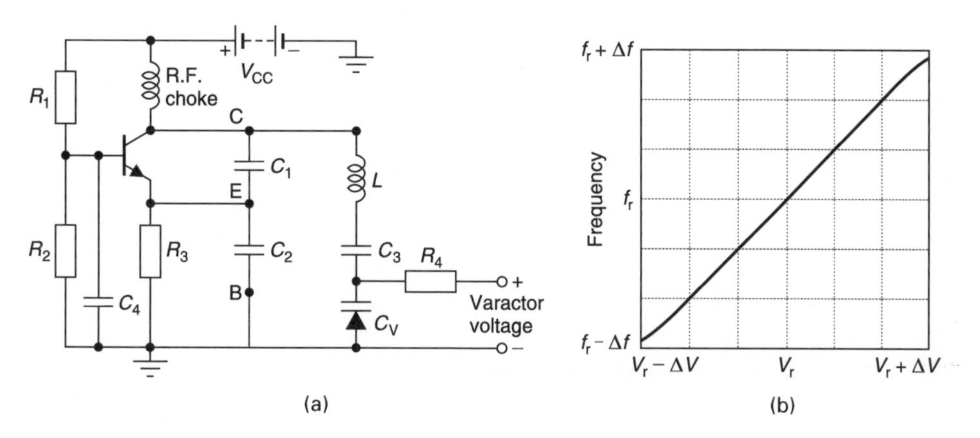

Fig. 8.11 (a) Voltage-controlled oscillator; (b) oscillator frequency (f_r) vs varactor (V_r)

$$k_o = \frac{\Delta f}{\Delta V} \tag{8.36}$$

Equation 8.36 is important because it tells us what voltage must be applied to the varactor to alter the oscillator frequency. In this case, the sensitivity is positive because the slope $\Delta f/\Delta v$ is positive.

However, if the varactor in Figure 8.11(a) is connected in the opposite direction then a negative voltage would be needed for control and, in this case, the sensitivity slope will be negative. We will be returning to the question of frequency sensitivity when we discuss phase locked loops.

Summing up Equation 8.31 determines the frequency of oscillation when you substitute C_v for C_T. Equation 8.35 determines the minimum gain of the amplifier for oscillation. Equation 8.36 describes the frequency sensitivity or frequency gain of the oscillator.

8.11 Comparison of the Hartley, Colpitts, Clapp and voltage-controlled oscillators

- The Hartley oscillator is very popular and is used extensively in low powered oscillators where the inductor value can be increased by winding the coil on ferrite cored material. It is used extensively as the local oscillator in domestic superhet radio receivers.
- The Colpitts oscillator is used in cases where a piezo-electric crystal is used in place of the inductor.
- The Clapp oscillator finds favour in electronically controlled circuits. It is more frequency stable than the other two oscillators and can be easily adapted for crystal control oscillators.
- The voltage-controlled oscillator is ideal for varying the frequency of an oscillator electronically. This is particularly true in phase locked loops and frequency synthesizers which will be explained shortly.

8.12 Crystal control oscillators

8.12.1 Crystals

Crystals are electromechanical circuits made from thin plates of quartz crystal or lead–zirconate–titanate. They are sometimes used in place of LCR-tuned circuits. The electrical symbol for a crystal[7] and its equivalent circuit are shown in Figure 8.12.

In this circuit, the capacitance between the connecting plates is represented by C_o, while L and C represent the electrical effect of the vibrating plate's mass stiffness, and R represents the effect of damping. The circuit has two main resonant modes; one when L, C and R are in series resonance and the other at a frequency slightly above the series resonant frequency when the total series combination is inductive and resonates with C_o to form a parallel resonant circuit. Engineers can use either of these resonant modes in their designs.

[7] It is common practice to abbreviate the term *crystal resonator* to *xtal.*

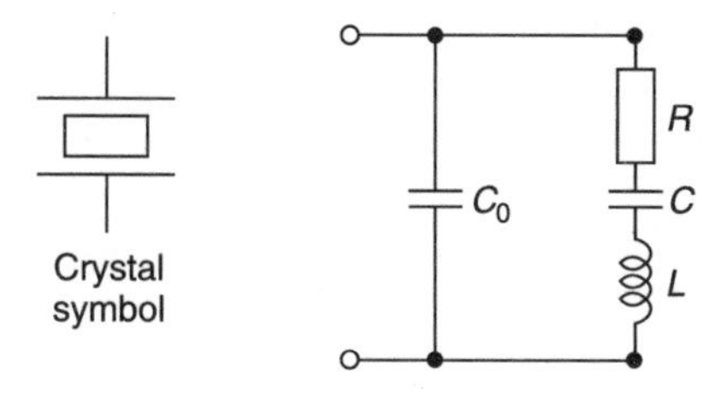

Fig. 8.12 Electrical equivalent circuit of a crystal resonator

Quartz and lead–zirconate–titanate are piezo-electric, i.e. they vibrate mechanically when an electrical signal is applied to them and vice-versa. The resonant frequencies at which they vibrate are dictated by their geometrical sizes, mainly plate thickness and angle of crystal cut with respect to the main electrical axes of the crystal. Crystals are normally cut to give the correct frequency in a specified oscillator circuit at a given temperature, nominally 15°C.

Plate thickness reduces with frequency and at the higher frequencies, plate thickness becomes so thin that the crystal is fragile. To avoid this condition and still operate at frequencies up to 200 MHz, manufacturers often resort to **overtone** operation where they cut the crystal to a lower frequency but mount it in such a manner that it operates at a higher harmonic. Overtone crystals which operate at *third*, *fifth* and *seventh* overtones are common.

The angle of cut determines the temperature stability of the crystal. Typical types of cuts are the AT cut, BT cut and SC cut. The frequency stabilities of these cuts against temperature are shown in Figure 8.13. The AT cut is popular because it is reasonably easy and cheap to make. It has a positive temperature coefficient (frequency increases) when the ambient temperature changes outside its design (turnover) temperature. The BT cut crystal has a negative temperature coefficient when the temperature is outside its turnover

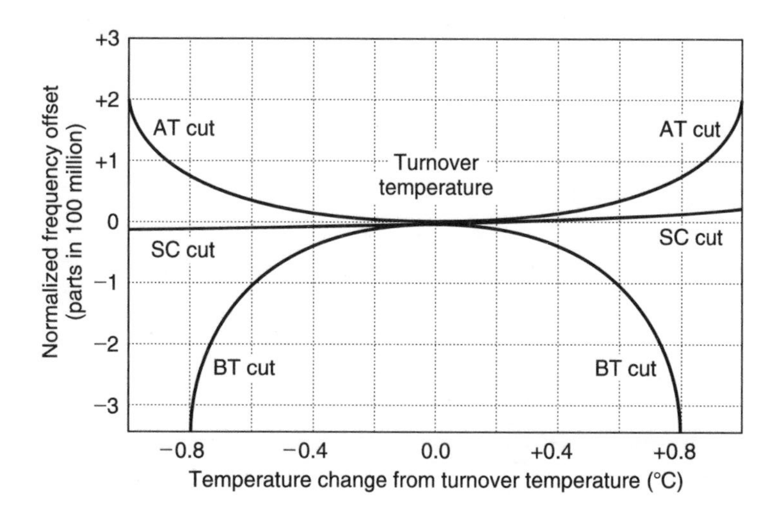

Fig. 8.13 Frequency stability against temperature

temperature. The SC cut crystal has an almost zero temperature coefficient but this crystal is difficult to make because it is first cut at an angle with respect to one axis and then rotated and cut again at a second angle to another axis.

From Figure 8.13, you should also note that changes of frequency for the crystal itself against temperature are very small. To put the matter in perspective, 1 part in 10^8 is equivalent to about 1 Hz in 100 MHz, but remember in a practical circuit, there are other parts (transistors, external capacitors, etc.) which will affect frequency as well.

Crystal resonators manifest very high Qs and values of Qs greater than 100 000 are common in 10 MHz crystals. Metallic plates are used to make electrical connections to the piezo material and the whole assembly is usually enclosed in a hermetically sealed can or glass bulb to minimise oxidation and the ingress of contaminants.

8.12.2 Crystal controlled oscillator circuits

Crystal oscillators are particularly useful because of their frequency stability and low noise properties. Figure 8.14 shows how a crystal is used in its parallel mode as an inductor in a Colpitts oscillator circuit. When used in this manner, it is sometimes called a Pierce oscillator. Such a circuit has the advantage of quick frequency changes by simply switching in different crystals. The resonance frequency of oscillation is determined by the equivalent circuit capacitance across the crystal and the equivalent inductance of the crystal. To order such a crystal from the manufacturer, it is essential to tell the manufacturer the model of the crystal required, its mode of operation (series or parallel), operating frequency and parallel capacitance loading. Typical capacitance loads are 10 pF, 30 pF and/or 50 pF.

Figure 8.15 shows how a crystal is used in its series resonance mode as an inductor in a Clapp type oscillator.

Crystal oscillators are used because they offer vastly superior frequency stability when compared with LC circuits. As crystal Qs are very high, crystal oscillators tend to produce much less noise than LC types. Typical values of stabilities and frequency ranges offered

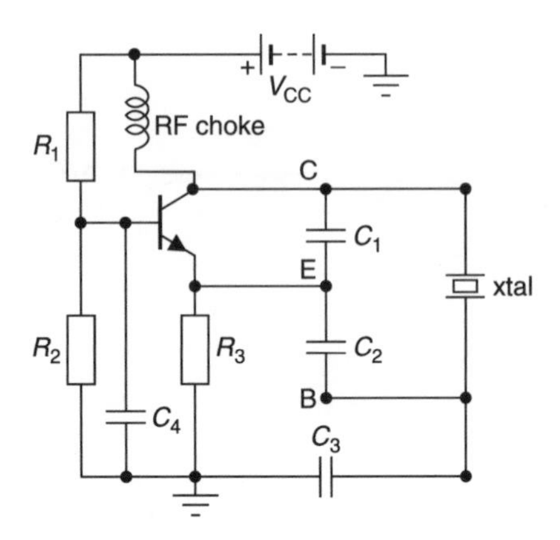

Fig. 8.14 Colpitts xtal oscillator

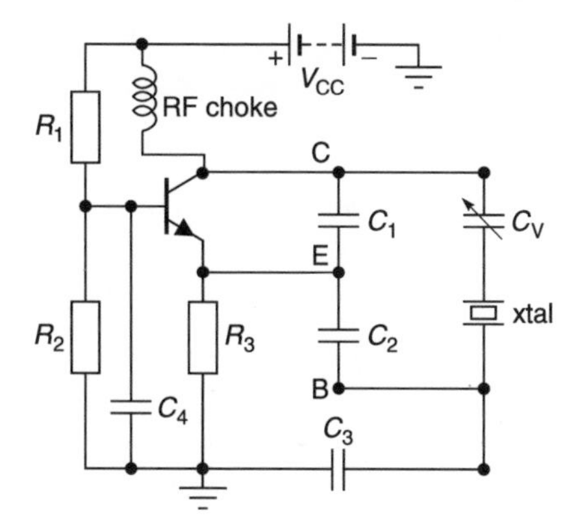

Fig. 8.15 Clapp xtal oscillator

by crystal oscillators are given in Table 8.1. Explanatory terms for some abbreviations are given below the table. Cathodeon and OSA are company names.

Table 8.1

Oscillator type	VCO	OCXO	TCXO (analogue)	TCXO (digital)
Type no	Cathodeon FS 5909	Cathodeon FS 5951	Cathodeon FS 5805	OSA DTCXO 8500
Frequency range	5–20 MHz	300kHz–40 MHz	5–15 MHz	1–20 MHz
Output	TTL	TTL	TTL or sine	Sine
Temperature range (°C) and stability	–25 to 80°C ±50 ppm on set frequency	0–60°C ±0.1 ppm or –40 to 70°C ±0.2 ppm	–10 to 55°C ±1 ppm or –20 to 70°C ±2 ppm	–40 to 85°C 0.5 ppm or –20 to 70°C ±0.3 ppm
Frequency adjust	External volt (50 ppm)	Internal trimmer	External resistor	External resistor
Ageing rate	2 ppm/year	3×10^{-9} ppm day	1 ppm/year	≤1 ppm/year
Oscillator supply	5–15 V	5 V	9 V	12 V
Oven supply		9–24 V		
Power oscillator	20 mA at 12 V	40/60 mA	9 mA	≤200 mW
Power oven		3–8 W		
Package size (mm)	36.1 × 26.7 × 15	36.1 × 26.8 × 25.4	36.1 × 26.8 × 19.2	35.33 × 26.9 × 7.19

VCO = voltage-controlled oscillator usually like that of Figure 8.11(a).
OCXO = oven-controlled crystal oscillator. This type of oscillator is usually mounted in an oven operating at 75°C. The frequency stability is extremely good because, as you can see from Figure 8.13, the frequency of a crystal is very stable. The disadvantages are additional bulk, size, weight and oven consumption of additional electrical power. The last is particularly undesirable in battery operated equipment.
TCXO (analogue) = temperature-compensated crystal oscillator which uses compensating circuits (usually thermistors) to correct frequency drifts with temperature.
TCXO (digital) = temperature-compensating oscillator which uses a microprocessor or look-up tables to correct frequency drifts with temperature.

8.12.3 Summary for crystal oscillators

In Section 8.12 you have gained an insight into quartz crystals and their properties. You have also seen how crystals are used in crystal oscillators, how they operate, their frequency stability with temperature, and factors which affect the ordering of crystals for use in oscillators.

8.13 Phase lock loops

8.13.1 Introduction

The necessity for stable, low noise, power oscillators at very high frequencies has led to many innovative oscillator design systems. Consider the case where a stable, low noise 3.6 GHz oscillator is required for a transmitter. Ideally we would like to use a crystal oscillator operating at 3.6 GHz. However, this is not possible because the maximum operating range for a crystal oscillator is about 300 MHz.

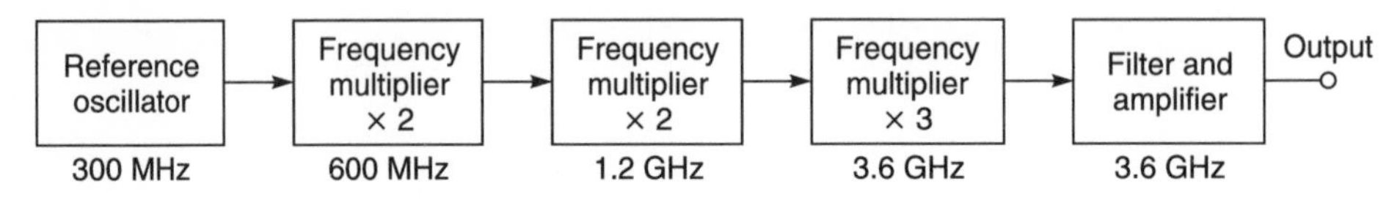

Fig. 8.16 Producing a crystal controlled high frequency signal

One way of producing the 3.6 GHz signal would be to use the scheme shown in Figure 8.16 where a 300 MHz oscillator is fed into a cascade of frequency multipliers[8] which amplify and select various harmonics of 300 MHz signal to produce the final 3.6 GHz. This method is expensive, requires a lot of circuit adjustment and is relatively inefficient. However, it can be used and is still in use particularly at frequencies which cannot be easily amplified. The great disadvantage of this method is that frequency multiplication increases unwanted f.m. noise and the oscillator output is generally noisy.

Another method of producing this signal would be to use a voltage-controlled LC oscillator at 3.6 GHz, divide its frequency, and compare its divided frequency against a reference crystal oscillator through a frequency or phase comparator which then emits a controlling voltage to shift the frequency back to the desired frequency. Such an arrangement is shown in Figure 8.17. This system is the basis of phase locked loop systems which we will now discuss more extensively.

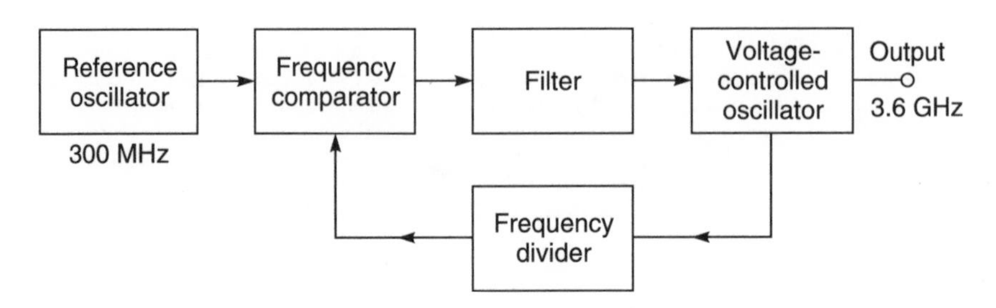

Fig. 8.17 A simple phase lock oscillator system

[8] A frequency multiplier can be made by over-driving an amplifier with a large signal so that the amplifer limits and produces a quasi-square waveform output. Fourier analysis tells us that a square waveform consists of many harmonics and we can select the desired harmonic to give the required signal.

8.13.2 Elements of a phase locked loop system

The block diagram of the basic phase locked loop (PLL) is shown in Figure 8.18. The phase detector, or phase comparator, compares the phase of the output waveform from the voltage-controlled oscillator with the phase of the r.f. reference oscillator. Their phase difference causes an output voltage from the phase detector, and this output voltage is fed to a low pass filter which removes frequency components at and above the frequencies of the r.f. input and the VCO. The filter output is a low frequency voltage which controls the frequency of the VCO.

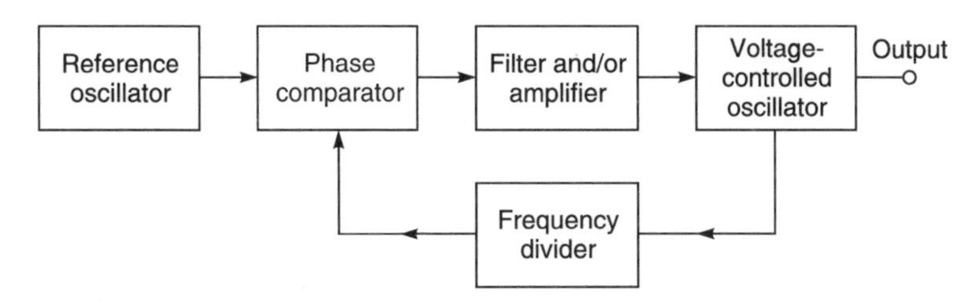

Fig. 8.18 The basic phase locked loop

When the loop is 'in lock', the phase difference has a steady value, which causes a d.c. voltage output from the filter. This d.c. voltage is sufficient to cause the VCO output frequency to become exactly equal to the input frequency. The two frequencies must be synchronised, otherwise there will be a continually-changing phase difference, the VCO input voltage will not be steady, and the loop will not be locked. Thus the loop 'locks onto' the reference frequency. Once the loop has locked, the reference frequency may vary, and the VCO output will follow it, over a range of frequencies called the **hold-in** range. That is the PLL will stay in lock, providing the output frequency does not fall outside the hold-in range.

In the following sections, we will look at each of the components of the PLL in turn, followed by the closed loop frequency response and step response.

8.13.3 The phase detector

The basic principle behind phase detection is signal multiplication. Figure 8.19 shows the principle using an ideal analogue multiplier. The VCO output voltage is represented by $v_v = \sin \omega_v t$ where ω_v is its angular frequency. The reference input signal is an unmodulated carrier $v_r \sin (\omega_r t + \phi)$ where ω_r is the input angular frequency and ϕ is its relative phase to v_v at $t = 0$. The multiplier output is

$$v_r v_v = \sin (\omega_r t + \phi) \sin \omega_v t \tag{8.37}$$

A little trigonometry[9] shows that

$$\begin{aligned} v_r v_v &= \sin (\omega_r t + \phi) \sin \omega_v t \\ &= 0.5 \{\cos[(\omega_r - \omega_v)t + \phi] - \cos (\omega_r + \omega_v)t + \phi\} \end{aligned}$$

[9] $\cos (A - B) = \cos A \cos B + \sin A \sin B$ and $\cos (A+B) = \cos A \cos B - \sin A \sin B$ and subtracting the two equations yields

$$\cos (A - B) - \cos (A + B) = 2 \sin A \sin B \text{ or } \sin A \sin B = 0.5 [\cos (A - B) - \cos (A + B)]$$

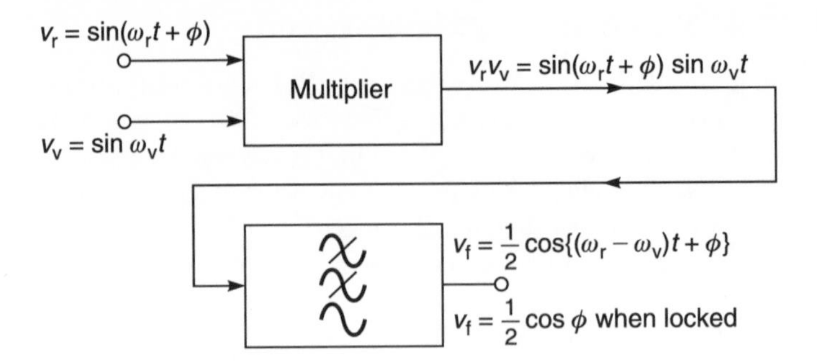

Fig. 8.19 Multiplication of two sine waves

The low pass filter removes the sum frequency ($\omega_r + \omega_v$) and the oscillator frequencies, leaving the difference–frequency component:

$$v_f = 0.5\{\cos[(\omega_r - \omega_v)t + \phi]\} \tag{8.38}$$

When the loop is in lock, the VCO frequency becomes equal to the reference input frequency and $\omega_r = \omega_v$ and Equation 8.38 becomes

$$v_f = 0.5[\cos \phi] \text{ when locked} \tag{8.39}$$

This is a d.c. level proportional to the cosine of the phase difference (ϕ) between the two signals at the input of the phase detector. Figure 8.20 shows the variation of this voltage with phase difference.

The sensitivity of Figure 8.20 is defined as rate of filter d.c./phase difference. It is maximum at points A and B when the phase difference between the two signals is ±90° and minimum at point C where the phase difference is zero. It follows that if we want maximum sensitivity, then the reference oscillator and the VCO should be out of phase by 90°. For the sake of clarity, let us choose point A. For this point we see:

- filtered d.c. output is zero for $\phi = -90°$
- filtered d.c. output is positive for $-90° < \phi < 0°$
- filtered d.c. output is negative for $-180° < \phi < -90°$

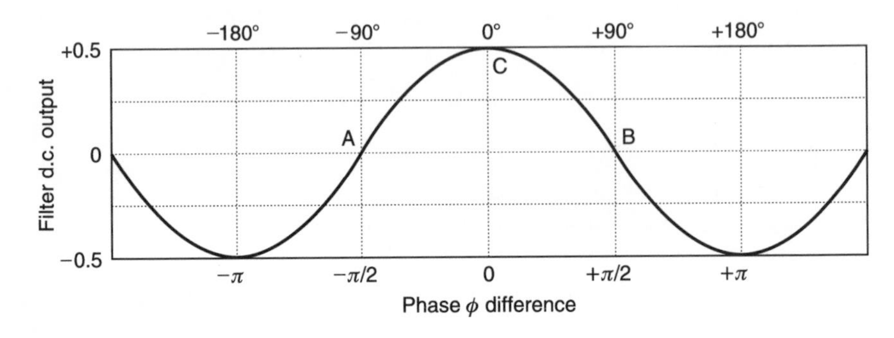

Fig. 8.20 Phase detector output when locked

Assume that the reference input frequency is constant. When the VCO drifts so that the relative phase shift $-90° < \phi < 0°$, a positive voltage will be generated for its frequency correction. When the VCO drifts in the opposite direction so the phase shift is $-180° < \phi < -90°$, a negative voltage will be generated for frequency correction. If the VCO is designed for the right sense of correction, it follows that the filter d.c. output voltage will keep the phase and hence the frequency constant.

8.13.4 Types of phase detectors

In practice, analogue multipliers are seldom used as phase detectors, because there are simpler circuits which can achieve the same overall result cheaper and faster. However, the theory of the analogue multiplier applies to these circuits too. We will now examine two basic phase detectors: the analogue switch, and the digital type.

Analogue switch type

One typical analogue switch phase detector is shown in Figure 8.21. In this circuit, the principle of operation described earlier is carried out by multiplying the two signals as described earlier. V_r remains the reference input signal, but part of the VCO signal is used to produce a square wave which switches the diodes ON and OFF when the square waveform is 1 and 0 respectively. Since a square wave is composed of a series of sine waves, it is apparent that the multiplication process is obtained and if a low pass filter is used after v_o, we will get the required d.c. term for VCO control as before.

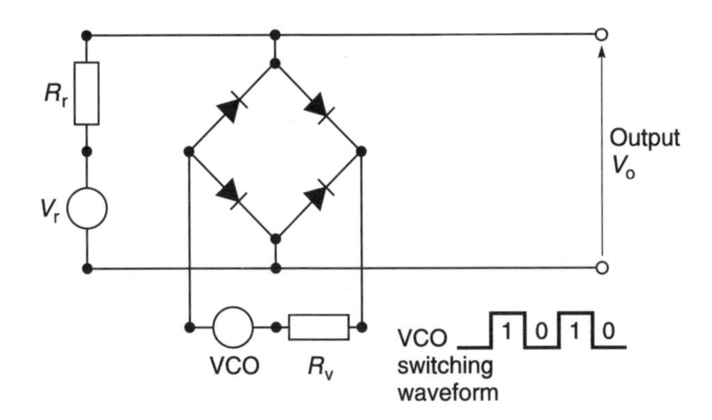

Fig. 8.21 Analogue type phase detector

Digital phase detectors

Digital phase detectors have both inputs in the form of digital waveforms. Typically, they use digital logic circuitry, such as TTL or CMOS.

The AND gate type

The most obvious digital equivalent of the analogue multiplier is an AND gate, as shown in Figure 8.22(a). For the AND gate, with logical inputs A and B, the output is given by $Y = A.B$. So, when the two input square waves have the same frequency and are in phase, the average output voltage is maximum and equal to half the logic 1 output voltage (V_o).

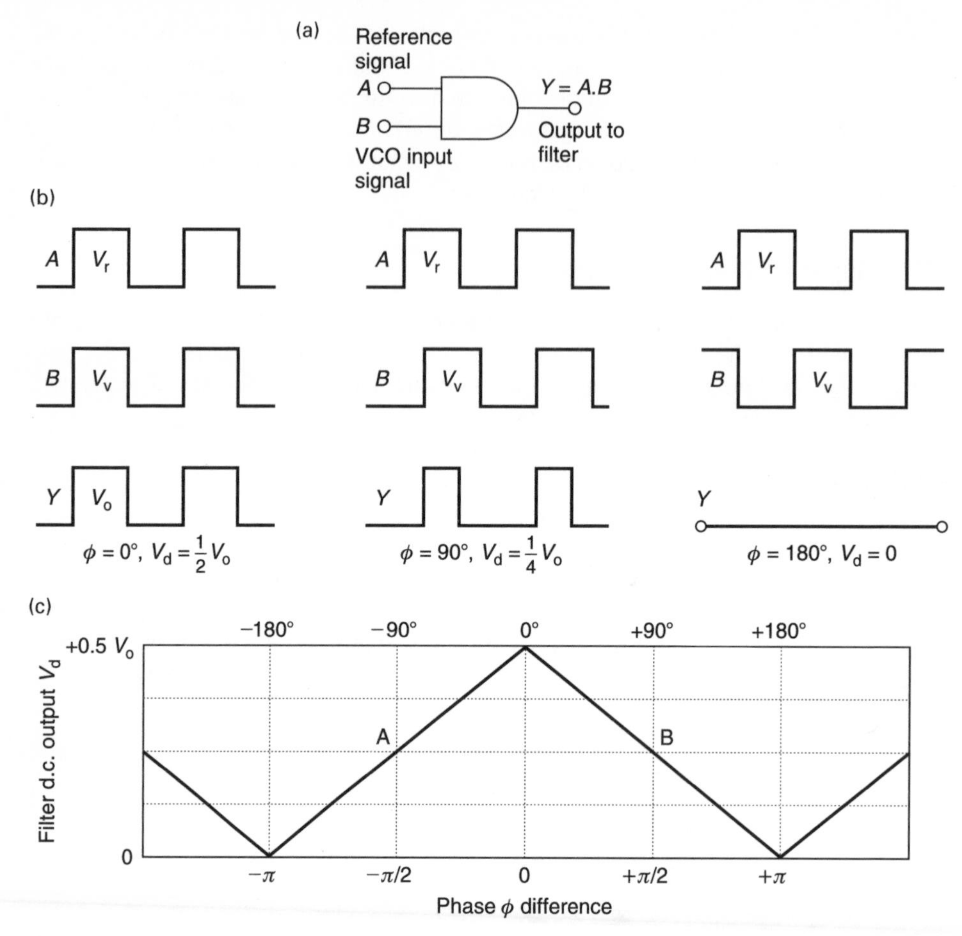

Fig. 8.22 The AND gate digital phase detector: (a) AND gate phase detector; (b) input and output waveforms for different phase shifts; (c) filter d.c. output versus phase difference, with locked loop

Figure 8.22(b) shows input square waves with various amounts of phase shift and corresponding output waveforms.

Figure 8.22(c) shows that, with input signals of the same frequency, the average output voltage varies linearly with the phase (ϕ). The filter output[10] voltage (v_d), when the loop is locked, is

$$v_d = (V_1/2)(1 + \phi/\pi) \text{ for } -\pi < \phi < 0 \text{ (see slope side A)}$$

and

$$v_d = (V_1/2)(1 - \phi/\pi) \text{ for } 0 < \phi < \pi \text{ (see slope side B)}$$

[10] The filter output voltage (v_d) is defined as the output voltage from the detector after removal of the oscillator and sum frequencies.

This function, which peaks at $\phi = 0$, is analogous to the cosine function of the analogue multiplier, which also peaks at $\phi = 0$.

Note: Although an AND gate has been featured in this case, you should be aware that it is also possible to use a NAND gate or an exclusive-OR gate as a phase detector but expect phase inversion in the output signal.

Gain sensitivity of the AND gate

The gain sensitivity or gain (k_ϕ) of a phase detector is defined as output voltage change (v_d)/phase difference (θ_e) change or

$$k_\phi = \frac{\Delta v_d}{\Delta \theta_e} \tag{8.40}$$

Using the same definition for Figure 8.22 we obtain for an AND gate

$$k_\phi = \frac{0.5V_o}{\pi} \tag{8.40a}$$

The sign of k_ϕ is dependent on the point used for the definition. If you use point A in Figure 8.22(c), you will get a positive slope and hence a positive value for k_ϕ whereas point B will give a negative slope and a negative value for k_ϕ.

The flip-flop type phase detector

In some applications of the phase locked loop, one or the other of the digital inputs to the phase detector may not be a square wave. For instance one input may come from the output of a counter used as a frequency divider, whose output waveform has a mark:space ratio which changes when the division factor is changed. Such a waveform is not suitable for use with the AND gate or the EX-OR gate. An alternative, which avoids this disadvantage, is the flip-flop. This is shown in Figure 8.23(a). Here, rising edges of one input set the output (Q) to logical 1, and rising edges of the other input reset the output (Q) to logical 0, as shown in the waveforms of Figure 8.23(b). You should see that the average output v_d varies with ϕ as in Figure 8.23(c) and not with the mark: space ratios of the input waveforms.

Gain sensitivity of the flip-flop detector

The gain sensitivity or gain (k_ϕ) of Figure 8.23 is defined as

$$k_\phi = \frac{V_o}{2\pi} \tag{8.41}$$

The sign of k_ϕ is dependent on the point used for the definition. If you use slope B in Figure 8.23(c), you will get a positive slope and hence a positive value for k_ϕ whereas point A will give a negative slope and a negative value for k_ϕ.

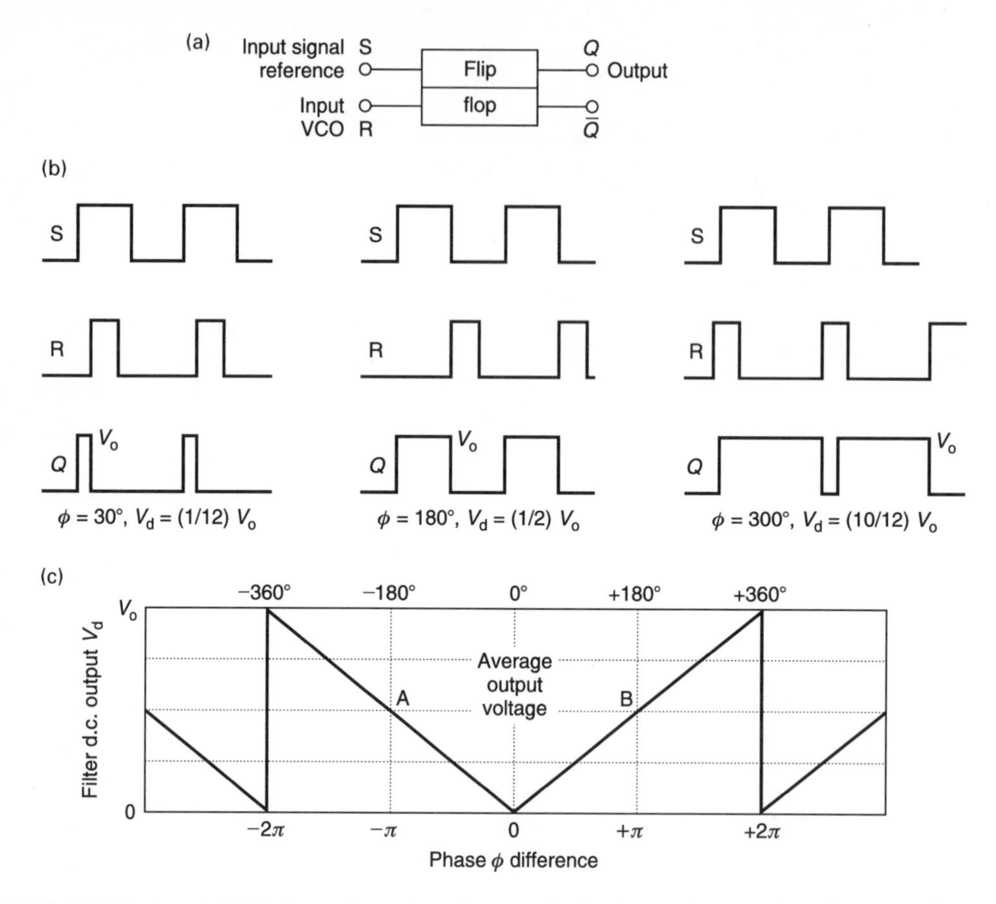

Fig. 8.23 A set–reset (SR) flip-flop used as a phase detector, with triggering on the rising edges of the input waveform: (a) set–reset flip-flop; (b) input and output waveforms; (c) average output voltage versus phase difference

8.13.5 The filter

The purpose of the filter is to remove the two oscillator frequencies f_r and f_{VCO} and the sum frequencies so that they do not cause instabilities in the loop system. The low pass filter is a simple RC network. In many applications it is a simple first-order single lag filter comprising just a series resistor followed by a shunt capacitor.

In some cases a lower frequency single lag filter is used for loop compensation. In other cases a slightly more complicated lag-lead RC network is used. Loop compensation is discussed further in the section on closed-loop response.

8.13.6 The d.c. amplifier

In many cases, the output from the phase detector is not sufficient to control the voltage-control led oscillator (VCO); therefore some form of d.c. amplification is necessary. Figure 8.24 shows three voltage gain amplifiers and their gain (volts out/volts in) parameters $k_A = v_o/v_i$.

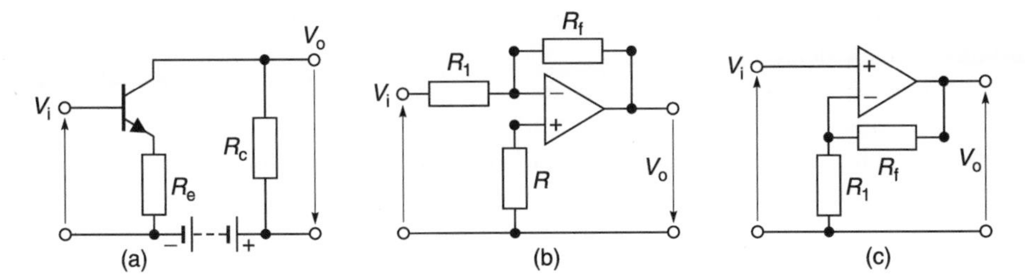

Fig. 8.24 (a) Common emitter; (b) inverting op-amp; (c) non-inverting op-amp

For Figure 8.24(a)

$$k_A \approx -R_c/R_e \tag{8.42}$$

For Figure 8.24(b)

$$k_A \approx -R_f/R_1 \tag{8.43}$$

For Figure 8.24(c)

$$k_A \approx (R_f + R_1)/R_1 \tag{8.44}$$

The bandwidth of the d.c. amplifier must be very high when compared to the loop bandwidth (explained later) otherwise loop instability will occur.

8.13.7 The voltage-controlled oscillator (VCO)

The sine wave VCO described in Section 8.10 is a suitable oscillator for use in phase locked loops, but other types such as the digital type oscillator shown in Figure 8.25 can also be used. The **sine wave** type is used as the oscillator in r.f. transmitters, and as local oscillators in radio and television receivers and as synthesised oscillators in mobile phones and signal generators. In these applications a pure sine wave is the ideal. With varactor diode type oscillators, the relation between the oscillator frequency and the control voltage is chiefly dependent on the chosen varactor and its direction of connection. Most varactors vary their capacitance as V^{α} where α can range from 0.5 to 3.

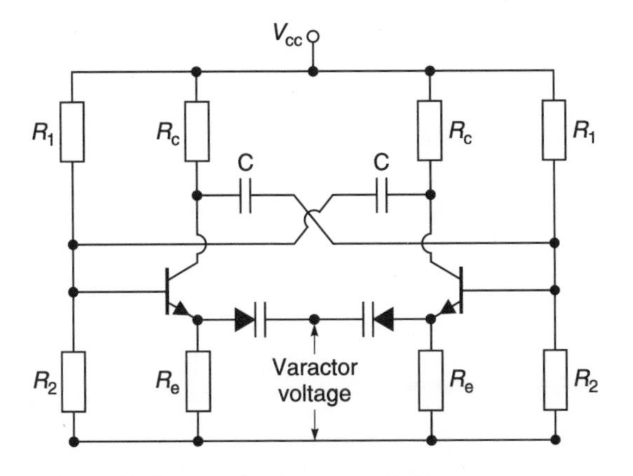

Fig. 8.25 Digital free-running multi-vibrator with voltage-controlled frequency

The linear, **square wave** type of VCO is suitable for use in a PLL system used for frequency demodulation. This is done at the receiver intermediate frequency, so no great strain is made on the operating frequency of the digital circuitry. If the loop is in lock, then the VCO frequency must be following the input frequency. In this case the VCO's control voltage, which is also the demodulated output from the loop, is linearly related to the frequency shift and, hence, to the original frequency modulation.

Gain sensitivity of a VCO

The gain sensitivity of a VCO is defined as

$$k_o = df/dv_c \tag{8.45}$$

where

df is the change in VCO frequency
dv_c is the change in control voltage v_c

Frequency of a VCO

If we define f_{FR} as the free-running frequency of a VCO when there is zero correction voltage from the phase detector, and assume oscillator control linearity, then by using Equation 8.45 we can describe the frequency (f) of the VCO as

$$f = f_{FR} + \Delta f = f_{FR} + k_o v_c \tag{8.46}$$

Some VCOs are designed for a control voltage centred on 0 V, at which point they generate their 'free-running' frequency, f_{FR}. This is shown in Figure 8.26(a). For a linear type VCO, the output frequency is given by

$$f = f_{FR} + k_o v_c \tag{8.46a}$$

Positive control voltages may increase the frequency, and negative ones decrease it, resulting in a positive value for k_o; or the converse may be true with k_o negative.

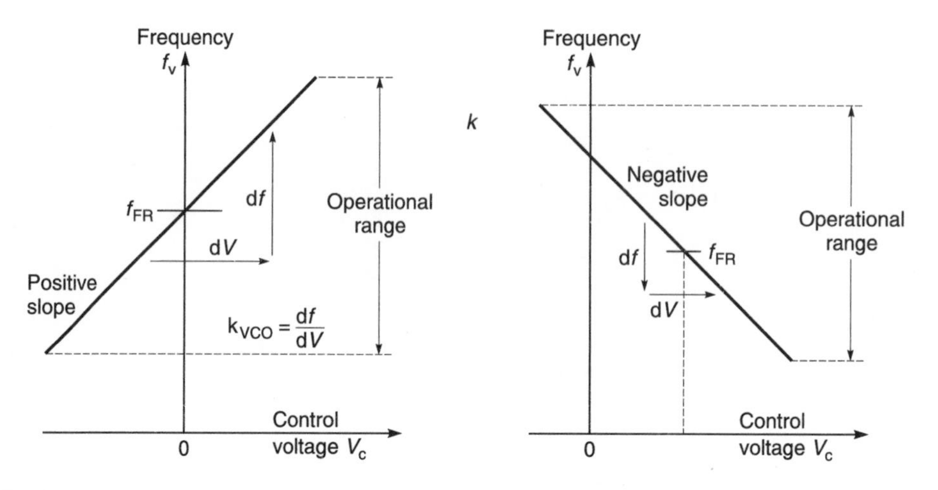

Fig. 8.26 Transfer characteristics of linear VCOs: (a) $v_c = 0$ for f_{FR} and a positive control slope; (b) $v_c \neq 0$ for f_{FR} and a negative control slope

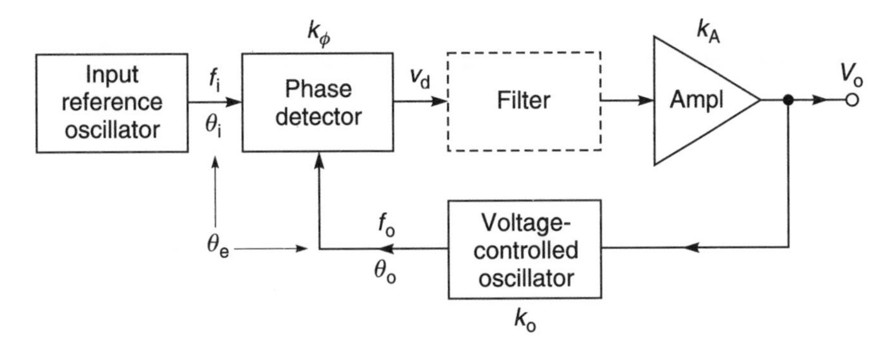

Fig. 8.27 Basic phase lock loop

Other VCOs have a control voltage range centred on a non-zero voltage. Figure 8.26(b) shows an example of this type of transfer characteristics. In this case and because of the negative slope

$$f = f_{FR} - k_o v_c \tag{8.46b}$$

Note: In each case, the slope of the characteristic $\partial f/\partial v_c$ gives k_o.

8.13.8 Loop gain and static phase error

At this stage, it would be prudent if we consolidate what we have discussed using the diagram shown in Figure 8.27. For clarity of understanding, we will consider the PLL initially as locked and tracked. Later we will consider how the PLL becomes locked. In Figure 8.27, each block has its own gain parameters. From Equation 8.40, we know that the phase detector develops an output parameter (v_d) in response to a phase difference (θ_e) between the reference input (f_i) and the VCO frequency (f_o). The transfer gain (k_ϕ) has units of volt/radians of phase difference.

At this locked stage, the main function of the filter is simply to remove f_i, f_o and the sum of these frequencies. We will temporarily ignore the parameters of the filter in the dashed block of Figure 8.27 because in the locked state, the filter has a parameter of unity.

The amplifier has a gain of k_A and its unit of gain (V_o/V_d) is dimensionless. Thus $V_o = k_A v_d$. The VCO free-running frequency is f_{FR}. The VCO frequency (f_o) will change in response to an input voltage change. The transfer gain (k_o) has units of Hertz/volt.

The loop gain (k_L) for this system is simply the product of the individual blocks:

$$k_L = k_\phi \, k_A k_o \tag{8.47}$$

The units for k_L = (v/rad) (v/v) (Hz/v) = Hz/rad.

In the diagram f_i is the input reference frequency to the phase detector and the system is in the locked stage. If the frequency difference *before* lock was $\Delta f = f_i - f_{FR}$, then a voltage $v_c = \Delta f/k_o$ is required to keep the VCO frequency equal to f_i. The amplifier must supply this v_c so its input must be v_c/k_A and this is the output voltage (V_d) required from the phase detector.

Summing up

$$v_d = v_c/k_A = (\Delta f/k_o)/k_A = \Delta f/(k_o k_A) \tag{8.48}$$

To produce v_d, we would need a phase difference error of θ_e radians between f_i and f_o because $k_\phi = v_d/\theta_e$ or $v_d = k_\phi\theta_e$ and substituting this in Equation 8.48 yields

$$\theta_e = \Delta f/(k_o k_A k_\phi)$$

and using Equation 8.47, we get

$$\theta_e = \frac{\Delta f}{k_o k_A k_\phi} = \frac{\Delta f}{k_L} \tag{8.49}$$

8.13.9 Hold-in range of frequencies

The hold-in range of frequency (Δf in Equation 8.49) is defined as the frequency range within which the VCO can drift before it becomes out of lock and stops being controlled by f_i. In practical circuits, the hold-in frequency range is limited by the operating range of the phase detector because the VCO has a much wider frequency range of operation.

In all of the types of phase detector described in Section 8.13.4, except for the flip-flop type, the useful phase–difference range is limited to $\pm\pi/2$. Outside this range, the slope of the phase detector's characteristic changes, altering the loop feedback from negative to positive, and causing the loop to lose lock. So the static phase error θ_e is limited to $\pm\pi/2$. Re-arranging Equation 8.49 for θ_e we get

$$\Delta f = \theta_e k_L \tag{8.50}$$

For the *AND gate* type phase detector, where θ_e has its maximum possible value of $\pm\pi/2$, Δf is the maximum possible deviation, $\Delta f_{max} = (\pm\pi/2)(k_L)$ and its hold-in range = $\pm\pi/2 \times$ (d.c. loop gain). For the *flip-flop* type, the static phase error (θ_e) is limited to $\pm\pi$, and its hold-in range = $\pm\pi \times$ (d.c. loop gain).

Again from Equation 8.50, you can see that if you want a wide hold-in frequency range, then you should increase the loop gain (k_L). This is true but care should be exercised in doing this because too high a loop gain will result in loop instability. Let us now consolidate our thoughts by carrying out Example 8.5.

Example 8.5

Example 8.5 summarises much of the information acquired on PLL at this stage. Figure 8.28 provides enough information to calculate the static behaviour of a phase locked loop. Calculate (a) the voltage gain (k_a) for the op-amp, (b) the loop gain (k_L) in units of second^{-1} and in decibels ($\omega = 1$ rad/s). (c) With S_1 open as shown, what is observed at v_o with an oscilloscope? (d) When the loop is closed, determine: (i) the VCO output frequency; (ii) the static phase error (θ_e) at the phase comparator output; (iii) V_o. (e) Calculate the hold-in range Δf (assume that the VCO and op-amp are not saturating). (f) Determine the maximum value of v_d.

Solution

(a) Using Equation 8.44

$$k_A = (R_f/R_1) + 1 = (9\ \text{k}\Omega/1\ \text{k}\Omega) + 1 = 10$$

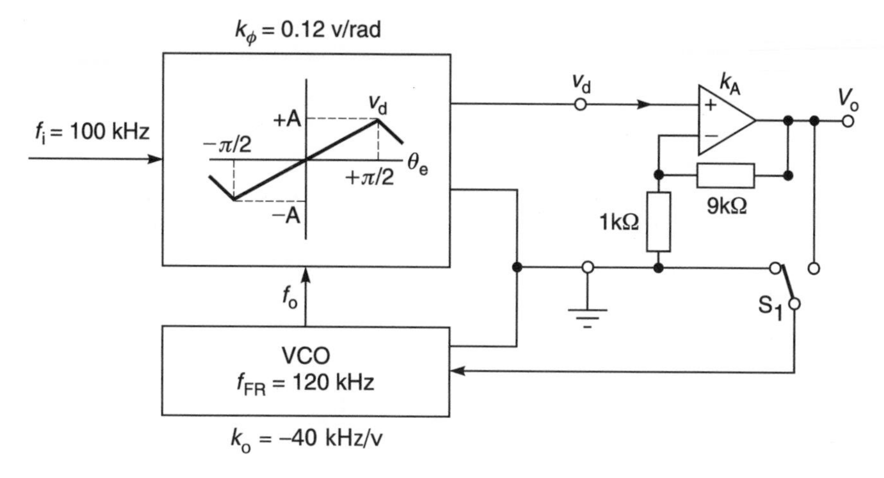

Fig. 8.28 Closed loop system

(b) Using Equation 8.47

$$k_L = k_\phi k_A k_o = (0.12 \text{ v/rad})(10)(-40 \text{ kHz/v}) = -48\,000 \text{ Hz/rad}$$

$$= \frac{48\,000 \text{ cycles/sec}}{\text{rad}} = \frac{48\,000 \text{ cycles/sec}}{\text{cycles}/(2\pi)}$$

$$= [2\pi \times 48\,000] \text{ s}^{-1} = 301\,593 \text{ s}^{-1}$$

and in terms of dB at 1 rad, we have

$$k_L = 20 \log (301\,593) \approx 109.6 \text{ dB at 1 rad/s}$$

(c) With S_1 open, there is no phase lock. If we assume that f_o, f_i and the sum frequencies have been removed by the filter, then all that will be seen is the beat or difference frequency $f_o - f_i$ = 120 kHz – 100 kHz = 20 kHz.

(d) (i) When the loop is closed and phase locked, then by definition $f_o = f_i$ and since f_i = 100 kHz, it follows that f_o = 100 kHz. There is no frequency error but there is a phase error between the two signals.

(ii) The free-running frequency f_{FR} of the VCO = 120 kHz. For the VCO output to be 100 kHz, we transpose Equation 8.45 to give

$$V_o = \Delta f / k_o = (100 - 120) \text{ kHz}/(-40 \text{ kHz/V}) = 0.500 \text{ V}$$

We want v_d, the input to the amplifier whose gain = 10. Using Equation 8.48

$$v_d = v_o / k_A = 0.5 \text{ V}/10 = 0.050 \text{ V}$$

Finally, using Equation 8.40 and transposing it, we obtain

$$\theta_e = v_d / k_\phi = 0.050 \text{ V}/0.12 \text{ V/rad} = 0.417 \text{ radians}$$

Alternatively, we could have used Equation 8.49 where

$$\theta_e = \frac{\Delta f}{k_o k_A k_\phi} = \frac{(100 - 120)\ \text{kHz}}{(-40\ \text{kHz/V})(0.12\ \text{V/rad})\ (10)} = 0.417\ \text{rad}$$

(iii) V_o was calculated in (ii) as 0.500 V d.c.

(e) Since the VCO and op-amp are assumed not to be saturating, then the limitation will obviously depend on the phase detector output. Clearly v_d can only increase until $v_d \to v_{max} = A$, at which point $\theta_e = \pi/2$. Beyond this point, v_d decreases for increasing static phase error, and the phase detector simply cannot produce more output voltage to increasing f_o, and the loop breaks lock. The total hold-in range is $\pm\pi/2$ or π radians. The total hold-in frequency range between these two break-lock points can be found by using Equation 8.50:

$$\Delta f = \theta_e k_L = (\pi)\ (-40\ \text{kHz/V})(0.12\ \text{V/rad})(10) = 48.00\ \text{kHz}$$

In the answer, I have dropped the minus sign because we are only interested in the frequency range.

(f) At the frequency where $\theta_e = \pi/2$, we have $v_{d(max)} = A$. Therefore

$$v_d = k_\phi \theta_e = 0.12\ \text{V/rad} \times \pi/2\ \text{rad} = 0.188\ \text{V d.c.}$$

This example shows clearly the conditions existing within a phase locked loop system when it is in lock.

Summary. Example 8.5 has shown clearly what happens in a phase locked loop when it is in lock. Certain facts are required to make a PLL function correctly.

- The sensitivity of the phase locked loop detector (k_ϕ) must be known. It can be obtained from measurement or calculation.
- The amplifier gain (k_A) must be known. This can be obtained by calculation or measurement.
- The sensitivity of the VCO must be known. The usual way to obtain this is by measurement.
- The hold-in range (Δf) must be known or measured.
- If the oscillator drifts outside this range either due to noise, instability, temperature, etc., then the phase locked loop will be erratic and will break lock and behave like a free-running oscillator.

The above conditions are basic to a phase locked loop when it is in a hold-in range situation.

Lock-in range

The lock-in range is defined as the range of input frequencies over which an unlocked loop will acquire lock. If f_h and f_L are respectively the highest and lowest frequencies at which the loop will attain lock, then

$$\text{lock-in range} = f_h - f_L$$

The lock-in range is usually smaller than the hold-in range. In practice when loop-lock is lost, the PLL generates a saw-tooth wave to sweep the VCO in the hope that lock-in may be re-captured.

There are many problems to be considered in acquiring frequency-lock. These include frequency sweep range, step response time, loop bandwidth frequency response, loop bandwidth gain, and the response of the individual components to the sweep range. The VCO phase response is important because its phase gain falls off with an increase in frequency sweep. The loop filter is also extremely important because it determines the step response time and hence the settle-time of the VCO to its new frequency.

The PLL is a very complicated system and to properly design such a system, the designer has to take into account many of the problems mentioned above. We do not propose to do it here because we have achieved our aim of showing the principles of a phase lock loop system. However, if you want an extensive source of material that covers phase lock loop design techniques, consult F. M. Gardner's *Phaselock Techniques*.[11]

8.14 Frequency synthesizers

8.14.1 Introduction

A frequency synthesizer is a variable-frequency oscillator with the frequency stability of a crystal-controlled oscillator. Synthesizers are used in radio transmitters and receivers because of their output signal stability which is essential in today's narrow band communication systems. In fact, modern communication systems cannot exist without them.

Two basic approaches are used in the design of synthesizers. They are the *direct* method and the *indirect* method. The direct method generates the output signal by combining one or more crystal-controlled oscillator outputs with frequency dividers/comb generators, filters and mixers. The indirect method utilises a spectrally pure VCO and programmable phase locked circuitry. Although slower than the direct approach and susceptible to f.m. noise on the VCO, indirect frequency synthesis using the phase lock loop principle is less expensive, requires less filtering, and offers greater output power with lower spurious harmonics.

8.14.2 The direct type synthesizer

This type uses no PLLs or VCOs, but only harmonic multipliers, dividers and filters. It may use only one crystal oscillator, with multiple harmonic multipliers and dividers, or it may use several crystal oscillators. An example of the single crystal type is shown in Figure 8.29. The 1 MHz crystal oscillator is followed by a harmonic multiplier, or comb generator. This is a circuit which 'squares-up' the signal from the crystal oscillator to generate a train of pulses, rich in harmonics of the crystal frequency. It is called a comb generator because its frequency spectrum resembles a comb. Harmonic selector filter 1 (HSF 1) allows harmonics of the 1 MHz signal to appear at 1 MHz intervals. Hence, by adjusting the filter, it is possible to get frequencies of 1, 2, 3, 4, etc. MHz. This selection of 1 MHz is rather a coarse adjustment.

The second output of the 1 MHz crystal oscillator is divided down to 100 kHz, fed to another comb generator and into harmonic selector filter 2 (HSF 2). This filter allows 100 kHz harmonics to pass throught it. The harmonic selected is dependent on the setting of the filter.

The two outputs from harmonic selector filter 1 and harmonic selector filter 2 are then mixed to produce sum and difference frequencies, amplified and filtered through the

[11] Gardner's book (1966) is published by John Wiley & Sons. It is old but it remains the classic on phase lock techniques.

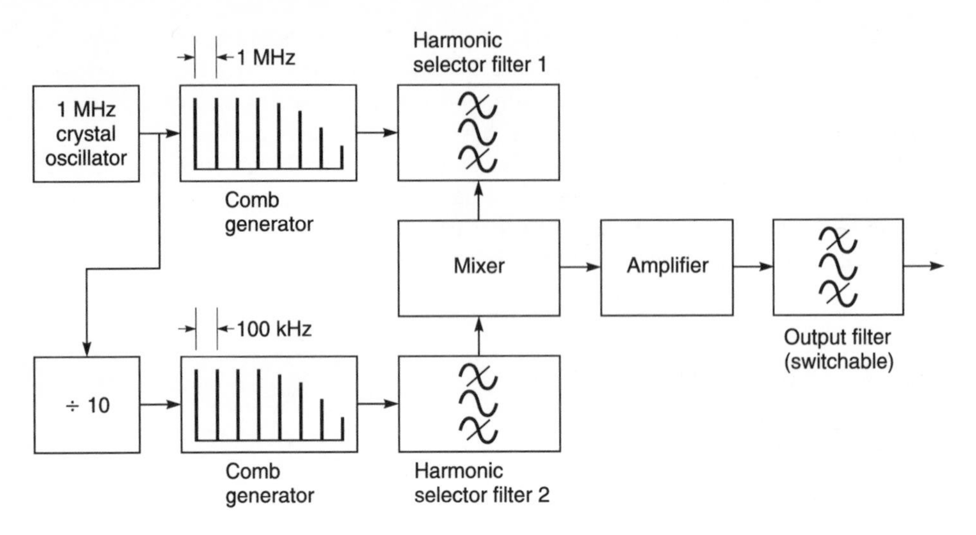

Fig. 8.29 An early direct frequency synthesizer

output filter. For example, if an output frequency of 6.5 MHz is required. The 6th harmonic of 1 MHz will be selected by (HSF 1), i.e. 6 MHz, and the 5th harmonic of 100 kHz will be selected by (HSF 2), i.e. 500 kHz. The two frequencies are fed into the mixer to produce sum 6.5 MHz and difference frequency 5.5 MHz. The four signals, 500 kHz, 5.5 MHz, 6 MHz and 6.5 MHz, are amplified but the switchable output filter rejects all but the desired 6.5 MHz signal.

The frequency resolution can be improved still further by adding further divider, comb generator and filter sections. For instance, a second decade divider and comb generator would provide 10 kHz steps, selectable by a second filter. Its output would be mixed with the output of the 100 kHz step filter in a second mixer, whose output would be filtered to select the sum or difference frequency. This would then be mixed with the selected 1 MHz step in the first mixer. Thus this circuit would provide frequencies up to 10.99 MHz, with 10 kHz resolution.

The biggest disadvantage of such a system is that it places stringent requirements on the output filter. This is because, in some cases, the sum and difference frequencies can differ by very little and selecting one frequency and rejecting the other means extremely steep filter slopes.

Example 8.6

A synthesizer with four decade divider, comb generator and harmonic filter sections has a crystal oscillator frequency of 10 MHz. State the frequency which could be selected by each section to produce a final output at 75.48 MHz, assuming the sum frequency is selected from each mixer output. State also the frequencies at the output of each mixer, and the frequency selected by the filter following each mixer.

Solution. Assuming that the sum frequency is selected from each mixer output, the frequencies are:

Section 1:	70 MHz
Section 2:	5 MHz
Section 3:	400 kHz
Section 4:	80 kHz

Mixer 3:	(400 ± 80) kHz = 480 kHz and 320 kHz
After filter 3:	480 kHz
Mixer 2:	(5 ± 0.48) MHz = 5.48 MHz and 4.52 MHz
After filter 2:	5.48 MHz
Mixer 1:	(70 ± 5.48) MHz = 75.48 MHz and 64.52 MHz
After filter 1:	75.48 MHz

8.14.3 Direct digital waveform synthesis

The system described in Section 8.14.2 is clumsy and is seldom used today. This is a more recent technique which synthesises a sine wave digitally. The basic principle is illustrated in Figure 8.30. Values of the sine function, at regularly-spaced angular intervals over one complete cycle, are stored digitally in a look-up table, typically in ROM. The values are clocked sequentially out of the look-up table to a digital-to-analogue convertor and via a filter to the output. The filter removes clock-frequency components.

The output frequency $f_o = f_c/S$, where f_c is the clock frequency and S is the number of samples per cycle stored in the look-up table. So the output frequency is determined by the clock frequency. This can be changed by changing the division factor of the programmable counter, which divides the crystal oscillator frequency f_{ref} by the factor N. Thus the output frequency is $f_o = f_{ref}/S\,N$. The value of S must be at least 4, but preferably 10 or so, unless complicated tuneable analogue filtering is used, which would defeat the object of the cheap digital chip.

Clearly, this type of synthesizer cannot produce frequencies much higher than, say, 100 MHz because the clock must run at S times the output frequency, and the fast test digital

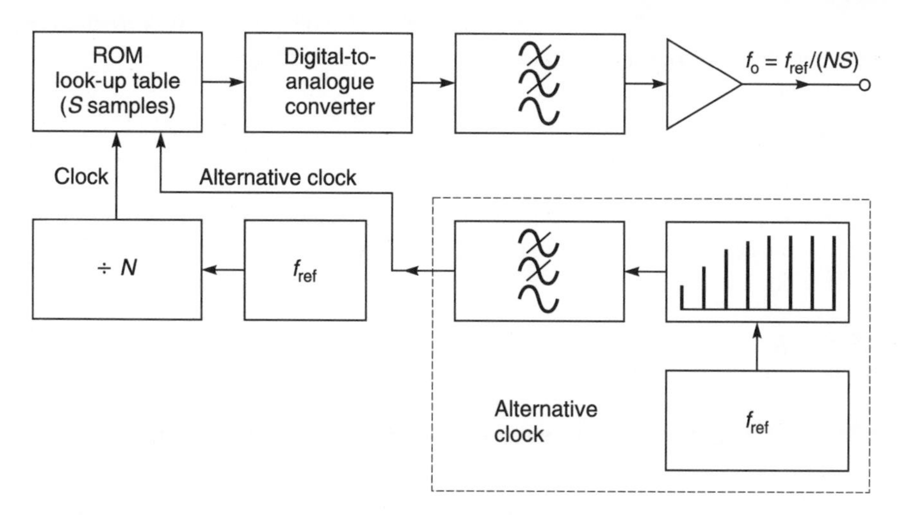

Fig. 8.30 Basic direct synthesizer using sinusoidal waveform synthesis

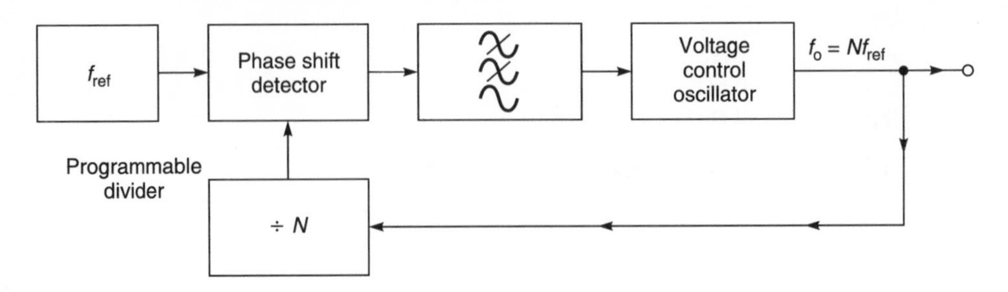

Fig. 8.31 Basic PLL synthesizer

circuits limit is currently about 2 GHz. In practice, a look-up table in such fast circuitry is currently too expensive, and the digital technique's cost advantage is realised only at lower frequencies.

8.14.4 Indirect synthesizers (phase-lock types)

These synthesizers use phase lock loops to control voltage-controlled oscillators, with good spectral purity, at frequencies locked to harmonics or sub-harmonics of crystal oscillators.

Single loop type

The simplest example is the single loop type of Figure 8.31 where a digital counter type frequency divider, set to divide by a programmable factor *N*, follows the VCO. The loop keeps the divider's output frequency equal to the crystal frequency (f_{ref}) so the output frequency from the VCO is an integer multiple (harmonic) of the crystal frequency: $f_o = Nf_{ref}$.

The output frequency can be changed by changing the division factor *N* of the divider. The highest output frequency is $f_o = N_{max} f_{ref}$, where N_{max} is the maximum division capacity of the counter. The frequency resolution, which is the minimum output frequency step size, is equal to the crystal frequency f_{ref}.

All the previous analysis of PLLs applies to this loop, but with the added complication that the loop gain is divided by the factor *N*, so the loop gain changes as the output frequency is changed. Because of this, the loop filter must be designed to maintain loop stability in the worst case. As the output frequency is raised, *N* is increased, which lowers the loop gain and the loop crossover frequency. So, if a lead-lag filter is used, its break points must be chosen well below the lowest loop crossover frequency, which is obtained at the highest output frequency.

The simple single loop synthesizer of Figure 8.31 cannot produce output frequencies any higher than can he handled by the digital divider. With moderately-priced TTL-variant programmable counters, this limits the frequency to the order of 100 MHz.

Pre-scaling

A simple modification called pre-scaling enables higher frequency VCOs to be used. Figure 8.31 shows an example. This synthesizer is used to generate the local-oscillator frequency for a UHF television receiver. The broadcast vision carriers[12] have frequencies

[12] Television channels and television channel spacing differ in different parts of the world and you should only take this frequency as representative.

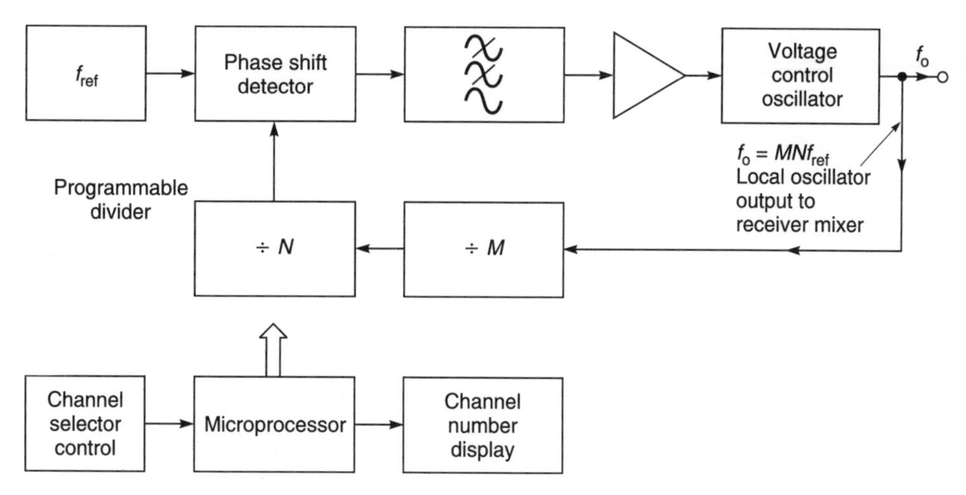

Fig. 8.32 A PLL synthesizer using pre-scaling, for generating the local-oscillator signal for a television receiver

from 471.25 to 847.25 MHz, spaced at exactly 8 MHz intervals. With a receiver vision i.f. of 39.75 MHz, the local oscillator must be tuneable over the range 511 to 887 MHz, assuming it works above the carrier frequency.

The pre-scaler is a high speed divider using ECL or Gallium Arsenide (GaAs) circuitry, which can work at these frequencies. It divides by a fixed factor M. The output frequency is now $f_o = MNf_{ref}$, and the frequency resolution is increased to Mf_{ref}. We still have f_o = ($N \times$ resolution), as with the simple loop.

The poorer resolution is not a problem in this case, because the TV channels are spaced 8 MHz apart. However, 8 MHz does not divide integrally into the required local-oscillator frequencies, so a frequency resolution must be chosen which does. The highest such frequency is 1 MHz, so this is the choice for the resolution, although of course the channel control logic will restrict N to selecting just channel frequencies at 8 MHz intervals. A value of 64 is chosen for M, to reduce the highest local-oscillator frequency down to less than 20 MHz so that a cheap, low power TTL or CMOS chip can be used for the variable divider. Since resolution = Mf_{ref}, we have 1 MHz = $64f_{ref}$, making f_{ref} = 15.625 kHz. This is best obtained from a cheap, higher frequency crystal and divider, such as a 1 MHz crystal with a divide-by-64 counter.

Example 8.7

Calculate the values of N to produce the lowest and the highest required local-oscillator frequencies.

Solution

$$f_o = MNf_{ref} = \text{resolution}$$

Also

$$f_o = N \times \text{resolution} = N \times 1\ \text{MHz and } N = f_o/1\ \text{MHZ}$$

The lowest local-oscillator frequency is 511 MHz, so $N = 511$. The highest local-oscillator frequency is 887 MHz, so $N = 887$.

8.14.5 Translation loops: frequency offset using heterodyne down-conversion

Another technique for avoiding the need for a VHF programmable divider is to translate the output frequency down to a lower frequency by mixing with a stable 'offset' oscillator. Figure 8.33 shows the principle. The balanced mixer output contains the sum and difference frequencies, $f_o + f_{osc}$ and $f_o - f_{osc}$, is of the same order as f_o then $(f_o + f_{osc})$ is much greater than $(f_o - f_{osc})$, and a simple low pass filter following the mixer can remove the sum easily, leaving only the difference frequency at the input to the divider. The loop containing the frequency-translation circuitry is called a **translation loop**.

In the example shown in Figure 8.33, the synthesizer is used as the local oscillator for a VHF FM receiver. The carrier signal band is 88.0–108.0 MHz, and the i.f. is 10.7 MHz so, with the local-oscillator frequency above the received frequency, the synthesizer must tune from 98.7 to 118.7 MHz. With $f_{osc} = 80$ MHz, the divider input frequency $(f_o - f_{osc})$ ranges from 18.7 to 38.7 MHz. The required resolution is 50 kHz, so the reference frequency is chosen as 50 kHz. This could be obtained by dividing from a 1 MHz crystal. In that case the 80 MHz source would be a separate crystal. Alternatively, one crystal oscillator running at 16 MHz could have its frequency multiplied and the appropriate harmonic (the fifth in this case) selected for the offset source, and its frequency divided for the reference source.

Example 8.8

Calculate the lowest and highest values of the division factor N of the divider in the loop.

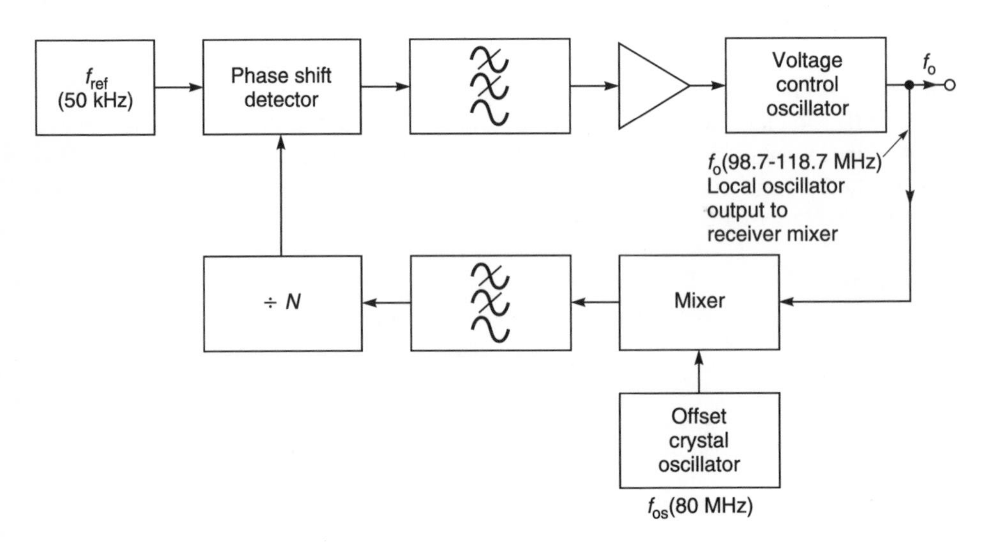

Fig. 8.33 A PLL synthesizer using an offset oscillator to generate the local oscillator for a VHF FM receiver

Solution. The lowest frequency at the divider input is 18.7 MHz and the divider output is 50 kHz, so in this case N = 18.7 MHz/50 kHz = 374. The highest frequency at the divider input is 38.7 MHz, so in this case N = 774.

8.15 Summary

This part has been mainly devoted to the more popular types of oscillators. The configurations discussed have included the Wien, phase shift, Hartley, Colpitts, Clapp, crystal and the phase locked loop. Frequency synthesizers and their basic operating configurations have also been shown. Oscillator design, particularly phased lock loop synthesizers, is a specialist subject but the information presented in this chapter has provided sufficient background information to allow further study.

9

Further topics

9.1 Aims

The primary aim of this chapter is to provide an introduction to signal flow graph analysis so that you will be able to analyse more complicated networks and to follow more advanced publications and papers. The secondary aim of this chapter is to offer you some comments on the use of small software packages in high frequency and microwave engineering.

9.2 Signal flow graph analysis

9.2.1 Introduction

Occasions often arise where a network is extremely cumbersome and difficult to understand, and it is hard to solve the circuit parameters by algebraic means. In such cases, **signal flow graph analysis** is used to help understanding, and to reduce circuit complexity until it can be handled easily by more conventional algebraic methods.

Signal flow analysis is used as a means of writing and solving linear microwave network equations. It is direct and relatively simple; variables are represented by points and the inter-relations between points are represented by directed lines giving a direct picture of signal flow. The easiest way to understand signal flow manupulation is by examples and in the following sections, we will show you several methods of applying signal flow analysis.

You are already familiar with Figure 9.1 which was used when we introduced you to

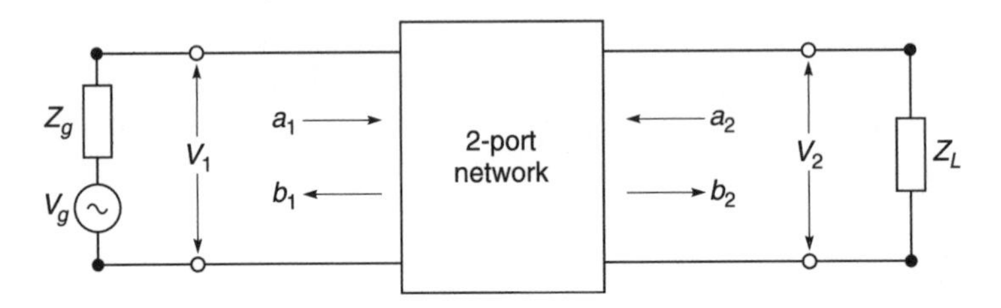

Fig. 9.1 General microwave two-port network showing incoming and outgoing waves

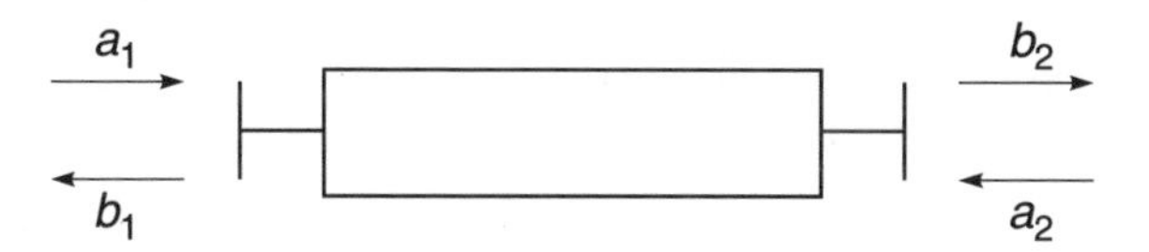

Fig. 9.2 Alternative signal flow graph for a two port network

scattering parameters. Figure 9.1 gives a semi-pictorial view of s-parameters. An alternative representation of Figure 9.1 is shown in Figure 9.2. This figure is often used in signal flow analysis because of its greater simplicity. Figure 9.2 shows a two-port network with wave a_1 entering port 1 and wave a_2 entering port 2. The emerging waves from the corresponding ports are represented by b_1 and b_2.

Figure 9.3 is a signal flow graph representation of Figure 9.2. In Figure 9.3 each port is represented by two nodes. Node a_n represents the wave coming into the device from another device at port n and node b_n represents the wave leaving the device at port n. A **directed** branch runs from each a node to each b node within the device. Each of these branches has one or more scattering coefficients associated with it. The coefficient/s shows how an incoming wave gets changed to become an outgoing wave at the node b. Scattering coefficients are complex quantities because they represent both amplitude and phase changes associated with a branch. The value of a wave at a b node when waves are coming in at both a nodes is the **superposition** of the individual waves arriving at b from each of the separate a nodes.

As you already know the relationship of the emergent waves to incident waves is written as the linear equations

$$b_1 = s_{11}a_1 + s_{12}a_2 \tag{9.1}$$

$$b_2 = s_{21}a_1 + s_{22}a_2 \tag{9.2}$$

These are called scattering equations and the s_{mn}s are the **scattering coefficients**.

By comparing Equations 9.1 and 9.2 and Figure 9.3 it is seen that s_{11} is the reflection coefficient looking into port 1 when port 2 is terminated in a perfect match ($a_2 = 0$). Similarly s_{22} is the reflection coefficient looking into port 2 when port 1 is terminated in a perfect match ($a_1 = 0$).

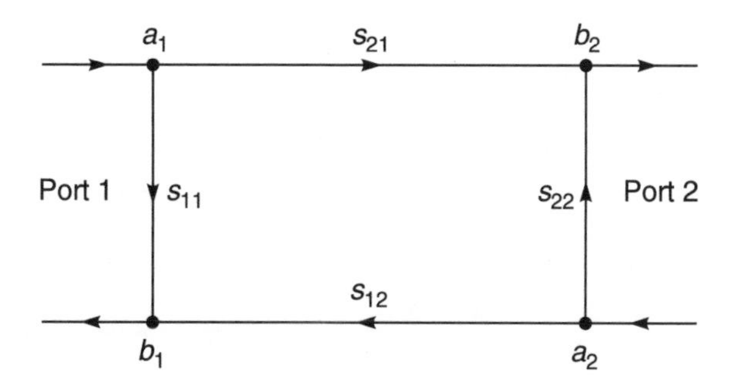

Fig. 9.3 Signal flow representation of Figure 9.2

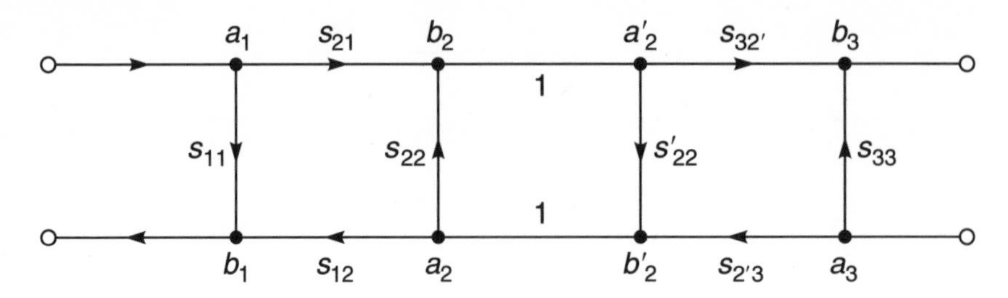

Fig. 9.4 Cascading of networks

The parameter s_{21} is the transmission coefficient from port 1 to port 2 when port 2 is terminated in a perfect match ($a_2 = 0$), and s_{12} is the transmission coefficient from port 2 to port 1 when port 1 is terminated in a perfect match ($a_1 = 0$).

Networks are cascaded by joining their individual flow graphs as in Figure 9.4. Note how a'_2 is synonymous with b_2 and how b'_2 is synonymous with a_2. This can be shown by a connecting branch of unity. See Figure 9.4.

9.2.2 Signal flow representation of elements

Some examples of transmission line elements and their equivalent signal flow graphs are shown in Figures 9.5 to 9.10.

In Figure 9.5, ρ represents the magnitude of the reflection coefficient. The phase change produced by the termination is shown as $e^{j\phi_L}$ where ϕ_L represents the load phase change in radians.

Fig. 9.5 Signal flow graph representation of a termination

Figure 9.6 shows a length of lossless transmission line which has no reflection coefficient. When compared to the flow graph of Figure 9.7, the s_{11} and s_{22} branches have the value zero or can be left out entirely. The term $e^{-j\phi_L}$ represents the phase change within the line.

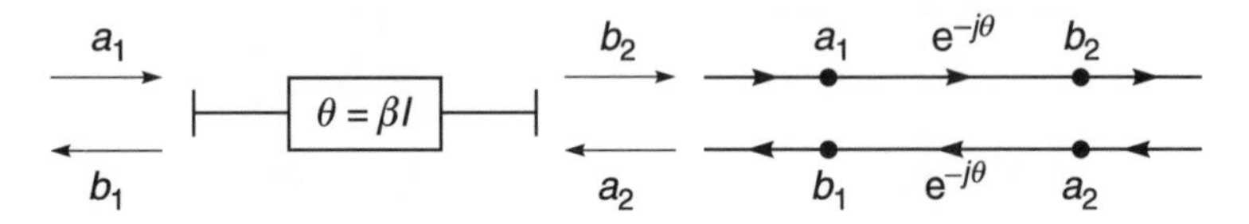

Fig. 9.6 Signal flow graph of a length of lossless transmission line

Figure 9.7 shows a detector and k denotes the scalar conversion efficiency relating the incoming wave amplitude to a meter reading. The meter reading M is assumed calibrated to take into account the detector law so k is independent of level.

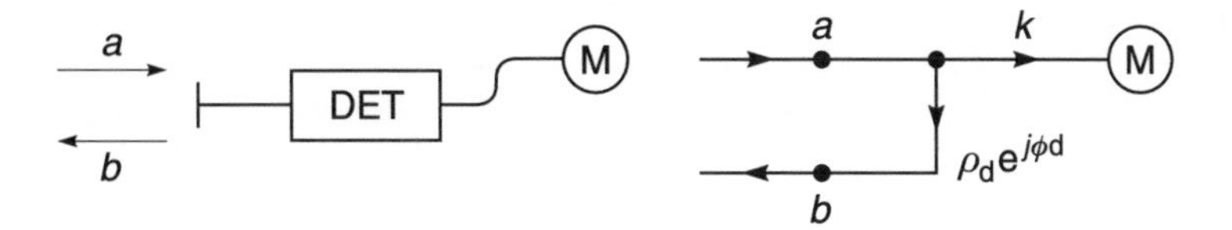

Fig. 9.7 Signal flow graph of a detector

Figure 9.8 shows the signal flow graph for a shunt admittance.

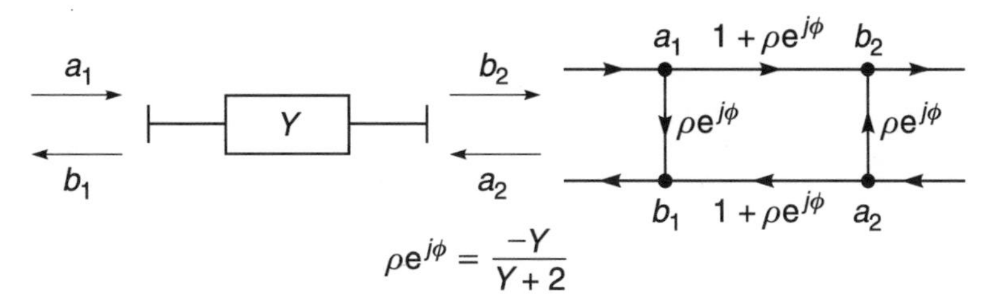

Fig. 9.8 Signal flow graph of a shunt admittance

Figure 9.9 shows the signal flow graph for a series impedance.

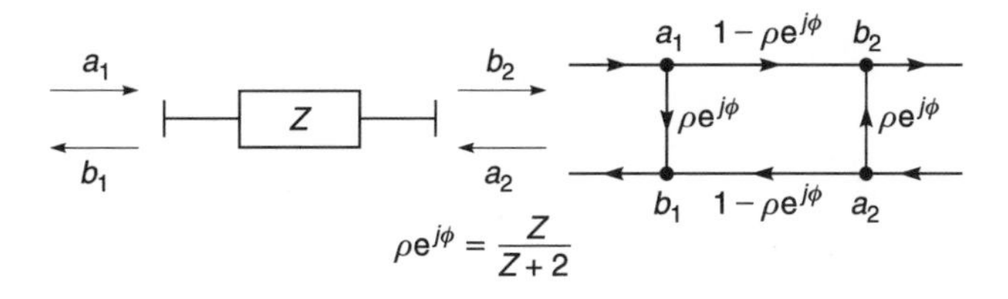

Fig. 9.9 Signal flow graph of a series impedance

Figure 9.10 shows a flow graph representation of a generator. In microwave systems, it is generally more convenient to think of a generator as a constant source of outward travelling wave with a reflection coefficient looking back into the generator output.

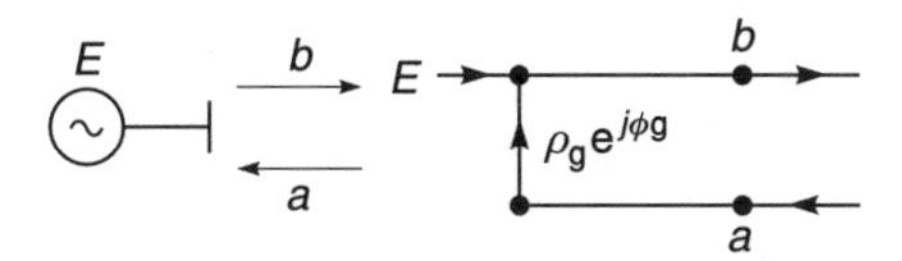

Fig. 9.10 Signal flow graph of a generator

9.2.3 Topological manipulation of signal flow graphs

The method of finding the value of a wave at a certain node may be arrived at with a series of topological manipulations which reduce a flow graph to simpler and simpler forms until the answer is apparent or until such a stage is reached when Mason's non-touching rule can be used easily. Mason's rule will be explained later. Four rules are given for easier understanding of signal flow manipulations.

Rule I: Branches in series (where the common node has only one incoming and one outgoing branch) may be combined to form a single branch whose coefficient is the product of the coefficients of the individual branches. A typical example of this is shown in Figure 9.11 where

$$E_2 = S_{21}E_1 \qquad E_3 = S_{32}E_2$$

making

$$E_3 = S_{32}S_{21}E_1 \tag{9.3}$$

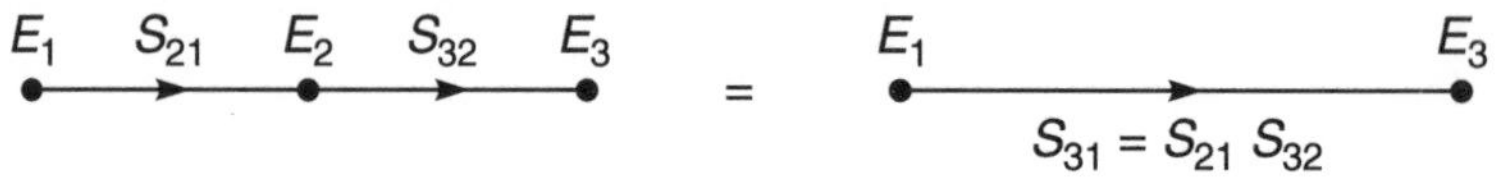

Fig. 9.11 Branches in series

Rule II: Two branches pointing from a common node to another common node (branches in parallel) may be combined into a single branch whose coefficient is the *sum* of the individual coefficients. A typical example of this is shown in Figure 9.12 where

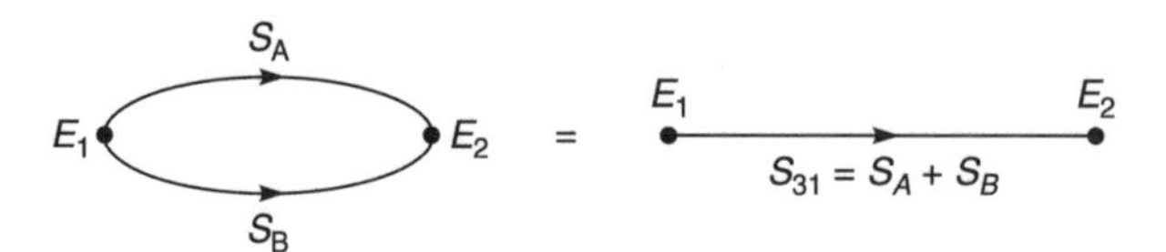

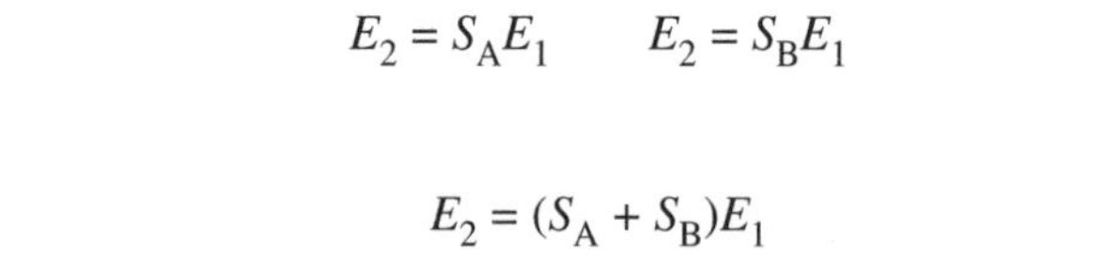

Fig. 9.12 Branches in parallel

$$E_2 = S_A E_1 \qquad E_2 = S_B E_1$$

making

$$E_2 = (S_A + S_B)E_1 \tag{9.4}$$

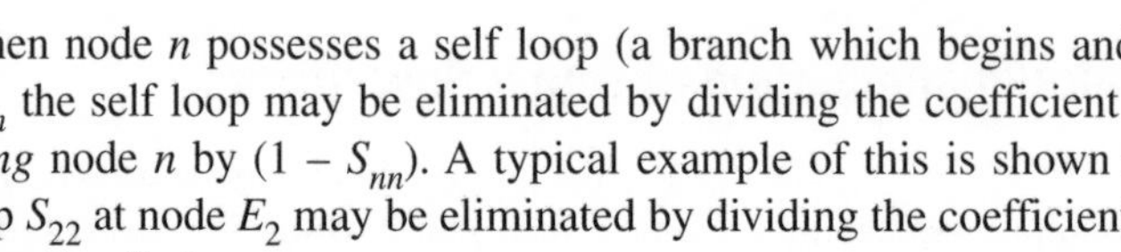

Fig. 9.13 Reduction of a feedback loop

Rule III: When node n possesses a self loop (a branch which begins and ends at n) of coefficient S_{nn} the self loop may be eliminated by dividing the coefficient of every other branch *entering* node n by $(1 - S_{nn})$. A typical example of this is shown in Figure 9.13 where the loop S_{22} at node E_2 may be eliminated by dividing the coefficient (S_{21}) entering the node E_2, by $(1 - S_{22})$.

Rule IV: A node may be duplicated, i.e. split into two nodes which may be subsequently treated as two separate nodes, providing the resulting signal flow graph contains, once and only once, each combination of separate (not a branch which forms a self loop) input and output branches which connect to the original node. Any self loop attached to the original node must also be attached to each of the nodes resulting from duplication. A typical example is shown in Figure 9.14(a) to (e) where a complicated signal flow graph is reduced to that of a far simpler one.

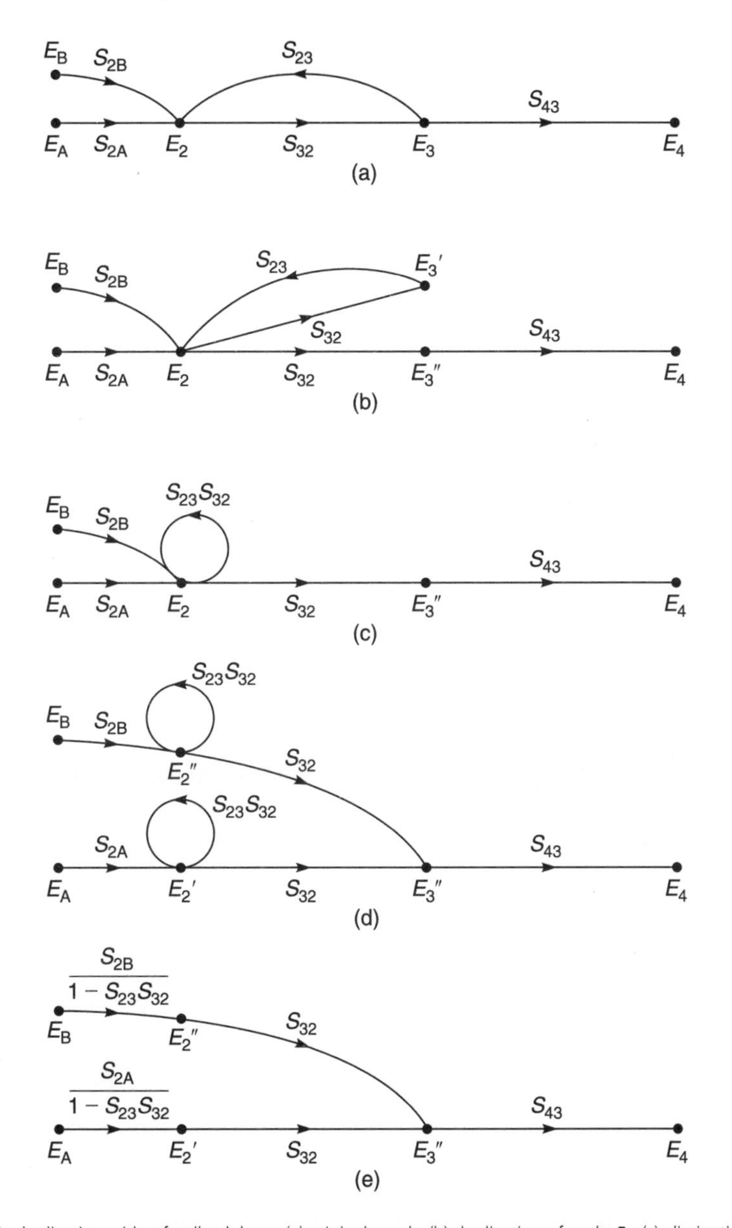

Fig. 9.14 Node duplication with a feedback loop: (a) original graph; (b) duplication of node E_3; (c) elimination of node E'_3 to form a self loop; (d) duplication of node E_2 with self loop; (e) elimination of self loops

9.2.4 Mason's non-touching loop rule

Mason's non-touching loop rule is extremely useful for calculating the wave parameters in a network. At first glance, Mason's expression appears to be very frightening and formidable but it can be easily applied once the fundamentals are understood. Some of the fundamentals relating to this rule have already been discussed in the preceding sections but for the sake of clarity, some of the material will be repeated in the application of Mason's rule to the example given in Figure 9.15 which shows the flow diagram of a network cascaded between a generator (E) and a load (Γ_L).

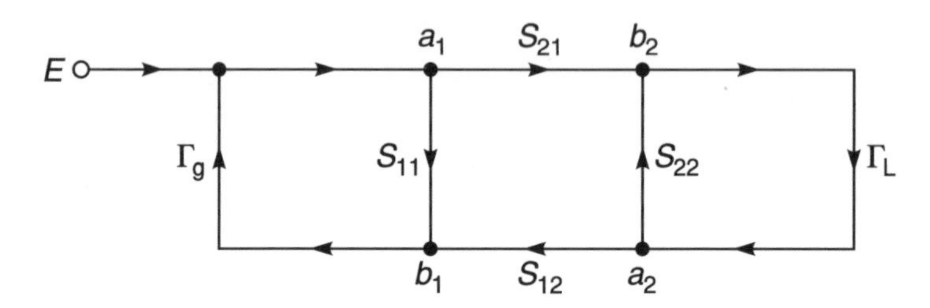

Fig. 9.15 Signal flow representation of a network

When networks are cascaded it is only necessary to cascade the flow graphs since the outgoing wave from the earlier network is the incoming wave to the next network. In Figure 9.15, the system has only one independent variable, the generator amplitude (E). The flow graph contains paths and loops.

A **path** is a series of directed lines followed in sequence and in the same direction in such a way that no node is touched more than once. The value of the path is the product of all coefficients encountered en route. In Figure 9.15, there is one path from E to b_2. It has the value S_{21}. There are two paths from E to b_1, namely S_{11} and $S_{21}\Gamma_L S_{12}$.

A **first-order loop** is a series of directed lines coming to a closure when followed in sequence and in the same direction with no node passed more than once. The value of the loop is the product of all coefficients encountered en route. In Figure 9.15, there are three first-order loops, namely $\Gamma_g S_{11}$, $S_{22}\Gamma_L$, and $\Gamma_g S_{21}\Gamma_L S_{12}$. A **second-order loop** is the product of any two first-order loops that do not touch at any point. In Figure 9.15, there is one second-order loop, namely $\Gamma_g S_{11} S_{22}\Gamma_L$. A **third-order loop** is the product of any three first-order loops which do not touch. There is no third-order loop in Figure 9.15. An ***n*th-order loop** is the product of any n first-order loops which do not touch and so on.

The solution of a flow graph is accomplished by application of Mason's non-touching loop rule[1] which, written symbolically, is

$$T = \frac{P_1[1 - \Sigma L(1)^{(1)} + \Sigma L(2)^{(1)} - \Sigma L(3)^{(1)} + \ldots] + P_2[1 - \Sigma L(1)^{(2)} + \Sigma L(2)^{(2)} \ldots] + P_3[1 - \Sigma L(1)^{(3)} + \ldots]}{1 - \Sigma L(1) + \Sigma L(2) - \Sigma L(3) + \ldots} \tag{9.5}$$

[1] Do not panic! Detailed examples follow. If you really want to know more about this topic see Mason, S.J., 'Feedback theory – some properties of signal flow graphs'. *Proc. IRE* (**41**) 1144–56, Sept 1953. See also Mason, S.J., 'Feedback theory – further properties of signal flow graphs'. *Proc. IRE* (**44**) 920–26, July 1956.

Here, P_1, P_2, P_3, etc. are the values of all the various paths which can be followed from the independent variable node to the node whose value is desired.

$\Sigma L(1)$ denotes the sum of all first-order loops. $\Sigma L(2)$ denotes the sum of all second-order loops and so on. $\Sigma L(1)^{(1)}$ denotes the sum of all first-order loops which do not touch P_1 at any point, and so on. $\Sigma L(2)^{(1)}$ denotes the sum of all second-order loops which do not touch P_1 at any point, the superscript (1) denoting path 1. Similarly, $\Sigma L(1)^{(2)}$ denotes the sum of all first-order loops which do not touch P_2 at any point and so on. In other words, each path is multiplied by the factor in brackets which involves all the loops of all orders which that path does not touch.

T is a general symbol representing the ratio between the dependent variable of interest and the independent variable. This process is repeated for each independent variable of the system and the results are summed.

As examples of the application of the rules in Figure 9.15, the transmission b_2/E and the reflection coefficient b_1/a_1 are written as follows:

$$\frac{b_2}{E} = \frac{S_{21}}{1 - \Gamma_g S_{11} - S_{22}\Gamma_L - \Gamma_g S_{21}\Gamma_L S_{12} + \Gamma_g S_{11} S_{22}\Gamma_L} \tag{9.6}$$

and

$$\frac{b_1}{a_1} = \frac{S_{11}(1 - S_{22}\Gamma_L) + S_{21}\Gamma_L S_{12}}{1 - S_{22}\Gamma_L} = S_{11} + \frac{S_{12}S_{21}\Gamma_L}{1 - S_{22}\Gamma_L} \tag{9.7}$$

Note that the generator flow graph is unnecessary when solving for b_1/a_1, and the loops associated with it are deleted when writing this solution. It is worth mentioning at this point that second- and higher-order loops can quite often be neglected while writing down the solution if one has orders of magnitude for the components in mind.

9.2.5 Signal flow applications

Signal flow graphs are best understood by some illustrative examples.

Example 9.1

Figure 9.16 illustrates a simple system where a generator is connected to a detector. The signal flow graph for the system is illustrated in Figure 9.17. It is made up from the basic

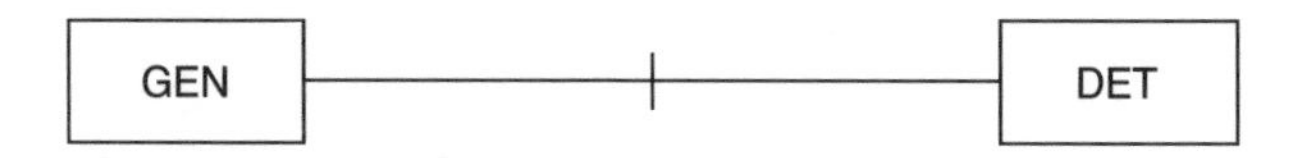

Fig. 9.16 Generator and detector system

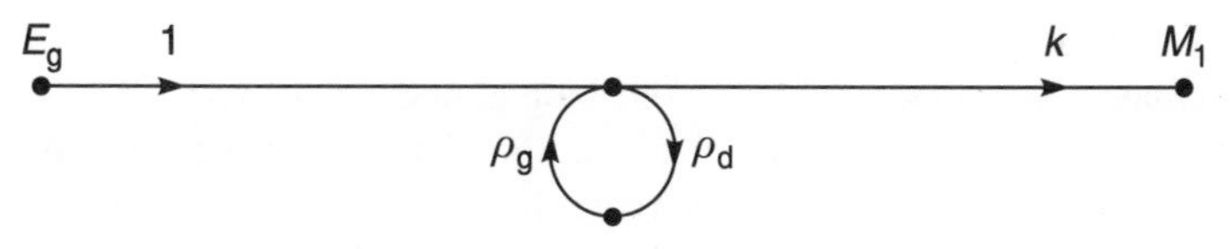

Fig. 9.17 Signal flow graph for system of Figure 9.16

building blocks of the generator and detector illustrated in Figures 9.7 and 9.10. The phase of the generator is not considered for ease of understanding.

By inspection, and by using Rules I and III, you can readily see that

$$M_1 = E_g \left[\frac{1}{1 - \rho_g \rho_d} \right] k \tag{9.8}$$

Example 9.2

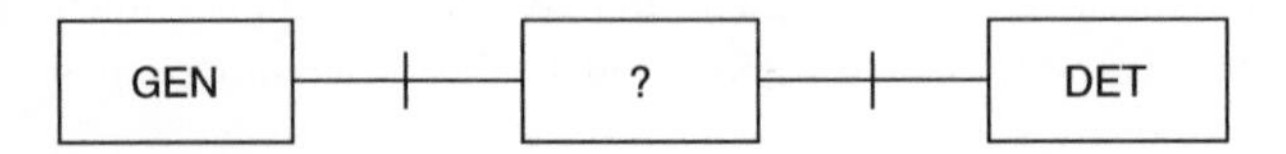

Fig. 9.18 Generator, two-port network and detector

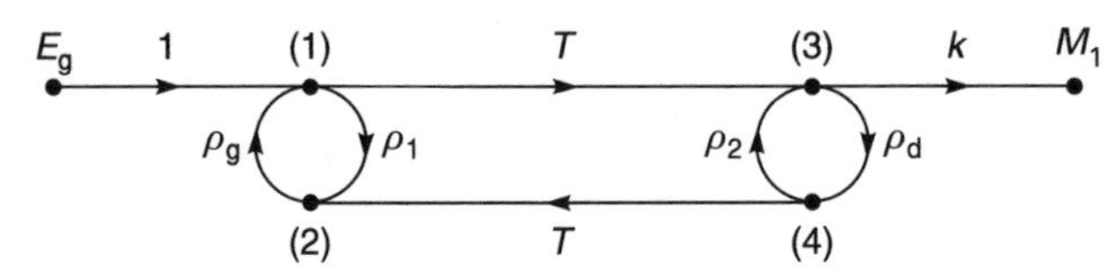

Fig. 9.19 Signal flow graph of system of Figure 9.18

Figure 9.18 depicts the case where a two port network is placed between the generator and the detector. The signal flow graph of Figure 9.19 is again made up from the basic building blocks of Figures 9.3, 9.7 and 9.10. Note particularly that we have used the signal flow graph of Figure 9.3 for the two-port network but for ease of working have replaced the *S*-parameters, s_{11}, s_{21}, s_{22}, s_{12}, with ρ_1, T, ρ_2, T, respectively.

In Figure 9.20, nodes (2) and (4) have been duplicated into nodes (2′), (2″), (4′), (4″) by Rule IV.

Figure 9.21 follows from the sequence of manipulations below.

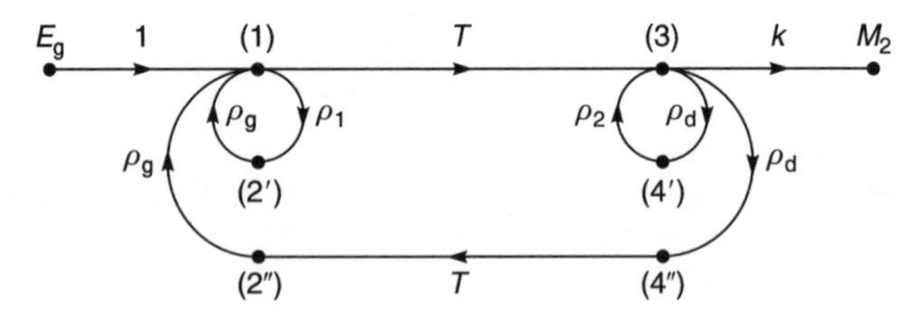

Fig. 9.20 Simplification process 1

1 Eliminate node (4′) by Rule I giving a self loop at node (3) of the value $\rho_d \rho_2$.
2 Eliminate this self loop by Rule III changing the value of the branch node (1) to node (3) from T to $T/(1 - \rho_d \rho_2)$.
3 Eliminate node (2′) by Rule I giving a self loop at node (1) of value $\rho_g \rho_1$.
4 Eliminate this self loop by Rule III changing the value of the branch leading from the generator to $1/(1 - \rho_g \rho_1)$ and also changing the branch from node (2″) to node (1) to $\rho_g/(1 - \rho_g \rho_1)$.

5 Eliminate nodes (2″) and (4″) giving a branch from node (3) to node (1) of value $(T\rho_g\rho_d)/(1 - \rho_g\rho_1)$.

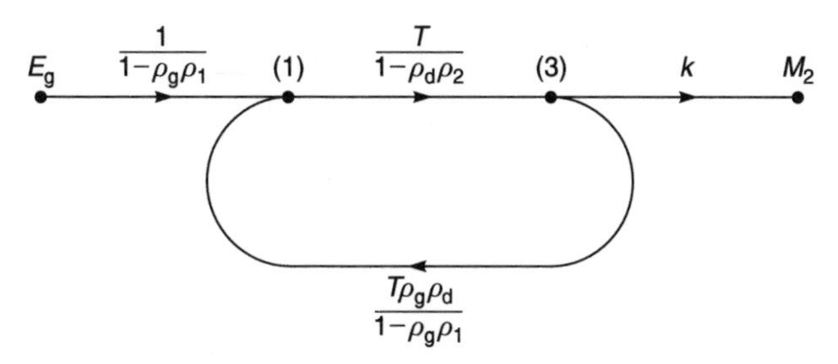

Fig. 9.21 Simplification process 2

Figure 9.22 shows the duplication of node (3) into nodes (3′) and (3″).

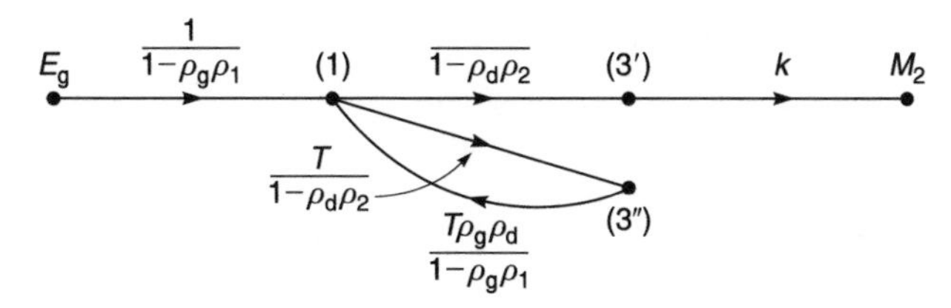

Fig. 9.22 Simplification process 3

Figure 9.23 shows the signal flow graph which results from eliminating node (3″) by Rule I and then eliminating the resulting self loop at node (1) by Rule III.

E_g (1) $\frac{T}{1-\rho_d\rho_2}$ (3′) k M_2

$$\frac{1}{\left[1-\rho_g\rho_1\right]\left[1-\frac{T^2\rho_g\rho_d}{(1-\rho_d\rho_2)(1-\rho_g\rho_1)}\right]}$$

Fig. 9.23 Simplification process 4

In Figure 9.23, there now exists a single path from E_g to M_2 and nodes (1) and (3) can be eliminated by Rule 1 yielding:

$$M_2 = \frac{E_gTk}{(1-\rho_g\rho_1)\left[1-\frac{T^2\rho_g\rho_d}{(1-\rho_d\rho_2)(1-\rho_g\rho_1)}\right](1-\rho_d\rho_2)}$$

which simplifies to

$$M_2 = E_gkT\left[\frac{1}{1-\rho_g\rho_1-\rho_d\rho_2-T^2\rho_g\rho_d+\rho_g\rho_d\rho_1\rho_2}\right] \tag{9.9}$$

9.2.6 Summary on signal flow graphs

The chief advantages of signal flow graphs over matrix algebra in solving cascaded networks are the convenient pictorial representations and the painless method of proceeding directly to the solution with approximations being obvious in the process. As your study in microwave engineering continues, you will come across more and more examples on the use of *S*-parameters and signal flow techniques.

9.3 Small effective microwave CAD packages

9.3.1 Introduction

The subject of software is a volatile one because new programs are being constantly introduced, old programs are constantly being updated, and last but not least, personal preferences come into the choice of a particular program.

Excellent radio engineering computer programs such as Hewlett Packard Design System 85150 series, EEsof's Libra*, Touchstone*, Super Compact*, and Academy* have been available for many years. These programs are excellent and extremely versatile. However, these facilities cost money and require quality computers with large RAM and disk facilities. If you or your company can afford these systems then by all means go for one or more of these software packages.

If you have an average personal computer with average RAM and only a little money, consider intermediate software programs such as ARRL (American Radio Relay League) 'ARRL Radio Designer', or Barnard's Microwave System's 'Wavemaker' or Number One System's 'Z match for Windows (Professional)'. The ARRL's program is a subset of 'Super Compact' and is quite powerful. If your frequency usage is not confined to microwave systems, then consider general purpose programs such as 'PSpice', 'HSpice' and 'Spice Age' which incorporate limited Smith chart facilities.

If money is nearly non-existent, then look in the Internet for free programs. Some of these programs are excellent. You can also go to the large commercial firms and enquire about the possibility of hiring software programs for a limited period. In some cases, it is possible to obtain reduced rates for educational establishments. Some firms such as Hewlett Packard offer internet sites where universities send in examples of their teaching and research work and may offer free copies of the programs they have developed. Some firms may offer you the free use of their small programs for a limited period.

Many people and some small radio engineering courses cannot afford even these costs. In long distance learning situations, software expenditure becomes doubly important because each student must be provided with programs which can be run on their home computers. Therefore low cost, small, powerful packages requiring *minimum* storage and RAM requirements are vitally important.

For personal and home use, I use inexpensive but relatively powerful programs such as Hewlett Packard's AppCAD, CalTech's PUFF and Motorola's Impedance Matching Program (MIMP) to investigate and produce simple designs. AppCAD is useful for the calculation of individual components, losses and gains. PUFF is valuable in calculating two port parameters, input and output matches, gains and losses and frequency responses and has plotting facilities for layout artwork. MIMP provides real time facilities for narrow and broadband matching using Smith charts and rectangular plots.

9.3.2 Hewlett-Packard's computer aided design program – AppCAD

Hewlett-Packard's Applications Computer Aided Design Program AppCAD is a collection of software tools or modules, which aid in the design of RF (radio frequency) and microwave circuits. AppCAD also includes a selection guide for Hewlett Packard RF and Microwave semiconductors. The modules considered in AppCAD are listed in Figure 9.24. Each main program is listed on the left table while the highlighted item is described on the right table.

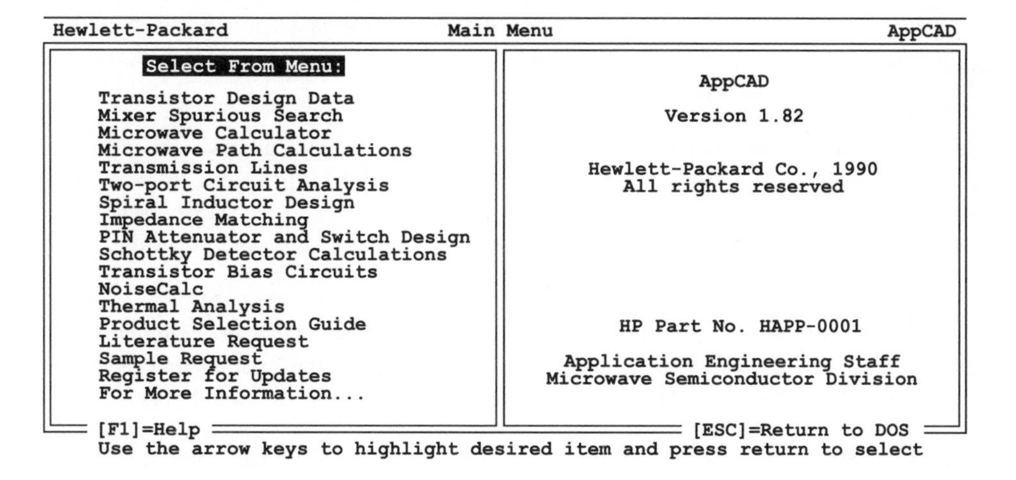

Fig. 9.24 Screen print-out of AppCAD

AppCAD modules

The modules considered in AppCAD are as follows.

(i) Transistor design data. This module computes various gain and stability data from *S*-parameters. Input *S*-parameters can be read from a TOUCHSTONE formatted file or entered manually. *S*-parameters input from a TOUCHSTONE formatted file may also be modified manually via an edit function.

The program

- calculates stability circles;
- converts *S*-parameters to *Y*-parameters, *Z*-parameters or *H*-parameters;
- calculates Γ_{ms} and Γ_{ml}.

(ii) Mixer spurious search. Mixers will generate harmonics of the RF input signal and the local-oscillator frequencies. This means that there exists a wide range of frequencies, many of which may lie outside the desired input passband, which will give unwanted responses in the IF band. This module will calculate all spurious responses for user chosen frequencies, passbands, and harmonics of the local oscillator and signal.

(iii) Microwave calculator. This is a computerised version of the famous Hewlett Packard 'Reflectometer' slide rule calculator. Calculations include reflection coefficients, standing wave ratios, mismatch loss, return loss, mismatch phase error, coupler directivity uncertainty and maximum standing wave ratio from mismatches.

(iv) Microwave path calculations. This module calculates the signal-to-noise (*S*/*N*) performance resulting from the following factors: receiver noise figure, antenna gain, transmitter power, path distance, frequency and line losses. The systems covered are one-way (communication) and two-way (radar).

(v) Transmission lines. From physical dimensions, this module calculates the following properties: characteristic impedance, effective dielectric constant, electrical line length and coupling factor. The structures covered include microstrip, stripline, coplanar waveguide with and without ground plane and coaxial lines.

(vi) Two-port circuit analysis. This easy to use linear circuit analysis program can include lumped or distributed circuit elements, which may be represented by an *S*-parameter file in TOUCHSTONE format. The calculated *S*-parameters can be displayed on the screen and printed. The data can also be graphed on to either a Smith chart or a linear *X*–*Y* plot, which can be printed to a Epson compatible printer.

(vii) Spiral inductor design. This module calculates the inductance of a circular spiral from its number of turns, conductor width, substrate height, inner radius and dielectric constant. Inductance is calculated for two cases, with and without a ground plane.

(viii) Impedance matching. This module determines both lumped and distributed elements for impedance matching of a source impedance and load impedance. The lumped or distributed elements determined are those for either an L-section, T-section, pi-section, transmission line transformer or tandem 3/8 wavelength transformer.

(ix) Pin attenuator and switch design. Insertion loss and isolation are calculated from PIN diode characteristics for both the series and shunt configurations. A built-in menu automatically returns the diode parameters from a menu selection of part numbers. Alternatively, custom diode series resistance and junction capacitance may be used. The program calculates the required resistance values for both Pi and bridged T attenuators from the desired attenuation in decibels (dB).

(x) Schottky detector calculations. This module calculates the effect of video amplifier characteristics, RF bypassing, amplifier input resistance and voltage sensitivity on pulse response, detected video bandwidth and TSS.

(xi) Transistor bias circuits. This module examines bias networks for microwave bi-polar transistors. Bias network resistors are calculated for a given collector current and voltage. The change in collector current with temperature is also computed for each network. Networks covered include non-stabilised, voltage feedback and voltage feedback constant base current.

(xii) Noise calculations. This module calculates the cascade noise figure and other performance parameters for a sub-system block diagram such as a receiver. This type of analysis allows system planning for the tradeoffs of important characteristics such as noise figure (sensitivity), gain distribution, dynamic range, signal levels and intermodulation products. Provision is made for system analysis with temperature.

(xiii) Thermal analysis. A general introduction to heat transfer is presented with emphasis on applications to semiconductors. This module includes a tutorial section, thermal resistance calculations for semiconductors, thermal analysis of MIC (hybrid) circuits and a table of thermal conductivities.

Details of programs

A program listed in Figure 9.24 usually sub-divides into other programs. For example, the transmission lines program of Figure 9.25 sub-divides into seven other types of transmission line. See the right column of Figure 9.25 for a description of the program. Figure 9.25 further sub-divides into another screen for the calculation of seven types of lines (Figure 9.26).

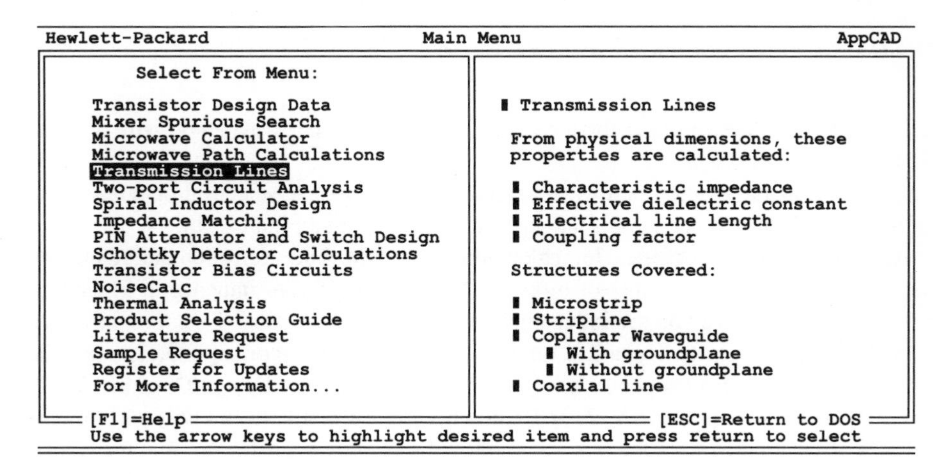

Fig. 9.25 Selection of transmission lines program

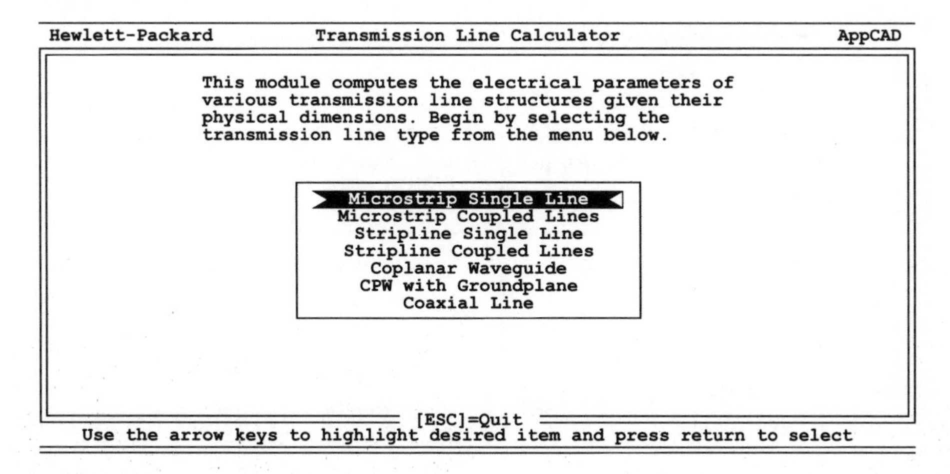

Fig. 9.26 Sub-division of the transmission lines program

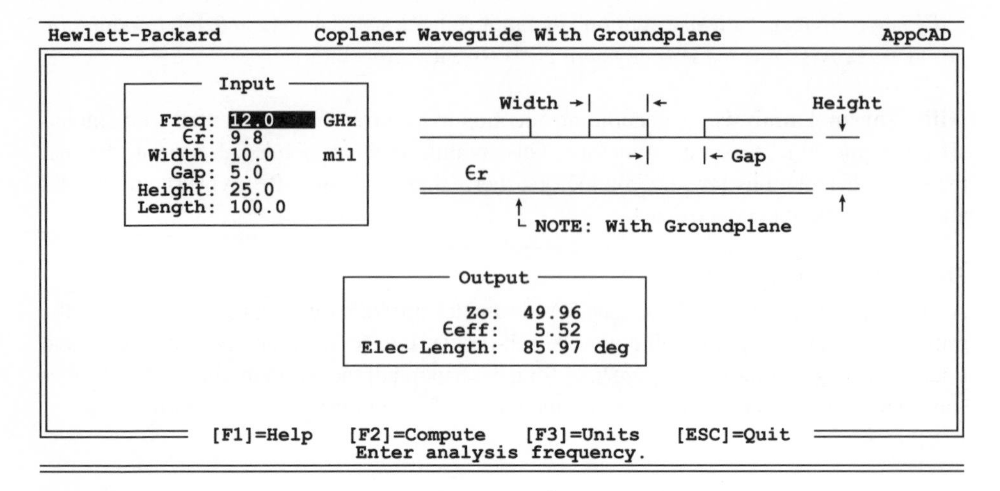

Fig. 9.27 Calculation of co-planar waveguide with ground plane

Figure 9.27 shows the case for the calculation of co-planar waveguide with a ground plane.

The same is true of the microwave calculator program. The selection of this program leads to further sub-division and four separate programs. See Figure 9.28.

AppCAD also has facilities for calculating the inductance of spiral inductors. You merely have to state the number of turns and dimensions of your inductor in Figure 9.29 and APPCAD will give you the inductance in the result box.

AppCAD is also extremely useful for conversion from one set of parameters to another. Figure 9.30 shows what happens when you select the 'Transistor Design Data' program. If you now select the Convert to Y, Z, or H and press the **return** key. You will now be given a choice of what type of parameters you require. See Figure 9.31. Ensure that the *Convert*

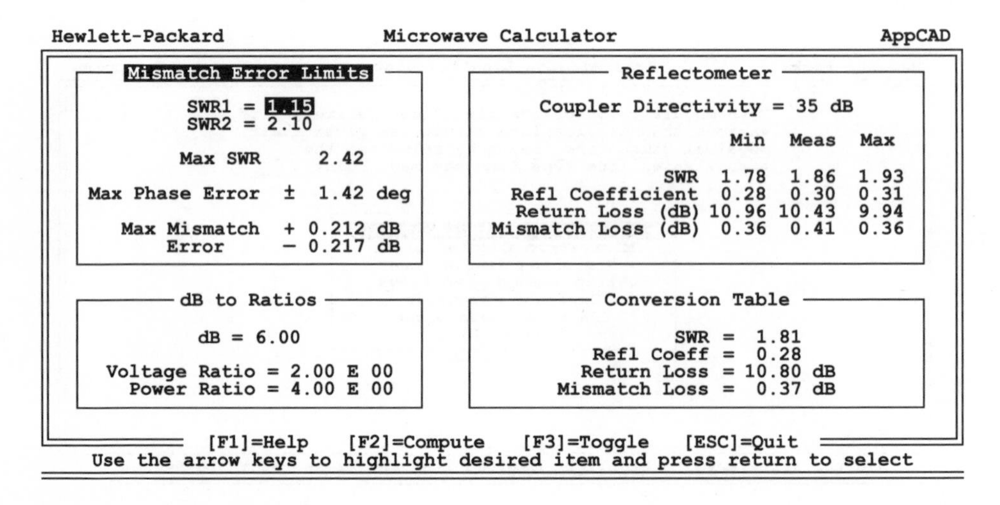

Fig. 9.28 AppCAD's calculator program

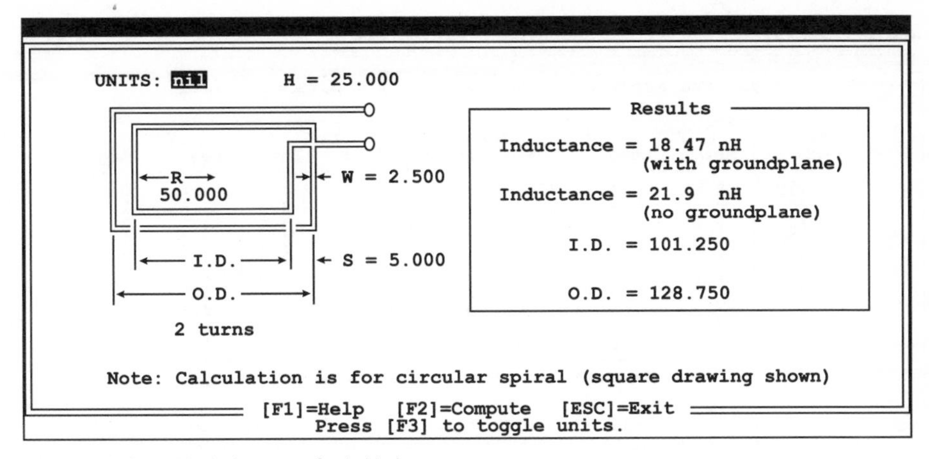

Fig. 9.29 Calculating the inductance of spiral inductors

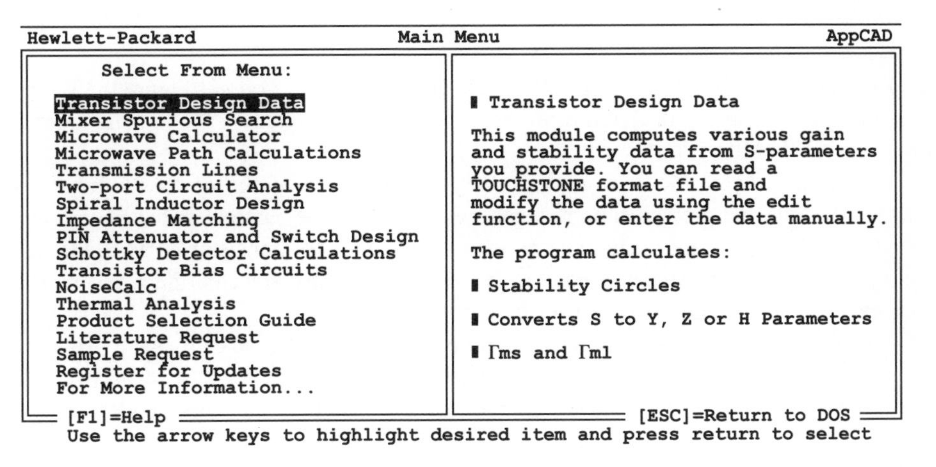

Fig. 9.30 Selection of transistor design data

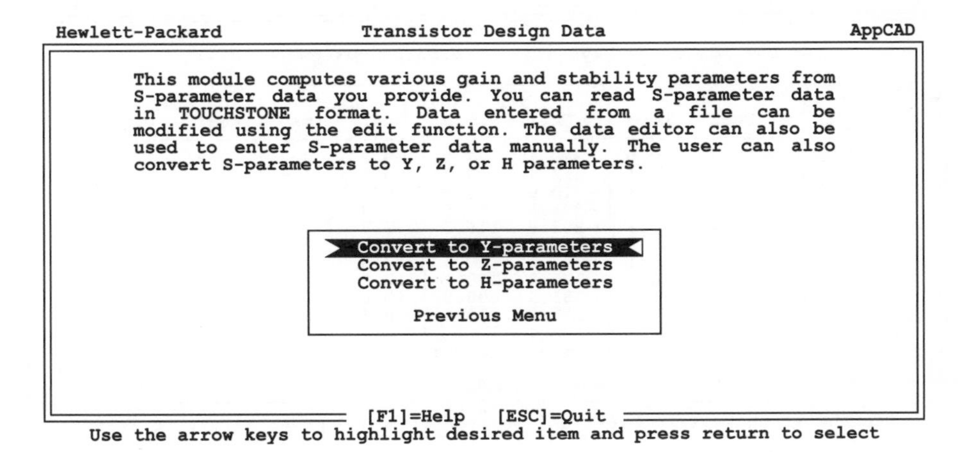

Fig. 9.31 Selection of *Y*-, *Z*- or *H*-parameters

Hewlett-Packard Computed Y-Parameters AppCAD

* Use the arrow keys (↑↓) to scroll Y-parameter data

Freq GHz	Y11 Real	Y11 Imag	Y21 Real	Y21 Imag	Y12 Real	Y12 Imag	Y22 Real	Y22 Imag
0.100	8.6	0.4	97.7	−6.9	−2.8	−0.1	3.4	0.9
0.200	8.5	0.9	97.0	−12.6	−2.8	−0.3	3.6	1.7
0.300	8.5	1.1	95.3	−18.5	−2.8	−0.4	3.5	2.3
0.400	8.3	1.4	93.0	−24.1	−2.8	−0.6	3.7	3.1
0.500	7.9	1.8	88.8	−27.7	−2.7	−0.7	3.8	3.5
0.600	7.8	2.2	87.1	−32.2	−2.7	−0.8	3.8	4.3
0.700	7.6	2.8	84.0	−36.6	−2.7	−1.0	3.9	5.0
0.800	7.7	3.5	81.6	−40.4	−2.7	−1.1	3.9	5.8
0.900	7.5	4.0	78.7	−44.4	−2.7	−1.3	4.0	6.5
1.000	7.5	4.6	75.9	−48.5	−2.7	−1.4	4.1	7.2

Y-parameter units are milli-mhos

[ESC]=Quit

Use the arrow keys to highlight desired item and press return to select

Fig. 9.32 *Y*-parameters of transistor HPMA0285.S2P

to Y-parameters line is highlighted. Press the **return** key and you get Figure 9.32. These parameters have been calculated from the *s*-parameters supplied by the manufacturer for transistor HPMA0285.S2P.

If you had wanted the Z-parameters, then you would have chosen the *Convert to Z-parameters* line and pressed **return** to get Figure 9.33. These parameters have been calculated from the *s*-parameters supplied by the manufacturer for transistor HPMA0285.S2P. Similarly if you had wanted *H*-parameters, you would have selected the *Convert to H-parameters* and pressed **return** to get Figure 9.34.

So you can see for yourself how much time and effort can be saved by using AppCAD. In most cases, you might be able to get the program free from Hewlett Packard who own the copyright. The hardware requirements for the program are very modest and the

Hewlett-Packard Computed Z-Parameters AppCAD

* Use the arrow keys (↑↓) to scroll Z-parameter data

Freq GHz	Z11 Real	Z11 Imag	Z21 Real	Z21 Imag	Z12 Real	Z12 Imag	Z22 Real	Z22 Imag
0.100	11.4	2.9	−320.1	27.2	9.2	0.3	28.2	1.1
0.200	11.9	5.1	−313.1	49.5	9.2	0.6	27.5	2.1
0.300	12.0	7.1	−310.8	75.6	9.3	1.0	28.0	2.4
0.400	12.7	9.6	−304.0	96.5	9.4	1.4	27.8	3.2
0.500	13.9	11.6	−303.8	116.0	9.7	1.7	27.8	4.4
0.600	14.6	13.8	−291.6	138.8	9.8	2.0	27.9	5.3
0.700	15.7	16.1	−280.2	159.4	10.0	2.4	27.7	6.8
0.800	16.4	18.7	−270.5	180.6	10.2	2.6	29.0	8.7
0.900	17.8	21.0	−255.9	200.9	10.4	3.0	28.8	10.1
1.000	19.0	23.0	−237.9	218.4	10.6	3.3	29.4	11.6

Z-parameter units are ohms

[ESC]=Quit

Use the arrow keys to highlight desired item and press return to select

Fig. 9.33 *Z*-parameters of transistor HPMA0285.S2P

Hewlett-Packard — Computed H-Parameters — AppCAD

* Use the arrow keys (↑↓) to scroll H-parameter data

Freq GHz	H11 Mag	H11 Ang	H21 Mag	H21 Ang	H21 Mag	H21 Ang	H22 Mag	H22 Ang
0.100	116.2	−3.8	11.4	−7.0	0.33	−0.0	0.035	−2.2
0.200	117.4	−5.9	11.5	−13.3	0.34	−0.3	0.036	−4.3
0.300	117.0	−7.6	11.4	−18.6	0.33	1.4	0.036	−4.9
0.400	118.5	−9.6	11.4	−24.2	0.34	1.8	0.036	−6.6
0.500	124.2	−12.6	11.6	−29.9	0.35	1.1	0.036	−9.0
0.600	122.6	−15.9	11.4	−36.2	0.35	0.8	0.035	−10.8
0.700	123.4	−19.9	11.3	−43.4	0.36	−0.4	0.035	−13.8
0.800	118.1	−24.2	10.8	−50.5	0.35	−2.5	0.033	−16.8
0.900	117.7	−28.1	10.6	−57.5	0.35	−3.5	0.033	−19.4
1.000	113.5	−31.4	10.2	−64.0	0.35	−4.0	0.032	−21.4

H11 units are in ohms and H22 units are in mhos.

[ESC]=Quit

Use the arrow keys to highlight desired item and press return to select

Fig. 9.34 *H*-parameters of transistor HPMA0285.S2P

DOS version will even run with an 8086 processor. There is also a Windows version of AppCAD available on the Internet.

9.3.3 PUFF Version 2.1

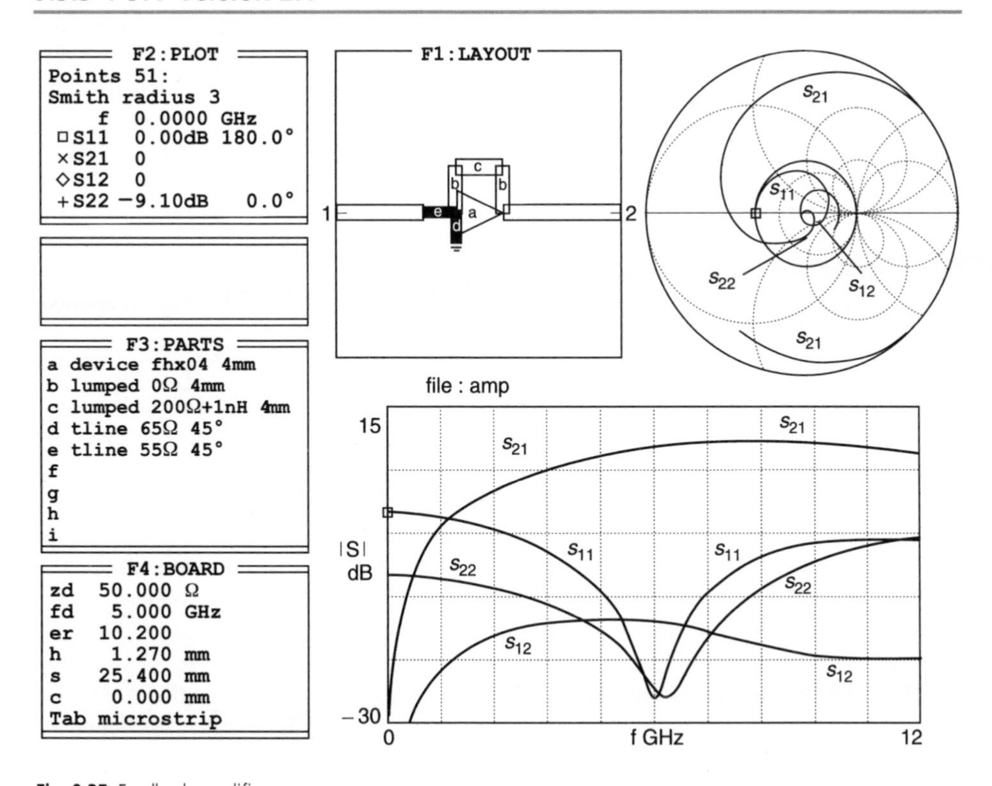

Fig. 9.35 Feedback amplifier response

Puff Version 2.1 was chosen for this book because of: (i) its ease of use – PC format, (ii) its versatility, (iii) its computer requirement flexibility – ≈290 KB on a floppy disk, any processor from an 8080 to Pentium, choice of display, CGA, EGA or VGA, and (iv) choice of printer, dot-matrix, bubble jet or laser and (v) its low costs. As you have seen for yourself, it can be used for (i) lumped and distributed filter design, evaluation, layout and fabrication, (ii) evaluation of *s*-parameter networks, (iii) lumped and distributed matching techniques, layout and construction, (iv) amplifier design layout and construction and (v) determination of input impedance and admittance of networks.

There are also many features to PUFF which have not been used in this book. For example, PUFF can be used for the design of oscillators; it has compressed Smith chart facilities. An example of this is shown in Figure 9.35.

If you want further information on PUFF, I suggest you contact PUFF Distribution, CalTech in Pasadena, California, USA. They can supply you with the source code, a manual for the program, and lecture notes for carrying out more advanced work with PUFF.

9.3.4 Motorola's Impedance Matching Program (MIMP)

MIMP is excellent for narrow-band and wide-band matching of impedances. It has facilities for matching complex source and load impedances and designing lumped or distributed circuits with the desired Q graphically. This program can be explained by an example. Consider the case where the output impedance, ≈(20 + j0) Ω, of a transmitter operating between 470 MHZ and 500 MHz is to be matched to an antenna whose nominal impedance is 50 Ω. A return loss of ≈20 dB is required. The conditions are entered into MIMPs as in Figure 9.36. Three frequencies are used to cover the band and the load and source impedances are also entered into the figure. When Figure 9.36 is completed, the **ESC** key is pressed to move on to Figure 9.37 where a network is chosen. For this case, a T network has been chosen. At this stage, the exact values of the components and the Q of the matching network are unknown so nominal values are inserted. The exact values will be derived later. For clarity, Z_{in} and the load have been annotated in Figure 9.37.

After completion of Figure 9.37, the **ESC** key is pressed to enter Figure 9.38. Starting from the top left line in Figure 9.37, we have SERIES CAP and up/down arrows which allow any component to be selected. C_1 is shown in this case. Capacitors are shown in this box with its arrow keys. The next right block shows values of inductors. Adjustment is provided by up/down arrow keys. The next right box with its arrow keys is the Q selection box. $Q = 3$ has been selected. The next box after the logo is the line impedance (Z_o) box. Its default position is 10 Ω but it can be changed and the Smith chart plot values will automatically change accordingly. The FREQ box allows selection of frequency. It has been set to mid-point, i.e. 485 MHz. The remaining three boxes are self-evident.

The middle left-hand box provides a read-out of impedance at points to the 'right' of a junction. The Smith chart return loss circle size is determined by the Return loss boundary set in the lower left box. Arcs AB, BC and CD are adjusted by components C_1, C_2 and L_1 respectively until the desired matching is obtained.

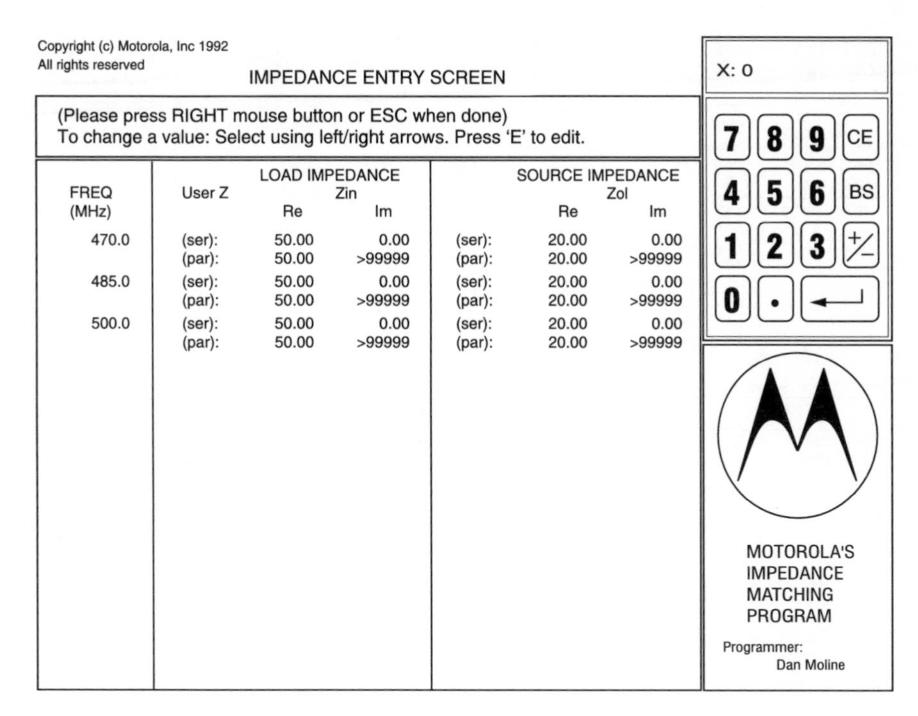

Fig. 9.36 Input data for Motorola's MIMP

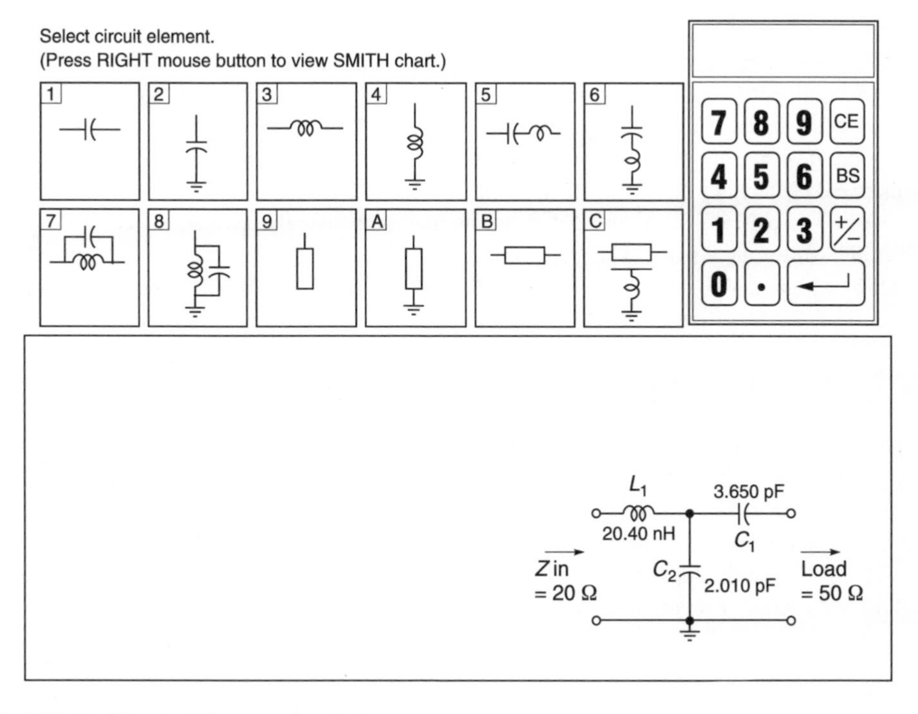

Fig. 9.37 Matching circuit for Motorola's MIMP

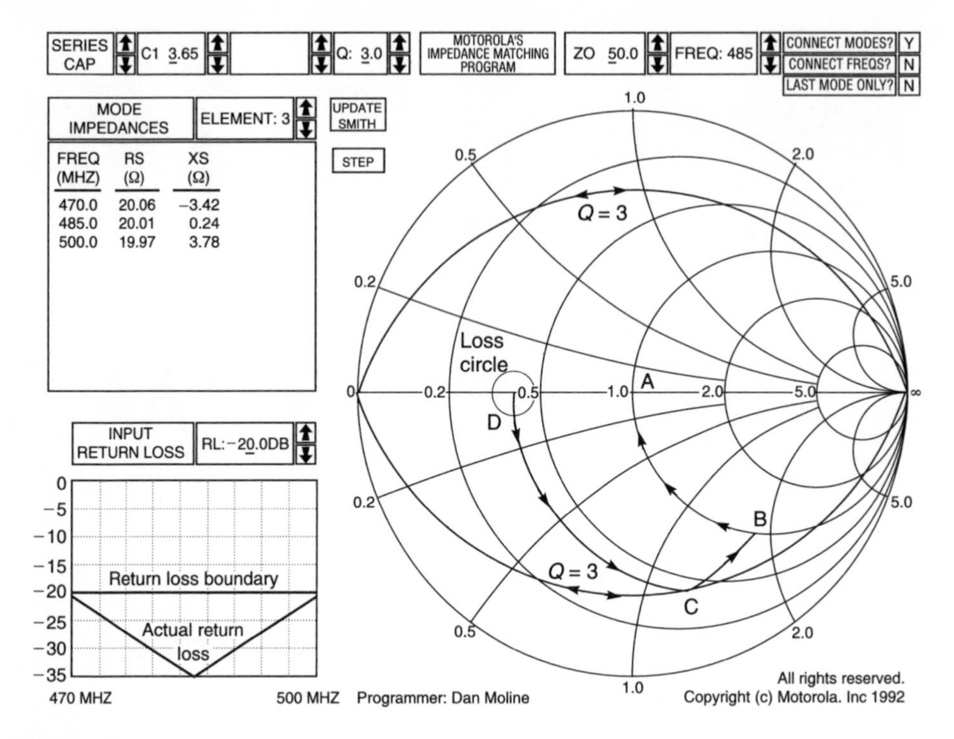

Fig. 9.38 Smith chart and input return loss

The great advantage of MIMP is that matching is carried out electronically and quickly. There are no peripheral scales on it like a conventional paper Smith chart but this is unnecessary because each individual point on the chart can be read from the information boxes. A good description of how this program works can be found in 'MIMP Analyzes Impedance Matching Network' *RF Design*, Jan 1993, 30–42.

MIMP is a Motorola copyright program but it is usually available free from your friendly Motorola agents. The program requirements are very modest with processor 80286 or higher, VGA graphics and 640k RAM.

9.4 Summary of software

The above computer aided design programs, namely AppCAD, PUFF and MIMP, are extremely low cost and provide a very good cross-section of theoretical and practical constructional techniques for microwave radio devices and circuits. AppCAD was used extensively for checking the bias and matching circuits. MIMP was used extensively for checking the Smith chart results in the book. I am sure that these programs will be useful additions to your software library.

References

These references are provided as a guide to readers who want more knowledge on the main items discussed in this book. They have been compiled into seven categories. These are circuit fundamentals, transmission lines, components, computer aided design, amplifiers, oscillators, and signal flow diagrams. The references are in alphabetical order in each section.

References soon become antiquated and to keep up with developments, it is best to read material, such as the IEEE Transactions on various topics, IEE journals, Microwave Engineering journal, etc. These journals are essential because they provide knowledge of the latest developments in the high frequency and microwave world. Attending conferences is also very important and many large firms like Hewlett Packard, Motorola, and Texas Instruments often provide free study seminars to keep engineers up to date on amplifier, oscillator, CAD and measurement techniques.

Many large firms such as Hewlett Packard also provide education material on the **Internet**. For example, many universities put their experimental work on http://www.hp.com/info/college_lab101.

Circuit Fundamentals

Avantek 1982: *High frequency transistor primer, Part 1.* Santa Clara CA: Avantek.

Festing, D. 1990. Realizing the theoretical harmonic attenuation of transmitter output matching and filter circuits, *RF Design*, February.

Granberg, H. O. 1980. Good RF construction practices and techniques. *RF Design*, September/ October.

Hewlett Packard. S parameters, circuit analysis and design. *Hewlett Packard Application Note 95*, Palo Alto CA: Hewlett Packard.

Johnsen, R.J. Thermal rating of RF power transistors. *Application Note AN790.* Phoenix Az: Motorola Semiconductor Products.

Jordan, E. 1979: *Reference data for engineers: radio, electronics, computers and communications.* Seventh Edition. Indianapolis IN: Howard Sams & Co.

Motorola. Controlled Q RF technology – what it means, how it is done. *Engineering bulletin EB19*, Phoenix AZ: Motorola Semiconductor Products Sector.

Motorola 1991. *RF data book DL110*, Revision 4, Phoenix AZ: Motorola Semiconductor Sector.

Saal, R. 1979: *Handbook of filter design.* Telefunken Aktiengesellschaft, 715 Backnang (Wurtt), Gerberstrasse 34 PO Box 129, Germany.

Transistor manual, Technical Series SC12, RCA, Electronic Components and Devices, Harrison NJ, 1966.

Transmission Lines

Babl, I.J. and Trivedi, D.K. 1977. A designer's guide to microstrip. *Micorwaves*, May.

Chipman, R.A. 1968: *Transmission lines.* New York NY: Schaum, McGraw-Hill.

Davidson, C.W. 1978: *Transmission lines for communications*. London: Macmillan.
Edwards, T.C. 1992: *Foundations for microstrip circuit design*. Second Edition. John Wiley & Sons.
Ho, C.Y. 1989. Design of wideband quadrature couplers for UHF/VHF. *RF Design*, 58–61, November (with further useful references).
Smith, P.H. 1944. An improved line calculator. *Electronics*, January, 130.

Components

Acrian Handbook 1987. *Various Application Notes*. The Acrian Handbook, Acrian Power Solutions, 490 Race Street, San Jose CA.
Blockmore, R.K. 1986. Practical wideband RF power transformers, combiners and splitters. *Proceedings of RF Expo West*, January.
Fair-Rite. Use of ferrities for wide band transformers. *Application Note*. Fair-Rite Products Corporation.
Haupt, D.N. 1990. Broadband-impedance matching transformers as applied to high-frequency power amplifiers. *Proceedings of RG Expo West*, March.
Myer, D. 1990. Equal delay networks match impedances over wide bandwidths. *Microwaves and RF*, April.
Phillips, 1969–72. On the design of HF wideband transformers, parts I and II. *Electronic Application Reports ECO69007 & ECOP7213*. Phillips Discrete Semiconductor Group.

Computer aided design

CAD Roundtable 1996. Diverse views on the future of RF design. *Microwave Engineering Europe Directory*, 20–26.
Da Silva, E. 1997: *Low cost microwave packages*. Fourth International Conference, Computer Aided Engineering Education CAEE97, Krakow, Poland.
Da Silva, E. 1997: *Low cost radio & microwave CAL packages*. EAEEIE, Eighth Annual conference, Edinburgh, Scotland.
Davis, F. Matching network designs with computer solutions. *Application Note AN267*. Phoenix AZ: Motorola Semiconductor Sector.
Edwards, T.C. 1992: *Foundations for microstrip circuit design*. Second Edition. John Wiley & Sons.
Gillick, M., Robertson, I.D. and Aghvami, A.H. 1994. Uniplanar techniques for MMICs. *Electronic and Communications Engineering Journal*, August, 187–94.
Hammerstad, E.O. 1975. Equations for microstrip circuit design. *Proceedings Fifth European Conference*, Hamburg.
Kirchning, N. 1983. Measurement of computer-aided modelling of microstrip discontinuities by an improved resonator method. *IEEE Trans MIT, International Symposium Digest*, 495–8.
Koster, W., Norbert, H.L. and Hanse, R.H. 1986. The microstrip discontinuity; a revised description. *IEEE MTT* **34** (2), 213–23.
Matthei, G.L., Young, L. and Jones, E.M.T. 1964: *Microwave filters, impedance matching networks and coupling structures*. New York NY: McGraw-Hill.
MMICAD (for IBM PCs). Optotek, 62 Steacie Drive, Kanata, Ontario, Canada, K2K2A9.
Moline, D. 1993. MIMP analyzes impedance matching networks. *RF Design*, January, 30–42.
Nagel, L.W. and Pederson, D.O. 1973. Simulation program with integrated circuit emphasis. *Electronics Research Lab Rep No ERL-M382*, University of Calif, Berkeley.
PSpice by MicroSim Corporation, 20 Fairbands, Irvine CA 92718.
Rutledge, D. 1996: *EE153 Microwave Circuits*. California Institute of Technology.
SpiceAge. Those Engineers Ltd, 31 Birbeck Road, Mill Hill, London, England.
Wheeler, H.A. 1977. Transmission line properties of a strip on a dielectric sheet on a plane. *IEEE MTT* **25** (8), 631–47.

Amplifiers

Bowick, C. 1982: *RF circuit design*. Indianapolis IN: Howard Sams.

Carson, R.S. 1975: *High frequency amplifiers*. New York NY: John Wiley and Sons.

Dye, N. and Shields, M. Considerations in using the MHW801 and MHW851 series power modules. *Application Note AN-1106*. Phoenix AZ: Motorola Semiconductor Sector.

Froehner, W.H. 1967. Quick amplifier design with scattering parameters. *Electronics*, October.

Gonzales, G. 1984: *Microwave transistor amplifier analysis and design*. Englewood Cliffs NJ: Prentice Hall.

Granberg, H.O. A two stage 1 kW linear amplifier. *Motorola Application Note A758*. Phoenix AZ: Motorola Semiconductor Sector.

Granberg, H.O. 1987. Building push-pull VHF power amplifiers. *Microwave and RF*, November.

Hejhall, R. RF small signal design using two port parameters. *Motorola Application Report AN 215A*.

Hewlett Packard. S parameter design. *Application Note 154*. Palo Alto CA: Hewlett Packard Co.

ITT Semiconductors. VHF/UHF power transistor amplifier design. *Application Note AN-1-1*, ITT Semiconductors.

Liechti, C.A. and Tillman, R.L. 1974. Design and performance of microwave amplifiers with GaAs Schottky-gate-field-effect transistors. *IEEE MTT-22*, May, 510–17.

Pengelly, R.S. 1987: *Microwave field effect transistors theory, design and applications*. Second Edition. Chichester, England: Research Studies Press, division of John Wiley and Sons.

Rohde, U.L. 1986. Designing a matched low noise amplifier using CAD tools. *Microwave Journal*, October 29, 154–60.

Vendelin, G., Pavio, A. and Rohde, U. *Microwave circuit design*. New York NY: John Wiley & Sons.

Vendelin, G.D., Archer, J. and Bechtel, G. 1974. A low-noise integrated s-band amplifier. *Microwave Journal*, February. Also *IEEE International Solid-state Circuits Conference*, February 1974.

Young, G.P. and Scalan, S.O. 1981. Matching network design studies for microwave transistor amplifiers. *IEEE MTT 29*, No 10, October, 1027–35.

Oscillators

Abe, H. A highly stabilized low-noise Ga-As FET integrated oscillator with a dielectric resonator in the C Band. *IEEE Trans MTT 20*, March.

Gilmore, R.J. and Rosenbaum, F.J. 1983. An analytical approach to optimum oscillator design using S-parameters. *IEEE Trans on Microwave Theory and Techniques MTT 31*, August, 663–9.

Johnson, K.M. 1980. Large signal GaAs FET oscillator design. *IEEE MTT-28*, No 8, August.

Khanna, A.P.S. and Obregon, J. 1981. Microwave oscillator analysis. *IEEE MTT-29*, June, 606–7.

Kotzebue, K.L. and Parrish, W.J. 1975. The use of large signal S-parameters in microwave oscillator design. *Proceedings of the International IEEE Microwave Symposium on circuits and systems*.

Rohde, Ulrich L. 1983: *Digital PLL frequency-synthesizers theory and design*. Englewood Cliffs NJ: Prentice Hall.

Vendelin, G.D. 1982: *Design of amplifiers and oscillators by the S-parameter method*. New York NY: John Wiley and Sons.

Signal flow

Chow, Y. and Cassignol, E. 1962: *Linear signal-flow graphs and applications*. New York NY: John Wiley and Sons.

Horizon House 1963: *Microwave engineers' handbook and buyers' guide*. Horison House Inc, Brookline Mass, T-15.

Hunton, J.K. 1960. Analysis of microwave measurement techniques by means of signal flow graphs. *Trans IRE MTT-8*, March, 206–12.

Mason, S. J. 1955. Feedback theory – some properties of signal flow graphs. *Proc IRE 41*, 1144–56, September.

Mason, S.J. 1955. Feedback theory – further properties of signal flow graphs. *Proc IRE 44*, 920–26, July.

Mason and Zimmerman. 1960: *Electronic circuits, signals and systems*. New York NY: John Wiley and Sons.

Montgomery et al 1948: *Principles of microwave circuits*. New York NY: McGraw-Hill Book Co.

Index